T0142824

Lecture Notes in Computational Science and Engineering

113

Editors:

Timothy J. Barth
Michael Griebel
David E. Keyes
Risto M. Nieminen
Dirk Roose
Tamar Schlick

More information about this series at http://www.springer.com/series/3527

Hans-Joachim Bungartz • Philipp Neumann •
Wolfgang E. Nagel

Editors

Software for Exascale Computing – SPPEXA 2013-2015

 Springer

Editors

Hans-Joachim Bungartz
Philipp Neumann
Institut für Informatik
Technische Universität München
Garching
Germany

Wolfgang E. Nagel
Technische Universität Dresden
Dresden
Germany

ISSN 1439-7358   ISSN 2197-7100   (electronic)
Lecture Notes in Computational Science and Engineering
ISBN 978-3-319-82123-8   ISBN 978-3-319-40528-5   (eBook)
DOI 10.1007/978-3-319-40528-5

Mathematics Subject Classification (2010): 65-XX, 68-XX, 70-XX, 76-XX, 85-XX, 86-XX, 92-XX

Cover illustration: Cover figure by courtesy of iStock.com/nadla

Printed on acid-free paper

This Springer imprint is published by Springer Nature
The registered company is Springer International Publishing AG Switzerland

Preface

One of the grand challenges with respect to current developments in high-performance computing (HPC) lies in exploiting the upcoming exascale systems, i.e. systems with 10^{18} floating point operations per second and beyond. Moore's law has proven to be astonishingly robust so far. However, these days, the respective progress can no longer be obtained via faster chips or higher clock rates, but only via a massive use of parallelism and by increasingly complex ways how this parallelism is arranged in the large systems.

Extreme-scale supercomputers will be made of heterogeneous hardware and will comprise millions of cores. This entails several challenges and adds new and severe requirements to large-scale (simulation) software, including efficient programming techniques for such kinds of hardware, fault-tolerance mechanisms to compensate the failure of single compute nodes or cores, or efficient algorithms able to cope with this massive amount of parallelism on heterogeneous systems. Collaborative, international, and multidisciplinary research efforts of mathematicians, computer scientists, application developers, and domain-specific research scientists are mandatory to prepare current research software for the upcoming exascale era on the one hand and to develop new approaches to exascale programming and software required by various disciplines on the other hand.

Globally, HPC is given utmost attention due to its scientific and economic relevance. Although recent initiatives also comprise simulation software, computational algorithms, and the complete underlying software stack, the primary focus, often, is still on the "race for the exascale systems", i.e. more on the racks than on the brains. Complementing related initiatives in other countries, the German Research Foundation (DFG) decided in 2011 to launch a strategic priority programme in that direction, i.e. a priority programme initiated by its Senate Committee on Strategic Planning and financed via special funds – the Priority Program 1648 "Software for Exascale Computing" (SPPEXA). After the standard reviewing process in 2012, 13 interdisciplinary consortia with, overall, more than 40 institutions involved started their research beginning of 2013. Each consortium addresses relevant issues in at least two of SPPEXA's six research domains: (1) computational algorithms, (2) system software, (3) application software, (4) data management and exploration, (5) programming, and (6) software tools.

At the end of the first of two 3-year funding phases, SPPEXA held a 3-day international symposium at Leibniz Supercomputing Centre and at the Department of Informatics of Technical University of Munich in Garching, Germany. The conference was structured in 3 invited plenary talks and 14 minisymposia – 1 organized by each of the 13 consortia of the first funding phase, plus 1 presenting the 4 new consortia which enter SPPEXA in the second phase.

An *overview of the project outcomes and particular research findings* from within SPPEXA have been collected in this book. *Further contributions* are made by other internationally recognized research groups who have participated in the SPPEXA Symposium. Due to the wide range of SPPEXA research related to exascale software and computing, the book at hand covers various topics. These include (but are definitely not limited to):

- novel algorithms for complex problems from science and engineering, such as scalable multigrid implementations or algorithms for high-dimensional problems,
- programming approaches for heterogeneous hardware devices,
- developments of exascale-enabling software systems, programming approaches, and tools,
- exascale-relevant applications from engineering; biology/chemistry, such as molecular dynamics; astrophysics; multiscale material science; or multi-physics problems,
- performance engineering, performance analysis, and performance prediction approaches,
- the preparation of existing PDE-frameworks for the upcoming exascale age.

As always, many people helped to make SPPEXA, the SPPEXA Symposium in January 2016, and this LNCSE volume a great success. First of all, we want to thank DFG, and in particular Dr. Marcus Wilms, for shaping and providing the frame for SPPEXA. It was not an easy path, since it was the first time that DFG put a strategic priority programme into life and since, for phase two having started in 2016, it was the first time that DFG and its partner agencies JST (Japan) and ANR (France) joined forces to allow for bi- and tri-national projects within a priority programme. However, SPPEXA's journey has been and is a rewarding endeavour. We further thank all helping hands who supported SPPEXA so far in terms of organizing and hosting events such as workshops, doctoral retreats, minisymposia, gender workshops, annual plenary meetings, and so forth. Moreover, concerning the preparation of this volume, we are grateful to Dr. Martin Peters and Mrs. Than-Ha Le Thi from Springer for their support – as in previous cases, it was again a pleasure to collaborate. Finally, we thank Michael Rippl for his support in compiling this book.

Looking forward to the second phase of SPPEXA with curiosity and high expectations.

Garching, Germany	Hans-Joachim Bungartz
Garching, Germany	Philipp Neumann
Dresden, Germany	Wolfgang E. Nagel

Contents

Part I
EXA-DUNE: Flexible PDE Solvers, Numerical Methods, and Applications

Hardware-Based Efficiency Advances in the EXA-DUNE Project

Peter Bastian, Christian Engwer, Jorrit Fahlke, Markus Geveler, Dominik Göddeke, Oleg Iliev, Olaf Ippisch, René Milk, Jan Mohring, Steffen Müthing, Mario Ohlberger, Dirk Ribbrock, and Stefan Turek

Abstract We present advances concerning efficient finite element assembly and linear solvers on current and upcoming HPC architectures obtained in the frame of the EXA-DUNE project, part of the DFG priority program 1648 *Software for Exascale Computing* (SPPEXA). In this project, we aim at the development of both flexible and efficient hardware-aware software components for the solution of PDEs based on the DUNE platform and the FEAST library. In this contribution, we focus on node-level performance and accelerator integration, which will complement the proven MPI-level scalability of the framework. The higher-level aspects of the EXA-DUNE project, in particular multiscale methods and uncertainty quantification, are detailed in the companion paper (Bastian et al., Advances concerning multiscale methods and uncertainty quantification in EXA-DUNE. In: Proceedings of the SPPEXA Symposium, 2016).

P. Bastian • S. Müthing (✉)
Interdisciplinary Center for Scientific Computing, Heidelberg University, Heidelberg, Germany
e-mail: peter.bastian@iwr.uni-heidelberg.de; steffen.muething@iwr.uni-heidelberg.de

C. Engwer • J. Fahlke • R. Milk • M. Ohlberger
Institute for Computational and Applied Mathematics, University of Münster, Münster, Germany
e-mail: christian.engwer@wwu.de; rene.milk@wwu.de; mario.ohlberger@wwu.de

D. Göddeke
Institute of Applied Analysis and Numerical Simulation, University of Stuttgart, Stuttgart, Germany
e-mail: dominik.goeddeke@mathematik.uni-stuttgart.de

M. Geveler • D. Ribbrock • S. Turek
Institute of Applied Mathematics, TU Dortmund, Dortmund, Germany
e-mail: markus.geveler@math.tu-dortmund.de; dirk.ribbrock@math.tu-dortmund.de; stefan.turek@math.tu-dortmund.de

O. Iliev • J. Mohring
Fraunhofer Institute for Industrial Mathematics ITWM, Kaiserslautern, Germany
e-mail: oleg.iliev@itwm.fraunhofer.de; jan.mohring@itwm.fraunhofer.de

O. Ippisch
Institut für Mathematik, TU Clausthal-Zellerfeld, Clausthal-Zellerfeld, Germany
e-mail: olaf.ippisch@tu-clausthal.de

© Springer International Publishing Switzerland 2016   3
H.-J. Bungartz et al. (eds.), *Software for Exascale Computing – SPPEXA 2013-2015*, Lecture Notes in Computational Science and Engineering 113, DOI 10.1007/978-3-319-40528-5_1

1   The EXA-DUNE Project

Partial differential equations (PDEs) – often parameterized or stochastic – lie at the heart of many models for processes from science and engineering. Despite ever-increasing computational capacities, many of these problems are still only solvable with severe simplifications, in particular when additional requirements like uncertainty quantification, parameter estimation or optimization in engineering applications come into play.

Within the EXA-DUNE[1] project we pursue three different routes to make progress towards exascale: (i) we develop new computational algorithms and implementations for solving PDEs that are highly suitable to better exploit the performance offered by prospective exascale hardware, (ii) we provide domain-specific abstractions that allow mathematicians and application scientists to exploit (exascale) hardware with reasonable effort in terms of programmers' time (a metric that we consider highly important) and (iii) we showcase our methodology to solve complex application problems of flow in porous media.

Software development, in the scope of our work for the numerical solution of a wide range of PDE problems, faces contradictory challenges. On the one hand, users and developers prefer flexibility and generality, on the other hand, the continuously changing hardware landscape requires algorithmic adaptation and specialization to be able to exploit a large fraction of peak performance.

A framework approach for entire application domains rather than distinct problem instances facilitates code reuse and thus substantially reduces development time. In contrast to the more conventional approach of developing in a 'bottom-up' fashion starting with only a limited set of problems and solution methods (likely a single problem/method), frameworks are designed from the beginning with flexibility and general applicability in mind so that new physics and new mathematical methods can be incorporated more easily. In a software framework the generic code of the framework is extended by the user to provide application specific code instead of just calling functions from a library. Template meta-programming in C++ supports this extension step in a very efficient way, performing the fusion of framework and user code at compile time which reduces granularity effects and enables a much wider range of optimizations by the compiler. In this project we strive to redesign components of the DUNE framework [1, 2] in such a way that hardware-specific adaptations based on the experience acquired within the FEAST project [18] can be exploited in a transparent way without affecting user code.

Future exascale systems are characterized by a massive increase in node-level parallelism, heterogeneity and non-uniform access to memory. Current examples include nodes with multiple conventional CPU cores arranged in different sockets. GPUs require much more fine-grained parallelism, and Intel's Xeon Phi design shares similarities with both these extremes. One important common feature of all

---

[1]http://www.sppexa.de/general-information/projects.html#EXADUNE

these architectures is that reasonable performance can only be achieved by explicitly using their (wide-) SIMD capabilities. The situation becomes more complicated as different programming models, APIs and language extensions are needed, which lack performance portability. Instead, different data structures and memory layouts are often required for different architectures. In addition, it is no longer possible to view the available off-chip DRAM memory within one node as globally shared in terms of performance. Accelerators are typically equipped with dedicated memory, which improves accelerator-local latency and bandwidth substantially, but at the same time suffers from a (relatively) slow connection to the host. Due to NUMA (non-uniform memory access) effects, a similar (albeit less dramatic in absolute numbers) imbalance can already be observed on multi-socket multi-core CPU systems. There is common agreement in the community that the existing MPI-only programming model has reached its limits. The most prominent successor will likely be 'MPI+X', so that MPI can still be used for coarse-grained communication, while some kind of shared memory abstraction is used within MPI processes at the UMA level. The upcoming second generation of Xeon Phi (Knight's Landing) will be available both as a traditional accelerator and as a standalone, bootable CPU, enabling new HPC architecture designs where a whole node with accelerator-like properties can function as a standalone component within a cluster. Combined with the ISA convergence between standard CPUs and the new Xeon Phi, this will allow for a common code base that only has to be parameterized for the different performance characteristics (powerful versus simplistic cores and the much higher level of intra-node parallelism of the Xeon Phi) with a high potential of vastly improved developer productivity. At the same time, Xeon Phi processors will contain special on-package RAM, bringing the highly segmented memory architecture of accelerator cards one step closer to general purpose CPUs. Similarly, NVIDIA has announced the inclusion of general purpose ARM cores in upcoming generations of their GPGPUs.

Our work within the EXA-DUNE project currently targets pilot applications in the field of porous media flow. These problems are characterized by coupled elliptic/parabolic-hyperbolic PDEs with strongly varying coefficients and highly anisotropic meshes. The elliptic part mandates robust solvers and thus does not lend itself to the current trend in HPC towards matrix-free methods with their beneficial properties in terms of memory bandwidth and/or Flops/degree of freedom (DOF) ratio; typical matrix-free techniques like stencil-based geometric multigrid are not suited to those types of problems. For that reason, we aim at algebraic multigrid (AMG) preconditioners known to work well in this context, and work towards further improving their scalability and (hardware) performance. Discontinuous Galerkin (DG) methods are employed to increase data locality and arithmetic intensity. Matrix-free techniques are investigated for the hyperbolic/parabolic parts.

In this paper we report on the current state of the lower-level components of the EXA-DUNE project, while the more application-oriented parts will be presented in a separate article in these proceedings [3]. Message passing parallelism is well established in DUNE (as documented by the inclusion of DUNE's solver library in the High-Q-Club[2]), and we thus concentrate on core/node level performance. After a short introduction to our UMA node concept in Sect. 2.1, the general structure comprises two major parts: Sect. 3 focuses on problem assembly and matrix-free solvers, in particular thread-parallel assembly (Sect. 3.1) and medium to high order DG schemes (Sect. 3.2), while Sect. 4 is concerned with linear algebra, where we show the integration of a modern, cross-platform matrix format (Sect. 4.1) as well as hardware-oriented preconditioners for the CUDA architecture (Sect. 4.2).

2   Hybrid Parallelism in DUNE

In the following, we introduce the 'virtual UMA node' concept at the heart of our hybrid parallelization strategy, and ongoing current steps to incorporate this concept into the assembly and solver stages of our framework.

2.1   UMA Concept

Current and upcoming HPC systems are characterized by two trends which greatly increase the complexity of efficient node-level programming: (i) a massive increase in the degree of parallelism restricts the amount of memory and bandwidth available to each compute unit, and (ii) the node architecture becomes increasingly heterogeneous. Consequently, on modern multi-socket nodes the memory performance depends on the location of the memory in relation to the compute core (NUMA). The problem becomes even more pronounced in the presence of accelerators like Xeon Phi or GPUs, for which memory accesses might have to traverse the PCIe bus, severely limiting bandwidth and latency. In order to demonstrate the performance implications of this design, we consider the relative runtime of an iterative linear solver (Krylov-DG), as shown in Table 1: an identical problem is solved with different mappings to MPI processes and threads, on a representative 4-socket server with AMD Opteron 6172 12-core processors and 128 GB RAM. On this architecture, a UMA domain comprises half a socket (6 cores), and thus, (explicit or implicit) multi-threading beyond 6 cores actually yields slowdowns. Note that all different configurations in this benchmark use all available cores; they only differ in how many MPI ranks (UMA domains) they allocate. This experiment validates our design decision to regard heterogeneous nodes as a collection of 'virtual UMA nodes' on the MPI level: internal uniform memory access characteristics are exploited by shared memory parallelism, while communication between UMA domains is handled via (classical/existing) message passing.

---

[2]http://www.fz-juelich.de/ias/jsc/EN/Expertise/High-Q-Club/_node.html

Table 1 Poisson problem on the unit cube, discretized by the DG-SIPG method, timings for 100 Krylov iterations. Comparison of different MPI/shared memory mappings for varying polynomial degree p of the DG discretization and mesh width h. Timings $t_{M/T}$ and speedups for varying numbers of MPI processes M and threads per process T

p	h^{-1}	$t_{48/1}[s]$	$t_{8/6}[s]$	$\frac{t_{48/1}}{t_{8/6}}$	$t_{4/12}[s]$	$\frac{t_{48/1}}{t_{4/12}}$	$t_{1/48}[s]$	$\frac{t_{48/1}}{t_{1/48}}$
1	256	645.1	600.2	1.07	1483.3	0.43	2491.7	0.26
2	128	999.5	785.7	1.27	1320.7	0.76	2619.0	0.38
3	64	709.6	502.9	1.41	1237.2	0.57	1958.2	0.36

3 Assembly

As discussed before, we distinguish between three different layers of concurrency. Below the classical MPI level, we introduce thread parallelization as a new layer of parallelism on top of the existing DUNE grid interface. The number of threads is governed by the level of concurrency within the current UMA node, as explained above.

The grid loop is distributed among the threads, which allows for parallel assembly of a finite element operator or a residual vector. In order to achieve good performance within an individual thread, two major problems need to be solved: (i) as the majority of operations during finite element assembly are memory bandwidth bound, a naive approach to multithreading will not perform very well and achieve only minor speedups. (ii) Vectorization (SIMD, ILP) is required to fully exploit the hardware of modern processors or many core systems. Even in memory bound contexts, this is important as it reduces the number of load and store instructions. Due to the much more complicated and problem-specific kernels that occur as part of problem assembly, a user-friendly integration of vectorization into assembly frameworks poses a much more difficult problem compared to linear algebra, where the number of kernels is much smaller.

Finally, these building blocks need to be optimized with regard to additional constraints like memory alignment and optimal cache utilization.

We believe that locally structured data is the key to achieve both of these goals. We follow two approaches to introduce local structure to a globally unstructured discretization:

1. Low order methods: we employ a structured refinement on top of an unstructured parallel mesh. The structured refinement is computed on the fly and leads to a well defined sparse local structure, i.e., band matrices.
2. Higher order methods: higher order DG methods lead to block structured data structures and access patterns, which allow for high computational efficiency. We pair these types of methods with a sum factorized assembly algorithm to achieve a competitive effort per DOF.

While both of these methods yield computational benefits, they are still very different, and while the higher amount of structure in a medium or high order DG method might yield a higher computational efficiency, these methods also require a solution with more regularity than a low order method to achieve good convergence, making the right choice of method strongly dependent on the problem at hand.

3.1 Thread Parallel Assembly

Thread parallel assembly mainly consists of thread parallel grid sweeps and concurrent writes to the residual vector or stiffness matrix. We identify two design decisions with a significant performance impact:

(A) *Partitioning:* an existing set of grid cells – with unknown ordering – has to be split into subsets for the different threads.
(B) *Access Conflicts:* each thread works on its own set of cells, but shares some DOFs with other threads, requiring a strategy to avoid those write conflicts.

In order to describe subsets of cells in an implementation agnostic manner, we introduce the concept of EntitySets. They encapsulate an iterator range, describing a set of grid objects, in this case a set of grid cells and provide a map from grid entities into a (globally) consecutive index range for data storage.

For a given mesh $\mathscr{T}(\Omega)$ we consider three different strategies, where the first two are directly based on the induced linear ordering of all mesh cells $e \in \mathscr{T}(\Omega)$. In [7] we presented performance tests to evaluate the different partitioning strategies.

strided: for P threads, each thread p iterates over the whole set of cells $e \in \mathscr{T}(\Omega)$, but stops only at cells where $e \bmod P = p$ holds.

ranged: $\mathscr{T}$ is split into consecutive iterator ranges of the size $|\mathscr{T}|/P$, using iterators over $\mathscr{T}$ to define the entry points.

general: general partitions, like those obtained from graph partitioning libraries like METIS or SCOTCH need to store copies of all cells in the EntitySet. This is the most flexible approach, but typically creates non-contiguous per-thread subsets of the EntitySet, which in turn leads to less cache-efficient memory access patterns.

To avoid *write conflicts* we consider three different strategies:

entity-wise locks are expected to give very good performance, at the cost of additional memory requirements.

batched write uses a global lock, but the frequency of locking attempts is reduced. Updates are collected in a temporary buffer and the lock is acquired when the buffer is full.

coloring avoids write conflicts totally, but requires a particular partitioning scheme.

The experiments performed in [7] indicate that ranged partitioning with entity-wise locking and coloring can be implemented with a low overhead in the thread parallelization layer and show good performance on classic multi-core CPUs and on modern many-core systems alike. In our test the performance gain from coloring was negligible (cf. Sect. 3.3), but the code complexity increased considerably, leading us to settle on the ranged partitioning strategy for the rest of this paper.

3.2 Higher Order DG Methods

As explained in the introduction, we focus on porous media applications to demonstrate the overall viability of our approach to extreme scale computing within our project and we initially consider the prototypical problem of density driven flow in a three-dimensional domain $\Omega = (0, 1)^3$ given by an elliptic equation for pressure $p(x, y, z, t)$ coupled to a parabolic equation for concentration $c(x, y, z, t)$:

$$- \nabla \cdot (\nabla p - c \mathbf{1}_z) = 0 , \tag{1}$$

$$\partial_t c - \nabla \cdot \left((\nabla p - c \mathbf{1}_z)c + \frac{1}{Ra} \nabla c \right) = 0 . \tag{2}$$

This system serves as a model for the dissolution of a CO_2 phase in brine, where the unstable flow behavior leads to enhanced dissolution. The system is formulated in non-dimensional form with the Raleigh number Ra as the only governing parameter. For further details, we refer to [4], where we introduce the problem in a more detailed fashion and describe our decoupled solution approach based on an operator splitting for the pressure and the concentration parts. In the following, we focus on the performance characteristics of the DG scheme used for the discretization of the transport equation and (optionally, instead of a finite volume scheme) the pressure equation.

DG methods are popular in the porous media flow community due to their local mass conservation properties, the ability to handle full diffusion tensors and unstructured, nonconforming meshes as well as the simple implementation of upwinding for convection dominated flows.

Due to the large computational effort per DOF, an efficient implementation of DG methods of intermediate and high order is crucial. In many situations it is possible to exploit the tensor product structure of the polynomial basis functions and the quadrature rules on cuboid elements by using sum factorization techniques. At each element the following three steps are performed: (i) evaluate the finite element function and gradient at quadrature points, (ii) evaluate PDE coefficients and geometric transformation at quadrature points, and (iii) evaluate the bilinear form for all test functions. The computational complexity of steps (i) and (iii) is reduced from $O(k^{2d})$, $k - 1$ being the polynomial degree and d the space dimension, to $O(dk^{d+1})$ with the sum factorization technique, see [14, 15]. This can be implemented with matrix–matrix products, albeit with small matrix dimensions. For the face terms, the complexity is reduced from $O(k^{2d-1})$ to $O(3dk^d)$. For practical polynomial degrees, $k \leq 10$, the face terms dominate the overall computation time, resulting in the time

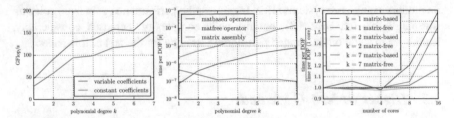

Fig. 1 Performance of the sum factorized DG assembly: GFlop/s rates for a matrix-free operator application *(left)*, time per DOF for matrix-based and matrix-free operator application as well as matrix assembly *(middle)*, relative overhead per DOF when weakly scaling from 1 to 16 cores *(right)*

per degree of freedom (DOF) to be independent of the polynomial degree. This can be seen in the second plot in Fig. 1, where the time per DOF for the matrix-free operator is almost constant starting from $k = 3$.

Our implementation of the DG scheme is based on exploiting the matrix–matrix product structure of the sum factorization kernels. We initially relied on compiler auto-vectorizers for the vectorization, but as can be seen in the results published in [4], this did not yield acceptable performance. We have thus reimplemented the kernels with explicit vectorization; for this purpose, we rely on the small external header library VCL [9] which provides thin wrappers around x86-64 intrinsics. In order to further improve and stabilize the performance of these kernels across different discretization orders k, we exploit the fact that our equation requires both the solution itself as well as its gradient, yielding a total of 4 scalar values per quadrature point, which fits perfectly with the 4-wide double precision SIMD registers of current CPUs, eliminating the need for complicated and costly data padding and or setup/tail loops. This scheme can be extended to wider architectures like AVX512 and GPUs by blocking multiple quadrature points together.

In the following, we present some results obtained with our new CPU implementation of this sum factorized DG scheme. For these benchmarks, we evaluate a stationary general convection diffusion reaction equation on the 3D unit cube. As many models assume the equation parameters to be constant within a single grid cell, our code has a special fast path that avoids parameter evaluation at each quadrature point, reducing the number of evaluations per cell from $O(k^d)$ to 1.

We performed our measurements on one CPU of a server with dual Xeon E5-2698v3 (Haswell-EP at 2.3 GHz, 16 cores, 32 hyper-threads, AVX2/FMA3, AVX clock 1.9 GHz, configured without TurboBoost and Cluster on Die, theoretical peak 486.4 GFlop/s) and 128 GB DDR4 DRAM at 2.13 GHz. All benchmarks were performed using thread pinning by first distributing hardware threads across the available cores before employing hyper-threading (if applicable). The same platform was used for all subsequent CPU benchmarks described in this paper except for Sect. 4.2. We investigated the scalability of our code within this UMA node according to our parallelization concept laid out in the introduction by means of a weak scalability study.

Figure 1 shows the overall GFlop/s rate achieved on all 16 cores during a complete matrix-free operator application for different polynomial degrees as well as the time required for these applications in relation to a matrix multiplication. As can be seen, the code already outperforms the matrix-based version for $k = 2$, without taking into account the large overhead of initial matrix assembly, which also becomes a severely limiting factor for the possible per-core problem sizes at larger core counts and higher discretization orders (for $k = 7$, we were only able to allocate 110,592 DOFs per core in order to stay within the available 4 GB RAM, which is already above average when considering current HPC systems). As can be seen, the larger amount of work per quadrature point in case of non-constant parameters directly translates into higher GFlop/s rates.

Finally, the third plot of Fig. 1 demonstrates the better scalability of the matrix-free scheme: it shows the relative overhead per DOF after weakly scaling to different numbers of active cores compared to the time per DOF required when running on a single core. While the computationally bound matrix-free scheme achieves almost perfect scalability, the matrix multiplication starts to saturate the 4 memory controllers of the CPU at between 4 and 8 active cores, causing a performance breakdown.

In order to gain further insight into the relative performance of the different assembly components, we instrumented our code to record separate timings and operation counts for the three parts of the sum factorized algorithm: calculation of the solutions at the quadrature points, per-quadrature point operations and the folding of the per-point integral arguments into the test function integrals. As can be seen in Fig. 2, the sum factorized kernels achieve very good GFlop/s rates due to their highly structured nature, especially for the 3D volume terms. In comparison, the face terms are slightly slower, which is due to both the lower dimension (less work per datum) and the additional work related to isolating the normal direction (the normal direction needs to be treated as the first/last direction in the sum factorization kernels, requiring an additional permutation step in most cases). While this step can be folded into the matrix multiplication, it creates a more complicated memory access pattern, reducing the available memory bandwidth due to less efficient prefetching, which is difficult to overcome as it involves scalar accesses spread over multiple cache lines. The lower amount of work also makes the face

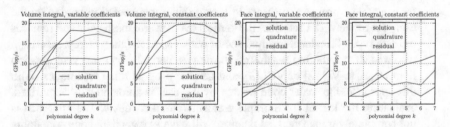

Fig. 2 GFlop/s rates for different parts of the sum factorized assembly. Rates are shown for a single core, benchmark was run with all 16 cores active

integrals more sensitive to the problem size, the residual calculation in particular hitting local performance peaks for $k = 3$ and $k = 7$, which translates into either 4 or 8 quadrature points per direction, exactly filling the 4-wide SIMD registers of the processor.

The calculations at the quadrature points do not achieve the same efficiency as the sum factorization, which is to be expected as they possess a more complex structure with e.g. parameter evaluations and (in the case of the face terms) branching due to the upwinding in the DG scheme. In order to improve performance in this area, we are currently investigating vectorization across multiple quadrature points.

3.3 Low Order Lagrange Methods

In contrast to the spectral DG methods, low order conforming methods have several major drawbacks regarding the possible performance:

 (i) The memory layout is much less favorable – assembly is performed cell wise, but the DOFs are attached to vertices (and edges etc. for polynomial degrees >1), leading to scattered memory accesses. Moreover, vertex DOFs are shared between multiple cells, increasing the size of access halos and the probability of write conflicts compared to DG.
(ii) The algorithmic intensity is very low and performance thus memory bandwidth bound rather than compute bound. While structured meshes allow to calculate a lot of information on-the-fly, reducing the amount of expensive memory transfers and increasing computational intensity, many real world applications do require unstructured meshes to correctly model complex geometries or for adaptive computations. We limit the costs of these unstructured meshes by combining globally unstructured coarse meshes with several levels of locally structured refinement on each cell to recover a minimum amount of local structure.

In the DUNE-PDELAB interface, users of the library must implement a local operator that contains the cell- and face-based integration kernels for the global operator. Vectorization has to be added at the innermost level to these kernels, i.e., at the level of cell operations, which is user code that has to be rewritten for every new problem. In order to lessen this implementation burden on the user, our framework vectorizes the kernels over several mesh cells and replaces the data type of the local residual vector with a special data type representing a whole SIMD vector. In C++ this can be done generically by using vectorization libraries, e. g. Vc[12] or VCL[9], and generic programming techniques. With this approach, the scalar code written by the user is automatically vectorized, evaluating the kernel for multiple elements simultaneously. The approach is not completely transparent, as the user will have to e.g. adapt code containing conditional branches, but the majority of user code can stay unchanged and will afterwards work for the scalar and the vectorized case alike.

Table 2 Matrix-based assembly performance: Poisson problem, Q_1 elements, assembly of Jacobian. *Left:* Xeon E5-2698v3 (cf. Sect. 3.2). *Right:* Xeon PHI 5110P (Knights Corner, 60 cores, 240 hyper-threads, peak 1011 GFlop/s)

SIMD	Lanes	Threads	Runtime	GFlop/s	%peak	SIMD	Lanes	Threads	Runtime	GFlop/s	%peak
None	1	1	38.626 s	3.01	0.6	None	1	1	43.641 s	0.17	0.02
None	1	16	2.455 s	47.28	9.7	None	1	60	2.974 s	2.44	0.24
None	1	32	3.426 s	33.88	7.0	None	1	120	1.376 s	5.27	0.52
AVX	4	1	16.570 s	4.95	1.0	Vect.	8	1	12.403 s	0.58	0.06
AVX	4	16	1.126 s	72.85	15.0	Vect.	8	60	1.474 s	4.92	0.49
AVX	4	32	2.271 s	36.12	7.4	Vect.	8	120	1.104 s	6.57	0.65

Table 3 Matrix-free assembly performance: Poisson problem, Q_1 elements, 10 iterations of a matrix-free CG. *Left:* Xeon E5-2698v3 (cf. Sect. 3.2). *Right:* Xeon PHI 5110P

SIMD	Lanes	Thread	Runtime	GFlop/s	%peak	SIMD	Lanes	Threads	Runtime	GFlop/s	%peak
None	1	1	56.19 s	0.10	0.02	None	1	1	139.61 s	0.12	0.01
None	1	16	6.84 s	0.82	0.17	None	1	60	14.74 s	1.09	0.11
None	1	32	6.13 s	0.91	0.19	None	1	120	10.50 s	1.53	0.15
AVX	4	1	44.55 s	0.09	0.02	Vect.	8	1	61.23 s	0.26	0.03
AVX	4	16	6.12 s	0.64	0.13	Vect.	8	60	12.47 s	1.29	0.13
AVX	4	32	5.50 s	0.72	0.15	Vect.	8	120	9.22 s	1.75	0.17

To evaluate the potential of vectorized assembly on structured (sub-)meshes, we present initial tests results in Table 2. The first test problem uses a conforming FEM Q_1 discretization. We measure the time to assemble a global stiffness matrix using numerical differentiation. Three levels of sub-refinement are applied and vectorization is employed across neighboring subelements. For the Xeon Phi, we include timings for 1, 60 and 120 threads. Here, the most real-world configuration involves 120 threads, as each of the 60 cores requires at least two threads to achieve full utilization. We do not include measurements for 180/240 threads, as our kernels saturate the cores at two threads per core and additional threads fail to provide further speedups. On the CPU we obtain a significant portion of the peak performance, in particular for low numbers of threads with less memory bandwidth pressure.

These results get worse if we switch to operations with lower algorithmic intensity or to many-core systems like the Xeon Phi 5110P. This is illustrated in the second example, where we consider the same problem but use a matrix free operator within an unpreconditioned CG solver, see Table 3. For such low order methods we expect this operation to be totally memory bound. In this case our benchmarks only show a very small speedup. This is in part due to bad SIMD utilization (cf. the single core results), but also due to the unstructured memory accesses, which are even more problematic on Xeon Phi due to its in-order architecture that precludes efficient latency hiding apart from its round-robin scheduling to multiple hardware threads per core; as a result, we are currently not able to leverage the performance potential of its wider SIMD architecture.

The scattered data issues can be reduced by employing special data structures for the mesh representation and the local stiffness matrix. Regarding the algorithmic intensity we expect a combination with strong smoothers, see Sect. 4.2, to improve the overall performance.

We will now discuss our modifications to the original data layout and data structures within PDELab aimed at making the data more streaming friendly. As general unstructured meshes are not suited for streaming and vectorization, we introduce a new layer, which we refer to as a *patch*. A patch represents a subset of an unstructured mesh with a local structured refinement, which is constructed on the fly and only used during assembly, which allows for a data layout which is well suited to accelerator units.

Each patch consists of a set of macro elements, made up of a number of elements extracted from the initial, unstructured mesh. We restrict ourselves to one type of macro element (i.e. either simplex or hypercube) per patch. In mixed type meshes a higher-level abstraction layer is expected to group the elements accordingly. This enables us to vectorize assembly across multiple macro elements of the patch. The macro elements are sub-refined on the fly in a structured way to a given level. For vectorized assembly, all lanes deal with corresponding subelements of different macro elements at the same time.

In the host mesh, a DOF may be associated with mesh entities of codimension > 0, which might form part of multiple patches. Thus, care must be taken to avoid data races when writing to global data structures. We circumvent this problem on the level of patches by provisioning memory for shared DOFs per patch macro element, enabling us to optimize the per-patch memory layout for vectorized access. Figure 3 illustrates the mapping of DOFs between global and per-patch storage. When preparing the assembly of a residual on a patch, the DOFs in the global coefficients vector are copied to a dedicated per-patch vector, and after the per-patch assembly we then need to accumulate the assembled data back into the layout imposed by the host mesh. While doing so we need to accumulate partial data for shared DOFs, taking care not to introduce races. The same issues and solution apply to Jacobian assembly.

This design trades increased data size for better access patterns, which creates its own set of trade-offs. In the case of the vertex coordinates used to describe the patch this should not have a big impact, because we apply virtual sub-refinement and the amount of storage for vertex coordinates should be much less than the amount of storage used for coefficient vectors and Jacobian matrices. In the case of coefficient vectors and Jacobian matrices we benefit not only from the improved access patterns, but also from the isolation of the per-patch data from the global state, reducing the need for locking or similar schemes to the patch setup/teardown layer.

The underlying DUNE interfaces, in particular the unstructured mesh, do not know about the virtual refinement. To allow reuse of existing components, we further provide a particular shape function implementation, which describes a

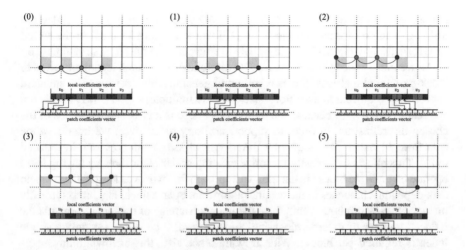

Fig. 3 Vectorized assembly for low order Lagrange discretization. Looping through consecutive elements in parallel and computing the local contributions for each local DOF. Each macro element is refined into 4 sub-cells and has 9 entries in the patch coefficients vector. Consecutive macro-cells are stored interleaved as vectors of doubles. This allows for fully automatic vectorization

refined Q_1 basis. This basis is used to encapsulate the additional intricacies of the virtual refinement layer and allows for re-use of existing DUNE components for visualization etc.

4 Linear Algebra

As laid out in the beginning, we are convinced that the problems from our domain of interest (porous media) will require a mix of matrix-free and matrix-based methods to be solved efficiently. The linear algebra part of these calculations will typically be memory bound, which makes it attractive to support moving these parts to accelerators and exploit their high memory bandwidth. In the following, we present some of our efforts in this direction.

4.1 Efficient Matrix Format for Higher Order DG

Designing efficient implementations and realizations of solvers effectively boils down to (i) a suitable choice of data structures for sparse matrix–vector multiply, and (ii) numerical components of the solver, i.e., preconditioners.

DUNE's initial matrix format, (block) compressed row storage, is ill-suited for modern hardware and SIMD, as there is no way to efficiently and generally expose a block structure that fits the size of the SIMD units. We have thus extended the SELL-C-σ matrix format introduced in [13] which is a tuned variant of the sorted ELL format known from GPUs, to be able to efficiently handle block structures [16].

As we mostly focus on solvers for DG discretizations, which lend themselves to block-structured matrices, this is a valid and generalizable decision. The standard approach of requiring matrix block sizes that are multiples of the SIMD size is not applicable in our case because the matrix block size is a direct consequence of the chosen discretization. In order to support arbitrary block sizes, we interleave the data from N matrix blocks given a SIMD unit of size N, an approach introduced in [6]. This allows us to easily vectorize existing scalar algorithms by having them operate on multiple blocks in parallel, an approach that works as long as there are no data-dependent branches in the original algorithm. Sparse linear algebra is typically memory bandwidth bound, and thus, the main advantage of the block format is the reduced number of column block indices that need to be stored (as only a single index is required per block). With growing block size, this bandwidth advantage quickly approaches 50 % of the overall required bandwidth.

So far, we have implemented the SELL-C-σ building blocks (vectors, matrices), and a (block) Jacobi preconditioner which fully inverts the corresponding subsystem; for all target architectures (CPU, MIC, CUDA). Moreover, there is an implementation of the blocked version for multi-threaded CPUs and MICs. While the GPU version is implemented as a set of CUDA kernels, we have not used any intrinsics for the standard CPU and the MIC – instead we rely on the auto-vectorization features of modern compilers without performance penalty [16]. Due to the abstract interfaces in our solver packages, all other components like the iterative solvers can work with the new data format without any changes. Finally, a new backend for our high-level PDE discretization package enables a direct assembly into the new containers, avoiding the overhead of a separate conversion step. Consequently, users can transparently benefit from our improvements through a simple C++ typedef.

We demonstrate the benefits of our approach for a linear system generated by a 3D stationary diffusion problem on the unit cube with unit permeability, discretized using a weighted SIPG DG scheme [8]. Timings of 100 iterations of a CG solver using a (block) Jacobi preconditioner on a Xeon E5-2698v3 (cf. Sect. 3.2, no hyperthreading), on a NVIDIA Tesla C2070 for the GPU measurements and on an Intel Xeon Phi 7120P, are presented in Fig. 4, normalized per iteration and DOF.

As can be seen, switching from MPI to threading affords moderate improvements due to the better surface-to-volume ratio of the threading approach, but we cannot expect very large gains because the required memory bandwidth is essentially identical. Accordingly, switching to the blocked SELL-C-σ format consistently yields good improvements due to the lower number of column indices that need to be loaded, an effect that becomes more pronounced as the polynomial degree grows due to larger matrix block sizes. Finally, the GPU and the MIC provide a further speedup of 2.5–5 as is to be expected given the relative theoretical peak memory bandwidth figures of the respective architectures, demonstrating that our

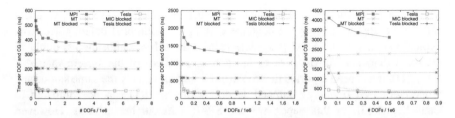

Fig. 4 Normalized execution time of the (block) Jacobi preconditioned CG solver for polynomial degrees $p = 1, 2, 3$ (*left* to *right*) of the DG discretization. The multithreaded (MT) and MIC versions use a SIMD block size of 8. The Tesla GPU versions use a SIMD block size of 32. Missing data points indicate insufficient memory

code manages to attain a constant fraction of the theoretically available memory bandwidth across all target architectures.

4.2 GPU Accelerated Preconditioners and Strong Smoothers

The promising results from enhancing the sparse matrix–vector multiply (SpMV) and therefore the whole DUNE-ISTL in DUNE by using the SELL-c-σ and BELL-c-σ storage formats lead to the idea of using this kernel in the linear solver more heavily by employing sparse approximate inverse preconditioners. Preconditioning with approximate inverses means direct application of a $M \approx A^{-1}$ that is, left-multiplying with a preassembled (sparse) preconditioner that approximates the inverse of the matrix A when solving $Ax = \mathbf{b}$. One potent representative of this family of preconditioners is the Sparse Approximate Inverse (SPAI) Algorithm initially proposed by Grote and Huckle [11] and recently applied very successfully in smoothers for Finite Element Multigrid solvers on the GPU within the FEAST software family [10]. The SPAI algorithm can briefly be described as follows:

$$\| I - M_{\mathrm{SPAI}}A \|_F^2 = \sum_{k=1}^{n} \| \mathbf{e}_k^{\mathrm{T}} - \mathbf{m}_k^{\mathrm{T}}A \|_2^2 = \sum_{k=1}^{n} \| A^{\mathrm{T}}\mathbf{m}_k - \mathbf{e}_k \|_2^2$$

where $\mathbf{e}_k$ is the k-th unit-vector and $\mathbf{m}_k$ is the k-th column of M_{SPAI}. Therefore it follows that for n columns of M we solve n independent and small *least-squares* optimization problems to construct $M = [\mathbf{m}_1, \mathbf{m}_2, \ldots \mathbf{m}_n]$:

$$\min_{m_k} \| A^{\mathrm{T}}\mathbf{m}_k - \mathbf{e}_k \|_2, \quad k = 1, \ldots n .$$

The resulting M_{SPAI} can then typically be used to accelerate a Richardson Iteration,

$$\mathbf{x}^{k+1} \leftarrow \mathbf{x}^k + \omega M_{\mathrm{SPAI}}(\mathbf{b} - A\mathbf{x}^k)$$

which can be employed as a stronger smoother in a multigrid scheme, or alternatively be applied directly without the cost of the additional defect correction. Typical variants of the SPAI procedure restrict the fill-in within the assembly to the main diagonal (SPAI(0)) or to the non-zero pattern of the system matrix A (SPAI(1)). It has been reported that SPAI(0) has approximately the same smoothing properties as an *optimally* damped Jacobi, while SPAI(1) can be compared to Gauß-Seidel [5]. In this paper, we use SPAI(1) in our benchmarks. However, stronger preconditioning based on such techniques needs discussion, in particular with regard to performance relative to simple preconditioning such as Jacobi. Although both kernels ($\omega M_{\mathrm{JAC}}\mathbf{d}$ and $M\mathbf{d}$ with the former representing a component-wise vector multiply and the latter an SpMV with the approximate inverse) are generally memory bound, the computational complexity of the SPAI preconditioner application depends on the sparsity pattern of M_{SPAI} and the memory access patterns imposed by the sparse matrix storage on the respective hardware. The Jacobi preconditioner comes at significantly lower cost and can be executed many times before reaching the computational cost of a single SPAI application. In addition, the performance gain through higher global convergence rates offered by SPAI must amortize the assembly of M_{SPAI}, which is still an open problem especially considering GPU acceleration (also being addressed within EXA-DUNE but not yet covered by this paper). On the other hand, with SPAI offering a numerical quality similar to Gauß-Seidel there is justified hope that in combination with well-optimized SpMV kernels based on the SELL-c-σ and BELL-c-σ storage formats a better overall solver performance can be achieved (also compared to even harder to parallelize alternatives such as ILU). In addition, the effectiveness for the Jacobi preconditioning depends on a good choice of ω, while SPAI is more robust in this regard.

In order to show that the SPAI preconditioner can be beneficial, we compare the overall performance of a Conjugate Gradient solver, preconditioned with SPAI(1) and Jacobi (with different values for ω) respectively. Here, we adapt an existing example program from DUNE-PDELAB that solves a stationary diffusion problem:

$$\nabla \cdot (K\nabla u) = f \text{ in } \Omega \subset \mathbb{R}^3$$

$$u = g \text{ on } \Gamma = \partial\overline{\Omega}$$

with $f = (6 - 4|x|^2)\exp(-|x|^2)$ and $g = \exp(-|x|^2)$, discretized with the same SIPG DG scheme [8] already used in Sect. 4.1. We restrict our experiments to the unit cube $\Omega = (0, 1)^3$ and unit permeability $K = 1$.

From the construction kit that comes with a fast SpMV on the GPU and a kernel to preassemble the global SPAI matrix in DUNE-ISTL, three types of preconditioners are directly made possible: a standard scalar Jacobi preconditioner,

$$S_{\mathrm{JAC}}^{\omega} : \mathbf{x}^{k+1} \leftarrow \mathbf{x}^k + \omega M_{\mathrm{JAC}}(\mathbf{b} - A\mathbf{x}^k), k = 1, \dots, K \tag{3}$$

with M_{JAC} as defined above and a fixed number of iterations K (note, that this is in order to describe how the preconditioner is applied to a vector x in the PCG solver and that this iteration solves a defect correction already and thus here, $\mathbf{b}$ is the global defect). In addition, we can also precompute the exact inverse of each logical DG block in the system matrix, making good use of the BELL storage by switching to a block Jacobi preconditioner:

$$S_{\text{BJAC}} : \mathbf{x}^{k+1} \leftarrow \mathbf{x}^k + M_{\text{BJAC}}(\mathbf{b} - A\mathbf{x}^k), k = 1, \ldots, K \tag{4}$$

with $M_{\text{BJAC}} = \sum_i R_i^T A_i^{-1} R_i$ being the exact DG-block-inverse, precomputed by a LU decomposition (using cuBLAS on the GPU). Third, a direct application of the SPAI(1) matrix to the defect can be employed:

$$S_{\text{SPAI}} : \mathbf{x} \leftarrow M_{\text{SPAI}}\mathbf{x} \tag{5}$$

with M_{SPAI} as defined above. We use both the SELL-c-σ and BELL-c-σ storage formats in this case.

We perform all benchmarks on the GPU and the CPU: here, we make use of a Maxwell GPU in a NVIDIA GTX 980 Ti consumer card with roughly 340 GB/s theoretical memory bandwidth. The Maxwell architecture of the 980 Ti is the same as in the (at the time of writing this paper) most recent iteration of the Tesla compute cards, the Tesla M40. For comparison, we use a 4-core Haswell CPU (Core i5 4690K) with 3.5 GHz (turbo: 3.9 GHz) and roughly 26 GB/s theoretical memory bandwidth.

First, we demonstrate the sensitivity of Jacobi preconditioning to damping in order to identify fair comparison configurations for the damping-free competitors Block-Jacobi and SPAI. Figure 5 shows the variation of the solver iterations depending on the damping parameter ω. We sample the parameter space in steps of

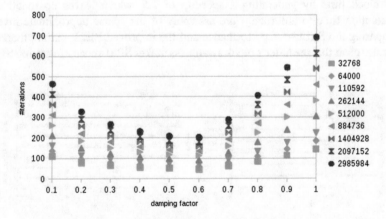

Fig. 5 Dependence of the Jacobi preconditioning on damping parameter

0.1 between 0 and 1. The measurements clearly show a 'sweet spot' around 0.6 and a worst case (expectedly) in the undamped case. Therefore, we employ $\omega = 0.6$ as an example of good damping and $\omega = 1.0$ as a bad damping coefficient. In addition, we consider a 'median' case of $\omega = 0.8$ in all following benchmarks. Note that in reality, ω is unknown prior to the solver run and has a huge impact on the overall performance which can be seen in the factor of more than 3 between the number of solver iterations for 'good' and 'bad' choices. In contrast to ω having an impact on the numerical quality of the preconditioner S_{JAC}^{ω} only and not on its computational cost, the parameter K is somewhat more complicated to take into account in the performance modelling process: here, a larger value for K produces a numerical benefit, but also increases computational cost due to the additional defect correction with each additional iteration. In many cases, the numerical benefit of increasing K does not amortize the additional cost beyond a certain value K_{opt}, as can be seen in Fig. 6 for the Jacobi preconditioner and $\omega = 0.5$. For the benchmark problem at hand, it is always beneficial to perform only 2 iterations. Note that this behavior also depends on the damping parameter and more importantly, that K is also unknown a priori. This makes both ω and K subject to *autotuning* in preconditioners that try to solve an unknown correction equation, which is also a research topic of the upcoming EXA-DUNE phase two.

Figures 7 and 8 show the timing results and numbers of iterations for first and second order DG discretizations and the preconditioners defined by Eqs. 3 (with different parameters for ω) through 5, where for the latter we employed both the SELL and BELL (labeled BSPAI) matrix storage techniques.

The first thing to notice here is that for the $p = 1$ case, the SPAI and BSPAI variants cannot beat the inexact Block-Jacobi solves, due to a better overall convergence behavior of the latter, although they come close (within 5 %). However, the assembly must be considered more expensive for both SPAI versions of M (see below). For higher order Finite Elements, the SPAI and especially the BSPAI preconditioning can beat the best (Block-) Jacobi ones concerning overall solver wall clock time by generating a speedup of 1.5, which leaves up to 50 % of the solution time to amortize a pre-assembly of the sparse approximate inverse. Comparing the SELL-c-σ performance and the improved BELL variant thereof it becomes clear that the latter's block awareness makes SPAI successful: using SELL,

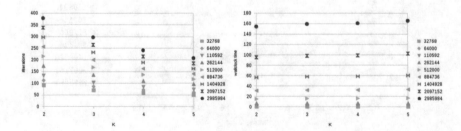

Fig. 6 K-Dependence of the Jacobi preconditioning. *Left*: Number of iterations. *Right*: Solver wall clock time

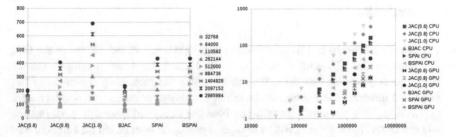

Fig. 7 Iteration count and wall clock times (logscale/logscale) of the PCG solver with different preconditioners for the benchmark problem using a first order DG discretization

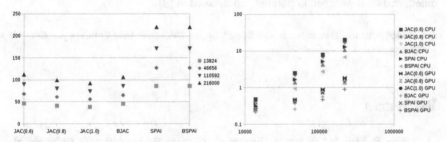

Fig. 8 Iteration count and wall clock times (logscale/logscale) of the PCG solver with different preconditioners for the benchmark problem using a second order DG discretization

SPAI is again only as good as Block-Jacobi. On the GPU, the solver performs 9.5 times better concerning wall clock time with the scalar SELL storage and 7.5 times with the BELL variant where in the former, the Haswell CPU can play out its sophisticated caches due to blocking. Here, the GPU cannot exploit the complete block structure due to the mapping of each row to one thread. Thus each thread can only exploit one row of each DG block. Both speedups are within good accordance of the factor between the theoretical memory bandwith of the respective architectures and a memory bound kernel.

Altogether, this shows that even for simple problems, the SPAI technique can be used to accelerate Krylov subspace solvers within DUNE especially for higher order Finite Elements. However, it must be stated that the overall feasibility of such Approximate Inverse techniques relies on being able to amortize the assembly time by means of faster application times of the preconditioner. In light of this, we are currently developing a GPU-based SPAI assembly based on fast QR decompositions with householder transforms on each column, which can be batched for execution similar to [17]. Also, SPAI(ϵ) (with more complex sparsity patterns for M) is being examined. Exploring the smoothing capabilities of SPAI-preconditioned iterations within DUNE's AMG schemes on the GPU is also currently under examination and expected to be finished within the remaining first phase of the EXA-DUNE project.

5   Outlook

The results presented in this contribution highlight some of the efforts of the first 2.5 years of the EXA-DUNE project. While these tools were developed mostly independently during that time, we intend to use the remaining 6 months of the project to integrate these tools into an initial demonstrator based on a porous media application. This demonstrator will combine the improved assembly performance and the faster linear algebra with a two-level preconditioner based on a matrix-free smoother for the DG level and an AMG-based subspace correction on a low order subspace, which we intend to combine with the multilevel methods and uncertainty quantification developed in parallel and detailed in [3].

Acknowledgements This research was funded by the DFG SPP 1648 *Software for Exascale Computing*.

References

1. Bastian, P., Blatt, M., Dedner, A., Engwer, C., Klöfkorn, R., Kornhuber, R., Ohlberger, M., Sander, O.: A generic grid interface for parallel and adaptive scientific computing. Part II: implementation and tests in DUNE. Computing **82**(2–3), 121–138 (2008)
2. Bastian, P., Blatt, M., Dedner, A., Engwer, C., Klöfkorn, R., Ohlberger, M., Sander, O.: A generic grid interface for parallel and adaptive scientific computing. Part I: abstract framework. Computing **82**(2–3), 103–119 (2008)
3. Bastian, P., Engwer, C., Fahlke, J., Geveler, M., Göddeke, D., Iliev, O., Ippisch, O., Milk, R., Mohring, J., Müthing, S., Ohlberger, M., Ribbrock, D., Turek, S.: Advances concerning multiscale methods and uncertainty quantification in EXA-DUNE. In: Proceedings of the SPPEXA Symposium 2016. Lecture Notes in Computational Science and Engineering. Springer (2016)
4. Bastian, P., Engwer, C., Göddeke, D., Iliev, O., Ippisch, O., Ohlberger, M., Turek, S., Fahlke, J., Kaulmann, S., Müthing, S., Ribbrock, D.: EXA-DUNE: flexible PDE solvers, numerical methods and applications. In: Lopes, L., et al. (eds.) Euro-Par 2014: Parallel Processing Workshops. Euro-Par 2014 International Workshops, Porto, 25–26 Aug 2014, Revised Selected Papers, Part II. Lecture Notes in Computer Science, vol. 8806, pp. 530–541. Springer (2014)
5. Bröker, O., Grote, M.J.: Sparse approximate inverse smoothers for geometric and algebraic multigrid. Appl. Numer. Math. **41**(1), 61–80 (2002)
6. Choi, J., Singh, A., Vuduc, R.: Model-driven autotuning of sparse matrix-vector multiply on GPUs. In: Principles and Practice of Parallel Programming, pp. 115–126. ACM, New York (2010)
7. Engwer, C., Fahlke, J.: Scalable hybrid parallelization strategies for the DUNE grid interface. In: Numerical Mathematics and Advanced Applications: Proceedings of ENUMATH 2013. Lecture Notes in Computational Science and Engineering, vol. 103, pp. 583–590. Springer (2014)
8. Ern, A., Stephansen, A., Zunino, P.: A discontinuous Galerkin method with weighted averages for advection-diffusion equations with locally small and anisotropic diffusivity. IMA J. Numer. Anal. **29**(2), 235–256 (2009)
9. Fog, A.: VCL vector class library, http://www.agner.org/optimize

10. Geveler, M., Ribbrock, D., Göddeke, D., Zajac, P., Turek, S.: Towards a complete FEM-based simulation toolkit on GPUs: unstructured grid finite element geometric multigrid solvers with strong smoothers based on sparse approximate inverses. Comput. Fluids **80**, 327–332 (2013)
11. Grote, M.J., Huckle, T.: Parallel preconditioning with sparse approximate inverses. SIAM J. Sci. Comput. **18**, 838–853 (1996)
12. Kretz, M., Lindenstruth, V.: Vc: A C++ library for explicit vectorization. Softw. Pract. Exp. **42**(11), 1409–1430 (2012)
13. Kreutzer, M., Hager, G., Wellein, G., Fehske, H., Bishop, A.R.: A unified sparse matrix data format for modern processors with wide SIMD units. SIAM J. Sci. Comput. **36**(5), C401–C423 (2014)
14. Kronbichler, M., Kormann, K.: A generic interface for parallel cell-based finite element operator application. Comput. Fluids **63**, 135–147 (2012)
15. Melenk, J.M., Gerdes, K., Schwab, C.: Fully discrete hp-finite elements: fast quadrature. Comput. Methods Appl. Mech. Eng. **190**(32–33), 4339–4364 (2001)
16. Müthing, S., Ribbrock, D., Göddeke, D.: Integrating multi-threading and accelerators into DUNE-ISTL. In: Numerical Mathematics and Advanced Applications: Proceedings of ENU-MATH 2013. Lecture Notes in Computational Science and Engineering, vol. 103, pp. 601–609. Springer (2014)
17. Sawyer, W., Vanini, C., Fourestey, G., Popescu, R.: SPAI preconditioners for HPC applications. PAMM **12**(1), 651–652 (2012)
18. Turek, S., Göddeke, D., Becker, C., Buijssen, S., Wobker, S.: FEAST – realisation of hardware-oriented numerics for HPC simulations with finite elements. Concurr. Comput.: Pract. Exp. **22**(6), 2247–2265 (2010)

Advances Concerning Multiscale Methods and Uncertainty Quantification in EXA-DUNE

Peter Bastian, Christian Engwer, Jorrit Fahlke, Markus Geveler,
Dominik Göddeke, Oleg Iliev, Olaf Ippisch, René Milk, Jan Mohring,
Steffen Müthing, Mario Ohlberger, Dirk Ribbrock, and Stefan Turek

Abstract In this contribution we present advances concerning efficient parallel multiscale methods and uncertainty quantification that have been obtained in the frame of the DFG priority program 1648 *Software for Exascale Computing* (SPPEXA) within the funded project EXA-DUNE. This project aims at the development of flexible but nevertheless hardware-specific software components and scalable high-level algorithms for the solution of partial differential equations based on the DUNE platform. While the development of hardware-based concepts and software components is detailed in the companion paper (Bastian et al., Hardware-based efficiency advances in the EXA-DUNE project. In: Proceedings of the SPPEXA Symposium 2016, Munich, 25–27 Jan 2016), we focus here on the development of scalable multiscale methods in the context of uncertainty

P. Bastian • S. Müthing
Interdisciplinary Center for Scientific Computing, Heidelberg University,
Heidelberg, Germany
e-mail: peter.bastian@iwr.uni-heidelberg.de; steffen.muething@iwr.uni-heidelberg.de

C. Engwer • J. Fahlke • R. Milk (✉) • M. Ohlberger
Institute for Computational and Applied Mathematics, University of Münster,
Münster, Germany
e-mail: christian.engwer@wwu.de; rene.milk@wwu.de; mario.ohlberger@wwu.de

D. Göddeke
Institute of Applied Analysis and Numerical Simulation, University of Stuttgart,
Stuttgart, Germany
e-mail: dominik.goeddeke@mathematik.uni-stuttgart.de

M. Geveler • D. Ribbrock • S. Turek
Institute for Applied Mathematics, TU Dortmund, Dortmund, Germany
e-mail: markus.geveler@math.tu-dortmund.de; dirk.ribbrock@math.tu-dortmund.de;
stefan.turek@math.tu-dortmund.de

O. Iliev • J. Mohring
Fraunhofer Institute for Industrial Mathematics ITWM, Kaiserslautern, Germany
e-mail: oleg.iliev@itwm.fraunhofer.de; jan.mohring@itwm.fraunhofer.de

O. Ippisch
Institut für Mathematik, TU Clausthal-Zellerfeld, Clausthal-Zellerfeld, Germany
e-mail: olaf.ippisch@tu-clausthal.de

© Springer International Publishing Switzerland 2016

25

H.-J. Bungartz et al. (eds.), *Software for Exascale Computing – SPPEXA*
2013-2015, Lecture Notes in Computational Science and Engineering 113,
DOI 10.1007/978-3-319-40528-5_2

quantification. Such problems add additional layers of coarse grained parallelism, as the underlying problems require the solution of many local or global partial differential equations in parallel that are only weakly coupled.

1 Introduction

Many physical, chemical, biological or technical processes can be described by means of partial differential equations. Due to nonlinear dynamics, interacting processes on different scales, and possible parametric or stochastic dependencies, an analysis and prediction of the complex behavior is often only possible by solving the underlying partial differential equations numerically on large scale parallel computing hardware. In spite of the increasing computational capacities plenty of such problems are still only solvable – if at all – with severe simplifications. This is in particular true if not only single forward problems are considered, but beyond that uncertainty quantification, parameter estimation or optimization in engineering applications are investigated. It has been proven that modern algorithmic approaches such as higher order adaptive modeling combined with efficient software design for highly parallel environments outperforms the pure gain of increasing compute power.[1] Hence, there is a need for algorithmic improvement, both concerning a reduction of the overall computational complexity and concerning parallel scalability of algorithms in order to exploit the computational resources of nowadays heterogeneous massively parallel architectures in an optimal manner.

Adaptive modeling, adaptive grid refinement, model reduction, multiscale methods and parallelization are important methods to increase the efficiency of numerical schemes. In the frame of the DFG priority program SPPEXA, the funded project EXA-DUNE [7] aims at the development of flexible but nevertheless hardware-specific software components and scalable high-level algorithms for the solution of partial differential equations based on the DUNE platform [3, 4] which uses state-of-the-art programming techniques to achieve great flexibility and high efficiency to the advantage of a steadily growing user-community.

While the development of hardware-based concepts and software components is detailed in [6], we focus here on the development of scalable multiscale methods in the context of uncertainty quantification which adds additional layers of coarse grained parallelism, as the underlying problems require the solution of many local or global partial differential equations in parallel that are only weakly coupled or not coupled at all (see [29] for preliminary results in this direction).

Our software concept for the efficient implementation of numerical multiscale methods in a parameterized setting is based on the general model reduction framework for multiscale problems that we recently presented in [33]. The framework covers a large class of numerical multiscale approaches based on an additive

---

[1] http://bits.blogs.nytimes.com/2011/03/07/software-progress-beats-moores-law/

splitting of function spaces into macroscopic and fine scale contributions combined with a tensor decomposition of function spaces in the context of multi query applications.

Numerical multiscale methods make use of a possible separation of scales in the underlying problem. The approximation spaces for the macroscopic and the fine scale are usually defined a priori. Typically piecewise polynomial functions are chosen on a relatively coarse and on a fine partition of the computational domain. Based on such discrete function spaces, an additive decomposition of the fine scale space into coarse parts and fine scale corrections is the basis for the derivation of large classes of numerical multiscale methods. A variety of numerical multiscale methods can be recovered by appropriate selection of such decomposed trial and test functions, the specific localizations of the function space for the fine scale correctors and the corresponding localized corrector operators. For a detailed derivation of the multiscale finite element method [14, 22], the variational multiscale method [24, 31], and the heterogeneous multiscale method [13], [19, 20, 32] in such a framework we refer to the expositions in [25] and [21].

Algorithmically, multiscale methods lead to a decomposition of the solution process into a large number of independent local so called cell problems and a global coarse problem associated with the solution of the macro scale. As the cell problems can be based on virtual locally structured meshes (only minimal memory needed) and do not need communication between each other, these methods have a high level of asynchronicity which can be realized very efficiently on accelerator cards or GPUs. The remaining global coarse problem still may be of large size and can be treated e.g., by applying AMG-type solvers.

Complex multiscale and multiphysics applications are naturally related to uncertainty quantification and stochastic modeling, as input data is often only given stochastically. Particular competing numerical methods for such problems are stochastic Galerkin/collocation methods on the one hand and Monte Carlo/Quasi-Monte Carlo methods on the other hand. While the first class of methods results in very high dimensional deterministic problems, in the second class of approaches usually very large numbers of deterministic problems have to be solved for suitable realizations of the stochastic input. Here we focus on multi-level Monte Carlo (MLMC) which was first introduced in [18] for high-dimensional or infinite-dimensional integration, was then used for stochastic ODEs [17] and has recently been applied successfully to PDEs with stochastic coefficients [2, 10, 15, 35]. The main idea of multi-level Monte Carlo is to consider the quantity of interest at different accuracy levels. At each level a different number of samples is used to compute the expected value of this quantity. In particular, very few samples are used at the finest level where the computation for each realization is expensive, while more samples are used at the coarsest level which is inexpensive to compute. By selecting the number of realizations at each level carefully one can thus decrease the computational costs enormously.

The article is organized as follows. In Sect. 2, the considered multiscale finite element method is introduced and pure MPI, as well as hybrid MPI/SMP parallelization concepts are discussed. The concept for an efficient parallel implementation of multi-level Monte Carlo methods is given in Sect. 3. Finally, results of numerical experiments are detailed in Sect. 4.

2 Numerical Multiscale Methods: A Case of Generality

Our software concept for the efficient implementation of numerical multiscale methods in a parameterized setting is based on the general model reduction framework for multiscale problems, recently presented in [33]. The framework consists of classical numerical multiscale approaches based on an additive splitting of function spaces into macroscopic and fine scale contributions combined with a tensor decomposition of function spaces in the context of multi query applications. In detail, let U, V denote suitable function spaces over a domain $\Omega \subset \mathbb{R}^d$ and let us look at solutions $u_\mu^\epsilon \in U$ of parameterized variational problems of the form

$$R_\mu^\epsilon[u_\mu^\epsilon](v) = 0 \qquad \forall v \in V \tag{1}$$

with an ϵ and μ-dependent mapping $R_\mu^\epsilon : U \to V'$ where ϵ denotes a parameter that indicates the multiscale character of the problem, and $\mu : \Omega \to \mathbb{R}^p, p \in \mathbb{N}$ denotes a vector of parameter functions that do not depend on ϵ.

Numerical multiscale methods make use of a possible separation of scales in the underlying problem. The macroscopic scale is defined by a priori chosen macroscopic approximation spaces $U_H \subset U, V_H \subset V$, typically chosen as piecewise polynomial functions on a uniform coarse partition $\mathcal{T}_H$ of Ω. The fine scale in the multiscale problem is usually defined by a priori chosen microscopic approximation spaces $U_h \subset U, V_h \subset V$, also typically chosen as piecewise polynomial functions on a uniform fine partition $\mathcal{T}_h$ of Ω. For suitable choices of polynomial degrees and meshes the spaces should satisfy $U_H \subset U_h \subset U$, and $V_H \subset V_h \subset V$, respectively. In this setting, let us denote with $\pi_{U_H} : U \to U_H, \pi_{V_H} : V \to V_H$ projections into the coarse spaces. We then define fine parts of U_h, or V_h through

$$U_{f,h} := \{u_h \in U_h : \pi_{U_H}(u_h) = 0\}, V_{f,h} := \{v_h \in V_h : \pi_{V_H}(v_h) = 0\}.$$

The discrete solution $u_{\mu,h}^\epsilon \in U_h$ is then defined through its decomposition $u_{\mu,h}^\epsilon = u_H + u_{f,h} \in U_H \oplus U_{f,h}$, satisfying

$$R_\mu^\epsilon[u_H + u_{f,h}](v_H) = 0 \qquad \forall v_H \in V_H, \tag{2}$$

$$R_\mu^\epsilon[u_H + u_{f,h}](v_{f,h}) = 0 \qquad \forall v_{f,h} \in V_{f,h}. \tag{3}$$

In a further step, a localization of the fine scale correction $u_{f,h}$ is obtained. Thus, let a coarse partition $\mathcal{T}_H$ of Ω and macroscopic discrete function spaces $U_H(\mathcal{T}_H), V_H(\mathcal{T}_H)$ be given, e.g., by choosing globally continuous, piecewise polynomial finite element spaces on $\mathcal{T}_H$. Furthermore, we choose quadrature rules $(\omega_{T,q}, x_{T,q})_{q=1}^Q$ for $T \in \mathcal{T}_H$ and associate with each quadrature point $x_{T,q}$ a local function space U^δ_{f,xT_q} which might for example be given as

$$U^\delta_{f,xT_q} := \{u_{f,xT_q} = u_{f,h}|_{Y^\delta(xT_q)} : u_{f,h} \in U_{f,h})\}$$

where $Y^\delta(xT_q)$ is an appropriate discrete δ-environment of xT_q that can be decomposed with elements from the fine mesh $\mathscr{T}_h$. Local function spaces V^δ_{f,xT_q} are defined analogously.

Next, we define local corrector operators $Q_{xT_q} : U_H \rightarrow U^\delta_{f,xT_q}$ through an appropriate localization of (3), e.g.,

$$R^\epsilon_\mu[u_H + Q_{xT_q}(u_H)](v_{f,xT_q}) = 0 \qquad \forall v_{f,xT_q} \in V^\delta_{f,xT_q} . \tag{4}$$

A corresponding local reconstruction operator $\mathscr{R}_{xT_q}$ is then given as

$$\mathscr{R}_{xT_q}(u_H) = u_H + Q_{xT_q}(u_H) \tag{5}$$

and we obtain the overall method using numerical quadrature in the coarse scale Eq. (2) and by replacing $u_H + u_{f,h}$ in (2) by the localized reconstruction $\mathscr{R}_{xT_q}(u_H)$.

Depending on the choice of trail and test functions, and on the choice of specific localizations of the function space for the fine scale correctors and by choosing corresponding localized corrector operators a variety of numerical multiscale methods can be recovered. For a detailed derivation of the multiscale finite element method, the variational multiscale method, and the heterogeneous multiscale method in such a framework we refer to the expositions in [25] and [21].

Concerning the structure of the solution spaces and the resulting discrete approximation schemes, for all numerical multiscale methods that fit into the above framework the global solution is decomposed into local solutions on merely structured sub-refinements of coarse grid blocks, and block-wise sparse global solutions. Hence, the general mathematical concept gives rise to the development of a unified interface based software framework for an efficient parallel implementation of such schemes.

As a particular realization within this concept we are considering here the multiscale finite element method (MsFEM) (see [14] for an overview) which we now detail, following the exposition in [22].

2.1 The Multiscale Finite Element Method for Multiscale Elliptic Equations

For simplicity, we will consider now the heterogeneous diffusion model problem:

$$\text{find} \quad u^\epsilon \in \overset{\circ}{H}{}^1(\Omega) : \quad \int_\Omega A^\epsilon \nabla u^\epsilon \cdot \nabla \Phi = \int_\Omega f\Phi \quad \forall \Phi \in \overset{\circ}{H}{}^1(\Omega) . \tag{6}$$

Here, $\Omega \subset \mathbb{R}^n, n \in \mathbb{N}_{>0}$ denotes a domain with a polygonal boundary and we define $\overset{\circ}{H}{}^1(\Omega) := \overline{C^\infty(\Omega)}^{\|\cdot\|_{H^1(\Omega)}}$. Furthermore, we assume that $A^\epsilon \in (L^\infty(\Omega))^{n \times n}$ and $f \in L^2(\Omega)$. For A^ϵ, we also suppose ellipticity, i.e. there exists some $\alpha \in \mathbb{R}_{>0}$ with

$$A^\epsilon(x)\xi \cdot \xi \geq \alpha|\xi|^2 \quad \forall \xi \in \mathbb{R}^n \text{ and for a.e. } x \in \Omega .$$

We note that the parameter ϵ does not have a particular value nor does it converge to zero, it just indicates that a certain quantity (such as A^ϵ or u^ϵ) exhibits microscopic features.

In order to formulate the method in a general way, we let $\mathscr{T}_H$ denote a regular partition of Ω with elements T and $\mathscr{T}_h$ a nested refinement of $\mathscr{T}_H$. Let $U_H := S_0^1(\mathscr{T}_H) \subset U_h := S_0^1(\mathscr{T}_h) \subset \overset{\circ}{H}{}^1(\Omega)$, denote associated piecewise linear finite element spaces. We assume that U_h is sufficiently accurate, i.e. we have a condition $\inf_{v_h \in U_h} \|u^\epsilon - v_h\|_{H^1(\Omega)} \leq \text{TOL}$. Furthermore, we define $\overset{\circ}{U}_h(\omega) := U_h \cap \overset{\circ}{H}{}^1(\omega)$ for $\omega \subset \Omega$. By A_h^ϵ we denote a suitable approximation of A^ϵ.

Definition 1 (Admissible environment) For $T \in \mathscr{T}_H$, we call $U(T)$ an *admissible environment* of T, if it is connected, if $T \subset U(T) \subset \Omega$ and if it is the union of elements of $\mathscr{T}_h$, i.e.

$$U(T) = \bigcup_{S \in \mathscr{T}_h^*} S, \quad \text{where } \mathscr{T}_h^* \subset \mathscr{T}_h .$$

Admissible environments will be used for oversampling. In particular T is an admissible environment of itself.

Now, we state the MsFEM in Petrov-Galerkin formulation with oversampling. The typical construction of an explicit multiscale finite element basis is already indirectly incorporated in the method. Also note that for $U(T) = T$ we obtain the MsFEM without oversampling.

Definition 2 Let $\mathscr{U}_H = \{U(T) | T \in \mathscr{T}_H\}$ denote a set of admissible environments of elements of $\mathscr{T}_H$. We call $\mathscr{R}_h^\epsilon(u_H) \in U_h \subset \overset{\circ}{H}{}^1(\Omega)$ the MsFEM-approximation of u^ϵ, if $u_H \in U_H$ solves:

$$\sum_{T \in \mathscr{T}_H} \int_T A_h^\epsilon \nabla \mathscr{R}_h^\epsilon(u_H) \cdot \nabla \Phi_H = \int_\Omega f \Phi_H \quad \forall \Phi_H \in U_H . \tag{7}$$

For $\Phi_H \in U_H$, the reconstruction $\mathscr{R}_h^\epsilon(\Phi_H)$ is defined by $\mathscr{R}_h^\epsilon(\Phi_H)_{|T} := \tilde{Q}_h^\epsilon(\Phi_H) + \Phi_H$, where $\tilde{Q}_h^\epsilon(\Phi_H)$ is obtained in the following way: first we solve for $Q_{h,T}^\epsilon(\Phi_H) \in \overset{\circ}{U}_h(U(T))$ with

$$\int_{U(T)} A_h^\epsilon \left(\nabla \Phi_H + \nabla Q_{h,T}^\epsilon(\Phi_H)\right) \cdot \nabla \phi_h = 0 \quad \forall \phi_h \in \overset{\circ}{U}_h(U(T)), \ \forall T \in \mathscr{T}_H . \tag{8}$$

Since we are interested in a globally continuous approximation, i.e. $\mathscr{R}_h^\epsilon(u_H) \in U_h \subset \overset{\circ}{H}^1(\Omega)$, we still need a conforming projection $P_{H,h}$ which maps the discontinuous parts $Q_{h,T}^\epsilon(\Phi_H)_{|T}$ to an element of U_h. Therefore, if

$$P_{H,h} : \{\phi_h \in L^2(\Omega)|\ \phi_h \in U_h(T)\ \forall T \in \mathscr{T}_H\} \longrightarrow U_h$$

denotes such a projection, we define

$$\tilde{Q}_h^\epsilon(\Phi_H) := P_{H,h}(\sum_{T \in \mathscr{T}_H} \chi_T Q_{h,T}^\epsilon(\Phi_H))$$

with indicator function χ_T.

$P_{H,h}$ might be for instance constructed by using a local average on the edges of T, i.e.

$$\tilde{Q}_h^\epsilon(\Phi_H)(x) = P_{H,h}(\sum_{T \in \mathscr{T}_H} \chi_T Q_{h,T}^\epsilon(\Phi_H)) := \frac{1}{\sharp \mathscr{T}_H(x)} \sum_{T \in \mathscr{T}_H(x)} Q_{h,T}^\epsilon(\Phi_H)(x) , \qquad (9)$$

where we defined $\mathscr{T}_H(x) := \{T \in \mathscr{T}_H|\ x \in \overline{T}\}$. For a more detailed discussion and analysis of this method we refer to [22].

2.2 Implementation and Parallelization

Parallelization of the MsFEM algorithm can yield two major benefits. The first is to increase the amount of available memory, and thereby the ability to solve ever larger problems in higher grid resolutions. The other benefit is minimizing time-to-solution for a given problem with fixed resolution. Our implementation of the general framework for multiscale methods (DUNE-MULTISCALE, [26]) is continually improving effort since the inception of the EXA-DUNE project and is built using the DUNE Generic Discretization Toolbox (DUNE-GDT, [34]) and DUNE-STUFF [28]. The implementation also relies on the DUNE Core Modules (-common, -geometry, -localfunctions, -grid) [5] and DUNE-PDELAB [1]. All of these modules are available as free and open source software and individually licensed under the GNU General Public License version 2 (GPLv2) or the BSD 2-Clause License (BSD2).

We will now describe our parallelization approach on an abstract compute cluster. This cluster shall consist of a set of N processors $\mathscr{P} = \{P_0, \ldots, P_N\}$, which we will assume as being interconnected over some reasonable fast interface (e.g., Infiniband). Each processor will run its own instance of our code, with communication between processors provided by an implementation of the Message-Passing Interface Standard (MPI [30], e.g., OpenMPI) and abstracted for our use by dune-grid.

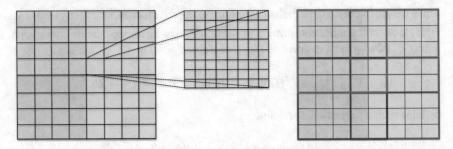

Fig. 1 Non-overlapping macro grid distribution of $\mathscr{T}_{\mathscr{H}}$ for $\mathscr{P} = P_0, \cdots, P_3$ and fine scale substructure *(left)*. Overlapping macro grid distribution of $\mathscr{T}_{\mathscr{H}}$ for $\mathscr{P} = P_0, \cdots, P_3$ *(right)*

The fundamental idea to distribute work across $\mathscr{P}$ is akin to domain decomposition. Given a coarse partition $\mathscr{T}_H$ of Ω dune-grid internally distributes subsets $\mathscr{T}_{H,P_i} \subset \mathscr{T}_H$ to each P_i. This distribution is called non-overlapping if $\bigcap_{i=0}^{N} \mathscr{T}_{H,P_i} = \emptyset$ (Fig. 1). Since we are interested in globally continuous solutions in U_H however, we require an overlapping distribution where cells can be present on multiple P_i. Consequently we will call $\mathscr{I}_i \subset \mathscr{T}_{H,P_i}$ the set of inner elements, if for all $T_H \in \mathscr{I}_i \Rightarrow T_H \notin \mathscr{I}_j$ for all i, j with $i \neq j$. Programmatically $\mathscr{I}_i$ and $\mathscr{T}_{H,P_i}$ both are provided to us by dune-grid as instances of the `Dune::GridView` interface. Let us note that for $\mathscr{P} = \{P_0\}$ our parallel implementation degrades to serial execution within the same code path, without the need for special casing, and that the current implementation is based on cubical grids on both coarse and fine scale.

The next step in the multiscale algorithm is to solve the cell corrector problems (8) from Definition 2 for all $U(T_H), T_H \in \mathscr{I}_i$, over which we iterate sequentially using the `Dune::Stuff::Grid::Walker` facility. For each T_H we create a new structured `Dune::YaspGrid` to cover $U(T_H)$. Next we need to obtain $Q_{h,T}^{\epsilon}(\Phi_H)$ for all J coarse scale basis function. After discretization this actually means assembling only one linear system matrix and J different right hand sides. The assembly is delegated to the `Dune::GDT::SystemAssembler` which is fed the appropriate elliptic operator `GDT::Operators::EllipticCG` and corresponding right hand side functionals `GDT::LocalFunctional::Codim0Integral`. The `SystemAssembler`, a class derived from `Stuff::Grid::Walker`, is designed to stack cell-local operations of any number of input functors, allowing us to complete all assembly in one single sweep over the grid.

Since our cell problems usually only contain up to about one hundred thousand elements it is especially efficient for us to factorize the assembled system matrix once and then backsolve for all right hand sides. For this we employ the `UMFPACK`[11] direct solver from the SuiteSparse library[2] and its abstraction through DUNE-ISTL [8] and DUNE-STUFF respectively. We now apply the

---

[2]http://faculty.cse.tamu.edu/davis/suitesparse.html

projections $P_{H,h}$ to get $\tilde{Q}_h^\epsilon(\Phi_H)$ and with that discretize Eq. (8), which yields a linear system in the standard way. Since this is a system with degrees of freedom (DoF) distributed across all P_i we need to select an appropriate iterative solver. Here we use the implementation of the bi-conjugate gradient stabilized method (BiCGSTAB) in DUNE-ISTL, preconditioned by an Algebraic Multigrid (AMG) solver[3] with symmetric successive overrelaxation (SSOR) smoothing or using an Incomplete LU-decomposition with thresholding (ILUT) as preconditioner. The necessary communication and cross-process DoF mapping pattern necessary for the BiCGSTAB we get from DUNE-PDELAB. For a detailed explanation and analysis of the distributed solver we refer to [9]. We note that the application of the linear system solver is the only step in our algorithm that requires a non-trivial communication.

2.3 Hybrid MPI/SMP Implementation

In contrast to the simplified cluster setup proposed in Sect. 2.2 modern compute servers contain multi-core processors, with the core count still trending up. Comparing the November 2010[4] and 2015[5] editions of the Top 500 © list of supercomputer sites we see an immense rise of systems using processors with six or more cores from 19% to 96.8%. It is evidently clear to us that efficiently using modern hardware means exploiting this thread-level, shared-memory parallelism (SMP) in addition to the message passing parallelism between processors. Employing a multithreading approach allows us to maximize CPU utilization by dynamically load balancing work items inside one CPU without expensive memory transfer or cross-node communication. Compared with the pure MPI case, we are effectively reducing the communication/overlap region of the coarse grid in a scenario with a fixed number of available cores. This also slightly reduces memory usage. Within EXA-DUNE we decided to use Intel's Thread Building Blocks (TBB) library as our multithreading abstraction.

Therefore, let us now consider a modified abstract compute cluster that is comprised of a set of processors $\mathscr{P}$, but now each P_i has an associated set of cores $C_{P_i} = \{C_{P_i}^j\}$ and a set of, hardware or software, threads $t_{C_j} = \{t_{C_j}^k\}$. For simplicity, we assume here that $j = k$ across $\mathscr{P}$. The most obvious place for us to utilize SMP is solving the cell corrector problems. These are truly locally solvable in the sense of data independence with respect to neighboring coarse cells. The idea of working on more than one cell problem in parallel comes naturally. We utilize extensions to the dune-grid module made within EXA-DUNE, presented in [16], that allow us to partition a given `GridView` into, amongst other options, connected ranges

---

[3]`Dune::Amg::AMG`, dune-istl/dune/istl/paamg/amg.hh
[4]http://www.top500.org/lists/2010/11/highlights/
[5]http://www.top500.org/lists/2015/11/highlights/

Fig. 2 Hybrid macro grid
distribution with two cores
per rank and fine scale
sub-structure of $U_{h,T}$ for
$U(T) = T$

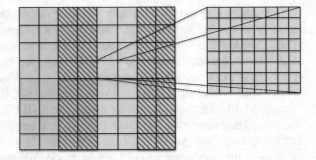

of cells. We have modified the `Stuff::Grid::Walker` accordingly to address
these partitions such that multiple threads may each iterate over one such range at a
time (Fig. 2), managed by TBB.

Without extra effort we gain thread parallel assembly of the coarse scale
system, since the `GDT::SystemAssembler` is derived from `Stuff::Grid::
Walker`. Only the application of the projections, which however have negligible
share in algorithm complexity and run time, and the coarse linear system solve have
yet to benefit from thread parallelism.

3   The Multi-level Monte-Carlo Method

MsFEM is an efficient numerical scheme for solving problems with data varying
on both, microscopic and macroscopic level. Usually, this kind of problems do not
show up in a fully determined way, but we can only guess the statistical distribution
of the underlying data. For instance, consider fluid flow through a given sector
of the ground. The complete permeability field is inaccessible, but assuming a
certain type of covariance function, we can at least determine its parameters from
a limited number of measurements. This means, however, that interesting aggregate
quantities, e.g., the total flux through the sector, can also be characterized only by
a stochastical distribution. In order to find moments of this distribution we have to
integrate the aggregate quantity with respect to the probability density of the random
parameters.

This kind of uncertainty quantification (UQ) is usually performed by Monte
Carlo methods (MC) as the dimension of the parameter space is too high to be
addressed by alternative approaches such as stochastic Galerkin schemes. Note that
UQ multiplies the effort for solving a single PDE by the combinations of random
parameters required to make some stochastic integrals converge. As solutions
for different parameters can be found independently, UQ turns out to be one of
the applications which do both, really require and benefit most from exascale
computing.

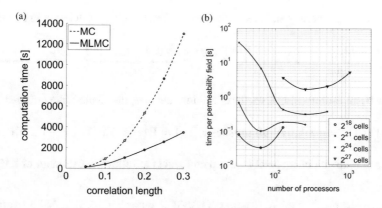

Fig. 3 Evaluation of MLMC and scaling of FFTW3. (**a**) MLMC versus MC. (**b**) Scaling of FFTW3

Unfortunately, standard MC methods are characterized by slow convergence rates. The multi-level Monte Carlo method (MLMC) is intended to attenuate this drawback [10, 15, 17]. The basic idea is to split the expected value of the aggregate quantity into two parts: the expected result of an inaccurate but fast method and the expected difference with respect to an accurate but slow solver. If the results of the fast method vary heavily for different random data, while fast and slow schemes give similar results when applied to the same data, then MLMC becomes considerably more efficient than standard MC. This is illustrated in Fig. 3a. It refers to the total flux through a unit cube the permeability of which is randomly distributed with some correlation length λ, cf. [29] for details. The higher the correlation length the more volatile is the total flux, while the deviation of coarse and fine solutions remains the same. As expected, the benefit of MLMC increases with λ.

In the above situation the two empirical mean values of coarse result and difference can be made equally accurate (similar variance) using many coarse samples, which are cheap, and only a few expensive differences of fine and coarse results. Note that MLMC can also be used to compute higher statistical moments and may comprise more levels.

3.1 Principle

Before explaining our parallel MLMC-module we have to repeat the basic principle of MLMC and introduce some notation.

Let ω be a random field characterizing different instances of the problem class. Assume we have $L + 1$ different numerical methods available to approximate the aggregate quantity $Q(\omega)$ by values $Q_l(\omega)$, $l = 0, \ldots, L$. Let the methods be ordered by increasing accuracy and cost. Then we can rewrite $Q(\omega)$ as telescoping sum:

$$Q(\omega) = \underbrace{Q_0(\omega)}_{Y_0(\omega)} + \underbrace{Q_1(\omega) - Q_0(\omega)}_{Y_1(\omega)} + \cdots + \underbrace{Q_L(\omega) - Q_{L-1}(\omega)}_{Y_L(\omega)} + \underbrace{Q(\omega) - Q_L(\omega)}_{Z_L(\omega)} .$$

Let $\mathbf{w}$ be an n-dimensional vector of random fields ω_i distributed like ω. Then

$$Y_{ln}(\mathbf{w}) := \tfrac{1}{n} \sum_{i=1}^{n} Y_l(\omega_i) \text{ satisfies } \mathrm{E}\,[Y_{ln}] = \mathrm{E}\,[Y_l] \,, \ \mathrm{Var}\,[Y_{ln}] = \tfrac{1}{n}\mathrm{Var}\,[Y_l] \,.$$

Given n_l realizations on level l we can construct the following estimator of $\mathrm{E}\,[Q]$:

$$\hat{Q}\left(\mathbf{w}^0 \ldots \mathbf{w}^L\right) = \sum_{l=0}^{L} Y_{ln_l}\left(\mathbf{w}^l\right) \text{ with } \mathrm{E}\big[Q - \hat{Q}\big] = \mathrm{E}\,[Z_L] \,, \ \mathrm{Var}\big[\hat{Q}\big] = \sum_{l=0}^{L} \tfrac{1}{n_l}\mathrm{Var}\,[Y_l] \,.$$

Let method L be chosen so accurate that $|\mathrm{E}\,[Z_L]| \leq \varepsilon$. Our goal is to have

$$\mathrm{E}\big[\big(\hat{Q} - \mathrm{E}\,[Q]\big)^2\big] = \mathrm{Var}\big[\hat{Q}\big] + \big(\mathrm{E}\big[\hat{Q} - Q\big]\big)^2 \leq 2\,\varepsilon^2 \text{ following from } \sum_{l=0}^{L} \tfrac{1}{n_l}\mathrm{Var}\,[Y_l] = \varepsilon^2 .$$
$$(10)$$

This condition may be achieved by different combinations of numbers n_l and we choose the one with minimal CPU time. Let $v_l = \mathrm{Var}\,[Y_l]$, t_l the mean time computing difference Y_l once, and $T = \sum_{l=0}^{L} n_l t_l$ the total time computing $\hat{Q}$. Minimizing T under constraint (10) and turning to integers gives

$$n_l = \text{ceil}\left[\alpha \sqrt{v_l/t_l}\right] \text{ with Lagrangian multiplier } \alpha = \tfrac{1}{\varepsilon^2} \sum_{l=0}^{L} \sqrt{v_l t_l} \,. \quad (11)$$

In practice, the n_l are computed based on empirical estimates of t_l and v_l.

3.2 Implementation

Multi-level Monte Carlo is quite a general framework for variance reduction within uncertainty quantification. In order to keep this generality, we have added a MLMC-module to DUNE [27] which does not depend on certain geometries, discretizations or solvers. The module simply expects classes providing the aggregate quantity computed by a coarse solver and classes providing the difference of results by two methods of increasing accuracy for the same set of random coefficients. The hierarchy of solvers may be related to different levels of grid refinement or different orders of a Karhunen-Loève expansion of a permeability field, for instance. But comparing completely different solution algorithms is permitted as well: in Sect. 4 we employ a one-step MsFEM-solver on the coarse level and standard FEM as fine solver. While a first version of the algorithms has shortly been sketched in [29] we now illustrate the present version in full detail.

Implementing our parallel MLMC framework we had to answer two main questions: how many processors p_l are used per sample on a given level and how can we minimize communication estimating mean times and variances? If the solvers scaled perfectly we could solve only one problem at a time using all cores in parallel. In this situation we could avoid any communication for estimating variances. In practice, however, there is an optimal number of processors p_l minimizing the product of p_l and the solution time t_l for a single problem on level l. Testing the MLMC-module we considered single phase flow through a cube. In this application the limiting factor is parallel creation of the random permeability field [29]. The underlying algorithm, circulant embedding, which is both fast and accurate, requires libraries on fast Fourier transform. Unfortunately, even the best library available, FFTW3, does not even scale monotonously, cf. Fig. 3b.

The second problem is estimating mean times and variances with a minimal communication overhead. This is done by an outer loop i over a few breaks, e.g., $n_b = 3$, and an inner loop over the levels. Here, Y_l are computed many times in parallel by groups of p_l processors until time T_l^i when statistical moments are exchanged between groups. At the beginning of a new outer loop we compute optimal n_l as in Eq. (11) and new stopping times as $T_l^i = (n_l - n_l') t_l \frac{p_l}{p} \frac{i}{n_b}$, where n_l' denotes the number of samples on level l created so far and p is the total number of processors.

Why do we feed the processor groups with stopping times rather than sample numbers? Processors may differ in performance and solution times may depend on the random data. Therefore, synchronizing the groups by sample numbers leads to undesired idle time.

Figure 4 illustrates the algorithm underlying our MLMC-module for a very simple setting with 8 processors and two levels.

1. Find the optimal number of processors per coarse sample problem (here 1).
2. Find the optimal number of processors per fine sample problem (here 4).
3. Compute the times T_1^1 and T_2^1 of the first two interrupts.
4. Coarse problems are run repeatedly on groups of one core without any communication until T_1^1 is reached.
5. Local moments (actual runtime, number of samples, sum of results, sum of squared results) are sent to master and combined to global moments of coarse results.
6. Fine problems are run repeatedly on groups of four cores without any communication until T_2^1 is reached.
7. Local moments of fine results are sent to the master and combined to global moments. New interrupt times T_1^2 and T_2^2 are computed from updated variances and estimated times per sample. The times are distributed to the groups and new independent runs are started.
8. The last pair of runs is prepared in the same way.
9. Procedure stops when sum of empirical variances is smaller than squared tolerance.

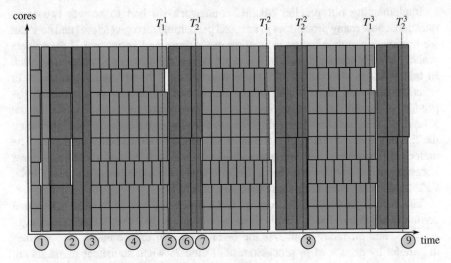

Fig. 4 Implemented multi-level Monte Carlo procedure

Up to now, only processor numbers equal to powers of 2 are supported. As illustrated in Fig. 7 the MLMC-module scales almost perfectly when applied to stationary porous media flow with random permeability as discussed above.

4 Numerical Experiments

To evaluate our MsFEM implementation we consider the following elliptic multi-scale benchmark problem.

Definition 3 (Testcase setup) $\Omega = [0, 1]^3$, with boundary conditions $u(x) = 0$ on $\partial\Omega \setminus x_3 \in \{0, 1\}$ and Neumann-zero elsewhere. With diffusion A^ϵ and source f^ϵ given as:

$$A^\epsilon(x_1, x_2, x_3) := \frac{1}{8\pi^2} \begin{pmatrix} 2\left(2 + \cos\left(2\pi\frac{x_1}{\epsilon}\right)\right) & 0 & 0 \\ 0 & 1 + \frac{1}{2}\cos\left(2\pi\frac{x_1}{\epsilon}\right) & 0 \\ 0 & 0 & 0 \end{pmatrix}$$

$$f^\epsilon(x) := -\nabla \cdot (A^\epsilon(x)\, \nabla v^\epsilon(x))$$

$$v^\epsilon(x_1, x_2, x_3) := \sin(2\pi x_1)\sin(2\pi x_2)$$

$$+ \frac{\epsilon}{2}\cos(2\pi x_1)\cos(2\pi x_2)\sin\left(2\pi\frac{x_1}{\epsilon}\right).$$

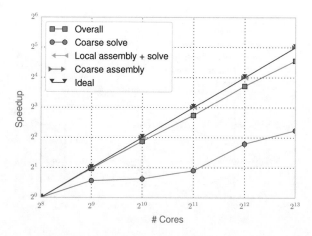

Fig. 5 Pure MPI MsFEM, 64^3 coarse cells, 8^3 fine cells per coarse cell

For the purely MPI parallel case (Fig. 5) we ran our code on the SuperMUC Petascale System (Phase 2) of the Leibniz Supercomputing Centre (LRZ). The compute nodes there each contain two Intel Xeon E5-2697 v3 (Haswell) processors with fourteen cores each and 64 GB of RAM. The uniformly refined, structured coarse grid $\mathcal{T}_H$ was created with $64^3 = 262,144$ cubes, whereas the cell problems were solved on grids with $8^3 = 512$ cubes. As the coarse scale linear solver we selected the ILUT preconditioned BiCGSTAB variant, for the local problems we always use the direct sparse solver UMFPACK. We began our strong scaling test with 256 MPI ranks on 10 nodes, with one core assigned to each rank, and continued up to 8192 MPI ranks on 293 nodes, which is eight times as many cores used as in previous results published in [7]. This means that the baseline presents 1024 coarse cells per rank, while only 32 cells per rank remain at 8192, which explains the rather poor scaling behavior of the BiCGSTAB. The overall scaling is diminished by this, as the remaining major parts of the algorithm scale near the linear ideal.

The hybrid SMP/MPI strong scaling (Fig. 6) test we performed on the CHEOPS infrastructure of the RRZK in Cologne with nodes holding two Intel Xeon X5650 (Westmere) processors with six cores each and 24 GB RAM. Since we have fewer total processors available here, we change the grid setup to 8^3 cells on the coarse grid and 32^3 cells in each local corrector problem's grid. In this case we spawn one MPI rank per processor and let TBB launch up to six threads per rank. We begin the speedup test with 16 ranks and continue up till 128 ranks, meaning we scale from using 96 cores to 786. The observed overall scaling of 7.83 is very close to the ideal scaling of 8. This is expected since the local problems are of higer dimensionality and the coarse cell per rank ratio is more favorable.

Testing the MLMC-module with respect to flexibility and parallel scaling we get by with a simple model problem: stationary single phase flow through a unit

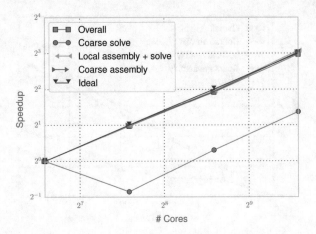

Fig. 6 Hybrid MPI/SMP MsFEM, 8^3 coarse cells, 32^3 fine cells per coarse cell

cube with random permeability field and a constant pressure difference between left and right end faces. The aggregate quantity is the total flux through the cube, which is proportional to an effective constant permeability. Permeability fields are characterized by the following type of correlation function [23]:

$$\mathrm{E}\left[\log\left(k\left(x,\cdot\right)\right)\log\left(k\left(y,\cdot\right)\right)\right] = \sigma^2 \exp\left(-\|x-y\|_2/\lambda\right), \quad x,y \in [0,1]^d, \quad (12)$$

where $k(x,\omega)$ is the permeability at position x for a vector ω of independent normally distributed random variables the number of which equals the number of cells in the discretization. As Karhunen-Loève expansions do not decay rapidly enough for our applications we use a parallel version of the circulant embedding algorithm [12], which is exact down to grid size, still fast, and also works for non-factoring covariance functions as in (12). It is based on the observation that a wide class of random permeability fields may be represented as superposition of Fourier basis functions with scaled independent normally distributed coefficients, where the scaling incorporates the correlation function. 3D permeability fields with up to 134 million cells and random variables have been computed on up to 1024 processors, cf. Fig. 3b.

In order to demonstrate flexibility, we have applied both, a three level scheme with FEM-solvers using different mesh refinements (Q_1-elements, AMG) and a two level scheme with MsFEM as coarse solver and standard FEM as fine solver. The three level scheme uses 1, 4, and 32 processors per realization and grid sizes of 2^{-5}, 2^{-6}, and 2^{-7}, respectively. The results illustrated in Fig. 7 have been generated for $\sigma = 1.0$, $\lambda = 0.2$, and tolerances $\varepsilon = 0.03$ for the 2-level-scheme and $\varepsilon = 0.005$ for the 3-level scheme.

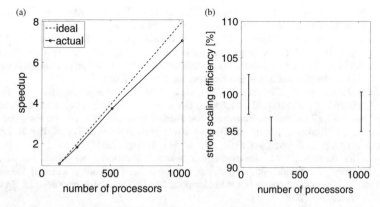

Fig. 7 Scaling of a 2-level MsFEM/FEM and a 3-level FEM MLMC. (**a**) 2-level, MsFEM/FEM. (**b**) 3-level, variable mesh size

5 Conclusion

In this contribution we introduced a software design for efficient parallelization of multiscale methods and their usage in parallel multi-level Monte Carlo methods. For the multiscale finite element method we have shown promising scaling behavior of our implementation, both in pure distributed MPI mode and in hybrid SMP/MPI mode. Furthermore, we demonstrated that our MLMC framework allows combining different kinds of algorithms per level, in particular our MsFEM implementation as a coarse solver.

Acknowledgements This research was funded by the DFG SPP 1648 'Software for Exascale Computing' under contracts IL 55/2-1, and OH 98/5-1. The authors gratefully acknowledge the Gauss Centre for Supercomputing e.V. (www.gauss-centre.eu) for funding this project by providing computing time on the GCS Supercomputer SuperMUC at Leibniz Supercomputing Centre (LRZ, www.lrz.de). We also gratefully acknowledge compute time provided by the RRZK Cologne, with funding from the DFG, on the CHEOPS HPC system under project name "Scalable, Hybrid-Parallel Multiscale Methods using DUNE".

References

1. DUNE pdelab. https://www.dune-project.org/pdelab/ (Nov 2015)
2. Barth, A., Schwab, C., Zollinger, N.: Multi-level Monte Carlo finite element method for elliptic PDEs with stochastic coefficients. Numer. Math. **119**(1), 123–161 (2011)
3. Bastian, P., Blatt, M., Dedner, A., Engwer, C., Klöfkorn, R., Kornhuber, R., Ohlberger, M., Sander, O.: A generic grid interface for parallel and adaptive scientific computing. Part II: implementation and tests in DUNE. Computing **82**(2–3), 121–138 (2008)
4. Bastian, P., Blatt, M., Dedner, A., Engwer, C., Klöfkorn, R., Ohlberger, M., Sander, O.: A generic grid interface for parallel and adaptive scientific computing. Part I: abstract framework. Computing **82**(2–3), 103–119 (2008)

5. Bastian, P., Blatt, M., Dedner, A., Engwer, C., Fahlke, J., Gräser, C., Klöfkorn, R., Nolte, M., Ohlberger, M., Sander, O.: DUNE Web page. http://www.dune-project.org (2011)
6. Bastian, P., Engwer, C., Fahlke, J., Geveler, M., Göddeke, D., Iliev, O., Ippisch, O., Milk, R., Mohring, J., Müthing, S., Ohlberger, M., Ribbrock, D., Turek, S.: Hardware-based efficiency advances in the EXA-DUNE project. In: Proceedings of the SPPEXA Symposium 2016. Lecture Notes in Computational Science and Engineering. Springer (2016)
7. Bastian, P., Engwer, C., Göddeke, D., Iliev, O., Ippisch, O., Ohlberger, M., Turek, S., Fahlke, J., Kaulmann, S., Müthing, S., Ribbrock, D.: Exa-dune: flexible PDE solvers, numerical methods and applications. In: Euro-Par 2014: Parallel Processing Workshops. Euro-Par 2014 International Workshops, Porto, 25–26 Aug 2014, Revised Selected Papers, Part II. Lecture Notes in Computer Science, vol. 8806, pp. 530–541. Springer (2014)
8. Blatt, M., Bastian, P.: The iterative solver template library. In: Kagstrom, B., Elmroth, E., Dongarra, J., Waśniewski, J. (eds.) Applied Parallel Computing. State of the Art in Scientific Computing. Lecture Notes in Computer Science, vol. 4699, pp. 666–675. Springer, Berlin/Heidelberg (2007)
9. Blatt, M., Bastian, P.: On the generic parallelisation of iterative solvers for the finite element method. Int. J. Comput. Sci. Eng. 4(1), 56–69 (2008)
10. Cliffe, K., Giles, M., Scheichl, R., Teckentrup, A.L.: Multilevel Monte Carlo methods and applications to elliptic PDEs with random coefficients. Comput. Vis. Sci. 14(1), 3–15 (2011)
11. Davis, T.A.: Algorithm 832: Umfpack v4.3 – an unsymmetric-pattern multifrontal method. ACM Trans. Math. Softw. 30(2), 196–199 (2004)
12. Dietrich, C., Newsam, G.N.: Fast and exact simulation of stationary Gaussian processes through circulant embedding of the covariance matrix. SIAM J. Sci. Comput. 18(4), 1088–1107 (1997)
13. Engquist, W.E.B.: The heterogeneous multiscale methods. Commun. Math. Sci. 1(1), 87–132 (2003)
14. Efendiev, Y., Hou, T.Y.: Multiscale Finite Element Methods: Theory and Applications. Surveys and Tutorials in the Applied Mathematical Sciences, vol. 4. Springer, New York (2009)
15. Efendiev, Y., Iliev, O., Kronsbein, C.: Multilevel monte carlo methods using ensemble level mixed MsFEM for two-phase flow and transport simulations. Comput. Geosci. 17(5), 833–850 (2013)
16. Engwer, C., Fahlke, J.: Scalable hybrid parallelization strategies for the dune grid interface. In: Numerical Mathematics and Advanced Applications: Proceedings of ENUMATH 2013. Lecture Notes in Computational Science and Engineering, vol. 103, pp. 583–590. Springer (2014)
17. Giles, M.B.: Multilevel Monte Carlo path simulation. Oper. Res. 56(3), 607–617 (2008)
18. Heinrich, S.: Multilevel Monte Carlo methods. In: Margenov, S., Wasniewski, J., Yalamov, P. (eds.) Large-Scale Scientific Computing 2001 (LSSC 2001). Lecture Notes in Computer Science, vol. 2179, pp. 58–67. Springer (2001)
19. Henning, P., Ohlberger, M.: The heterogeneous multiscale finite element method for elliptic homogenization problems in perforated domains. Numer. Math. 113(4), 601–629 (2009)
20. Henning, P., Ohlberger, M.: The heterogeneous multiscale finite element method for advection-diffusion problems with rapidly oscillating coefficients and large expected drift. Netw. Heterog. Media 5(4), 711–744 (2010)
21. Henning, P., Ohlberger, M.: A Newton-scheme framework for multiscale methods for nonlinear elliptic homogenization problems. In: Proceedings of Algoritmy 2012, Conference on Scientific Computing, Vysoke Tatry, Podbanske, 9–14 Sept 2012, pp. 65–74. Slovak University of Technology in Bratislava, Publishing House of STU (2012)
22. Henning, P., Ohlberger, M., Schweizer, B.: An adaptive multiscale finite element method. Multiscale Model. Sim. 12(3), 1078–1107 (2014)
23. Hoeksema, R.J., Kitanidis, P.K.: Analysis of the spatial structure of properties of selected aquifers. Water Resour. Res. 21(4), 563–572 (1985)

24. Hughes, T.J.R.: Multiscale phenomena: Green's functions, the Dirichlet-to-Neumann formulation, subgrid scale models, bubbles and the origins of stabilized methods. Comput. Methods Appl. Mech. Eng. **127**(1–4), 387–401 (1995)
25. Målqvist, A.: Multiscale methods for elliptic problems. Multiscale Model. Simul. **9**(3), 1064–1086 (2011)
26. Milk, R., Kaulmann, S.: DUNE multiscale. http://dx.doi.org/10.5281/zenodo.34416 (Nov 2015)
27. Milk, R., Mohring, J.: DUNE mlmc. http://dx.doi.org/10.5281/zenodo.34412 (Nov 2015)
28. Milk, R., Schindler, F.: DUNE stuff. http://dx.doi.org/10.5281/zenodo.34409 (Nov 2015)
29. Mohring, J., Milk, R., Ngo, A., Klein, O., Iliev, O., Ohlberger, M., Bastian, P.: Uncertainty quantification for porous media flow using multilevel Monte Carlo. In: Large-Scale Scientific Computing. Lecture Notes in Computer Science, vol. 9374, pp. 145–152. Springer (2015)
30. MPI Forum: MPI: A Message-Passing Interface Standard. Version 3.1 (Nov 2015). Available at: http://www.mpi-forum.org (June 2015)
31. Nordbotten, J.M., Bjørstad, P.E.: On the relationship between the multiscale finite-volume method and domain decomposition preconditioners. Comput. Geosci. **12**(3), 367–376 (2008)
32. Ohlberger, M.: A posteriori error estimates for the heterogeneous multiscale finite element method for elliptic homogenization problems. Multiscale Model. Simul. **4**(1), 88–114 (2005)
33. Ohlberger, M.: Error control based model reduction for multiscale problems. In: Proceedings of Algoritmy 2012, Conference on Scientific Computing, Vysoke Tatry, Podbanske, 9–14 Sept 2012, pp. 1–10. Slovak University of Technology in Bratislava, Publishing House of STU (2012)
34. Schindler, F., Milk, R.: DUNE generic discretization toolbox. http://dx.doi.org/10.5281/zenodo.34414 (Nov 2015)
35. Teckentrup, A., Scheichl, R., Giles, M., Ullmann, E.: Further analysis of multilevel monte carlo methods for elliptic PDEs with random coefficients. Numer. Math. **125**(3), 569–600 (2013)

Part II
ExaStencils: Advanced Stencil-Code Engineering

Systems of Partial Differential Equations in ExaSlang

Christian Schmitt, Sebastian Kuckuk, Frank Hannig, Jürgen Teich, Harald Köstler, Ulrich Rüde, and Christian Lengauer

Abstract As HPC systems are becoming increasingly heterogeneous and diverse, writing software that attains maximum performance and scalability while remaining portable as well as easily composable is getting more and more challenging. Additionally, code that has been aggressively optimized for certain execution platforms is usually not easily portable to others without either losing a great share of performance or investing many hours by re-applying optimizations. One possible remedy is to exploit the potential given by technologies such as domain-specific languages (DSLs) that provide appropriate abstractions and allow the application of technologies like automatic code generation and auto-tuning. In the domain of geometric multigrid solvers, project ExaStencils follows this road by aiming at providing highly optimized and scalable numerical solvers, specifically tuned for a given application and target platform. Here, we introduce its DSL ExaSlang with data types for local vectors to support computations that use point-local vectors and matrices. These data types allow an intuitive modeling of many physical problems represented by systems of partial differential equations (PDEs), e.g., the simulation of flows that include vector-valued velocities.

1 Introduction

The solution of PDEs is a part of many problems that arise in science and engineering. Often, a PDE cannot be solved analytically but must be solved numerically. As a consequence, the first step towards a solution is to discretize the equation, which results in a system of (linear) equations. However, depending on the size of the problem and the targeted numerical accuracy, the systems can grow quite large and result in the need for large clusters or supercomputers. These

C. Schmitt (✉) • S. Kuckuk • F. Hannig • J. Teich • H. Köstler • U. Rüde
Department of Computer Science, Friedrich-Alexander University Erlangen-Nürnberg, Erlangen, Germany
e-mail: christian.j.schmitt@fau.de

C. Lengauer
Faculty of Informatics and Mathematics, University of Passau, Passau, Germany

© Springer International Publishing Switzerland 2016
H.-J. Bungartz et al. (eds.), *Software for Exascale Computing – SPPEXA 2013-2015*, Lecture Notes in Computational Science and Engineering 113, DOI 10.1007/978-3-319-40528-5_3

execution platforms are increasingly heterogeneous for reasons such as performance and energy efficiency. Today, a compute cluster consists of hundreds of nodes, where each node may contain multiple CPU cores—sometimes even of different type—and one or more accelerators, e.g., a GPU or some other manycore accelerator such as the Xeon Phi.

A common approach to enabling performance portability for a variety of platforms is the separation of algorithm and implementation via a domain-specific language (DSL). In a DSL, domain experts can specify an algorithm to solve a certain problem without having to pay attention to implementation details. Instead, they can rely on the DSL compiler to generate a program with good—or even near optimal—performance. Usually, the execution of hand-written program code is faster than that of automatically generated code. However, rather than putting hardware and optimization knowledge individually into each application's implementation in isolation, in the DSL approach, all these efforts are put into the compiler and, consequently, every program benefits. Thus, for a new execution platform, only the compiler must be adapted, not individual application programs. This enhances performance portability.

In contrast, library-based approaches require updates of the library to make use of new technologies. This may potentially break backward compatibility and incur changes to the application program which often lead to laborious re-programming efforts. Often, a new technology comes with a new programming paradigm which is not easily captured using a library that was developed with a previous technology and paradigm in mind.

An additional advantage of DSLs is that users can be more productive by composing a new algorithm much more quickly, since it requires only a short specification. Yet another advantage of generative approaches is the ability to validate models. By providing language elements with corresponding constraints, a great number of invalid models become non-specifiable. Furthermore, since the DSL compiler has some knowledge about the application domain and works at a much higher level of abstraction, it can perform semantic validations, avoiding the generation of invalid programs and helping end-users in the correction of errors.

2 Multigrid Methods

In this section, we give a short introduction to multigrid methods. For a more in-depth review, we refer to the respective literature [6, 23].

In scientific computing, multigrid methods are a popular choice for the solution of large systems of linear equations that stem from the discretization of PDEs. The basic multigrid method cycle is shown in Algorithm 1. Here, by modifying the parameter γ that controls the number of recursive calls, one can choose between the V-cycle ($\gamma = 1$), and the W-cycle ($\gamma = 2$). There exist additional cycle types that provide higher convergence rates for certain problems [23].

Algorithm 1 Recursive multigrid algorithm to solve $u_h^{(k+1)} = \mathrm{MG}_h\left(u_h^{(k)}, A_h, f_h, \gamma, \nu_1, \nu_2\right)$

---

if coarsest level **then**
    solve $A_h u_h = f_h$ exactly or by many smoothing iterations
else
    $\bar{u}_h^{(k)} = \mathscr{S}_h^{\nu_1}\left(u_h^{(k)}, A_h, f_h\right)$ $\triangleright$ pre-smoothing
    $r_h = f_h - A_h \bar{u}_h^{(k)}$ $\triangleright$ compute residual
    $r_H = R r_h$ $\triangleright$ restrict residual
    for $j = 1$ to γ **do**
        $e_H^{(j)} = \mathrm{MG}_H\left(e_H^{(j-1)}, A_H, r_H, \gamma, \nu_1, \nu_2\right)$ $\triangleright$ recursion
    $e_h = P e_H^{(\gamma)}$ $\triangleright$ interpolate error
    $\tilde{u}_h^{(k)} = \bar{u}_h^{(k)} + e_h$ $\triangleright$ coarse grid correction
    $u_h^{(k+1)} = \mathscr{S}_h^{\nu_2}\left(\tilde{u}_h^{(k)}, A_h, f_h\right)$ $\triangleright$ post-smoothing

---

In the pre- and post-smoothing steps, high-frequency components of the error are damped by smoothers such as the *Jacobi* or the *Gauss-Seidel* method. In Algorithm 1, ν_1 and ν_2 denote the number of smoothing steps that are applied. Low-frequency components are transformed to high-frequency components by restricting them to a coarser level, making them good targets for smoothers once again.

At the coarsest level, the small number of unknowns makes a direct solution for the remaining unknowns feasible. In the special case of a single unknown, the single smoother iteration corresponds to solving the problem directly. When moving to large-scale clusters, parallel efficiency can be potentially improved by stopping at a few unknowns per compute node and relying on specialized coarse grid solvers such as the conjugate gradient (CG) and generalized minimal residual (GMRES) methods.

3 The ExaStencils Approach

ExaStencils[1] [9] is a basic research project focused on a single application domain: geometric multigrid. The implementation of large simulations involving a great diversity of different mathematical models or complex work flows is out of Exa-Stencils' scope. The project's goal is to explore how to obtain optimal performance on highly heterogeneous HPC clusters automatically. By employing a DSL for the specification of algorithms and, therefore, separating it from the implementation, we are able to operate on different levels of abstraction that we traverse during code generation. As a consequence, we can apply appropriate optimizations in every code refinement step, i.e., algorithmic optimizations, parallelization and communication

---

[1] http://www.exastencils.org/

optimizations down to low-level optimizations, resulting in a holistic optimization process. One key element in this optimization chain, working mainly at the algorithmic level, is local Fourier analysis (LFA) [2, 25] to obtain a-priori convergence predictions of iterative schemes. This helps to select adequate solver components—if not specified by the user—and to fine-tune numerical parameters. Another central feature of the ExaStencils approach is software product line (SPL) technology [21], which treats an application program not as an individual but as a member of a family with commonalities and variabilities. Based on machine learning from previous code-generation and benchmark runs, this supports the automatic selection of the optimization strategy that is most effective for the given combination of algorithm and target hardware. Embedded into the ExaStencils compiler, the techniques of LFA and SPL are sources of domain knowledge that is available at compile time.

4   The ExaStencils DSL ExaSlang

When creating a new programming language—especially a DSL—it is of utmost importance to pay attention to the user's experience. A language that is very complex will not be used by novices, whereas a very abstract language will not be used by experts. For our DSL ExaSlang—short for ExaStencils language—we identified three categories of users: domain experts, mathematicians, and computer scientists.

Each category of users focuses on a different aspect of the work flow resulting in the numerical solver software, starting with the system of equations to be solved. Whereas the domain expert cares about the underlying problem, and to some extent, about the discretization, the mathematician focuses on the discretization and components of the multigrid-based solver implementation. Finally, the computer scientist is mainly interested in the numerical solver implementation, e.g., parallelization and communication strategies.

The following subsections highlight a number of concepts and features of ExaSlang. A more detailed description can be found elsewhere [18].

4.1   Multi-layered Approach

As pictured in Fig. 1, ExaSlang consists of four layers that address the needs of the different user groups introduced previously. We call them ExaSlang 1–4; higher numbers offer less abstraction and more language features.

In ExaSlang 1, the problem is defined in the form of an energy functional to be minimized or a partial differential equation to be solved, with a corresponding computational domain and boundary definitions. In any case, this is a continuous description of the problem. We propose this layer for use by scientists and engineers that have little or no experience in programming. The problem specification might be on paper or also in LATEX or the like.

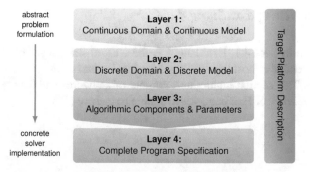

Fig. 1 Multi-layered approach of ExaSlang [18]

In ExaSlang 2, details of the discretization of the problem are specified. We deem this layer suitable for more advanced scientists and engineers as well as mathematicians.

In ExaSlang 3, algorithmic components, settings and parameters are modeled. Since they build on the discretized problem specified in ExaSlang 2, this is the first layer at which the multigrid method is discernible. At this layer, it is possible to define smoothers and to select the multigrid cycle. Computations are specified with respect to the complete computational domain. Since this is already a very advanced layer in terms of algorithm and discretization details, we see mainly mathematicians and computer scientists working here.

In ExaSlang 4, the most concrete language layer, user-relevant parts of the parallelization become visible. Data structures can be adapted for data exchange and communication patterns can be specified via simple statements. We classify this layer as semi-explicitly parallel and see only computer scientists using it. A detailed description of its key elements is given in the next subsection. Note that, even though this is the least abstract layer, it is still quite a bit more abstract than the solver implementation generated in, e.g., C++.

Orthogonal to the functional program description is the target platform description language (TPDL), which specifies not only the hardware components of the target system such as CPUs, memory hierarchies, accelerators, and the cluster topology, but also available software such as compilers or MPI implementations.

Unavailable to the user and, thus, not illustrated in Fig. 1 is what we call the intermediate representation (IR). It forms a bridge between the code in ExaSlang 4 and the target code in, e.g., C++ and contains elements of both. This is the stage at which most of the compiler-internal transformations take place, i.e., parallelization efforts such as domain partitioning, and high-level and low-level optimizations such as polyhedral optimizations and vectorization. Finally, the IR is transformed to target source code, e.g., in C++, that is written to disk and available for the user to transfer to the designated hardware to compile and run.

4.2   Overview of ExaSlang 4

As explained in Sect. 4.1, ExaSlang 4 is the least abstract layer of ExaSlang and has been extended to host the data types for local vectors that form a crucial part of ExaSlang 3. This section highlights a number of keywords and data types. A more thorough overview of ExaSlang 4 is available elsewhere [18].

4.2.1   Stencils

Stencils are crucial for the application domain and approach of project ExaStencils. They are declared by specifying the offset from the grid point that is at the center of the stencil and a corresponding coefficient. Coefficients may be any numeric expression, including global variables and constants, binary expressions and function calls. Since access is via offsets, the declarations of coefficients do not need to be ordered. Furthermore, unused coefficients, which would have a value of 0, can be omitted. An example declaration using constant coefficients is provided in Listing 1.

```
1   Stencil Laplace@all {
2       [ 0,   0,   0] =>   6.0
3       [ 1,   0,   0] => -1.0
4       [-1,   0,   0] => -1.0
5       [ 0,   1,   0] => -1.0
6       [ 0,  -1,   0] => -1.0
7       [ 0,   0,   1] => -1.0
8       [ 0,   0,  -1] => -1.0
9   }
```

Listing 1 Example 3D stencil declaration

4.2.2   Fields and Layouts

From the mathematical point of view, fields are vectors that arise, for example, in the discretization of functions. Therefore, a field may form the right-hand side of a partial differential equation, the unknown to be solved, or represent any other value that is important to the algorithm, such as the residual. As such, different boundary conditions can be specified. Currently, Neumann, Dirichlet, and no special treatment are supported. Values of fields may either be specified by the users via constants or expressions, or calculated as part of the program. Multiple copies of the same fields can be created easily via our slotting mechanism that works similarly to a ring buffer and can be used for intuitive specifications of Jacobi-type updates and time-stepping schemes. To define a field, a layout is mandatory. It specifies a data type and location of the discretized values in the grid, e.g., grid nodes or cells, and communication properties such as the number of ghost layers. In case the special field declaration `external Field` is detected, data exchange functions are generated for linked fields. They can be used to interface generated solvers as part of larger projects.

4.2.3 Data Types, Variables, and Values

As a statically-typed language, ExaSlang 4 provides a number of data types which are grouped into three categories. The first category are *simple data types*, which consist of Real for floating-point values, Integer for whole numbers, String for the definition of character sequences, and Boolean for use in conditional control flow statements. Additionally, the Unit type is used to declare functions that do not return any value. The second category are *aggregate data types*, a combination of simple data types, namely for complex numbers and the new data types for local vectors and matrices which are introduced in Sect. 6. Finally, there are *algorithmic data types* that stem from the domain of numerical calculations. Apart from the aforementioned data types stencil, field and layout, the domain type belongs to this category and is used to specify the size and shape of the computational domain.

Note that variables and values using algorithmic data types can only be declared globally. Other data types can also be declared locally, i.e., inside functions bodies or nested local scopes such as conditional branch bodies or loop bodies. Additionally, to keep variable content in sync across program instances running on distributed-memory parallel systems, they can be declared as part of a special global declaration block.

The syntax of variable and constant declarations is similar to that of Scala, with the keywords `Variable` and `Value` or, in short, `Var` and `Val`. Followed by the user-specified name, both definitions require specification of the data type, which can be of either simple or aggregate. Optionally for variables—mandatory for values—an initial value is specified via the assignment operator.

4.2.4 Control Flow

Functions can take an arbitrary number of parameters of simple or aggregate types and return exactly one value of a simple or aggregate type, or nothing. In the latter case, the return type is `Unit`. If the compiler detects a function with the signature `Function Application() : Unit`, a C++ function `main()` is generated and the compilation process is switched to the generation of a standalone program. A lot of the ExaSlang 4 syntax is like Scala, but there are additional features. In ExaSlang 4, functions are introduced with the keyword `Function`, or shorter, `Func`. An example declaration, which additionally uses the concept of level specifications presented later, is depicted in Listing 2.

The syntax and semantics of conditionals in ExaSlang 4 corresponds to Scala.

An important concept in ExaSlang 4 are loops, which are available in two main types: temporal (i.e., sequential in time) and spatial (i.e., parallel across the

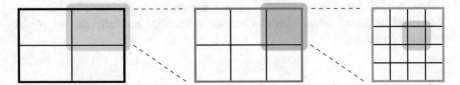

Fig. 2 Example partitioning of the computational domain into 4 blocks (*green*) of 6 fragments (*brown*) each, with 16 data values (*blue*) per fragment

computational domain). The temporal loop has the keyword `repeat` and comes as a post-test loop (`repeat until <condition>`) or a counting loop (`repeat <integer> times`). Spatial loops iterate across the computational domain. Since ExaSlang 4 is explicitly parallel, a `loop over <field>` can be nested inside a `loop over fragments` loop. Fragments are entities that arise during domain partitioning. Fragments aggregate to blocks, which in turn form the computational domain. This hierarchy is depicted in Fig. 2, where the computational domain is divided into four blocks, each consisting of six fragments. Each fragment consists of 16 data values at chosen discretization locations. The reasoning behind this strategy is to connect primitives with different parallelization concepts such as distributed- and shared-memory parallelism. One example is to map blocks to MPI ranks and fragments to OpenMP threads.

4.2.5 Level Specifications

Function, as well as layout, field, stencil, variable and value declarations can be postfixed by an @ symbol, followed by one or more integers or keywords. We call them level specifications, as they bind a certain program entity to one or several specific multigrid levels. This feature is unique to ExaSlang 4. A common usage example is to end the multigrid recursion at the coarsest level, as depicted in Listing 2. Level specifications support a number of keywords that are mapped to discrete levels during code generation. To write ExaSlang 4 programs without the explicit definition of the multigrid cycle size—and, thus, enable the application of domain knowledge at compile time—aliases such as `coarsest` and `finest` can reference bottom and top levels of the multigrid algorithm. For declarations, the keyword `all` marks an element to be available at all multigrid levels. Inside a function, relative addressing is possible by specifying `coarser` and `finer`, or by specifying simple expressions. Here, decreasing level numbers corresponds to decreasing (coarsening) the grid size, with 0 being the coarsest level if not defined otherwise by the compiler's domain knowledge. Structures at the current multigrid level are referenced by `current`. Level specifications are resolved at compile time. Thus, general specifications such as `all` are overridden by more specific ones. For example, line 5 of Listing 2 could also be declared as `Function VCycle@all`, since the definition at the coarsest level would be overridden by the definition on line 1.

```
1  Function VCycle@coarsest () : Unit {
2    // solve on coarsest grid
3  }
4
5  Function VCycle@((coarsest + 1) to finest) () : Unit {
6    // standard V-cycle
7  }
```

Listing 2 Specifying direct solving on the coarsest multigrid level to exit recursion using level specifications

5 Code Generation

Our transformation and code generation framework, which forms the basis for all transformations that drive the compilation process towards the various target platforms, is written in Scala [13, 19]. Because of its flexible object-functional nature, we deem Scala a suitable language for the implementation of DSLs and corresponding compilers. Scala features the powerful technique of pattern matching that is used to identify object instances based on types or values at run time, making it easy and elegant to find and replace parts of the program during the compilation process.

Since ExaStencils is meant to support high-performance computing, our target platforms include clusters and supercomputers such as SuperMUC, TSUBAME and JUQUEEN. We used especially the latter to validate our scaling efforts [18]. However, while scalability is one thing, run-time performance is what users are interested in. Thus, during code generation, a number of high-level optimizations based on polyhedral transformations are applied [8], such as loop tiling to enable parallelization. Another optimization is the increase of data locality by tiling and modifying the schedule of loops. Additionally, low-level optimizations such as CPU-specific vectorization have been implemented. Furthermore, we demonstrated that our compilation framework and code generation approach is flexible enough to generate specialized hardware designs from the abstract algorithm description given in ExaSlang 4 [20].

6 Data Types for Systems of Partial Differential Equations

This section highlights the advantages of local vectors and matrices for systems of PDEs and sketches their usage in ExaSlang 3 and ExaSlang 4.

```
1  Layout flowLayout < ColumnVector<Real, 3>, Node> @all {
2    ghostLayers = [ 0, 0, 0 ]
3    duplicateLayers = [ 1, 1, 1 ]
4  }
5  Field Flow < global, flowLayout, Neumann >[2]@all
```

Listing 3 Definition of layout and field of vectors with Neumann boundary conditions

6.1 Motivation

Systems of PDEs can always be expressed in ExaSlang 4 by splitting up components since, this way, only scalar data types are required. However, to implement computations of coupled components, data structures require multiple scalar values per point. We call such data types vectors or matrices, respectively, and have just recently incorporated them in ExaSlang 4, as a preparation step for code specified in ExaSlang 3.

One added benefit of specialized data types for the specification of systems of PDEs is the much increased readability of the source code—for us, of ExaSlang 4 code. Especially for domain experts, who should not have to be experts in programming, they correspond to a more natural representation of the mathematical problem which will help when checking or modifying ExaSlang 3 and 4 code that has been generated from more abstract layers of ExaSlang, i.e., ExaSlang 2.

6.2 The ExaSlang Data Types

In ExaSlang 4, the new data types `Vector` and `Matrix` belong to the category of aggregate data types and can be given a fixed dimensionality. Additionally, a `ColumnVector` (short: `CVector`) can be specified to explicitly set the vector type when the direction cannot be derived from assigned values. The element types of these aggregated data types can be simple numeric data types, i.e., integers, reals or complex numbers. As is the case for other declarations in ExaSlang 4, it is possible to use a short-hand notation by specifying the designated inner data type, followed by the corresponding number of elements in each direction. An example is shown in Listing 4, where lines 1 and 2 are equivalent.

Anonymous constant vectors default to row vectors. The suffix T transposes vector and matrix expression, thus defines the second vector to be a column vector expression in line 3 of Listing 4.

As part of the optimization process, the ExaStencils compiler applies transformations such as constant propagation and folding also to expressions containing vectors and matrices. Beside the standard operators such as addition, subtraction and multiplication that consider the vector and matrix data types in a mathematical sense, there are element-wise operators. Example calculations are depicted in Listing 4, for both vector-wise and element-wise operators. Of course, vector and matrix

```
1  Var a : Matrix<Real, 3, 3> = { {1,2,3}, {4,5,6}, {7,8,9} }
2  Var b : Real<3, 3> = { {1,2,3}, {4,5,6}, {7,8,9} }
3  Var c : Real = {1,2,3} * {1,2,3}T
4  Var d : Vector<Real, 3>
5  print("Matrix scaling: ", 7 * b)
6  print("Vector addition: ", {1,2,3} + {3,4,5})
7  print("Matrix multiplication: ", b * {{1,2}, {3,4}, {5,6}})
8  print("Vector mult.: ", {1,2,3}T * {1,2,3}) // yields a 3x3
       matrix
9  print("Element-wise mult.: ", {1,2,3} .* {1,2,3}) // yields
       {1,4,9}
```

Listing 4 Example declarations and calculations using vectors and matrices

entries do not need to be constant, but can be any valid expression evaluating to a numeric value.

7   Modifications to the Code Generator

In ExaSlang 4, the dimensionality of a field equals the dimensionality of the problem. That is, fields may have up to three dimensions. However, with our new data types, each grid point in the field may have a non-zero dimensionality as well. At present we work with vectors and 2D matrices, but our implementation can also handle higher dimensionalities.

In order to support the new data types in the generated C++ code, one could simply store multiple scalar values inside a structure to represent a local vector (or matrix) at a certain grid point, such that a field becomes an array of structures. However, arrays of structures potentially introduce run-time overhead caused, for one, by the dynamic memory management applied to instances of the structure and, for another, because custom compilers like for CUDA and OpenCL generate inferior target code if they can handle arrays of structures at all. Also, high-level synthesis (HLS) tools, which emit hardware descriptions for FPGAs, provide limited or no support for arrays of structures.

To overcome these limitations and to enable optimizations such as tiling a field for parallelization or, for hybrid target platforms, distribution across different devices, we linearize fields and the structures in them. This exposes the size and memory layout of a field and provides the possibility to modify them. The dimensionality of the array that represents a field with non-scalar grid points is the sum of the dimensionality of the field and that of the structure at the grid points. For example, a three-dimensional field of 2×2 matrices becomes a five-dimensional array. At each grid point, one matrix consisting of four scalar values is stored, resulting in a total of $4 \cdot n \cdot m$ values for a field of size $n \times m$. Each value has five coordinates as depicted in Fig. 3: the d_i denote the coordinates of the grid point, the c_i those of the structure at the grid point.

During code generation, special care is necessary for non-scalar variables that appear on both sides of an assignment involving a multiplication such as $A = A B$

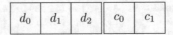

Fig. 3 Access to one element of three-dimensional matrix field consists of the index to field's grid point (d_i) and of the matrix element (c_i)

with A and B being matrices or vectors of appropriate shape. The assignment is refined to two separate assignments: first, the operation is applied and the result is saved into a temporary variable ($A' = A B$), then the original variable is reassigned ($A = A'$). This guarantees that no intermediate result is used to calculate subsequent vector or matrix entries while, at the same time, resulting in code that can be vectorized easily.

8 Example Application

To demonstrate the application of the new data types, we choose the calculation of the optical flow detection. In contrast to solving the incompressible Navier-Stokes equations, it does not need specialized smoothers and also exhibits acceptable convergence rates when solved without the use of systems of PDEs, making it an excellent example to compare code sizes of the two approaches. The optical flow approximates the apparent motion of patterns such as edges, surfaces or objects in a sequence of images, e.g., two still images taken by a camera or a video stream. Note that this approximation does not need to necessarily describe the physical motion; the actual motion of an object is not always reflected in intensity changes in the images. To be more precise, we actually calculate the displacement field between two images.

8.1 Theoretical Background

Among the numerous approaches to approximate the optical flow, we opt for a multigrid-based algorithm [7].

Our goal is to approximate the 2D motion field (u, v) between two images that are part of an image sequence **I**. An image point $I(x, y, t)$ has, aside from the two spatial coordinates x and y, a temporal coordinate t. As an example, a certain value of t can correspond to one frame of a video stream. We assume that a moving object does not change in intensity in time, i.e., we neglect changes in illumination. We call this the *constant brightness assumption*, which can be written as follows:

$$\frac{d\mathbf{I}}{dt} = 0 \ . \tag{1}$$

For small movements, i.e., for small time differences between two images, the movement of an intensity value at a pixel (x, y, t) can be described by:

$$I(x, y, t) = I(x + dx, y + dy, t + dt) . \tag{2}$$

Taylor expansion of this term around (x, y, t) and reordering results in:

$$\frac{\partial \mathbf{I}}{\partial x}\frac{dx}{dt} + \frac{\partial \mathbf{I}}{\partial y}\frac{dy}{dt} + \frac{\partial \mathbf{I}}{\partial t} \approx 0 . \tag{3}$$

We now can define the partial image derivatives $I_x := \frac{\partial \mathbf{I}}{\partial x}$, $I_y := \frac{\partial \mathbf{I}}{\partial y}$, $I_t := \frac{\partial \mathbf{I}}{\partial t}$, the spatio-temporal gradient $\nabla_\theta \mathbf{I} := \left(I_x, I_y, I_t\right)^{\mathsf{T}}$ and the optical flow vector $(u, v) := \left(\frac{dx}{dt}, \frac{dy}{dt}\right)$.

After more transformation steps, we end up with a two-dimensional system of PDEs:

$$-\alpha \Delta u + I_x(I_x u + I_y v) = -I_x I_t \tag{4}$$

$$-\alpha \Delta v + I_y(I_x u + I_y v) = -I_y I_t . \tag{5}$$

After discretization using finite differences (FD) for constant coefficient operators and image derivatives and finite volumes (FV) for variable operators, we obtain the following linear system:

$$\left(\begin{matrix} \alpha + \begin{pmatrix} & -1 & \\ -1 & 4 & -1 \\ & -1 & \end{pmatrix} + I_x^2 & I_x I_y \\ I_x I_y & \alpha + \begin{pmatrix} & -1 & \\ -1 & 4 & -1 \\ & -1 & \end{pmatrix} + I_y^2 \end{matrix} \right) \begin{pmatrix} u \\ v \end{pmatrix} = \begin{pmatrix} -I_x I_t \\ -I_y I_t \end{pmatrix} . \tag{6}$$

For simplification purposes, we disregard the time gradient I_t and fix it to 1. After more transformations, we obtain the following 5-point stencil to use in our iterative scheme:

$$\left(\begin{matrix} & \begin{pmatrix} -1 \\ -1 \end{pmatrix} & \\ \begin{pmatrix} -1 \\ -1 \end{pmatrix} & \begin{pmatrix} 4\alpha + I_x^2 & I_x I_y \\ I_x I_y & 4\alpha + I_y^2 \end{pmatrix} & \begin{pmatrix} -1 \\ -1 \end{pmatrix} \\ & \begin{pmatrix} -1 \\ -1 \end{pmatrix} & \end{matrix} \right) . \tag{7}$$

An extension in 3D space to detect the optical flow of volumes is trivial and omitted here because of space constraints.

```
1   Stencil SmootherStencil@all {
2     [ 0, 0] => { { 4.0 * alpha + GradX@current * GradX@current,
3                     GradX@current * GradY@current },
4                   { GradX@current * GradY@current,
5                     4.0 * alpha + GradY@current * GradY@current
      } }
6     [ 1, 0] => { { -1.0, 0.0 }, { 0.0, -1.0 } }
7     [-1, 0] => { { -1.0, 0.0 }, { 0.0, -1.0 } }
8     [ 0, 1] => { { -1.0, 0.0 }, { 0.0, -1.0 } }
9     [ 0,-1] => { { -1.0, 0.0 }, { 0.0, -1.0 } }
10  }
```

Listing 5 Declaration of the smoothing stencil for the optical flow in 2D

```
1   Function Smoother@all () : Unit {
2     loop over Flow@current {
3       Flow[next]@current = Flow[active]@current + (
4           ( inverse ( diag ( SmootherStencil@current ) ) ) *
5           ( RHS@current -
6             SmootherStencil@current * Flow[active]@current )
7         )
8     }
9     advance Flow@current
10  }
```

Listing 6 Smoother definition using slots for the flow field

8.2   Mapping to ExaSlang 4

Mapping the introduced algorithm to ExaSlang 4 is straight-forward thanks to the new local vector data types. In Listing 5, code corresponding to (7) is depicted. Here, we first defined the central coefficient, followed by the four directly neighboring values with offsets ± 1 in x and y direction. Each stencil coefficient consists of two components, as our system of PDEs is to be solved for the velocities in x and y direction of the image.

The smoother function using the previously introduced stencil is shown in Listing 6. As we will also use the smoother for coarse-grid solution, it has been defined for all multigrid levels using @all. For the computations, we loop over the flow field, calculating values based on the active field slot and writing them into the next slot. After calculations are done, we set the next field slot to be active using advance. Effectively, both slots are swapped, as only two slots have been defined.

Note the function calls inverse(diag(SmootherStencil@current)) which are used to invert the 2×2 matrix that is the central stencil element without further user intervention.

In Listing 7, the ExaSlang 4 implementation of a V(3,3)-cycle is depicted. This corresponds to Algorithm 1 with parameters $\gamma = 1$ and $v_1 = v_2 = 3$. The function has been defined for all multigrid levels except the coarsest one, with a separate function declaration a few lines below for the coarsest level. This function exits

```
1   Function VCycle@((coarsest + 1) to finest) () : Unit {
2     repeat 3 times {
3       Smoother@current ()
4     }
5     UpResidual@current ()
6     Restriction@current ()
7     SetSolution@coarser (0)
8     VCycle@coarser ()
9     Correction@current ()
10    repeat 3 times {
11      Smoother@current ()
12    }
13  }
14
15  Function VCycle@coarsest () : Unit {
16    Smoother@current ()
17  }
```

Listing 7 V(3,3)-cycle function in ExaSlang 4

the multigrid recursion by omitting the recursive call. As highlighted previously, it calls the smoother once to solve the system of PDEs on the coarsest grid.

In our optical flow implementation, application of stencils on coarser grids works by coarsening the gradient fields using full weighting restriction. Then, the discrete stencil is composed based on the coefficients—including level-dependent accesses to fields—specified by the user.

One big advantage of the local vector data types is that many existing multigrid component implementations can be re-used. For example, in this application no changes are needed for inter-grid operators such as restriction and prolongation, as they are based on scaling or adding values at discretization points regardless of whether these are represented by scalars or local vectors. During code generation, our framework detects the underlying data type the operators are working on and emits corresponding code. Consequently, it is very easy to adapt existing solver implementations to the new data types: Most often, only field layout definitions and stencils computing components of the system of PDEs need to be changed.

8.3 Results

In Fig. 4, the resulting flow field for the standard example of a rotating sphere is depicted. Fig. 5 shows the optical flow of a driving car. Because the scene has not been filmed using a fixed camera, there is also a movement of the background. In both result plots, a number of vectors has been omitted to improve clearness and reduce file size.

Fig. 4 Optical flow of rotating sphere

Fig. 5 Optical flow of an image sequence showing a driving car

Figure 6 shows the code sizes in lines of code for a few optical flow implementations, among them the implementation yielding the depicted flow fields. Both serial and parallel version have been generated from the exact same ExaSlang file. OpenMP has been used as the underlying parallelization technology. For both 2D cases, using local vectors instead of computing each component separately reduces the ExaSlang 4 program sizes by around 16 %. In 3D, stencils are larger and a third component must be computed, so with the use of the new data types, the savings increase to around 28 %. Consequently, the generated C++ source is smaller, since fewer loops are generated. However, expressions involving the new data types are not yet being optimized by our code generation framework. For the driving car test case, the average time per V(3,3)-cycle using Jacobi smoothers on an Intel i7-3770 increases from 31.6 ms to 36.3 ms with the new data types, due to slightly higher efforts at run time and optimization steps still missing. For two OpenMP threads using the new data types, average time decreases to 18.9 ms. As memory bandwidth seems to be already saturated, adding more threads does not yield further speedup. Input images are 512×512 pixels large, which results in a V-cycle consisting of nine levels, each with three pre- and post-smoothing steps. For the solution on the coarsest grid consisting of one unknown, another smoother iteration is applied. As our focus is on the introduction of the new data types and their advantages with

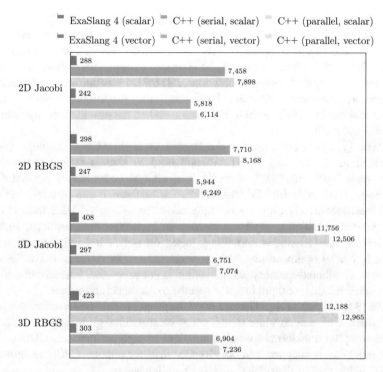

Fig. 6 Comparison of code sizes in lines of code LoC of user-specified ExaSlang 4 and generated code for different implementations of optical flow detection using a V(3,3)-cycle with Jacobi resp. red-black Gauss-Seidel (RBGS) smoothers and direct solution on the coarsest grid level (one unknown) by a single smoother iteration

respect to modeling of algorithms, we deliberately postpone the dissemination and discussion of further performance results.

9 Related Work

In previous work, the benefits of domain-specific optimization have been demonstrated in various domains. The project closest in spirit to ExaStencils has been *SPIRAL* [14], a widely recognized framework for the generation of hard- and software implementations of digital signal processing algorithms (linear transformations, such as FIR filtering, FFT, and DCT). It takes a description in a domain-specific language and applies domain-specific transformations and auto-tuning techniques to optimize run-time performance specifically for a given target hardware platform. Since it operates at the level of linear algebra, it directly supports vectors and matrices.

Many languages and corresponding compilers have been customized for the domain of stencil computations. Examples include *Liszt* [4], which adds abstractions to Java to ease stencil computations for unstructured problems, and *Pochoir* [22], which offers a divide-and-conquer skeleton on top of the parallel C extension Cilk to make stencil computations cache-oblivious. *PATUS* [3] uses auto-tuning techniques to improve performance. Other than ExaStencils, they support only vectors of fixed lengths, operate at a lower level of abstraction and do not provide language support for multigrid methods.

SDSLc [17] is a compiler for the Stencil DSL (SDSL), a language that is embedded in C, C++ and MATLAB, and used to express stencil expressions. Given such input, the SDSL compiler can emit shared-memory parallel CPU code and CUDA code for NVIDIA GPUs. Furthermore, it can generate FPGAs-based hardware descriptions by emitting code for a C-based HLS tool. During code generation, SDSLc applies a number of high-level optimizations, such as data layout transformations and tiling, based on polyhedral transformations, to enable low-level optimizations such as vectorization. In contrast to ExaStencils, automatic distributed-memory parallelization is not supported. Furthermore, SDSL is an embedded DSL without features specific to multigrid algorithms.

Mint [24] and *STELLA* (STEncil Loop LAnguage) [5] are DSLs embedded in C, respectively C++, and consider stencil codes on structured grids. Mint's source-to-source compiler transforms special annotations to high-performance CUDA code, whereas STELLA supports additionally OpenMP for parallel CPU execution. At present, neither offers distributed-memory parallelization.

In the past, several approaches to the generation of low-level stencil code from abstract descriptions have been pursued. However, to the best of our knowledge, most do not target multigrid methods for exascale machines.

Julia [1] centers around the multiple dispatch concept to enable distributed parallel execution. It builds on a just-in-time (JIT) compiler and can also be used to write stencil codes in a notation similar to Matlab. It works at a level of abstraction lower than ExaStencils.

HIPAcc [11] is a DSL for the domain of image processing and generates OpenCL and CUDA from a kernel specification embedded into C++. It provides explicit support for image pyramids, which are data structures for multi-resolution techniques that bear a great resemblance to multigrid methods [12]. However, it supports only fixed length vectors of size four and only supports 2D data structures. Furthermore, it does not consider distributed-memory parallelization such as MPI.

The finite element method library *FEniCS* [10] provides a Python-embedded DSL, called Unified Form Language (UFL), with support of vector data types. Multigrid support is available via PETSc, which provides shared-memory and distributed-memory parallelization via Pthreads and MPI, as well as support for GPU accelerators. The ExaStencils approach and domain-specific language aim at another class of users and provide a much more abstract level of programming.

PyOP2 [16] uses Python as the host language. It targets mesh-based simulation codes over unstructured meshes and uses FEniCS to generate kernel code for different multicore CPUs and GPUs. Furthermore, it employs run-time compilation and scheduling. FireDrake [15] is another Python-based DSL employing FEniCS' UFL and uses PyOP2 for parallel execution. While PyOP2 supports vector data types, it does not feature the extensive, domain-specific, automatic optimizations that are the goals of project ExaStencils.

10 Future Work

In future work, we will embed the data types introduced here in our code generator's optimization process in order to reach the same performance as existing code. For example, the polyhedral optimization stages must be aware of the sizes of data of these types when calculating calculation schedules. Consequently, low-level optimizations, especially data pre-fetching and vectorization transformations, must be adapted.

Additionally, we will showcase more applications using the new local vector and matrix data types. One application domain that will benefit greatly is that of solvers for coupled problems occurring, e.g., in computational fluid dynamics such as the incompressible Navier-Stokes equations. Here, the components of a vector field can be used to express unknowns for various physical quantities such as velocity components, pressure and temperature. The vector and matrix data types will greatly simplify the way in which such problems, and their solvers, can be expressed. Furthermore, not solving for each component separately but for the coupled system in one go allows for increased numerical stability and faster convergence. Of course, this may require the specification of specialized coarsening and interpolation strategies for unknowns and stencils. Moreover, specialized smoothers, such as Vanka-type ones, are crucial for optimal results.

11 Conclusions

We reviewed ExaSlang 4, the most concrete layer of project ExaStencils' hierarchical DSL for the specification of geometric multigrid solvers. To ease description of solvers for systems of PDEs, we introduced new data types that represent vector and matrices. The benefits of these data types, such as increased programmer productivity and cleaner code, were illustrated by evaluating program sizes of an example application computing the optical flow.

The new data types are also a big step towards an implementation of ExaSlang 3, since functionality that is available at the more abstract ExaSlang layers must be available at the more concrete layers as well. Furthermore, they expand the application domain of project ExaStencils, e.g., towards computational fluid dynamics.

Acknowledgements This work is supported by the German Research Foundation (DFG), as part of Priority Program 1648 "Software for Exascale Computing" in project under contracts TE 163/17-1, RU 422/15-1 and LE 912/15-1.

References

1. Bezanson, J., Karpinski, S., Shah, V.B., Edelman, A.: Julia: a fast dynamic language for technical computing. CoRR (2012). arXiv:1209.5145
2. Brandt, A.: Rigorous quantitative analysis of multigrid, I: constant coefficients two-level cycle with L_2-norm. SIAM J. Numer. Anal. **31**(6), 1695–1730 (1994)
3. Christen, M., Schenk, O., Burkhart, H.: PATUS: A code generation and autotuning framework for parallel iterative stencil computations on modern microarchitectures. In: Proceedings of IEEE International Parallel & Distributed Processing Symposium (IPDPS). pp. 676–687. IEEE (2011)
4. DeVito, Z., Joubert, N., Palaciosy, F., Oakleyz, S., Medinaz, M., Barrientos, M., Elsenz, E., Hamz, F., Aiken, A., Duraisamy, K., Darvez, E., Alonso, J., Hanrahan, P.: Liszt: A domain specific language for building portable mesh-based PDE solvers. In: Proceedings of the Conference on High Performance Computing Networking, Storage and Analysis (SC). ACM (2011), paper 9, 12pp.
5. Gysi, T., Osuna, C., Fuhrer, O., Bianco, M., Schulthess, T.C.: STELLA: a domain-specific tool for structured grid methods in weather and climate models. In: Proceedings of the International Conference for High Performance Computing, Networking, Storage and Analysis (SC), pp. 41:1–41:12. ACM (2015)
6. Hackbusch, W.: Multi-Grid Methods and Applications. Springer, Berlin/New York (1985)
7. Köstler, H.: A multigrid framework for variational approaches in medical image processing and computer vision. Ph.D. thesis, Friedrich-Alexander University of Erlangen-Nürnberg (2008)
8. Kronawitter, S., Lengauer, C.: Optimizations applied by the ExaStencils code generator. Technical Report, MIP-1502, Faculty of Informatics and Mathematics, University of Passau (2015)
9. Lengauer, C., Apel, S., Bolten, M., Größlinger, A., Hannig, F., Köstler, H., Rüde, U., Teich, J., Grebhahn, A., Kronawitter, S., Kuckuk, S., Rittich, H., Schmitt, C.: ExaStencils: advanced stencil-code engineering. In: Euro-Par 2014: Parallel Processing Workshops. Lecture Notes in Computer Science, vol. 8806, pp. 553–564. Springer (2014)
10. Logg, A., Mardal, K.A., Wells, G.N. (eds.): Automated Solution of Differential Equations by the Finite Element Method. Lecture Notes in Computational Science and Engineering, vol. 84. Springer, Berlin/New York (2012)
11. Membarth, R., Reiche, O., Hannig, F., Teich, J., Körner, M., Eckert, W.: HIPAcc: a domain-specific language and compiler for image processing. IEEE T. Parall. Distr. (2015), early view, 14 pages. doi:10.1109/TPDS.2015.2394802
12. Membarth, R., Reiche, O., Schmitt, C., Hannig, F., Teich, J., Stürmer, M., Köstler, H.: Towards a performance-portable description of geometric multigrid algorithms using a domain-specific language. J. Parallel Distrib. Comput. **74**(12), 3191–3201 (2014)
13. Odersky, M., Spoon, L., Venners, B.: Programming in Scala, 2nd edn. Artima, Walnut Creek (2011)
14. Püschel, M., Franchetti, F., Voronenko, Y.: SPIRAL. In: Padua, D.A., et al. (eds.) Encyclopedia of Parallel Computing, pp. 1920–1933. Springer (2011)
15. Rathgeber, F., Ham, D.A., Mitchell, L., Lange, M., Luporini, F., McRae, A.T.T., Bercea, G.T., Markall, G.R., Kelly, P.H.J.: Firedrake: automating the finite element method by composing abstractions. CoRR (2015). arXiv:1501.01809
16. Rathgeber, F., Markall, G.R., Mitchell, L., Loriant, N., Ham, D.A., Bertolli, C., Kelly, P.H.: PyOP2: A high-level framework for performance-portable simulations on unstructured meshes.

In: Proceedings of the 2nd International Workshop on Domain-Specific Languages and High-Level Frameworks for High Performance Computing (WOLFHPC), pp. 1116–1123. IEEE Computer Society (2012)

17. Rawat, P., Kong, M., Henretty, T., Holewinski, J., Stock, K., Pouchet, L.N., Ramanujam, J., Rountev, A., Sadayappan, P.: SDSLc: A multi-target domain-specific compiler for stencil computations. In: Proceedings of the 5th International Workshop on Domain-Specific Languages and High-Level Frameworks for High Performance Computing (WOLFHPC). pp. 6:1–6:10. ACM (2015)

18. Schmitt, C., Kuckuk, S., Hannig, F., Köstler, H., Teich, J.: ExaSlang: A Domain-Specific Language for Highly Scalable Multigrid Solvers. In: Proceedings of the 4th International Workshop on Domain-Specific Languages and High-Level Frameworks for High Performance Computing (WOLFHPC), pp. 42–51. ACM (2014)

19. Schmitt, C., Kuckuk, S., Köstler, H., Hannig, F., Teich, J.: An evaluation of domain-specific language technologies for code generation. In: Proceedings of the International Conference on Computational Science and its Applications (ICCSA), pp. 18–26. IEEE Computer Society (2014)

20. Schmitt, C., Schmid, M., Hannig, F., Teich, J., Kuckuk, S., Köstler, H.: Generation of multigrid-based numerical solvers for FPGA accelerators. In: Größlinger, A., Köstler, H. (eds.) Proceedings of the 2nd International Workshop on High-Performance Stencil Computations (HiStencils), pp. 9–15 (2015)

21. Siegmund, N., Grebhahn, A., Apel, S., Kästner, C.: Performance-influence models for highly configurable systems. In: Proceedings of the European Software Engineering Conference and ACM SIGSOFT Symposium on the Foundations of Software Engineering (ESEC/FSE), pp. 284–294. ACM (2015)

22. Tang, Y., Chowdhury, R.A., Kuszmaul, B.C., Luk, C.K., Leiserson, C.E.: The Pochoir stencil compiler. In: Proceedings of the ACM Symposium on Parallelism in Algorithms and Architectures (SPAA), pp. 117–128. ACM (2011)

23. Trottenberg, U., Oosterlee, C.W., Schüller, A.: Multigrid. Academic, San Diego (2001)

24. Unat, D., Cai, X., Baden, S.B.: Mint: Realizing CUDA performance in 3D stencil methods with annotated C. In: Proceedings of the International Conference on Supercomputing (ISC), pp. 214–224. ACM (2011)

25. Wienands, R., Joppich, W.: Practical Fourier Analysis for Multigrid Methods. Chapman Hall/CRC Press, Boca Raton (2005)

Performance Prediction of Multigrid-Solver Configurations

Alexander Grebhahn, Norbert Siegmund, Harald Köstler, and Sven Apel

Abstract Geometric multigrid solvers are among the most efficient methods for solving partial differential equations. To optimize performance, developers have to select an appropriate combination of algorithms for the hardware and problem at hand. Since a manual configuration of a multigrid solver is tedious and does not scale for a large number of different hardware platforms, we have been developing a code generator that automatically generates a multigrid-solver configuration tailored to a given problem. However, identifying a performance-optimal solver configuration is typically a non-trivial task, because there is a large number of configuration options from which developers can choose. As a solution, we present a machine-learning approach that allows developers to make predictions of the performance of solver configurations, based on quantifying the influence of individual configuration options and interactions between them. As our preliminary results on three configurable multigrid solvers were encouraging, we focus on a larger, non-tivial case-study in this work. Furthermore, we discuss and demonstrate how to integrate domain knowledge in our machine-learning approach to improve accuracy and scalability and to explore how the performance models we learn can help developers and domain experts in understanding their system.

1 Introduction

Many real-world and scientific problems can be modeled using partial differential equations (PDEs). After performing an adequate approximation using a discretization based on finite differences, volumes, or elements, it is possible to use multigrid methods to solve the resulting symmetric, positive definite system matrices. Since

A. Grebhahn (✉) • N. Siegmund • S. Apel
University of Passau, Passau, Germany
e-mail: Alexander.Grebhahn@uni-passau.de; Norbert.Siegmund@uni-passau.de;
apel@uni-passau.de

H. Köstler
Friedrich-Alexander University Erlangen-Nürnberg, Erlangen, Germany
e-mail: harald.koestler@fau.de

© Springer International Publishing Switzerland 2016   69
H.-J. Bungartz et al. (eds.), *Software for Exascale Computing – SPPEXA*
2013-2015, Lecture Notes in Computational Science and Engineering 113,
DOI 10.1007/978-3-319-40528-5_4

their development in the 1970s, it has been proved that multigrid methods are among the most effective iterative algorithms to solve discretized PDEs [3, 7, 27, 28]. This is because of their runtime complexity being in the order of the number of unknowns.

Before applying a given multigrid implementation on a specific hardware, it is often unavoidable to re-implement parts of a system to achieve optimal performance. This process results in considerable maintenance costs and program errors. In many different areas, it has been shown that domain-specific program transformation and generation can be used to overcome this challenge [15, 20]. In our project *ExaStencils* [13], we follow this idea by developing a code generator that automatically generates multigrid code tailored to a given hardware platform, based on a domain-specific specification in the language *ExaSlang* [22]. Internally, the code generator uses program transformations in a stepwise refinement process, as proposed by Wirth et al. [29]. During this process, different choices need to be made, each leading to unique multigrid code. These choices include, for example, the selection of a specific smoother or the number of pre-smoothing and post-smoothing steps to be performed in a single iteration. Since these components of the multigrid system have a strong influence on the performance of the overall system [27], they have to be chosen carefully. However, this is a non-trivial task, because developers and users often lack knowledge about how their choices influence the performance. Worse, the influence of a choice might also depend on other decisions. We call choices that have to be made during this process *configuration options* and a valid selection of options a *configuration*, which gives rise to a *variant* of the system. In this light, we see the code generator as a configurable system that can be used in a black-box manner to tune its parameters to generate performance-optimal code for a specific hardware.

To identify the optimal variant arising from such a refinement process, auto-tuning or machine-learning techniques, such as genetic algorithms [1, 26], neuronal networks [14], or random forests [9] can be used. Although many of these techniques can identify optimal variants, they fail to make the influences of individual configuration options on the performance explicit. That is, they do not provide a comprehensible model that allows developers to reason about the performance influences of individual options and their interactions. For the same reason, it is hardly possible to integrate existing domain knowledge (e.g., that two ore more options interact with each other) in the tuning or learning process, prohibiting further optimizations.

To give domain experts the possibility of integrating and validating their knowledge, we employ a unique combination of forward feature selection and multivariate regression [24]. Based on the resulting *performance-influence model*, it is possible to predict the performance of all variants of a system and to determine the influence of the individual configuration options and their interactions on performance. To learn a performance-influence model, we first select and measure a set of variants of the system in question using structured sampling strategies (see Sect. 3.1), because the selection of a suitable set of variants is essential for model accuracy. In prior work, we already demonstrated that such an approach can be used to identify the influence of configuration options and their interactions on the performance, independent of the application domain and programming language [24]. There, we considered 6 configurable systems, including the *Java* garbage collector, the Single Assignment C compiler (SAC^1), the video encoder *x264*, and three customizable multigrid systems implemented for different purposes. For the multigrid systems, we are able to predict performance with an accuracy of 90 %, on average, and, for all systems, with an accuracy of 86 %, on average.

In this paper, we focus on the integration of domain knowledge into our approach and present first results on predicting performance of multigrid variants generated by our *ExaStencils* code generator [13]. To increase prediction accuracy of our approach and to reduce the number of measurements, domain knowledge is a valuable source of information, which can be integrated in the sampling and the learning procedure of our machine-learning algorithm. However, if the domain knowledge does not hold for an application, its integration might have negative effects on the prediction accuracy of our approach. To this end, we demonstrate the benefits and also the risks of exploiting this knowledge by means of our configurable multigrid system. Using this knowledge, we were able to reduce the number of measurements from 162 to 39 by increasing the prediction error by less than 1 %. Based on the results, we are able to see the existence of interactions between different smoothers and different coarse-grid solvers provided by the configurable multigrid system we use in the experiments. Additionally, we performed first experiments on predicting performance of multigrid solver variants generated by our *ExaStencils* code generator. There, we were able to predict performance of the variants with a median accuracy of 83 %.

In summary, our contributions are the following:

- We discuss which kind of domain knowledge can be used to tune the sampling and the learning process of our machine-learning algorithm and describe how it can be integrated.
- We demonstrate the benefits and the risks of integrating domain knowledge in our approach on a real-world multigrid system.
- We present how our approach can be used to test, whether domain knowledge holds for an application.
- We learn a performance-influence model to predict performance of different multigrid-solver variants generated by the *ExaStencils* code generator.

2 Configurable Multigrid Solvers and the ExaStencils Code Generator

A multigrid solver is an iterative algorithm working on a hierarchy of grids. In general, one iteration of the algorithm, which is also called a cycle, works as follows: First, the algorithm starts at the finest level, which arises from the

---

[1] http://www.sac-home.org/

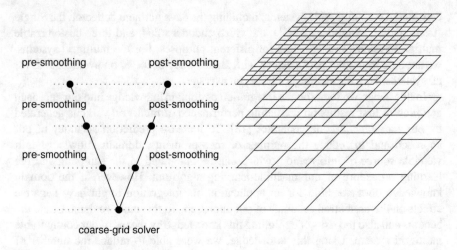

Fig. 1 The basic structure of a V-cycle iteration of a multigrid algorithm

discretization of the equation. The algorithm uses a smoother to transform high-frequency errors to lower-frequency errors in a predefined number of pre-smoothing steps. Subsequently, the residual is computed and restricted to a coarser grid. This process is performed recursively until the coarsest grid is reached. On this grid, the equation is solved using a coarse-grid solver, and the solution is propagated back to a finer level. In this propagation process, the error is again smoothed with a number of post-smoothing steps. Subsequently, the solution is recursively propagated to finer levels. This process is repeated until a given convergence is reached. In Fig. 1, we give an overview of the structure of this process by means of the example of a V-cycle, who describes in which order the different levels have to be visited. Beside the V-cycle, there are also other cycle types, such as the W-cycle. Although the process is relatively simple, it can be parametrized, using a large number of different algorithms, which again have a large number of parameters (i.e., configuration options). For example, there exists a whole array of different smoothers, such as the Jacobi smoother or the Gauss-Seidel smoother. Additionally, it is possible to use many code optimization strategies on a code representing this process, such as loop unrolling, temporal blocking, [17] and parallelization, to achieve optimal performance.

To take advantage of the variability in multigrid algorithms, we developed a code generator in our project *ExaStencils*[2] that aims at generating code that is tailored to a specific problem and a specific hardware platform. To give different groups of users (e.g., application scientists or computation scientists), the possibility of working with the code generator efficiently, the *ExaSlang* language—the input language of the generator—offers four layers of abstraction [22].

---

[2]http://exastencils.org/

The generator transforms a domain-specific specification of a problem in a stepwise manner to produce optimized C++ code. During this refinement process, several choices have to be made, all having an effect on the performance of the generated code. Overall, we have about 100 choices in our generator, ranging from the selection of the number of nodes used in the implementation, over the selection of a smoother, the selection of the number of levels of the multigrid system, to choosing optimization strategies, such as temporal blocking or loop unrolling. However, the optimal combination of these options highly depends on the problem and the underlying hardware, which complicates the task of identifying the optimal set of options.

3 Performance Prediction

To predict the performance of the variants of a configurable system, we use a machine-learning approach that learns an empirical model based on measuring a sample of variants. The resulting model describes the influence of the individual configuration options on performance. To learn such a performance-influence model, we first select and measure a sample set of variants and use them as input for our learning procedure. Although using all variants during learning would lead to the best results regarding accuracy, this is infeasible for highly-configurable systems, because the number of variants grows exponentially with the number of options. Hence, we select only a subset of variants for learning based on structured sampling approaches (Sect. 3.1). Based on this sample set, we learn a performance-influence model in a stepwise manner, to identify the most relevant configuration options first (Sect. 3.2). To speed up the performance of our approach and to reduce the number of measurements, we exploit domain knowledge about some characteristics of influences of configuration options on performance (Sect. 3.3).

3.1 Sampling

When selecting a sample set of variants for learning, we have to consider binary and numeric configuration options. Because of the different value domains of the two kinds of options, it is necessary to handle them differently during sampling [24]. To this end, we apply different structured sampling strategies to select a representative sample set of configurations. For binary options, we apply sampling heuristics from the software-product-line domain, which have been developed to detect individual influences and interactions between options [25]. For numeric options, we draw on the established theory of experimental designs [16]. After sampling the binary and numeric configuration spaces, we compute the Cartesian product of the two sets of partial configurations, to create a set of variants used as input for the learning process.

3.1.1 Binary Sampling Heuristics

For binary configuration options, we use two different sampling heuristics.

The *Option-wise heuristic (OW)* aims at identifying the influences of all individual configuration options on a non-functional property (i.e., performance), for all variants of a configurable system. To this end, it selects one configuration for each binary option. In each of these configurations, all options are disabled and only the one being considered is enabled. Additionally, we select a configuration with no option enabled. Based on these configurations, we can identify the influence of the individual options. This heuristic requires a linear number of configurations with respect to the number of binary options.

The *Pair-wise heuristic (PW)* aims at identifying interactions between pairs of configuration options. To this end, one configuration for each pair of options is selected. In each of these configurations, the options being considered are enabled and all other options are disabled. The number of selected configurations is quadratic with the number of options.

3.1.2 Experimental Designs

Sampling in the presence of numeric options is a widely researched area, known under the umbrella of *experimental designs* [16]. Although a large number of different designs have been proposed, we focus only on two designs in this work. In prior work, we systematically compared different designs and found the Plackett-Burmann design to perform best in our setting [24].

The *Plackett-Burman Design (PBD)* was developed for configuration spaces in which the strength of interactions compared to individual influences can be neglected [18]. However, it is still possible to identify relevant interactions between options. In contrast to other designs, where the number of configurations grows quickly with the number of options considered, the PBD defines specific seeds that define the number of configurations that are selected. A seed also defines how many different values of the numeric options are considered during sampling. In Fig. 2, we give an example for the configurations that are selected using a PBD

Fig. 2 Plackett-Burman Design for 8 configuration options using a seed defining that 9 configurations are selected and 3 values of the options are considered in the sampling

options

0	1	2	2	0	2	1	1
1	0	1	2	2	0	2	1
1	1	0	1	2	2	0	2
2	1	1	0	1	2	2	0
0	2	1	1	0	1	2	2
2	0	2	1	1	0	1	2
2	2	0	2	1	1	0	1
1	2	2	0	2	1	1	0
0	0	0	0	0	0	0	0

Mapping
0 -> minimal value
1 -> center value
2 -> maximal value

with a seed defining that 9 configurations are selected and 3 different values of the options are considered. Each line of the table defines the values of options in a single configuration. In the first line, we see the original seed, which is shifted to the right in each subsequent configuration. Internally, we map the values 0, 1, and 2, used in the seed, to the minimal, center, and maximal, value of the corresponding numeric option. Beside the presented seed, other seeds are defined, which consider, for example, five or seven different values of the options and a larger number of configurations.

It is also possible to use a *Random Design (RD)* selecting a defined number of configurations randomly distributed over the whole value domain. This is possible because the value domains of the options as well as constraints between the options are known. Since this design is very simple and often effective, it is often used in literature [2, 4].

3.2 Performance-Influence Models

To describe the influences of configuration options and their interactions on the performance, we learn a *performance-influence model* Π [24]. Mathematically, Π is a mapping from a configuration $c \in C$ to its performance in $\mathbb{R}$. The different influences of the options and their interactions are described by individual *terms* π_i of a model. In the following, we give a simplified example of a performance-influence model consisting of four terms:

$$\overbrace{86}^{\pi_1} + \overbrace{27 * \textit{pre-smoothing}}^{\pi_2} + \overbrace{323 * \textit{GaussSeidel} * \textit{pre-smoothing}}^{\pi_3} - \overbrace{1.8 * \textit{cgsAmg}}^{\pi_4}$$

Based on the model, the options *pre-smoothing*, *GaussSeidel*, and *cgsAmg* have an influence on performance, where option *pre-smoothing* has an independent influence (π_2), but also interacts with the *GaussSeidel* option (π_3). Additionally, we can see a general runtime that can not be dedicated to a specific option described through term π_1. To predict the performance of a configuration, we compute the sum of the influences represented by the terms. During the generation of the model, we consider and model only the relevant influences on the performance to keep the model as simple as possible. To consider only these relevant influences, we use a unique combination of *forward feature selection* and *multivariate regression* to learn performance-influence models in an iterative manner.

The basic idea of our machine-learning approach is to start with an empty model and to expand it with those terms that increase the prediction accuracy at most (forward feature selection). In Algorithm 1, we show the basic structure of the learning algorithm. We start the learning process with an empty model, which does not consider the influence of any configuration option (Line 3). Then, we create a set of candidate models, where each model covers the terms of the original model (i.e., the model from the previous round) and one additional term that is not covered in

---

Algorithm 1 Forward feature selection

1: **Data:** measurements, options
2: **Result:** model
3: model = Ø, error = ∞
4: **repeat**
5:     lastError = error
6:     bestModel = ⊥
7:     candidates = generateCandidates(model, options)
8:     **for each** candidate **in** candidates **do**
9:         refinedModel = learnFunction(candidate, measurements)
10:        modelError = computeError(refinedModel, measurements)
11:        **if** modelError < error **then**
12:            error = modelError, bestModel = candidate
13:     **if** bestCandidate ≠ ⊥ **then**
14:         model = bestModel
15: **until** (lastError − error < margin) ∨ (error < threshold)
16: **return** learnFunction(model, measurements)

---

the original model (Line 7). This term considers either an individual influence or an interaction. In more detail, for each term in the original model, we generate a set of candidate terms, where each of the candidate terms covers the options of the term of the original model and one additional option. So, terms covering interactions with three options are created, if the model contains a term with two options.

To identify the most promising candidate model, we use multivariate regression on the sampled configurations to determine the coefficients of all terms of the model (Line 9). The coefficients describe the influence of terms on the performance variation in the set of system variants. After adding one term to the model, we have to recompute the coefficients of all terms of a model to correctly determine the influence of the options of the terms. Then, we predict the performance of configurations using the candidates and compare the predicted and the measured performance. Based on the prediction error of the candidates, we select the most accurate model and use it as the basis for the next iteration of the learning algorithm. To consider only relevant interactions, we probe only for existence of interactions where, at least, one option has an individual influence or contributes to an interaction with a smaller degree. We iteratively expand this model until predefined thresholds are reached or until no candidate model further increases accuracy.

3.3   Integration of Domain Knowledge

In what follows, we discuss which kinds of domain knowledge can be integrated in our machine-learning approach to, (1) minimize the number of configurations used during the learning process, to (2) minimize runtime of the learning process without causing a loss of prediction accuracy, and to (3) increase prediction accuracy of the model.

3.3.1 Shrinking the Configuration Space

Domain knowledge can be helpful to reduce the size of the configuration space to search for the optimal variant. For example, it is not necessary to consider all configuration options if it is known that some of them do not have an influence on the performance of the system. These options can be ignored during sampling and the subsequent learning process without loss of accuracy. Additionally, it is not necessary to consider the whole value domain of numeric options. If we know that the optimal performance can be achieved by setting only a subset of the values in the whole range, we can simply modify the option's value range.

3.3.2 Domain Knowledge on Interactions

It is also useful to exploit domain knowledge about the existence and absence of interactions between options in our machine-learning algorithm. First, if two or more configuration options do not interact with each other, the sampling strategies can be modified to reduce the number of measurements. For example, to identify whether there is an interaction between two binary options, it is necessary to measure one configuration with both options enabled. If it is known that an interaction does not exist, it is possible to omit the corresponding configuration during the sampling process, without losing accuracy in the learned model. During the generation of the candidate models, it is also possible to use this knowledge to reject candidates that consider this interaction. Second, if it is known that two or more options interact with each other, configurations that cover this interaction should be selected during the sampling. This is of special importance when a large number of options interact with each other. Based on the sampling strategy used, it is not possible to identify such interactions. To support this, Siegmund et al. proposed further heuristics aiming at identifying these higher-order interactions [23]. However, these heuristics require a large number of configurations when applying them on many configuration options. For this reason, they should be used only if it is known that higher-order interactions exist. Additionally, knowledge on existing interactions can be integrated in the learning process. This way, it is possible to start the learning process with a model considering all of these interactions instead of starting with an empty model.

3.3.3 Independent Sampling Strategies and Independent Models

So far, we considered only a single sampling strategy for all configuration options of one kind (e.g., the PBD for numeric option or the PW heuristic for binary options). However, it is also possible to use different sampling strategies for different sets of options, if there are, for example, two or more independent groups of options with no interactions between them. For instance, it is possible to use the PBD with a specific seed for one set of numeric options and a PBD with another seed, and a completely different design for another set of options. During sampling of one group of options, it is necessary to not modify the values of options of other groups. This is similar to group-based sampling [21]. Based on the multiple sets of options, we can learn different models and combine these models subsequently to a single model. This combination is possible because of the additive structure of performance-influence models.

3.3.4 Integration of Analytical Models

Based on theoretical knowledge about the performance behavior of parts of an algorithm, it is possible to create an analytical model. Although the creation process of such a model is error prone, it has been shown that it is possible to perform accurate predictions using this kind of models [8, 10].

In our setting, analytical models can simply be used as a starting point to learn more accurate and more complex models. This can also be done if only a subset of all options is considered in the analytical model. Additionally, the selection of the sampling strategy should be adapted to be able to learn the coefficients for the analytical model correctly. However, it might not be possible to use existing experimental designs, because the designs assume that the influence of an option can be described with a function of a small polynominal degree. So, the sampling strategy should be selected carefully and combined with other sampling strategies, if necessary.

3.3.5 Models for Disjoint Parts of a System

Rather learning a model to describe the influence of all options on the whole runtime of an application, it is also possible to learn a set of models for disjoint parts (e.g., one model for the initialization and one model for the computation) of a system, if the execution time of one part does not have an influence on the execution time of another part. With this set of *local* models, it might be possible to perform more accurate predictions compared to using a single *global* model. This is because interactions with a small influence on the overall performance, but strong influences on the performance of a part of the system, can be better identified. However, learning different models leads to some drawbacks regarding the performance of the learning process, because candidates considering the same influences are used to predict performance variations of different parts of the system and thus have to be tested several times.

4 Evaluation

In our evaluation, we focus on three research questions:

RQ1 Can we observe significant benefits in terms of reducing the number of sampled configurations if we integrate domain knowledge in the sampling and learning process of our machine-learning algorithm?

RQ2 Are we able to predict the runtime of multigrid-solver variants generated by our code generator with an accuracy comparable to comparable experiments without using domain knowledge?

RQ3 Is it beneficial to learn a set of local models compared to learning a global model to predict the runtime of multigrid variants produced by our code generator?

4.1 Leveraging Domain Knowledge

To demonstrate the usefulness of integrating domain knowledge in our machine-learning approach and to answer RQ1, we performed a number of experiments based on the customizable multigrid system HSMGP, which solves a finite differences discretization of the Poisson's equation. HSMGP was developed for testing various algorithms and data structures on high-performance computing systems, such as *JuQueen* at the Jülich Supercomputing Centre, Germany [12]. We use HSMGP because of its simplicity and because we are familiar with the system. In our experiments, we consider a configuration space with six different smoothers, three different coarse-grid solvers, and a variable number of pre-smoothing and post-smoothing steps. We also performed weak scaling experiments from 64 to 4096 nodes, where we see the number of nodes used by a variant as a further configuration option of the system. A more detailed description of the configuration space of the system is given elsewhere [5].

4.1.1 Experimental Setup

In the experiments, we focus on predicting the time needed to perform a single V-cycle iteration. In advance to our experiments, we performed an exhaustive search and measured all variants of the configuration space, to be able to determine prediction accuracy of unseen variants. The experiment consists of three phases. First, we select a set of configurations based on domain knowledge and the structured sampling strategies. Then, we use our machine-learning approach, to learn a performance-influence model. Last, we predict the performance of all variants of the system using the identified influences and compute prediction accuracy.

To demonstrate the usefulness of incorporating domain knowledge, we state the following assumptions on the influences of the configuration options, based on domain knowledge:

(1) The runtime of the coarse-grid solver is not affected by the selection of the smoother and the number of pre-smoothing and post-smoothing steps.
(2) One pre-smoothing sweep needs the same amount of time as one post-smoothing sweep.

Based on assumption 1, there are no interactions between the coarse-grid solver and the smoothers and the pre-smoothing and post-smoothing steps. So, it is not necessary to sample configurations to identify whether such an interaction exists. Additionally, it is not necessary to generate candidate models during learning, covering such an interaction. Thus, we can reduce the number of measurements and improve performance of the learning algorithm.

We can assume 2 because we perform the same operations in one pre-smoothing and one post-smoothing sweep on the same grid. Hence, it is possible to infer the influence of a post-smoothing sweep based on the identified influence of a pre-smoothing sweep. This allows us to ignore the post-smoothing option during sampling and learning entirely. However, during sampling, configurations have to be selected with no runtime caused by the post-smoothing steps. For that reason, we select configurations with zero post-smoothing steps.

We apply an iterative sampling process, considering a small number of options first and learn models for these options. Consecutively, we sample the other configuration options and learn models for them. This way, we first generate a model considering the different smoothers and the number of pre-smoothing and post-smoothing steps. Later, we generate models considering the coarse-grid solvers and the influence of the number of cores, and we combine these models. For assumption 2, we omit the post-smoothing steps and infer the influence of these options based on the identified influence of pre-smoothing.

Additionally, we learn a model based on a selection of the PW heuristic in combination with the PBD using a seed defining that 9 configurations are selected considering 3 values of the options and a model based on 50 randomly selected configurations.

4.1.2   Results and Discussion

In Table 1, we present our results using the different sampling strategies. In each row of the table, we present the number of configurations selected during sampling and the prediction accuracy of the learned performance-influence model. When making assumption 1, 69 configurations are used for learning. If assumptions 1 and 2 are used during sampling, 39 configurations are selected. In both experiments, we achieve the same median prediction error rate of 2.1 %. If we instead apply the PW heuristic, which also considers interactions between the different binary options, in combination with the PBD, we have a median prediction error of 1.4 %.

Table 1 Experimental results on using domain knowledge. *Heuristic*: different sampling strategies, $|e|$: number of measurements required for the heuristic, $|C|$: number of configuration of the configuration space, *Error-rate dist.*: error-rate distribution, σ: median error rate

Heuristic	$|e| / |C|$	Error-rate distribution	σ (%)
Assumption 1	69 / 3456		2.1
Assumption 1 and 2	39 / 3456		2.1
PW, PDB(9,3)	162 / 3456		1.4
Random	50 / 3456		11.7
		0 5 10 15 20 25	

However, to learn such an accurate model, we have to measure 162 configurations. For comparison, if we select a set of 50 randomly distributed configurations, we achieve a median error rate of 11.7 %

These results show that the use of domain knowledge during the sampling process can lead to a drastic reduction of the number of configurations needed for learning. However, we also observe a loss of prediction accuracy of the model, if the knowledge does not completely hold, as in the case of assumption 1. The analysis why assumption 1 does not hold reveals that there is an interaction between the coarse-grid solver and the smoother performing the pre-smoothing and post-smoothing steps. This is because we apply the coarse-grid solver until the error on the coarsest grid is reduced by a certain factor. Thus, smoothers, such as Jacobi and Gauss-Seidel, have different convergence properties, which has an influence on the error of the input. This has an influence on the runtime, because when starting with a larger error it can be reduced by a larger factor in the first iterations of the coarse-grid solver.

As result, for RQ1, we can state that an integration of domain knowledge in the sampling and learning process has significant benefits as long as the domain knowledge holds for the system. For the two assumtions, we can state that assumption 2 holds for the system, while assumption 1 does not hold. We can also see that selection of configurations completely randomly does not lead to a high prediction accuracy.

4.2 Code Generator

In the second experiment, we aim at answering RQ2 and RQ3. To this end, we try to identify the influence of 20 configuration options (out of approximately 100) on the performance of the generated variants using the *ExaStencils* code generator [13]. During the manual pre-selection process, we selected options from different transformation steps, all having a strong influence on the performance of the resulting code. For this purpose, we consider various options, such as the

Table 2 Configuration options of the *ExaStencils* code generator that we consider in our experiments after performing a manual pre-selection, grouped in hardware-specific, multigrid-specific parameters, and optimization options

Hardware-specific	
• Ranks per node	• # of nodes
• Use fragment loops for each operation	• Use custom MPI data types
• # of OMP-threats x direction	• # of OMP-threats y direction
• Tile size x direction	• Dimension for ratio offset

Multigrid-specific	
• Smoother	• Recursive cycle calls
• # pre-smoothing steps	• # post-smoothing steps
• Use slots for Jacobi	• Min level

Optimizations	
• Optimize number of finest levels	• Parallelize loop over dimensions
• Use address pre-calculation	• Vectorize
• Loop unrolling	• Loop unrolling interleave

selection of a smoother and cycle type, but also optimization strategies, such as vectorization and loop unrolling [11]. In Table 2, we give an overview of the configuration options we consider in our experiment. The remaining options are set to their default value.

4.2.1 Experimental Setup

After defining the configuration space, we apply the OW and PW heuristics on the set of binary options and PBD on the numeric options to select configurations used during the learning process. Beside this set of configurations, we also select a set of random variants and measured their performance to predict the performance of configurations not used during the learning process.

During code generation, we included code in the variants to measure the runtime of different components of a single cycle iteration independently. That gave us the possibility to learn models representing the performance behavior of specific parts of a cycle iteration. This way, we are able to learn models, for example, for the residual update, for the restriction, and for the time needed to perform pre-smoothing and post-smoothing. After combining these models, it is possible to predict the whole execution time needed for a single iteration of a given variant.

To answer RQ2 and RQ3, we learn two different kinds of models. First, we learn a global model describing the influence of all options on the performance of a multigrid iteration. Second, we learn a set of local models to identify the influence of the options on smaller parts of the system. After learning these local models, we combine them and predict the runtime of an iteration.

We performed all measurements on *JuQueen* at the Jülich Supercomputing Centre, Germany. We considered a variable number of nodes from 16 to 64, because of the high number of configurations we measured. In each of the configurations, we measured the time for solving Poisson's equation in 2D with constant coefficients.

4.2.2 Results and Discussion

In Table 3, we present the prediction results of using the different performance-influence models. In each row of the table, we present the error distribution of the model when predicting the runtime of the variants and the median error rate. If we learn a global model, we can predict all measured configurations (configurations used for learning and unseen configurations) with a median error rate of 19.5 %. This means that we can predict runtime with an accuracy of 80 %. If we instead learn a set of local models, and combine them afterwards, we are able to predict performance with a median error rate of 16.2 %.

So, we can state that we are able to identify most of the relevant influences on the performance. Furthermore, learning a set of local models has benefits compared to learning a global model, which answers RQ3. Nevertheless, we have a median error rate of more than 15 % in both cases, which indicates that some relevant influences have not been detected or cannot be modeled using our approach, or that we overfitted the model. To learn whether we overfitted the model, we predict only the performance of the configurations used during the learning process with the set of local models. For these configurations, we got a median error rate of 16.5 %. This indicates that we do not overfit the model.

To conclude, we can predict the performance of the different variants with an accuracy of more than 83 %. But, we were not able to identify all existing influences. Regarding RQ2, we can state that we are able predict performance of the variants generated by our code generator with an accuracy comparable to previous experiments on other configurable systems. Regarding RQ3, we can conclude that learning a set of local models is more beneficial than learning a global model.

Table 3 Experimental results of predicting performance of multigrid solvers generated by the *ExaStencils* code generator using a global model or a set of local models. $|e|$: number of measurements required for the heuristic, $|C|$: number of all measured configurations, *Error-rate distribution*: error-rate distribution, σ: median error rate

Heuristic	$|e| / |C|$	Error-rate distribution	σ (%)
Global model	8131 / 10445		19.5
Local models	8131 / 10445		16.2
		0 5 10 15 20	

4.3   Threats to Validity

In our experiments, we considered only a subset of the existing experimental designs. There might be other designs that can lead to more accurate predictions. However, we selected the design leading to the most accurate results in previous case studies.

Furthermore, we considered only a subset of the configuration options of the code generator in our experiments. Based on prior knowledge, we selected options having a considerable influence on the performance of a multigrid solver. Moreover, we also selected hardware-specific options and optimization options.

Although, the prediction accuracy for the two systems are very different, the results are in line with the results of previous work [24]. The differences in accuracy come in line with the different complexity of the systems. While HSMGP has only a limited number of variants and a small number of configuration options, the code generator has an infinite number of variants and a high number of options that interact with each other. In the earlier work, we could also see that for complex systems, such as the Java garbage collector or in our case, the code generator, we are able only to predict performance with an accuracy of more than 70 %.

Regarding the generality of the integration of domain knowledge in our machine-learning approach, we can state, that it is possible to integrate domain knowledge also in the prediction process for other systems. This is, because it is easy possible to modify the sampling strategies and to enrich the learning process with domain knowledge.

Last, in both experiments, we only predict the time needed for one multigrid-cycle iteration. However, the time to solution can be computed using our approach in combination with local Fourier analysis [27].

5   Related Work

Program generation in combination with an integrated domain-specific optimization process is used in many different domains. Often, standard machine-learning approaches, such as evolutionary algorithms [26] and neural networks [14], are used to identify the performance-optimal variant of a specific system. Although many of these approaches can predict performance accurately, they are intransparent in describing the influence of options on the performance explicitly.

Agakov et al. [1] aim at identifying the optimal sequence of source-to-source transformations of iterative compiler optimizations to generate performance-optimal code. They use *Markov models* and *Independent identically distributed models* to support random search and genetic algorithms in identifying promising configurations. In their models, they quantify the probability of a transformation leading to a good performance. Using the Markov models, they are able to consider the influence of small sequences of transformations on performance, which is equal to an interaction between different options. However, they have to model each interaction explicitly before generating the model. Furthermore, it is not possible to model the influence of numeric options on the performance without discretizing the numeric option to a set of binary options. Finally, it is not possible to infer the absolute influence of an option based on the probability of this option leading to a code with good performance.

Another famous example for domain-specific optimization is SPIRAL [19, 20]. It generates implementations of signal-processing algorithms, such as FFTs, based on a domain-specific description. To identify optimal implementations for a specific hardware, they use a feedback-loop mechanism where previous runs are used to guide the search. To this end, they use a set of different machine-learning algorithms, such as hill climbing and evolutionary search algorithms. In contrast to us, they are interested only in the optimal configuration. Thus, they guide the search to sampling only areas with promising configurations, whereas we try to identify the influences of all options.

Ganapathi et al. use statistical machine learning to identify performance optimal stencil computations [4]. After selecting a random set of configurations, they use kernel canonical correlation analysis to identify the correlation between the performance and the different parameters. To identify the optimal configuration, they perform a nearest-neighbor search around the projection space of the performance-optimal configuration of the training set. Using properties of the projection space, it might be possible to get information about the influence of different options on the performance.

Random forests are another supervised learning technique that is often used. For example, Jain et al. use random forests to predict performance of an application using communication data [9]. As input for the learning procedure, they use properties of the implementations, such as the average number of bytes per link or the maximal dilation of a communication. To improve prediction accuracy, they also combine the different metrics to consider interactions among them. Guo et al. [6] use Classification and Regression Trees to identify the influence of configuration options on the performance of the system. In contrast to our approach, they focus only on binary options and do not consider interactions between options. In general, approaches based on a random forest or regression tree, cannot easily quantify the influence of individual configuration options.

6   Conclusion and Future Work

Multigrid solvers are the most efficient methods for solving discretized partial differential equations. However, to achieve a good performance, multigrid implementations have to be tailored by the means of configuration options to the characteristics of the given hardware platform and the problem at hand. Identifying the optimal configuration, is a non-trivial task, because the influence of the options on the performance of the solver is not known and options might also interact with each other. To identify the influence of the configuration options and interactions on performance, we use a combination of forward feature selection and multivariate regression. We tailored our approach such that the learned performance-influence models are easy to read and understand and that we can incorporate domain knowledge to fasten learning and improve accuracy of the models. To this end, we discuss and present how domain knowledge can be integrated in the approach and present first results how the domain knowledge effects prediction accuracy and measurement overhead. Here, we showed experiments with one system, where we were able to reduce the number of configurations used in our approach from 162 to 39 while increasing the mean prediction error by less then 1 %. Moreover, we also demonstrated that the whole approach can be used to predict performance of different variants generated by our *ExaStencils* code generator with a median prediction error of 16.2 %.

In future work, we will create a common interface for our approach to give users the possibility to easily integrate their domain knowledge in the sampling and the learning procedure. Additionally, based on our results in predicting the performance of different multigrid solvers created by our code generator, we see that it is necessary to consider more complex interactions between options to predict performance with a high accuracy. However, the additional candidates caused by considering complex interactions can substantially increase the search space for finding optimal performance-influence models. As result, it is necessary to identify heuristics that describe whether these candidates have to be considered. Last, we will focus on a combination of our approach with local Fourier analysis to predict performance of a whole multigrid solver variant.

Acknowledgements We thank the Jülich Supercomputing Center for providing access to the supercomputer JuQueen. This work is supported by the German Research Foundation (DFG), as part of the Priority Program 1648 "Software for Exascale Computing", under the contract RU 422/15-1 and AP 206/7-1. Sven Apel's work is also supported by the DFG under the contracts AP 206/4-1 and AP 206/6-1.

References

1. Agakov, F., Bonilla, E., Cavazos, J., Franke, B., Fursin, G., O'Boyle, M.F.P., Thomson, J., Toussaint, M., Williams, C.K.I.: Using machine learning to focus iterative optimization. In: Proceedings of the International Symposium on Code Generation and Optimization (CGO), Manhattan, pp. 295–305. IEEE (2006)
2. Bergstra, J., Pinto, N., Cox, D.: Machine learning for predictive auto-tuning with boosted regression trees. In: Proceedings of the Innovative Parallel Computing (InPar), San Jose, pp. 1–9. IEEE (2012)
3. Brandt, A.: Multi-level adaptive solutions to boundary-value problems. Math. Comput. **31**(138), 333–390 (1977)
4. Ganapathi, A., Datta, K., Fox, A., Patterson, D.: A case for machine learning to optimize multicore performance. In: Proceedings of the USENIX Conference on Hot Topics in Parallelism (HotPar), Berkeley, pp. 1–6. USENIX Association (2009)

5. Grebhahn, A., Kuckuk, S., Schmitt, C., Köstler, H., Siegmund, N., Apel, S., Hannig, F., Teich, J.: Experiments on optimizing the performance of stencil codes with SPL conqueror. Parallel Process. Lett. **24**(3), 19 (2014). Article 1441001
6. Guo, J., Czarnecki, K., Apel, S., Siegmund, N., Wasowski, A.: Variability-aware performance prediction: a statistical learning approach. In: Proceedings of the IEEE/ACM International Conference on Automated Software Engineering (ASE), Palo Alto, pp. 301–311. IEEE (2013)
7. Hackbusch, W.: Multi-grid Methods and Applications. Springer, Berlin (2003)
8. Ipek, E., de Supinski, B.R., Schulz, M., McKee, S.A.: An approach to performance prediction for parallel applications. In: Euro-Par 2005 Parallel Processing, Lisboa, pp. 196–205. Springer (2005)
9. Jain, N., Bhatele, A., Robson, M.P., Gamblin, T., Kale, L.V.: Predicting application performance using supervised learning on communication features. In: Proceedings of the International Conference on High Performance Computing, Networking, Storage and Analysis (SC), Denver, pp. 95:1–95:12. ACM (2013)
10. Kerbyson, D.J., Alme, H.J., Hoisie, A., Petrini, F., Wasserman, H.J., Gittings, M.: Predictive performance and scalability modeling of a large-scale application. In: Proceedings of the ACM/IEEE Conference on Supercomputing (SC), Denver, pp. 37–48. ACM (2001)
11. Kronawitter, S., Lengauer, C.: Optimizations Applied by the ExaStencils Code Generator. Technical report MIP-1502, Faculty of Informatics and Mathematics, p. 10. University of Passau (2015)
12. Kuckuk, S., Gmeiner, B., Köstler, H., Rüde, U.: A generic prototype to benchmark algorithms and data structures for hierarchical hybrid grids. In: Parallel Computing: Accelerating Computational Science and Engineering (CSE), pp. 813–822. IOS Press (2013)
13. Lengauer, C., Apel, S., Bolten, M., Größlinger, A., Hannig, F., Köstler, H., Rüde, U., Teich, J., Grebhahn, A., Kronawitter, S., Kuckuk, S., Rittich, H., Schmitt, C.: ExaStencils: advanced stencil-code engineering. In: Euro-Par 2014: Parallel Processing Workshops, Part II. Lecture Notes in Computer Science, Porto, vol. 8806, pp. 553–564. Springer (2014)
14. Magni, A., Dubach, C., O'Boyle, M.: Automatic optimization of thread-coarsening for graphics processors. In: Proceedings of the International Conference on Parallel Architectures and Compilation (PACT), Alberta, pp. 455–466. ACM (2014)
15. Membarth, R., Reiche, O., Hannig, F., Teich, J., Korner, M., Eckert, W.: Hipacc: a domain-specific language and compiler for image processing. IEEE Trans. Parallel Distrib. Syst. **PP**(99), 1–1 (2015)
16. Montgomery, D.C.: Design and Analysis of Experiments. Wiley, New York/Chichester (2006)
17. Nguyen, A., Satish, N., Chhugani, J., Kim, C., Dubey, P.: 3.5-d blocking optimization for stencil computations on modern cpus and gpus. In: Proceedings of the International Conference for High Performance Computing, Networking, Storage and Analysis (SC), New Orleans, pp. 1–13. IEEE (2010)
18. Plackett, R.L., Burman, J.P.: The design of optimum multifactorial experiments. Biometrika **33**(4), 305–325 (1946)
19. Püschel, M., Franchetti, F., Voronenko, Y.: Spiral. In: Encyclopedia of Parallel Computing, pp. 1920–1933. Springer (2011)
20. Püschel, M., Moura, J.M.F., Singer, B., Xiong, J., Johnson, J., Padua, D., Veloso, M., Johnson, R.W.: Spiral: a generator for platform-adapted libraries of signal processing algorithms. J. High Perform. Comput. Appl. **18**, 21–45 (2004)
21. Saltelli, A., Ratto, M., Andres, T., Campolongo, F., Cariboni, J., Gatelli, D., Saisana, M., Tarantola, S.: Global Sensitivity Analysis. The Primer. Wiley, New York/Chichester (2008)
22. Schmitt, C., Kuckuk, S., Hannig, F., Köstler, H., Teich, J.: Exaslang: a domain-specific language for highly scalable multigrid solvers. In: Proceedings of the International Workshop on Domain-Specific Languages and High-Level Frameworks for High Performance Computing (WOLFHPC), New Orleans, pp. 42–51. IEEE (2014)
23. Siegmund, N.: Measuring and predicting non-functional properties of customizable programs. Dissertation, University of Magdeburg (2012)

24. Siegmund, N., Grebhahn, A., Apel, S., Kästner, C.: Performance-influence models for highly configurable systems. In: Proceedings of the European Software Engineering Conference and the ACM SIGSOFT Symposium on the Foundations of Software Engineering (ESEC/FSE), Bergamo, pp. 284–294. ACM (2015)
25. Siegmund, N., Kolesnikov, S.S., Kästner, C., Apel, S., Batory, D., Rosenmüller, M., Saake, G.: Predicting performance via automated feature-interaction detection. In: Proceedings of the International Conference on Software Engineering (ICSE), Zürich, pp. 167–177. IEEE (2012)
26. Simon, D.: Evolutionary optimization algorithms. Wiley, New York/Chichester (2013)
27. Trottenberg, U., Oosterlee, C.W., Schüller, A.: Multigrid. Academic Press, Orlando (2001)
28. Wesseling, P.: An Introduction to Multigrid Methods. Wiley, New York/Chichester (1992)
29. Wirth, N.: Program development by stepwise refinement. Commun. ACM **14**(4), 221–227 (1971)

Part III
EXASTEEL: Bridging Scales
for Multiphase Steels

One-Way and Fully-Coupled FE^2 Methods for Heterogeneous Elasticity and Plasticity Problems: Parallel Scalability and an Application to Thermo-Elastoplasticity of Dual-Phase Steels

Daniel Balzani, Ashutosh Gandhi, Axel Klawonn, Martin Lanser, Oliver Rheinbach, and Jörg Schröder

Abstract In this paper, aspects of the two-scale simulation of dual-phase steels are considered. First, we present two-scale simulations applying a top-down one-way coupling to a full thermo-elastoplastic model in order to study the emerging temperature field. We find that, for our purposes, the consideration of thermo-mechanics at the microscale is not necessary. Second, we present highly parallel fully-coupled two-scale FE^2 simulations, now neglecting temperature, using up to 458,752 cores of the JUQUEEN supercomputer at Forschungszentrum Jülich. The strong and weak parallel scalability results obtained for heterogeneous nonlinear hyperelasticity exemplify the massively parallel potential of the FE^2 multiscale method.

D. Balzani (✉) • A. Gandhi (✉)
Faculty of Civil Engineering, Institute of Mechanics and Shell Structures, TU Dresden, Dresden, Germany
e-mail: daniel.balzani@tu-dresden.de; ashutosh.gandhi@tu-dresden.de

A. Klawonn (✉) • M. Lanser (✉)
Mathematisches Institut, Universität zu Köln, Köln, Germany
e-mail: axel.klawonn@uni-koeln.de; martin.lanser@uni-koeln.de

O. Rheinbach (✉)
Institut für Numerische Mathematik und Optimierung, Technische Universität Bergakademie Freiberg, Freiberg, Germany
e-mail: oliver.rheinbach@math.tu-freiberg.de

J. Schröder (✉)
Faculty of Engineering, Department of Civil Engineering, Institute of Mechanics, Universität Duisburg-Essen, Essen, Germany
e-mail: j.schroeder@uni-due.de

© Springer International Publishing Switzerland 2016   91
H.-J. Bungartz et al. (eds.), *Software for Exascale Computing – SPPEXA 2013-2015*, Lecture Notes in Computational Science and Engineering 113, DOI 10.1007/978-3-319-40528-5_5

1 Introduction

Advanced High Strength Steels (AHSS) provide a good combination of both, strength and formability and are therefore applied extensively in the automotive industry, especially in the crash relevant parts of the vehicle. One such AHSS which is widely employed is dual-phase (DP) steel. The excellent macroscopic behavior of this steel is a result of the inherent micro-heterogeneity and complex interactions between the ferritic and martensitic phases in the microstructure. The microstructural phases are affected by both, mechanical and thermal loads. The modeling of such steels poses a challenge because capturing all the mentioned effects leads to rather complex phenomenological models, which may still be valid for a limited number of loading scenarios.

A more promising modeling approach is the application of multiscale methods. The current contribution proposes a two-scale strategy to analyze the forming process of a DP steel sheet. In this context, the predictions of the overall mechanical response of phenomenological and multiscale-based approaches are compared. We also study the impact of considering pure mechanics versus thermo-mechanics at the microstructure on the quality of the results with view to a predictive mechanical response and the computational effort. Our scale-coupling approach for the two-scale computation of maximal stresses in largely deformed dual-phase steel sheets can be seen as a two-scale FE^2 approach with one-way coupling which consists of two steps. First, a single-scale macroscopic simulation of the deformed steel sheet based on a phenomenological material model representing the macroscopic material behavior is performed. Then, the macroscopic deformation gradient is stored at all Gauß points for each iterated load step. On the basis of macroscopic distributions of plastic strains or stresses, critical regions are identified. Second, microscopic boundary value problems are solved for all Gauß points within the critical regions. Here, the macroscopic deformation gradients are used to define the microscopic deformation-driven boundary conditions. In order to enable a higher efficiency of the scheme we propose to only compute the thermo-mechanical problem at the macroscale. Based on the temperature at each macroscopic Gauß point, we focus on a purely mechanical microscopic boundary value problem, where the temperature-dependent material parameters are updated in each load step according to the macroscopic temperature.

Compared to the high computational cost of the fully-coupled thermo-mechanical FE^2 scheme considering the temperature field at the macro- and microscale, the proposed method is clearly computationally cheaper. Furthermore, an estimator for the quality of the phenomenological macroscopic material model in the critical macroscopic region is obtained by comparing it to the homogenized material response from the microscopic computations. For the simulations including the temperature field we have made use of computing resources in Essen as well as of the CHEOPS cluster in Cologne.

The parallel scalability results for nonlinear hyperelasticity problems presented in this paper were obtained on the JUQUEEN supercomputer [30] at Forschungszentrum Jülich and make use of the FE2TI software package. The FE2TI package is a parallel implementation of the fully coupled FE^2 approach using FETI-DP (Finite Element Tearing and Interconnecting—Dual Primal) methods to solve the problems on the microscopic scale. The FE2TI package has qualified for the High-Q-Club[1] membership in 2015, and its parallel performance has previously been reported in [19, 20]. JUQUEEN is a 28,672 node 6-petaflops Blue Gene/Q system at Jülich Supercomputing Center (JSC, Germany), with a total number of 458,752 processor cores and a power consumption of 2.3 MW. It runs Linux and is ranked 11th on the TOP500 list of the world's fastest supercomputers of November 2015. It uses a Power BQC 16C 1.6 GHz processor with 16 cores and 16 GB memory per node.

The paper is organized in various sections. The material model and a numerical differentiation scheme based on complex step derivative approximation (CSDA) that has been used in the implementation of numerical examples of the one-way coupling FE^2 method are briefly discussed in Sect. 2. A short summary of the general FE^2 multiscale method and the one-way scale-coupling strategy introduced here to study the DP steel sheet response is given in Sect. 3. The details regarding the numerical example and the results obtained with the various strategies are then illustrated in Sect. 4. In Sect. 5 the parallel implementation of the FE^2 method is described, weak parallel scalability for production runs up to the complete JUQUEEN are presented, and strong parallel scalability results for up to 131,072 cores are reported. Finally, the conclusion is presented in Sect. 6.

2   Thermodynamic and Continuum Mechanical Framework

We now present the thermo-elastoplastic framework used in our one-way scale-coupling method. Thermo-mechanics at finite strains are governed by the balance equation of linear momentum and energy. In this section, we only recapitulate the main results of the formulation in the reference configuration, given as

$$- \operatorname{Div} \mathbf{F} \, \mathbf{S} - \mathbf{f} = \mathbf{0} \,, \tag{1}$$

$$\mathbf{S} \cdot \frac{1}{2} \dot{\mathbf{C}} + \rho_0 r - \operatorname{Div} \mathbf{q}_0 - \rho_0 (\dot{\Psi} + \overline{\dot{\theta}\eta}) = 0 \,, \tag{2}$$

and refer the interested reader to [15] for a detailed derivation of these equations and the corresponding weak forms required for the finite element implementation. In equation (2), the Legendre transformation $\Psi = U - \theta\eta$ has been performed, where Ψ, U, η and θ denote the Helmholtz free energy, the specific internal energy, the specific entropy and the temperature, respectively, cf. [33] and [28]. $\mathbf{S}$ denotes

---

[1] http://www.fz-juelich.de/ias/jsc/EN/Expertise/High-Q-Club/FE2TI/_node.html

the second Piola-Kirchoff stress tensor, $\mathbf{C} = \mathbf{F}^T\mathbf{F}$ represents the right Cauchy Green deformation tensor, $\mathbf{F} = \mathrm{Grad}\varphi$ is the deformation gradient and φ defines the nonlinear deformation map, which maps points $\mathbf{X}$ of the undeformed reference configuration $\mathcal{B}_0$ onto points $\mathbf{x}$ of the deformed (actual) configuration. Note that a simple dot notation is used in $\mathbf{S} \cdot \dot{\mathbf{C}}$ to express the full contraction of $\mathbf{S}$ and $\dot{\mathbf{C}}$. $\mathbf{q}_0$ is the heat flux through the body in the reference configuration, which is related to the Cauchy heat flux $\mathbf{q} = -k_\theta \mathrm{grad}\theta$ by $\mathbf{q}_0 = J\mathbf{F}^{-1}\mathbf{q}$. Herein, k_θ is the isotropic heat conduction coefficient and J is the determinant of $\mathbf{F}$. The operators $\mathrm{Grad}(\bullet)$ and $\mathrm{grad}(\bullet)$ denote the gradient with respect to coordinates in the reference and actual configuration. Also, $\mathbf{f}$, r and ρ_0 are the body force vector, internal heat source and the reference density of the body, respectively. Applying the standard Galerkin method, the weak forms of these balance equations can be derived, see e.g. [34] or [28]. Herein, approximations for the displacements in the sense of isoparametric finite elements are inserted. Thus the system to be solved can be written as

$$G_{\mathrm{u}} = G_{\mathrm{u}}^{\mathrm{int}} - G_{\mathrm{u}}^{\mathrm{ext}} \approx \sum_{e=1}^{n_{\mathrm{ele}}} (\delta \mathbf{d}_{\mathrm{u}}^e)^{\mathrm{T}} \left[\mathbf{r}_{\mathrm{u}}^{e,\mathrm{int}} - \mathbf{r}_{\mathrm{u}}^{e,\mathrm{ext}} \right] = 0 \,, \tag{3}$$

$$G_{\theta} = G_{\theta}^{\mathrm{int}} - G_{\theta}^{\mathrm{ext}} \approx \sum_{e=1}^{n_{\mathrm{ele}}} (\delta \mathbf{d}_{\theta}^e)^{\mathrm{T}} \left[\mathbf{r}_{\theta}^{e,\mathrm{int}} - \mathbf{r}_{\theta}^{e,\mathrm{ext}} \right] = 0 \,, \tag{4}$$

where 'G' denotes the weak forms, while the elemental residuals and degree of freedom vectors are introduced as $\mathbf{r}^e$ and $\mathbf{d}^e$ respectively. Here, the subscripts 'u' and 'θ' represent the mechanical and thermal contributions, respectively, and n_{ele} is the number of elements.

2.1 Incorporation of Thermo-mechanics

Since advanced high strength steels are fundamentally thermo-mechanical in nature, the study presented here employs a thermo-elastoplastic material model, as established in [33] and [28]. The main features of the implementation are briefly described in this section. The deformation gradient is multiplicatively decomposed into an elastic ($\mathbf{F}^e$) and a plastic part ($\mathbf{F}^p$) such that $\mathbf{F} = \mathbf{F}^e\mathbf{F}^p$. The isotropic free energy function, incorporating isotropic hardening, takes the form $\Psi = \Psi(\mathbf{b}^e, \theta, \alpha)$, where $\mathbf{b}^e = \mathbf{F}^e\mathbf{F}^{eT}$ and α represent the left Cauchy-Green deformation tensor and the equivalent plastic strain, respectively. The internal dissipation consists of a mechanical and a thermal contribution, $\mathcal{D}_{\mathrm{int}} = \mathcal{D}_{\mathrm{mech}} + \mathcal{D}_{\mathrm{therm}}$. The expressions for these are obtained using the principle of maximum dissipation, the evolution equations for the internal variables $\mathbf{b}^e$, α and the Kuhn-Tucker optimality conditions. For a von Mises type limit surface, the mechanical part reduces to $\mathcal{D}_{\mathrm{mech}} = \lambda \sqrt{\frac{2}{3}} y(\theta)$, where λ is the consistency parameter and $y(\theta)$ the temperature dependent initial yield stress. Exploiting the entropy inequality and Gauß-theorem in (1) and (2) leads upon discretization to the detailed form of the elemental residual vectors

$$\mathbf{r}_u^{e,\text{int}} = \sum_{I=1}^{nen} \int_{\mathscr{B}_0^e} (\mathbf{B}_u^I)^\text{T} \mathbf{S} \, dV , \tag{5}$$

$$\mathbf{r}_\theta^{e,\text{int}} = \sum_{I=1}^{nen} \int_{\mathscr{B}_0^e} \left((\mathbf{B}_\theta^I)^\text{T} \mathbf{q}_0 + N^I \rho_0 \theta \partial_{\theta\theta}^2 \Psi \dot{\theta} + N^I \rho_0 \theta \partial_{\theta\alpha}^2 \Psi \dot{\alpha} \right.$$
$$\left. + N^I \rho_0 \theta \partial_{\theta\mathbf{b}^e}^2 \Psi \cdot \dot{\mathbf{b}}^e + N^I \lambda \sqrt{\tfrac{2}{3}} y(\theta) \right) dV . \tag{6}$$

Here, $\mathbf{B}_u$ and $\mathbf{B}_\theta$ matrices hold the derivatives of the shape functions with respect to spatial coordinates, cf. [4], the $(\dot{\bullet})$ represents the material time derivative of $(\bullet)$ and nen is the number of nodes per element. Note that in the current work we use an additively decoupled, isotropic free energy function with a mechanical part $\Psi_{\text{vol}}^e + \Psi_{\text{iso}}^e + \Psi^p$, a thermo-mechanical coupling part Ψ^c and thermal part Ψ^θ, i.e. $\Psi = \Psi^e(\mathbf{b}^e) + \Psi^p(\alpha) + \Psi^c(\mathbf{b}^e, \theta) + \Psi^\theta(\theta)$, with the individual parts

$$\begin{aligned}
\Psi_{\text{vol}}^e &= \tfrac{\kappa}{\rho_0} \left[\tfrac{1}{2}(J^2 - 1) - \ln J \right] , \\
\Psi_{\text{iso}}^e &= \tfrac{\mu}{2\rho_0} \left[\text{tr}\mathbf{b}^e (\det \mathbf{b}^e)^{-1/3} - 3 \right] , \\
\Psi^p &= \tfrac{1}{2\rho_0} H \alpha^2 , \\
\Psi^c &= -\tfrac{3}{\rho_0} \alpha_t (\theta - \theta_0) \, \partial_J \Psi_{\text{vol}}^e , \\
\Psi^\theta &= -\rho_0 c \left(\theta \ln \tfrac{\theta}{\theta_0} - \theta + \theta_0 \right) .
\end{aligned} \tag{7}$$

For the yield stress $y(\theta)$, a linearly decreasing function in terms of the temperature is considered. Here, H is the linear isotropic hardening modulus for plasticity. The external residual vectors in Eqs. (3) and (4) are obtained on discretizing the external parts of the weak form consisting of the traction vectors and the surface heat fluxes for mechanical and thermal contributions respectively, cf. [34]. These are not discussed here for conciseness of the text; for further details on the algorithmic treatment of thermoplasticity see [28].

2.2 Implementation Using a Complex Step Derivative Approximation

Two widely used numerical differentiation schemes, namely the finite difference method (FD) and the Complex Step Derivative Approximation (CSDA) approach, have been employed to evaluate the stiffness matrix in nonlinear finite element simulations; cf. [22, 25]. Another successful approach is Automatic Differentiation (AD) and, interestingly, relations of CSDA to the forward mode of AD have been pointed out [13]. All these approaches eliminate the need to compute analytical linearizations of the weak forms, which is especially useful in the early development stage of elaborate material models. However, the FD approach leads to round-off errors for small step sizes. The CSDA based strategy overcomes this issue by applying perturbations (of size h) along the imaginary axis of the complex number (cf. [31]) and thus permits the choice of perturbations at the order of the machine precision. Thus, although the method is a (second order) approximation, as with AD, local quadratic convergence can be expected. The implementation of CSDA is simple, especially if a Fortran FD implementation is already available, since Fortran and Fortran libraries have handled complex numbers consistently for a long time. The computational cost, however, is typically larger than for the forward mode of AD.

Our implementation of CSDA was extended also to nonlinear thermo-mechanical problems, where again quadratic convergence rates were obtained; see [4]. A brief summary of this method is presented here. Considering conservative loading and the functional dependencies of the residuals, i.e., $\mathbf{r}_u^{e,\text{int}} := \mathbf{r}_u^{e,\text{int}}(\mathbf{d}_u^e, \mathbf{d}_\theta^e)$ and $\mathbf{r}_\theta^{e,\text{int}} := \mathbf{r}_\theta^{e,\text{int}}(\mathbf{d}_u^e, \mathbf{d}_\theta^e)$, the linearized increments are obtained by differentiating $\mathbf{r}_u^{e,\text{int}}$ and $\mathbf{r}_\theta^{e,\text{int}}$ with respect to both $\mathbf{d}_u^e$ and $\mathbf{d}_\theta^e$ and can be written as

$$\Delta G_u^{\text{int,h}} \approx \sum_{e=1}^{n_{\text{ele}}} (\delta \mathbf{d}_u^e)^{\text{T}} \left(\mathbf{k}_{uu}^e \, \Delta \mathbf{d}_u^e + \mathbf{k}_{u\theta}^e \, \Delta \mathbf{d}_\theta^e \right) , \tag{8}$$

$$\Delta G_\theta^{\text{int,h}} \approx \sum_{e=1}^{n_{\text{ele}}} (\delta \mathbf{d}_\theta^e)^{\text{T}} \left(\mathbf{k}_{\theta u}^e \, \Delta \mathbf{d}_u^e + \mathbf{k}_{\theta\theta}^e \, \Delta \mathbf{d}_\theta^e \right) . \tag{9}$$

Now the CSDA scheme can be used to evaluate the stiffness matrix contributions. The approximations of the k-th column vectors $\tilde{\mathbf{k}}_{uu(k)}^e$ and $\tilde{\mathbf{k}}_{\theta u(k)}^e$ in $\mathbf{k}_{uu}^e$ and $\mathbf{k}_{\theta u}^e$, respectively, and of the j-th column vectors $\tilde{\mathbf{k}}_{u\theta(j)}^e$ and $\tilde{\mathbf{k}}_{\theta\theta(j)}^e$ in $\mathbf{k}_{u\theta}^e$ and $\mathbf{k}_{\theta\theta}^e$, respectively, are given by

$$
\begin{aligned}
\tilde{\mathbf{k}}_{uu(k)}^e &:= \frac{\partial \mathbf{r}_u^{e,\text{int}}}{\partial \{d_u^e\}_k} \approx \frac{\Im \left[\mathbf{r}_u^e(\mathbf{d}_u^e + ih\tilde{\mathbf{d}}_{u(k)}^e, \mathbf{d}_\theta^e) \right]}{h}, \\[1mm]
\tilde{\mathbf{k}}_{u\theta(j)}^e &:= \frac{\partial \mathbf{r}_u^{e,\text{int}}}{\partial \{d_\theta^e\}_j} \approx \frac{\Im \left[\mathbf{r}_u^e(\mathbf{d}_u^e, \mathbf{d}_\theta^e + ih\tilde{\mathbf{d}}_{\theta(j)}^e) \right]}{h}, \\[1mm]
\tilde{\mathbf{k}}_{\theta u(k)}^e &:= \frac{\partial \mathbf{r}_\theta^{e,\text{int}}}{\partial \{d_u^e\}_k} \approx \frac{\Im \left[\mathbf{r}_\theta^e(\mathbf{d}_u^e + ih\tilde{\mathbf{d}}_{u(k)}^e, \mathbf{d}_\theta^e) \right]}{h}, \\[1mm]
\tilde{\mathbf{k}}_{\theta\theta(j)}^e &:= \frac{\partial \mathbf{r}_\theta^{e,\text{int}}}{\partial \{d_\theta^e\}_j} \approx \frac{\Im \left[\mathbf{r}_\theta^e(\mathbf{d}_u^e, \mathbf{d}_\theta^e + ih\tilde{\mathbf{d}}_{\theta(j)}^e) \right]}{h},
\end{aligned}
\tag{10}
$$

where the indices $k \in [1, tdof_u]$ and $j \in [1, tdof_\theta]$ on the left hand side of the equations represent the column index. On the right hand side these indices correspond to the individual perturbation vectors $\tilde{\mathbf{d}}^e$ whose components with indices $m \in [1, tdof_u]$ and $q \in [1, tdof_\theta]$, respectively, are defined as

$$\{\tilde{d}^e_{u(k)}\}_m = \delta_{(k)m} \quad \text{and} \quad \{\tilde{d}^e_{\theta(j)}\}_q = \delta_{(j)q} . \tag{11}$$

Herein, the Kronecker symbol is defined as $\delta_{ab} = 1$ for $a = b$ and $\delta_{ab} = 0$ otherwise. $tdof_u$ and $tdof_\theta$ are the total mechanical and thermal elemental degrees of freedoms, respectively, and $\Im$ is the imaginary operator.

3 Framework for Direct-Micro-Macro Computations

The direct micro-macro approach for computation of material behavior of micro-heterogeneous materials has been well-developed in the last 15 years, see e.g. [9–12, 23, 24, 29], see also [27]. For sake of completeness, in the following subsection this method is briefly recapitulated. Thereafter in Sect. 3.2, we discuss the multiscale treatment proposed here to model DP steel sheet behavior.

3.1 General Approach

The general FE^2 concept involves solving a microscopic boundary value problem at each macroscopic integration point during the solution of the macroscopic boundary value problem. These nested problems are defined on representative volume elements (RVEs) that describe the complex geometry of the microstructure adequately. Appropriate boundary conditions are applied to these in terms of, e.g., the deformation gradients at the macroscopic integration point. After solving the microscopic problem, suitable volume averages of microscopic stresses $\mathbf{P}$ and microscopic tangent moduli $\mathbb{A}$ are computed and returned back to the macroscopic integration point, which replaces the evaluation of a classical phenomenological macroscopic material law.

These averages are computed as

$$\overline{\mathbf{P}} = \frac{1}{V} \int_{\mathcal{B}_0} \mathbf{P} \, dV \quad \text{and} \quad \overline{\mathbb{A}} = \frac{1}{V} \int_{\mathcal{B}_0} \mathbb{A} \, dV - \frac{1}{V} \mathbf{L}^{\mathsf{T}} \mathbf{K}^{-1} \mathbf{L} , \quad \mathbf{L} = \int_{\mathcal{B}_0} \mathbf{B}^{\mathsf{T}} \mathbb{A} \, dV . \tag{12}$$

Here, $\overline{\mathbf{P}}$, $\overline{\mathbb{A}}$ represent the macroscopic first Piola-Kirchhoff stresses and the macroscopic material tangent moduli. The global stiffness matrix and the spatial derivatives of the shape functions of the microscopic boundary value problem are denoted by $\mathbf{K}$ and $\mathbf{B}$, respectively. This procedure eliminates the need of a phenomenological law at the macroscale. Furthermore, certain effects like anisotropy, and its evolution as well as kinematic hardening are automatically included through the solution of the micro-problem due to its heterogeneity.

Algorithm 1 Algorithmic description of the FE²TI approach. Overlined letters denote macroscopic quantities. This algorithm consists of the classical FE² scheme using (ir)FETI-DP for solving the microscopic boundary value problems. We consider all macroscopic as well as microscopic Newton iterations as converged, if the l_2-norm of the Newton update is smaller than $1e - 6$. The GMRES iteration in our FETI-DP methods is stopped, if a relative residual reduction of $1e-8$ is reached. This pseudocode is taken from [20]

Repeat until convergence of the Newton iteration:

1. Apply boundary conditions to RVE (representative volume element) based on macroscopic deformation gradient: Enforce $x = \overline{F}X$ on the boundary of the microscopic problem $\partial\mathscr{B}$ in the case of Dirichlet constraints.
2. Solve one microscopic nonlinear implicit boundary value problem for each macroscopic Gauß point using Newton-Krylov-(ir)FETI-DP or related methods.
3. Compute and return macroscopic stresses as volumetric average of microscopic stresses $\mathbf{P}^h$:

$$\overline{P}^h = \frac{1}{V}\sum_{T\in\tau}\int_T P^h dV .$$

4. Compute and return macroscopic tangent moduli as average over microscopic tangent moduli $\mathbb{A}^h$:

$$\overline{\mathbb{A}}^h = \frac{1}{V}\left(\sum_{T\in\tau}\int_T \mathbb{A}^h \, dV\right) - \frac{1}{V}L^T (K)^{-1} L .$$

5. Assemble tangent matrix and right hand side of the linearized macroscopic boundary value problem using $\overline{P}^h$ and $\overline{\mathbb{A}}^h$.
6. Solve linearized macroscopic boundary value problem.
7. Update macroscopic deformation gradient $\overline{F}$.

Note that, because of the two-scale procedure, the Newton linearization of the FE² method was not straight forward but rather a significant step in the development of the method. In our fully coupled FE² simulations, we use a consistent Newton linearization and thus can expect locally quadratic convergence of the fully coupled two-scale method, i.e., of the outermost loop in Algorithm 1.

For a brief description of the algorithm and an efficient parallel implementation, see also Sect. 5.1 and Algorithm 1.

3.2 Approaches for Multiphase-Steel Incorporating Thermo-mechanics

Two-scale analysis is performed to study the influence of macroscopic deformation on the microscopic mechanical fields to obtain more realistic simulations of sheet metal forming processes. For efficiency reasons, here, we focus on a one-way scale-coupling scheme, using efficient parallel algorithms to solve complex microscopic boundary value problems of DP steel microstructures. For that purpose we first perform a thermo-mechanical simulation of the macroscopic sheet metal forming process using a phenomenological thermo-elastoplastic material model at finite strains, as described in Sect. 2, which would be used in engineering practice. Then, in order to obtain more information of those mechanical fields at the microscale which are important for failure initialization analysis, the macroscopic regime with high plastic strains is identified. Only there, additional microscopic boundary value problems are solved which are driven by the macroscopic deformation gradients and temperatures computed at the macroscopic integration points. In detail, regarding the displacements, linear displacement boundary conditions are applied to the real DP steel microstructures and periodic boundary conditions are considered when using statistically similar RVEs (Representative Volume Elements) in the sense of [5]. The microstructure consists of two phases—ferrite as the matrix phase and martensite as the inclusion phase.

With respect to the temperature, we focus on different approaches: (i) the temperature calculated at the macroscopic integration point is applied to the boundary of the microscopic boundary value problem, where thermo-mechanics are considered and the temperature is free to evolve, and (ii) the microscopic thermal fluctuations are considered to be small due to small deviations of thermo-mechanical parameters for ferrite and martensite. Therefore, only mechanical boundary value problems taking into account temperature-dependent yield stresses are considered at the microscale. The latter approach enables more efficient computations since the temperature is not a degree of freedom in the microscopic calculations anymore. Simulations based on such one-way scale-couplings have two important advantages compared with purely macroscopic computations: first, they provide valuable information regarding those microscopic mechanical fields in the macroscopic domains where failure is expected to initialize. Second, an estimation of the quality of the macroscopic material model is obtained by comparing to the more accurate micro-macro computation.

4   Numerical Examples for the One-Way FE^2 Coupling

In the analysis performed here, we consider at the macroscale the extension of a DP steel sheet containing a regular arrangement of holes. The dimensions of the sheet and the diameter of the holes are $140 \times 140 \times 6$ mm and 20 mm, respectively. A displacement of 10 mm is applied in X-direction at the outer surface of the sheet such that the metal sheet is extended up to 14.28% nominal strain. The time considered for this deformation is 10 s. Due to the symmetry of the problem, we only simulate 1/8th of the plate, see Fig. 1a, and incorporate appropriate symmetry conditions. The plate is discretized with 10-noded tetrahedral finite elements. As a phenomenological description at the macroscale, we consider the thermo-

(a) (b)

(c) (d)

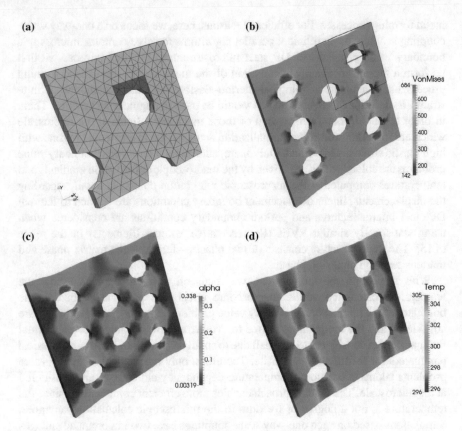

Fig. 1 (a) 1/8th geometry of the plate with tetrahedral finite element mesh, macroscopic (b) von Mises stress, (c) equivalent plastic strain and (d) temperature distributions in the deformed configuration of the sheet metal after applying full extension at the macro-level

mechanical formulation of [28], which was implemented using the new CSDA scheme. The initial yield stress as well as the linear hardening modulus were adjusted to yield curves calculated as volumetric averages of purely mechanical micro-macro computations of uni-axial tension tests. The hardening modulus was chosen such that the model response matches this yield curve at approximately 30 % strain. In the micro-macro computations the same thermo-mechanical framework was used as in the macroscopic computations. The resulting distributions of stress, equivalent plastic strains and temperature are as shown in Fig. 1b–d, respectively. As can be seen, the fluctuation of temperature is rather small although rather large plastic strains are obtained. However, in particular for the detection of necking, the incorporation of even small temperature deviations may be essential, cf. the findings in [28]. In Fig. 1a the eighth of the complete sheet metal considered for computation is depicted. Additionally, the outline of a subregion is marked which is considered as most critical for failure initialization since here the largest macroscopic stresses

(a)   (b)

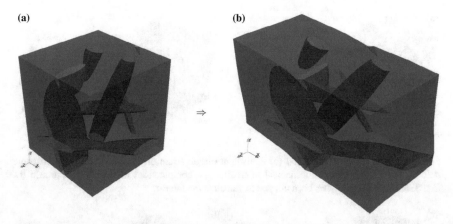

Fig. 2 (**a**) Undeformed SSRVE structure for evaluation of the microscopic problem at the point of interest indicated by the bullet and (**b**) deformed configuration of the SSRVE at full load

are found. This subregion is therefore considered as the most relevant regime, and detailed micro-macro computations are performed, here. For this purpose, there the deformation gradient and the temperature at each macroscopic integration point is stored for every load step in order to be applied in subsequent microscopic computations.

In order to analyze the influence of the two approaches (i) and (ii) we focus on statistically similar RVEs (SSRVEs) which were computed for DP steel in [5]; cf. Fig. 2a. For the analysis, we consider a macroscopic integration point within the critical region, where its position is marked by the bullet in Fig. 1a. The hardening modulus for the pure ferrite and the pure martensite is chosen such that the model response corresponds to the experimental yield stress in uni-axial tension at 10 % strains. The distributions of von Mises stresses and equivalent plastic strains as a result of the thermo-mechanical computations associated with approach (i) are depicted in Fig. 3. They indicate a higher development of stresses and negligible plastic strains in the martensitic inclusions. The ferritic matrix phase shows lower stresses and higher plastic strains are accumulated here due to the lower yield stress as compared to the martensite.

For the purely mechanical microscopic computation the temperature-dependent initial yield stress y is taken into account such that $y = \langle \omega(\theta - \theta_0) + y_0 - \tilde{y}_0 \rangle + \tilde{y}_0$, where the Macauley brackets $\langle(\bullet)\rangle = \frac{1}{2}[|(\bullet)| + (\bullet)]$ ensure a limiting yield stress $\tilde{y}_0$. Herein, θ, θ_0 and ω are the current temperature, the room temperature and a thermal softening parameter; y_0 denotes the initial yield stress at room temperature. When comparing the stresses and plastic strains resulting from the purely mechanical computation where the temperature-dependent yield behavior is incorporated (approach (ii)), see Fig. 4, we obtain quite similar distributions at the microscale. This is also observed for the macroscopic values: the macroscopic von

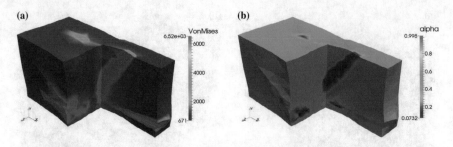

Fig. 3 (a) Von Mises stresses and (b) equivalent plastic strain distributions at full macroscopic deformation for the thermo-mechanical microstructure computations according to approach (i). The SSRVE, see Fig. 2b, has been clipped to visualize the interior

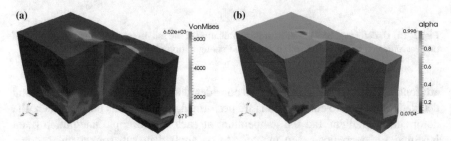

Fig. 4 (a) Von Mises stresses and (b) equivalent plastic strain distributions over the SSRVE at full macroscopic deformation for the purely mechanical calculations according to approach (ii). The SSRVE, see Fig. 2b, has been clipped to visualize the interior

Mises stress is computed from the volume averaged Cauchy stress and takes a value of 1050.4 MPa, whereas for the thermo-mechanical computation it is 1009.85 MPa.

Now, we compare the results of the micro-macro computations with the response of the purely macroscopic phenomenological model. Therefore, the macroscopic von Mises stress versus nominal extension at the bullet point in the sheet metal is plotted in Fig. 5a. As can be seen, the difference between the purely macroscopic computation and the micro-macro computation is rather large, compared to the difference between the two approaches (i) and (ii). Furthermore, Fig. 5b shows the temperature distribution at the microscale as a result of approach (i). A quite small fluctuation even below 1 K is observed. This indicates that the consideration of thermo-mechanics at the microscale is not necessarily required. In Fig. 5a also the response of a purely mechanical micro-macro computation is plotted, where not even the temperature dependency of the yield stress is taken into account. A small deviation from the computation including temperature-dependent yield stresses is observed. However, the incorporation of temperature-dependent evolving yield stresses may be important in order to accurately represent a potential necking at the microscale, cf. [28]. Therefore, in the following, the model based on approach (ii) is used to perform parallel micro-macro simulations of the entire critical region as seen in Fig. 1a which consists of 468 microscopic boundary value problems associated with the macroscopic integration points. We consider realistic microstructures with

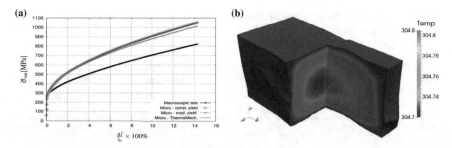

Fig. 5 **(a)** Comparison of the macroscopic von Mises stress vs. nominal strain curves resulting from the macroscopic phenomenological law and from the micro-macro computations and **(b)** bi-sectional view of the temperature distribution at the microscale as a result of approach (i)

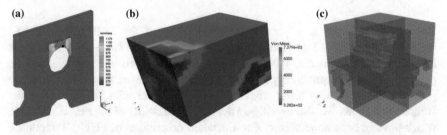

Fig. 6 **(a)** Von Mises stresses in the critical region after applying the full extension obtained from averaging the stress distributions from the microscopic boundary value problems with the **(c)** realistic microstructure; **(b)** von Mises stresses in the deformed configuration of one exemplary microscopic boundary value problem corresponding to the macroscopic integration point P marked with the *black dot* in Fig. 6a

206,763 degrees of freedom each, see Fig. 6c. For an efficient and fast solution we decomposed the microstructure in eight cubical subdomains and used a Newton-Krylov FETI-DP approach. All microscopic computations have been performed on eight cores of the CHEOPS cluster at the RRZK in Cologne. In Fig. 6b we present the von Mises stresses in the deformed configuration of one exemplary microscopic problem and in Fig. 6b the von Mises stresses in the complete critical region of the macroscopic problem. The von Mises stresses in the integration points of the macroscopic problem are obtained from a suitable volumetric average over the microscopic quantities. When comparing the results with the purely macroscopic computations shown in Fig. 1a, the qualitative distribution of the stresses looks similar. The quantitative results differ however more than 30 %, which shows the necessity to analyze scale-coupled computations. In addition to a more reliable prediction of stresses at the macroscale, microscopic stress distributions are available building the basis for failure initialization analysis. We additionally provide a brief summary of the RVE computations performed on CHEOPS and using FETI-DP for the solution of all linear systems; see Table 1.

Table 1 RVEs using the J2 plasticity material model in 3D. For the three dimensional micro structure; see Fig. 6c. *Average Newton It.* denotes the number of Newton iterations per RVE, summed up over all macroscopic load steps and averaged over all RVEs

Realistic RVEs with thermo-plasticity and realistic microstructure				
#RVEs	D.o.f. per RVE	FETI-DP subdomains per RVE	Average Newton It.	Total core× h
468	206,763	8	2113	9.28 h × 8 × 468

5   FE2TI: A Parallel Implementation of the Fully Coupled FE² Approach

The FE2TI software is a parallel implementation of the (fully coupled) FE² method using FETI-DP domain decomposition methods to solve the microscopic boundary value problems. We have reported on the software package and its parallel performance previously [19, 20]. In the current paper, for the first time, we provide weak scalability results for large production runs with parallel I/O on the complete machine. We also investigate the strong scalability of the FE2TI software, which has not been done before. For a detailed description of FETI-DP methods, see [8, 16–18].

5.1   Implementation Remarks

FE2TI is implemented using PETSc 3.5.2 [3], MPI, and hybrid MPI/OpenMP. Furthermore, we make use of the software libraries MUMPS [1, 2], UMFPACK [6], and PARDISO [26] as sequential (or parallel) direct solvers for subdomain problems. We also make use of inexact FETI-DP variants using BoomerAMG [14] from the hypre [7] package as a preconditioner of the FETI-DP coarse problem. On Blue Gene/Q, the software environment is compiled using the IBM XL C/C++ compilers using auto vectorization. When using UMFPACK as a direct solver for the subproblems, a large portion of the computing time is spent inside IBM's ESSL library, which implements efficient auto vectorization. In the computations presented here, we use piecewise linear brick elements (Q1) for all finite element discretizations. In our FE2TI implementation, an MPI_Comm_split is used to create subcommunicators for the computations on the RVEs (Representative Volume Elements). On Blue Gene/Q supercomputers, we use the environment variable

```
PAMID_COLLECTIVES_MEMORY_OPTIMIZED=1
```

to enable an efficient communicator split even for a large number of cores.

Each RVE is assigned to exactly one of the MPI subcommunicators, and the computations in 1. to 4. in Algorithm 1 can be carried out independently on each subcommunicator. This includes several parallel (inexact reduced) FETI-DP [17] setups and solves. Communication in between the several communicators is only necessary for the assembly of the linearized macroscopic problem (see 5. in Algorithm 1) and the update of the macroscopic variables (see 7. in Algorithm 1). The macroscopic solve (see 6. in Algorithm 1) is performed on each MPI rank redundantly, using a sparse direct solver. This is feasible due to the small macroscopic problem size. To assemble and solve the macroscopic problem on each MPI rank, the consistent tangent moduli and the averaged stresses in the macroscopic Gauß points have to be communicated to all ranks. Therefore, we have implemented a gather operation in two steps. First, the tangent moduli and stresses are averaged and collected on the master ranks of each RVE subcommunicator. This corresponds to an MPI_Reduce operation on each subcommunicator. In a second step, an MPI_Gather operation collects the data from the master ranks of the subcommunicators in the global master rank. This avoids a global all-to-all communication and only includes one MPI rank per RVE. Finally, we broadcast all tangent moduli and stresses from the global master rank to all MPI ranks. For some more details on the FE2TI implementation, see [19, 20].

A highly efficient parallel I/O strategy is also provided in the FE2TI package, based on HDF5 [32]. All data, as stresses and displacements on the RVEs, is written to one single parallel file, currently once every four macroscopic load steps. For a production run on the complete JUQUEEN, we have measured an I/O time of less than 2 % of the total runtime.

In all computations presented in this section, we consider two different Neo-Hooke materials. Note that for the results in this section, we do not use a CSDA approximation but rather the exact tangent. We have inclusions of stiff material ($E = 2100$, $\nu = 0.3$) in softer matrix material ($E = 210$, $\nu = 0.3$) and consider a realistic microstructure depicted in [20, Fig. 1]. The strain energy density function of the Neo-Hooke material W [15, 34] is given by

$$W(u) = \frac{\mu}{2}\left(\mathrm{tr}(\mathbf{F}^T\mathbf{F}) - 3\right) - \mu\ln(J) + \frac{\lambda}{2}\ln^2(J)$$

with the Lamé constants $\lambda = \frac{\nu E}{(1+\nu)(1-2\nu)}$, $\mu = \frac{E}{2(1+\nu)}$ and the deformation gradient $\mathbf{F}(x) := \nabla\varphi(\mathbf{x})$; here, $\varphi(\mathbf{x}) = \mathbf{x} + \mathbf{u}(\mathbf{x})$ denotes the deformation and $\mathbf{u}(\mathbf{x})$ the displacement of $\mathbf{x}$.

5.2 Production Runs on the JUQUEEN Supercomputer

First, we present three different production runs of different problem sizes in Table 2. Here, as a macroscopic problem, we discretize a thin plate with a rectangular hole with 8, 32, and finally 224 finite elements. This corresponds to a full simulation of 64, 256, and finally 1792 RVEs in the corresponding macroscopic Gauß integration points. In 40–41 load steps, we apply a deformation of the plate of approximately 8 %. A visualization of the results of the largest production run has been previously reported on in [20, Fig. 1]. Considering only a few macroscopic load steps and disabling I/O, e.g., for checkpointing, we have already shown nearly optimal weak scalability for the FE2TI package [19, 20]. Here, for the first time, we present weak scalability for the production runs using full I/O (using HDF5 [32]), many load steps, an unstructured mesh on the macroscale, and a realistic microstructure from dual phase steel.

In our multiscale approach, the size of the RVE must be determined such that it is representative of the microstructure (sufficient size) and that it must capture all important features of the microstructure (sufficient resolution). Once the type of discretization is chosen, the number of degrees of freedom for the RVE is thus fixed. Here, each RVE has 823,875 degrees of freedom. In our computation, a problem on an RVE is then solved iteratively and in parallel, using 512 MPI ranks running on 256 cores, by a FETI-DP domain decomposition method using 512 subdomains. This choice results in an appropriate workload for each core. Therefore, the largest multiscale production run on the complete JUQUEEN at Forschungszentrum Jülich (917,504 MPI ranks on 458,752 cores) makes use of a total number of 1,476,384,000 degrees of freedom (of course representing a much larger full scale problem).

Neglecting the fact that we use slightly different dimensions for the macroscopic plate in the three different production runs, this set of production runs can also be viewed as a weak parallel scaling test. In addition to the total time to solution, we have also reported on the average runtime for the solution of a nonlinear RVE problem in Table 2. Here, we have a slight increase in the runtime of approximately 10 % when scaling from 1 to 28 racks. This is partially due to a small increase in I/O time and also slightly higher numbers of GMRES iterations in the FETI-DP solver, probably due to the larger and more complicated macroscopic problem. Nevertheless, for a complete production run including parallel I/O, these scalability results are satisfying. In Table 2, we also provide timings for the macroscopic solve. Since the macroscopic problem is currently solved redundantly on each core, this phase of the method does not scale. But even for the largest production run, the cost

Table 2 Complete FE^2 production runs using the FE2TI software; realistic microstructure in the RVEs; nonlinear elasticity model; 32 MPI ranks per node. *Avg. RVE Solve* denotes the average runtime to solve the nonlinear microscopic boundary value problems; *Avg. Macro Solve* denotes the average runtime of a direct solve on the macroscale

JUQUEEN—Complete FE^2 runs for elasticity						
#Racks	#MPI ranks	#RVEs	#Load steps	Time (s)	Avg. RVE solve (s)	Avg. macro solve (s)
1	32,768	64 RVEs	41LS	16,899	101.13	0.06
4	131,072	256 RVEs	41LS	17,733	105.95	0.22
28	917,504	1792 RVEs	40LS	18,587	112.48	1.54

for the macroscopic problem is currently negligible, i.e., it contributes less than 1% to the total runtime.

5.3   Strong Scalability on JUQUEEN

For the first time, we also present strong scalability results for the FE2TI software for a nonlinear model problem; see Table 3. Let us first describe the model problem used here. On the macroscale, we use the geometry and discretization of the second largest production run presented before in Table 2. Thus we have 256 microscopic boundary value problems (RVEs). In contrast to the previous production runs, we now consider smaller RVEs with 107 K degrees of freedom each. Each RVE has one stiff, spherical inclusion and is decomposed into 512 FETI-DP subdomains. The subdomains are thus quite small, only consisting of 375 degrees of freedom. Let us note that this setup avoids memory problems on the smallest partition (1024 MPI ranks). Let us remark that we always use 32 MPI ranks per BlueGene/Q node and thus less than 512 MByte are available per rank. This setup was found to be most efficient in [19, 20].

In our strong scaling test, we simulate one macroscopic load step which converges in three Newton steps. In 222 of the 256 RVE problems, 9 microscopic Newton steps are necessary for convergence during the complete macroscopic load step. For the remaining 34 Gauß points only 8 microscopic Newton steps are

Table 3 Strong scaling of FE^2 using the FE2TI software; nonlinear elasticity model; 32 MPI ranks per node. Macroscopic problem with 256 Gauß integration points; in each macroscopic integration point an RVE with 107K degrees of freedom is solved using 512 FETI-DP subdomains. Simulation of one macroscopic load step. *Time to Solution* denotes the total time needed for one FE^2 load step; *Eff.* denotes the parallel efficiency, where the total time to solution on 1 024 ranks is chosen as a baseline; *Speedup* denotes the speedup compared to the runtime on 1 024 cores; *Avg. FETI-DP Setup Time* denotes the average runtime necessary for a FETI-DP setup for one linearized system on an RVE; *Avg. Ass. Time* denotes the average runtime of the assembly of one linearized system on an RVE; *Avg. Solve Time* denotes the average iteration time to solve one linearized system on an RVE; all averages consider all linearized systems occurring in all microscopic Newton steps

Strong scaling on JUQUEEN						
	Time			Avg. FETI-DP	Avg.	Avg.
MPI ranks	to solution (s)	Eff. (%)	Speedup	setup time (s)	ass. time (s)	solve time (s)
1024	1568.9	100	1.00	14.98	44.29	21.08
2048	822.0	95	1.91	6.97	22.13	11.54
4968	431.7	91	3.63	3.51	11.07	5.65
8192	225.0	87	6.97	1.85	5.54	3.05
16,384	138.7	71	11.39	1.04	2.77	2.32
32,768	90.0	54	17.42	0.61	1.38	1.43
65,536	40.4	61	38.81	0.39	0.69	0.67
131,072	35.6	34	44.10	0.29	0.35	0.65

performed. This means, depending on the RVE, we perform 11 or 12 FETI-DP setups including problem assembly, while 35 or 36 FETI-DP solves are necessary. Let us recall that we have to perform one FETI-DP setup and solve per microscopic Newton step. Additionally, after convergence on the microscale, we have to compute the consistent tangent moduli (see 4. in Algorithm 1). Therefore, for each of the three macroscopic Newton steps, one further FETI-DP setup and nine FETI-DP solves with different right hand sides are necessary. This sums up to the mentioned number of FETI-DP setups and solves on each microstructure. In average, we have 44.8 GMRES iterations for each FETI-DP solve. Let us remark that we consider all macroscopic as well as microscopic Newton iterations as converged, if the l_2-norm of the Newton update is smaller than $1e - 6$.

Since the lion's share of the runtime of the FE2TI package is spent in the assembly of the microscopic problems and in FETI-DP, the strong scalability is dominated by three phases: the problem assembly on the RVEs, the FETI-DP setup, and the FETI-DP solve; see also [21] for a detailed discussion on the strong scaling behavior of (ir)FETI-DP methods. Therefore, we provide detailed timings for those three phases in Table 3. We obtain, as it can be expected, perfect scalability for the assembly phase and also convincing results for the FETI-DP setup phase. The FETI-DP solution phase scales well up to 65 K ranks. Scaling further up to 131 K ranks the additional benefit is small. These results are also depicted in Fig. 7. All in all, this leads to a satisfying strong scaling behavior of the complete FE2TI package from 1 K up to 65 K ranks with 61 % parallel efficiency and a speedup of 38.8; see also Fig. 8. Let us finally remark that the FE2TI package can thus solve 256 times 36 linear systems with 107 K degrees of freedom in approximately 40 s on 65 K MPI ranks and 32 K BlueGene/Q cores.

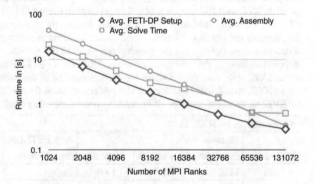

Fig. 7 Strong scalability of the FE2TI software. Figure corresponds to data from Table 3. Scalability of the different phases of the RVE solver FETI-DP

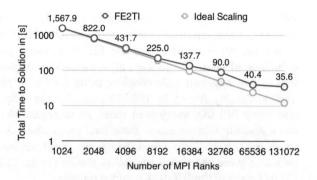

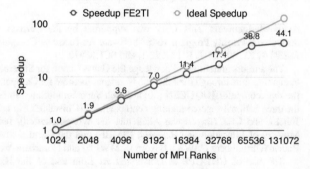

Fig. 8 Strong scalability of the FE2TI software. Figure corresponds to data from Table 3. *Top*: Total time to solution. *Bottom*: Speedup

6 Conclusion

We have presented two steps towards the realistic two-scale simulation of dual-phase steel. First, we have discussed our isotropic, thermodynamically-consistent, thermo-elastoplastic material model, based on [28], to be employed in the multiscale simulation of dual-phase steel sheets. A numerical differentiation scheme, which relies on the complex step derivative approximation approach, was used to compute the tangent stiffness matrices in the thermo-mechanical simulations. It allows to obtain locally quadratic convergence of Newton's method. Within this setting, a one-way coupling scheme is utilized to increase the efficiency in the multiscale analysis of the steel sheet subjected to inhomogeneous deformations. The multiscale analysis presented here indicates that the higher level of information involved in the micro-level computation leads to a more accurate assessment of critical states during the forming process. The resulting mechanical field distributions help to identify areas in the microstructure geometry where concentrations of stress or strains may lead to initialization of failure. This information is not accessible by purely phenomeno-logical material models and limits their predictive capabilities. Additionally, the comparison between various approaches at the micro-level show that, for DP steels, where the thermal properties of the phases are almost identical, for the considered nominal strain rates ($\dot{\bar{\varepsilon}} \approx 10^{-2}\,\text{s}^{-1}$), the thermo-mechanical consideration does not yield significantly different response than the purely mechanical one. Thus, for cases similar to the one presented here, considering only mechanics at the microscale can reduce computational effort substantially without significant loss of accuracy.

Second, we have presented the FE2TI software package for the two-scale simulation of steel. The package allows one-way coupling, as described above, as well as two-way, two-scale coupling using the FE^2 approach. We have discussed weak scalability for up to 458,752 cores for the fully coupled FE^2 production runs using full I/O, many load steps, an unstructured mesh on the macroscale, and a realistic microstructure from dual phase steel. As a result of the previous considerations, in these computations, we could neglect temperature effects. We have also presented strong scalability results for the FE2TI software using up to 131 072 cores of the JUQUEEN supercomputer.

Acknowledgements This work was supported by the German Research Foundation (DFG) through the Priority Program 1648 "Software for Exascale Computing" (**SPPEXA**), projects BA 2823/8-1, KL 2094/4-1, RH 122/2-1, and SCHR 570/19-1.

The authors gratefully acknowledge the Gauss Centre for Supercomputing (GCS) for providing computing time through the John von Neumann Institute for Computing (NIC) on the GCS share of the supercomputer **JUQUEEN** [30] at Jülich Supercomputing Centre (JSC). GCS is the alliance of the three national supercomputing centres HLRS (Universität Stuttgart), JSC (Forschungszentrum Jülich), and LRZ (Bayerische Akademie der Wissenschaften), funded by the German Federal Ministry of Education and Research (BMBF) and the German State Ministries for Research of Baden-Württemberg (MWK), Bayern (StMWFK) and Nordrhein-Westfalen (MIWF).

The use of **CHEOPS** at Universität zu Köln and of the **High Performance Cluster** at Technische Universität Bergakademie Freiberg are also gratefully acknowledged. Furthermore, the authors D. Balzani and A. Gandhi appreciate S. Prüger for helpful scientific discussions.

References

1. Amestoy, P.R., Duff, I.S., Koster, J., L'Excellent, J.Y.: A fully asynchronous multifrontal solver using distributed dynamic scheduling. SIAM J. Matrix Anal. Appl. **23**(1), 15–41 (2001)
2. Amestoy, P.R., Guermouche, A., L'Excellent, J.Y., Pralet, S.: Hybrid scheduling for the parallel solution of linear systems. Parallel Comput. **32**(2), 136–156 (2006)
3. Balay, S., Brown, J., Buschelman, K., Gropp, W.D., Kaushik, D., Knepley, M.G., McInnes, L.C., Smith, B.F., Zhang, H.: PETSc Web page. http://www.mcs.anl.gov/petsc (2014)
4. Balzani, D., Gandhi, A., Tanaka, M., Schröder, J.: Numerical calculation of thermo-mechanical problems at large strains based on complex step derivative approximation of tangent stiffness matrices. Comput. Mech. **55**, 861–871 (2015)
5. Balzani, D., Scheunemann, L., Brands, D., Schröder, J.: Construction of two- and three-dimensional statistically similar RVEs for coupled micro-macro simulations. Comput. Mech. **54**, 1269–1284 (2014)
6. Davis, T.A.: Direct Methods for Sparse Linear Systems. SIAM, Philadelphia (2006)
7. Falgout, R.D., Jones, J.E., Yang, U.M.: The design and implementation of hypre, a library of parallel high performance preconditioners. In: Bruaset, A.M., Bjorstad, P., Tveito, A. (eds.) Numerical Solution of Partial Differential Equations on Parallel Computers. Lecture Notes in Computational Science and Engineering, vol. 51, pp. 267–294. Springer, Berlin (2006). http://dx.doi.org/10.1007/3-540-31619-1_8

8. Farhat, C., Lesoinne, M., LeTallec, P., Pierson, K., Rixen, D.: FETI-DP: a dual-primal unified FETI method – part I: a faster alternative to the two-level FETI method. Int. J. Numer. Methods Eng. **50**, 1523–1544 (2001)
9. Feyel, F., Chaboche, J.: Fe^2 multiscale approach for modelling the elastoviscoplastic behaviour of long fibre SiC/Ti composite materials. Comput. Methods Appl. Mech. Eng. **183**, 309–330 (2000)
10. Fish, J., Shek, K.: Finite deformation plasticity for composite structures: computational models and adaptive strategies. Comput. Methods Appl. Mech. Eng. **172**, 145–174 (1999)
11. Geers, M., Kouznetsova, V., Brekelmans, W.: Multi-scale first-order and second-order computational homogenization of microstructures towards continua. Int. J. Multiscale Comput. **1**, 371–386 (2003)
12. Golanski, D., Terada, K., Kikuchi, N.: Macro and micro scale modeling of thermal residual stresses in metal matrix composite surface layers by the homogenization method. Comput. Mech. **19**, 188–201 (1997)
13. Griewank, A., Walther, A.: Evaluating Derivatives. Society for Industrial and Applied Mathematics, 2nd edn. (2008). http://epubs.siam.org/doi/abs/10.1137/1.9780898717761
14. Henson, V.E., Yang, U.M.: BoomerAMG: a parallel algebraic multigrid solver and preconditioner. Appl. Numer. Math. **41**, 155–177 (2002)
15. Holzapfel, G.A.: Nonlinear Solid Mechanics. A Continuum Approach for Engineering. John Wiley and Sons, Chichester (2000). http://opac.inria.fr/record=b1132727
16. Klawonn, A., Rheinbach, O.: Robust FETI-DP methods for heterogeneous three dimensional elasticity problems. Comput. Methods Appl. Mech. Eng. **196**(8), 1400–1414 (2007)
17. Klawonn, A., Rheinbach, O.: Highly scalable parallel domain decomposition methods with an application to biomechanics. ZAMM Z. Angew. Math. Mech. **90**(1), 5–32 (2010). http://dx. doi.org/10.1002/zamm.200900329
18. Klawonn, A., Widlund, O.B.: Dual-primal FETI methods for linear elasticity. Commun. Pure Appl. Math. **59**(11), 1523–1572 (2006)
19. Klawonn, A., Lanser, M., Rheinbach, O.: EXASTEEL – computational scale bridging using a FE^{2}TI approach with ex_nl/FE^2. Technical report FZJ-JSC-IB-2015-01, Jülich Supercomputing Center, Germany (2015). https://juser.fz-juelich.de/record/188191/files/FZJ-2015-01645. pdf. In: Frings, Brian J.N. Wylie (eds.) JUQUEEN Extreme Scaling Workshop 2015. Dirk Brömmel and Wolfgang
20. Klawonn, A., Lanser, M., Rheinbach, O.: FE2TI: Computational Scale Bridging for Dual-Phase Steels (2015). Accepted for publication to the proceedings of the 16th ParCo Conference, Edinburgh. To be published in Advances in Parallel Computing
21. Klawonn, A., Lanser, M., Rheinbach, O.: Towards extremely scalable nonlinear domain decomposition methods for elliptic partial differential equations. SIAM J. Sci. Comput. **37**(6), C667–C696 (2015)
22. Miehe, C.: Numerical computation of algorithmic (consistent) tangent moduli in large-strain computational inelasticity. Comput. Methods Appl. Mech. Eng. **134**, 223–240 (1996)
23. Miehe, C., Schröder, J., Schotte, J.: Computational homogenization analysis in finite plasticity. simulation of texture development in polycrystalline materials. Comput. Methods Appl. Mech. Eng. **171**, 387–418 (1999)
24. Moulinec, H., Suquet, P.: A numerical method for computing the overall response of nonlinear composites with complex microstructure. Comput. Methods Appl. Mech. Eng. **157**, 69–94 (1998)
25. Pérez-Foguet, A., Rodríguez-Ferran, A., Huerta, A.: Numerical differentiation for local and global tangent operators in computational plasticity. Comput. Methods Appl. Mech. Eng. **189**, 277–296 (2000)
26. Schenk, O., Gärtner, K.: Two-level dynamic scheduling in PARDISO: improved scalability on shared memory multiprocessing systems. Parallel Comput. **28**(2), 187–197 (2002)
27. Schröder, J.: A numerical two-scale homogenization scheme: the FE^2-method. In: J. Schröder, K. Hackl (eds.) Plasticity and Beyond – Microstructures, Crystal-Plasticity and Phase Transitions. CISM Lecture Notes 550. Springer, Wien (2013)

28. Simo, J., Miehe, C.: Associative coupled thermoplasticity at finite strains: formulations, numerical analysis and implementation. Comput. Methods Appl. Mech. Eng. **98**, 41–104 (1992)
29. Smit, R., Brekelmans, W., Meijer, H.: Prediction of the mechanical behavior of nonlinear heterogeneous systems by multi-level finite element modeling. Comput. Methods Appl. Mech. Eng. **155**, 181–192 (1998)
30. Stephan, M., Docter, J.: JUQUEEN: IBM Blue Gene/Q® Supercomputer System at the Jülich Supercomputing Centre. JLSRF 1, A1. http://dx.doi.org/10.17815/jlsrf-1-18 (2015)
31. Tanaka, M., Fujikawa, M., Balzani, D., Schröder, J.: Robust numerical calculation of tangent moduli at finite strains based on complex-step derivative approximation and its application to localization analysis. Comput. Methods Appl. Mech. Eng. **269**, 454–470 (2014)
32. The HDF Group: Hierarchical Data Format, version 5. http://www.hdfgroup.org/HDF5/ (1997-NNNN)
33. Wriggers, P., Miehe, C., Kleiber, M., Simo, J.: On the coupled thermomechanical treatment of necking problems via finite element methods. Int. J. Numer. Methods Eng. **33**, 869–883 (1992)
34. Zienkiewicz, O.C., Taylor, R.L.: The Finite Element Method for Solid and Structural Mechanics. Elsevier, Oxford (2005)

Scalability of Classical Algebraic Multigrid for Elasticity to Half a Million Parallel Tasks

Allison H. Baker, Axel Klawonn, Tzanio Kolev, Martin Lanser, Oliver Rheinbach, and Ulrike Meier Yang

Abstract The parallel performance of several classical Algebraic Multigrid (AMG) methods applied to linear elasticity problems is investigated. These methods include standard AMG approaches for systems of partial differential equations such as the unknown and hybrid approaches, as well as the more recent global matrix (GM) and local neighborhood (LN) approaches, which incorporate rigid body modes (RBMs) into the AMG interpolation operator. Numerical experiments are presented for both two- and three-dimensional elasticity problems on up to 131,072 cores (and 262,144 MPI processes) on the Vulcan supercomputer (LLNL, USA) and up to 262,144 cores (and 524,288 MPI processes) on the JUQUEEN supercomputer (JSC, Jülich, Germany). It is demonstrated that incorporating all RBMs into the interpolation leads generally to faster convergence and improved scalability.

1 Introduction

Classical Algebraic Multigrid (AMG) methods were originally designed for scalar partial differential equations (PDEs) and usually assume that the nullspace of the operator is one-dimensional and constant. This assumption does not hold for many systems of PDEs. For elasticity problems in particular, the nullspace consists of three (in 2D) or six (in 3D) rigid body modes (RBMs), which comprise translations

A.H. Baker (✉)
National Center for Atmospheric Research, Boulder, CO, USA
e-mail: abaker@ucar.edu

A. Klawonn (✉) • M. Lanser (✉)
Mathematisches Institut, Universität zu Köln, Köln, Germany
e-mail: axel.klawonn@uni-koeln.de; martin.lanser@uni-koeln.de

T. Kolev (✉) • U.M. Yang (✉)
Lawrence Livermore National Laboratory, Livermore, CA, USA
e-mail: tzanio@llnl.gov; umyang@llnl.gov

O. Rheinbach (✉)
Institut für Numerische Mathematik und Optimierung, Technische Universität Bergakademie
Freiberg, Freiberg, Germany
e-mail: oliver.rheinbach@math.tu-freiberg.de

© Springer International Publishing Switzerland 2016   113
H.-J. Bungartz et al. (eds.), *Software for Exascale Computing – SPPEXA*
2013-2015, Lecture Notes in Computational Science and Engineering 113,
DOI 10.1007/978-3-319-40528-5_6

and rotations. Classical AMG methods, including standard approaches modified to handle systems of PDEs, e.g., the unknown approach [23], interpolate translations but not rotations. This limitation will typically result in a loss of optimality and scalability for these approaches when applied to systems problems.

Different approaches to handle linear elasticity problems with AMG methods have been suggested in the last decades, e.g., smoothed aggregation [7, 27], unsmoothed aggregation [3, 4, 8, 19–21], AMGe [6], element-free AMGe [15], local optimization problems to incorporate the RBMs in the interpolation [13], or the global matrix (GM) and local neighborhood (LN) approaches [2].

In this paper, we provide a brief overview of AMG methods and AMG for systems in Sects. 2 and 3. In Sects. 4 and 5, we describe the GM and LN approaches, which were first introduced in [2]. These two approaches explicitly incorporate given smooth error vectors into the AMG interpolation in order to handle the correction of these error components in the coarse grid correction. We note that the descriptions of the AMG methods and interpolations in this paper are based on both [2] (which only considered sequential AMG) and on Chap. 4 of the dissertation [18]. In Sect. 6, we compare the performance of AMG approaches for systems of PDEs and show that the GM and LN approaches can improve convergence and scalability for elasticity problems. The parallel numerical results on up to half a million MPI processes presented in Sect. 6 are new and have not been published elsewhere (as [2] contained only serial results for small problems).

2 Algebraic Multigrid

We first give a brief overview of AMG. Consider the linear system $Au = f$, which is often generated from the discretization of a scalar PDE. We denote with u_i the i-th entry of u. As in geometric multigrid, one needs to define a hierarchy of coarser grids or levels, adequate smoothers or relaxation schemes for each level and restriction, and interpolation operators to move between levels. However, unlike geometric multigrid, algebraic multigrid methods are applied to the linear system without any geometrical or mesh-related information.

Because grid information is not given, one needs to use the linear system to define a "grid". The variables u_i are now the grid points and the non-zero entries a_{ij} of matrix A define the connections between the grid points. Because not all variables are equally important, one defines the concept of *strong dependence*. For a threshold $0 < \theta \leq 1$, a variable u_i *strongly depends* on the variable u_j if

$$- a_{ij} \geq \theta \max_{k \neq i} (-a_{ik}) \,. \tag{1}$$

To determine the coarse-grid variables, which are a subset of the variables u_i, one first eliminates all connections that do not fulfill (1). Then one applies a coarsening algorithm to the remaining "grid". For brevity, we do not describe any coarsening algorithms here, but note that descriptions of several common coarsening strategies and an investigation of their parallel performance can be found in [28] (e.g., Ruge-Stüben [23, 25], HMIS [11] and Falgout [16]).

In AMG, errors are reduced by two separate operations: the smoothing or relaxation steps and the coarse grid correction. For an optimal AMG method, the coarse grid correction and the relaxation strategy must be chosen carefully to complement each other. While simple pointwise relaxation methods such as Jacobi or Gauß-Seidel rapidly reduce errors in the directions of eigenvectors associated with large eigenvalues, the reduction in directions of eigenvectors associated with small eigenvalues is less optimal; see [6] for details. Errors that are poorly reduced by the smoothing steps are referred to as *smooth errors*. More precisely, *algebraic smooth errors* can be characterized by $Ae \approx 0$, where e is an eigenvector associated with a small eigenvalue. For an effective AMG method, the smooth error must be reduced by the coarse grid correction. Therefore, an interpolation operator P needs to be defined in such a way that the smooth errors are approximately in the range of P [6]. For additional details on interpolation operators, we refer the reader to various publications, e.g. [12, 23, 25, 26, 29]. The restriction operator R is often defined to be the transposed operator P^T, so that in the case of a symmetric positive definite matrix A, the coarse grid operator RAP is also symmetric positive definite. After interpolation, restriction, and coarse grid operators have been defined and a relaxation strategy has been determined, the solve phase can be performed.

For simplicity, consider the two-level case with one fine grid and one coarse grid. For an approximate solution u and the exact solution u^* of the system $Au^* = f$ on the fine grid, we have the relationship $Ae = r$, where $e := u^* - u$ is the error vector and $r := f - Au$ is the residual. One AMG cycle to correct (or update) u is as follows:

(1) Smooth ν_1 times on: $Au = f$
(2) Compute the residual: $r = f - Au$
(3) Solve on the coarse grid: $RAPe_c = Rr$
(4) Correct u: $u = u + Pe_c$
(5) Smooth ν_2 times on: $Au = f$.

T obtain a full multi-level AMG V-cycle, one needs to apply this algorithm recursively, as depicted in Fig. 1. For more details on classical AMG methods, see, e.g., [23, 25].

3   Algebraic Multigrid for Systems of PDEs

We now consider a linear system of equations $Au = f$ derived from the discretization of a system of PDEs with p scalar functions or unknowns. Now, each variable or degree of freedom (dof) of the linear system describes one physical quantity in

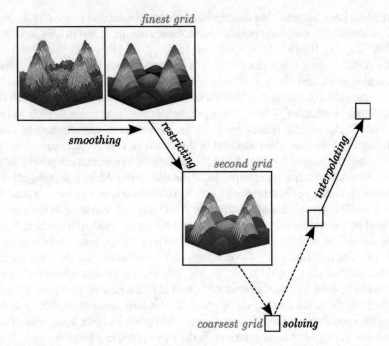

Fig. 1 One AMG V-cycle. Smoothing on the fine grid $\rightarrow$ Restricting to the coarsest grid $\rightarrow$ Solving on coarsest grid $\rightarrow$ Interpolating to the finest grid (Figure from [18])

a grid point or node. For example, in linear or nonlinear elasticity, we have one dof describing one spatial direction in each node. For simplicity, we restrict our presentation here to the two-dimensional case and consider an elasticity problem with two unknowns, x and y, representing the two spatial directions. A detailed three-dimensional description can be found in [2].

For algebraic multigrid methods, the two common approaches to treating systems of PDEs such as $Au = f$ are the *unknown approach (U-AMG)* and the nodal approach, e.g., [1, 10, 14, 22, 23, 25]. While U-AMG completely separates the different physical quantities, the nodal approach considers all unknowns belonging to the same node at once and thus acts on a nodal basis.

We first take a brief look at the U-AMG. Here, we assume an unknown-related ordering of the system matrix (i.e., first all dofs related to the unknown x followed by those associated with y):

$$A = \begin{bmatrix} A_{xx} \, A_{xy} \\ A_{yx} \, A_{yy} \end{bmatrix}. \tag{2}$$

One now applies classical AMG coarsening and interpolation strategies to the different variables separately, i.e., only to the diagonal blocks A_{xx} and A_{yy}. Note that this strategy ignores couplings between unknowns x and y, which are contained in A_{xy} and A_{yx}, and leads to an AMG interpolation matrix P that has the diagonal block structure

$$P = \begin{bmatrix} P_x & 0 \\ 0 & P_y \end{bmatrix}. \tag{3}$$

In general, U-AMG is often used to handle systems of PDEs and is quite effective for problems with weak coupling between the different unknowns. Of course, performance also strongly depends on the general quality of the chosen coarsening, interpolation, and smoothing techniques for the diagonal blocks A_{xx} and A_{yy}.

We now describe the nodal approach, which is often a more effective approach for problems with a stronger coupling between the different physical quantities. If we block all unknowns that share the same node and consider a node-related ordering, then the system matrix A can be written as

$$A = \begin{bmatrix} A_{11} & A_{12} & \cdots & A_{1N} \\ A_{21} & A_{22} & \cdots & A_{2N} \\ \vdots & \vdots & \ddots & \vdots \\ A_{N1} & A_{N2} & \cdots & A_{NN} \end{bmatrix}, \tag{4}$$

where the 2×2 blocks A_{ij} connect nodes i and j. Note that if we define N as the number of nodes or grid points, then A is a $N \times N$ block matrix. With the nodal approach, we consider strong dependence between two nodes i and j, instead of between two variables. Therefore, we now have to compare block entries, such as A_{ji} or A_{jj}. This comparison typically involves an appropriate norm such as the Frobenius norm $|| \cdot ||_F$ or the row-sum norm $|| \cdot ||_\infty$. Applying the norm to the blocks of the system matrix A results in a condensed $N \times N$ matrix with scalar entries

$$C = \begin{bmatrix} c_{11} & c_{12} & \cdots & c_{1N} \\ c_{21} & c_{22} & \cdots & c_{2N} \\ \vdots & \vdots & \ddots & \vdots \\ c_{N1} & c_{N2} & \cdots & c_{NN} \end{bmatrix} := \begin{bmatrix} ||A_{11}|| & ||A_{12}|| & \cdots & ||A_{1N}|| \\ ||A_{21}|| & ||A_{22}|| & \cdots & ||A_{2N}|| \\ \vdots & \vdots & \ddots & \vdots \\ ||A_{N1}|| & ||A_{N2}|| & \cdots & ||A_{NN}|| \end{bmatrix}. \tag{5}$$

The definition of strong dependence in Eq. (1) is based on A or C being an M-matrix, i.e., a matrix whose off-diagonal elements have the opposite sign of the diagonal elements. Therefore, we change the diagonal elements c_{ii} of C to $c_{ii} = -||A_{ii}||$ or

$$c_{ii} = - \sum_{j=1, j \neq i}^{N} ||A_{ij}||. \tag{6}$$

This approach as well as additional options for defining C are further discussed in [10]. In our experiments, we found the latter approach (i.e., the row-sum norm) to give better convergence, and we are using Eq. (6) in the numerical results presented in Sect. 6. The AMG coarse grids are now obtained by applying classical AMG coarsening techniques to the condensed matrix C. In the nodal coarsening approach, all unknowns on one grid point share the same set of coarse grids. Note the contrast with the unknown approach, which can result in completely different coarse meshes for each unknown. The interpolation matrix in the nodal approach can be obtained by applying scalar AMG interpolation techniques to the blocks (e.g., [14]). Another option, used in our experiments in Sect. 6, is to combine nodal coarsening with unknown-based interpolation. We call this approach the *hybrid approach (H-AMG)*.

4   The Global Matrix Approach

As mentioned in Sect. 2, smooth error vectors should be in the range of the interpolation operator. In the case of linear elasticity, the nullspace of the matrix A consists of the rigid body modes (RBMs), i.e., all rotations and translations of the domain. Since classical AMG interpolation P already interpolates constant vectors exactly, we only have to take care of rotations (i.e., in two dimensions, the single rotation $s(x, y) := [y, -x]$). A possible approach to incorporate an exact interpolation of smooth error vectors in the AMG interpolation is, as already mentioned, the *global matrix* (GM) approach, introduced in [2]. The following description is restricted to two levels. A generalization to the multilevel-case can be found in [2].

The GM approach is based on the idea of augmenting a given global AMG interpolation matrix P with several matrices Q^j. Each matrix Q^j is chosen to exactly interpolate a specified smooth error vector s_j. We designate the rotation $s := [y, -x]$ in two dimensions as the smooth error vector. We define s_C as the restriction of s onto the coarse grid and define a new interpolation matrix $\widetilde{P}$ by augmenting P:

$$\widetilde{P} := [P \quad Q], \text{ such that } s \in \text{range}(\widetilde{P}) . \tag{7}$$

There are several possibilities to define a matrix Q fulfilling Eq. (7) and also retain the sparsity of P. We will consider both variants suggested in [2]. For *Variant 1* or *GM1* we define $\widetilde{P}$ such that

$$\widetilde{P}\begin{bmatrix} 0 \\ 1 \end{bmatrix} = s , \tag{8}$$

whereas for *Variant 2* or *GM2*, $\widetilde{P}$ is defined such that

$$\widetilde{P}\begin{bmatrix} s_C \\ 1 \end{bmatrix} = s . \tag{9}$$

For GM1, the coefficients Q_{ij} of Q, where i is the index of a fine grid point and j the index of a coarse grid point, are then defined as

$$Q_{ij} := P_{ij}\left(\frac{s_i}{\sum\limits_{k \in C_i} P_{ik}}\right), \tag{10}$$

where C_i is the set of coarse points in the direct neighborhood of i, i.e., the indices of the columns with non-zero entries in row i of the interpolation P. For GM2, the entries Q_{ij}, are given by

$$Q_{ij} := P_{ij}\left(\frac{s_i}{(\sum\limits_{k \in C_i} P_{ik})} - (s_C)_j\right). \tag{11}$$

The unknown-based GM interpolation in two dimensions can then be written as

$$\widetilde{P} = \begin{bmatrix} P_x & 0 & Q_x \\ 0 & P_y & Q_y \end{bmatrix},$$

where Q_x and Q_y can be computed independently and have the same sparsity as P_x and P_y. Note that this leads to a coarse grid space with a larger number of degrees of freedom than the coarse grid space generated by the unknown-based or the hybrid approach. The increase in degrees of freedom is even further exacerbated in three dimensions, where one needs to add three rigid body modes. So, while we expect improved convergence, the new method is potentially significantly more expensive, and the increased complexities could prevent better performance. Therefore, to mitigate the increase in complexities, we also truncate the Q matrices (see also our numerical results in Sect. 6). Truncation of Q needs to be done independently from truncation of P, because P-truncation is normalized to interpolate constants whereas the truncated Q matrices need to interpolate the rotations. When truncating Q to $\tilde{Q}$, we adjust the weights of $\tilde{Q}$ so that the row sums of $\tilde{Q}$ equal those of Q.

Interestingly enough, the application of both variants beyond the first level leads to very different algorithms. GM1 needs to only interpolate constants after the first level, whereas GM2 needs to continue to interpolate coarser versions of the rigid body modes, thus requiring the storage of coarse grid versions of the rigid body modes as well as additional computations. More details are available in [2]. However, note that GM2 leads to coefficients of similar size, which is not the case for GM1. It is therefore much more difficult to effectively truncate the Q matrices generated in GM1. This difficulty will become evident in Sect. 6.

5   The Local Neighborhood Approach

We now consider an approach where the rigid body modes are incorporated locally. Because exact local interpolation leads to exact global interpolation, this approach should work at least as well as the global matrix approach. This approach requires looking at interpolation from a different angle. Assume that the error at the fine points, e_F, is interpolated by the error at the coarse points, e_C, such that

$$e_F = W_{FC}e_C . \tag{12}$$

Let $\tilde{C}$ be the set of new coarse points that have been introduced by adding new degrees of freedom to the coarse nodes. Further, s is a rigid body mode, s_C is s at the original coarse grid points, and s_F is s at the fine grid points. The idea for the local neighborhood approach is then to exactly interpolate the rigid body mode using an extension operator

$$e_F = W_{FC}e_C + W_{F\tilde{C}}e_{\tilde{C}} \quad s.t. \quad s_F = W_{FC}s_C + W_{F\tilde{C}}s_{\tilde{C}} , \tag{13}$$

where $s_{\tilde{C}} = 1$ at the new degrees of freedom in $\tilde{C}$. The LN interpolation matrix needs to be defined by harmonic extension based on the local extension $\tilde{W}_{FC} = [W_{FC}, W_{F\tilde{C}}]$. Let D_s be the matrix with diagonal s. Because W_{FC} interpolates constants, the following definition, which is similar to GM2, satisfies Eq. (13):

$$W_{F\tilde{C}} = [D_s^F W_{FC} - W_{FC}D_s^C] . \tag{14}$$

To allow for an arbitrary interpolation matrix P, the implementation of this approach performs a preprocessing step (cf. "iterative weight refinement" [9]) that results in $\bar{P}$ where

$$\bar{P}_{ij} = -\frac{1}{a_{ii}}\left(a_{ij} + \sum_{k\in F_i} a_{ik}w_{kj}\right) , \tag{15}$$

where F_i is the fine neighborhood of point i and

$$w_{kj} = \frac{P_{kj}}{\sum_{n\in C_i} P_{kn}} . \tag{16}$$

Now that $\bar{P}$ is based on harmonic extension, Q can be determined using the following formula

$$Q_{ij} = -\frac{1}{a_{ii}}\sum_{k\in F_i} a_{ik}w_{kj}(s_k - s_j) . \tag{17}$$

For k rigid body modes $s_1, \ldots, s_k$, the new LN interpolation operator is given by

$$\tilde{P} = [\ \bar{P}\ Q^1\ \ldots\ Q^k\ ] . \tag{18}$$

Note that this approach assumes that $As = 0$. However, the unknown-based interpolation is not generated from A, but from the block diagonal matrix A_D with block diagonals A_{xx} and A_{yy} in 2D (as well as A_{zz} in 3D). In this situation it is important to modify Eq. (17) by incorporating the non-zero residual. We refer to [2] for further details. In addition, like GM2, the LN approach requires the generation of Q on all coarse levels.

6   Numerical Results

In this section, we present numerical results that compare the performance of the previously described AMG approaches. AMG is here used as a preconditioner to either GMRES or CG. The parallel experiments were conducted on the Vulcan supercomputer (LLNL), except for those presented in Table 6, which were computed on the JUQUEEN supercomputer (JSC) [24]. JUQUEEN and Vulcan were ranked 11th and 12th respectively on the TOP500 list of the world's fastest supercomputers of November 2015. JUQUEEN is a 28,672 node 6 Petaflop Blue Gene/Q system at Jülich Supercomputing Center (JSC, Germany), with a total number of 458,752 processor cores. Vulcan is a 24,576 node 5 Petaflop Blue Gene/Q production system at Lawrence Livermore National Laboratory (USA) with a total number of 393,216 processor cores. Both Blue Gene/Q systems use a Power BQC 16C 1.6 GHz processor with 16 cores and 16 GB memory per node.

We use BoomerAMG [16], the unstructured algebraic multigrid solver in hypre version 2.10.0b [17], which now provides an efficient parallel implementation of the GM and the LN approaches. In hypre version 2.10.0b, the user now simply has to provide the smooth error vectors on the fine grid in addition to the linear system. In our case, we provide the rotations s_j, one in 2D, three in 3D. In order to make efficient use of the hardware threads for the 3D results in Tables 3 and 6, we use oversubscription with 2 MPI ranks for each core of the Power BQC processor. Note that no parallel results are given in [2], as a parallel implementation was not available at that time. To ensure a fair comparison of the different methods, we chose an AMG setup such that all components have shown the potential to scale up to large scales. In particular, for all methods, we use HMIS coarsening, introduced in [11], the *extended+i* interpolation method described in [12, 29] and symmetric SOR/Jacobi smoothing in a V(1,1)-cycle.

We consider the compressible linear elasticity problem

$$-2\mu \operatorname{div}(\epsilon(u)) - \lambda \operatorname{grad}(\operatorname{div}(u)) = f ,$$

where u is the unknown displacement and $\epsilon(u)$ is the strain. The parameters are $\lambda = \frac{vE}{(1+v)(1-2v)}, \mu = \frac{E}{2(1+v)}$ (cf. [5]), where the Young modulus is $E = 210$, and we vary the Poisson ratio v between 0.3 and 0.49.

More detailed descriptions of the various model problems in two and three dimensions are given in subsequent subsections. The finite element assembly is performed in PETSc, and we also use PETSc's GMRES/CG implementation. In all tables we use the abbreviations *U-AMG* for the unknown approach, *H-AMG* for the hybrid approach with the nodal coarsening strategy in Eq. (6) and the row-sum norm, and *H-AMG-GM1/GM2/LN* for the interpolation approaches GM1, GM2, and LN, respectively. Cop denotes the *operator complexity*, which is defined as the sum of the non-zeros of all matrices A_i on all levels divided by the number of non-zeros of the original matrix A. Operator complexity is an indication of memory usage and the number of flops per iteration and also affects setup times. In order to reduce Cop, we truncate P to at most Pmax non-zero elements per row and use a truncation factor of Q-th (absolute threshold) to truncate Q. In the tables, we mark the fastest time (for the sum of setup and solve) as well as the lowest number of iterations in **bold face**. As a baseline for our weak scalability tests, in order to avoid cache effects, we use the smallest problem which still makes use of at least a single full node.

6.1   Results in Two Dimensions

If a Dirichlet boundary condition is applied to a large portion of the boundary, standard nodal or unknown approaches are known to perform well, and we do not expect any additional benefit from the GM or LN approach. Therefore, we consider an elasticity problem on a rectangular domain $[0, 8] \times [0, 1]$ in 2D, fixed on one of the short sides. A volume force orthogonal to the longer sides is applied. We refer to this problem as *2D beam*, and a solution for a linear elastic material is presented in Fig. 2. We use piecewise quadratic finite elements on triangles in all experiments in two dimensions, and, by reordering the unknowns, we ensure that each MPI rank holds a portion of the beam of favorable shape, i.e., close to a square. We present weak scalability results for the *2D beam* in Tables 1 and 2, comparing the unknown

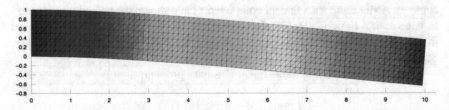

Fig. 2 Solution of the *2D beam* considering linear elasticity with $E = 210$ and $v = 0.3$. The color represents the norm of the displacement

Table 1 Weak scalability of the *2D beam* problem with $E = 210$ and $v = 0.3$; iterative solver: preconditioned GMRES; stopping tolerance for the relative residual: 1e-8; quadratic triangular finite elements; *Preconditioner* denotes the AMG approach (one V-cycle); *Pmax/Q-th* denotes the truncation of the interpolation operators for P (max non-zeros per row) and Q (absolute threshold); *It.* denotes the number of GMRES iterations and *(Cop)* the operator complexity; *Time GMRES* denotes the runtime of the AMG-GMRES solve phase; *Time BoomerSetup* denotes the time spent in the BoomerAMG setup; *Setup + Solve* denotes the total solution time spent in the AMG setup (BoomerSetup) and the AMG-GMRES (Time GMRES) solve. The fastest variant is marked in bold face

#MPI ranks (=#Cores)	Problem size	Preconditioner	Pmax/Q-th	It. (Cop)	Time GMRES (s)	Time BoomerSetup (s)	Time Setup + Solve (s)
32	643,602	U-AMG	– / –	23 (2.4)	4.5	1.1	5.6
		H-AMG	– / –	21 (2.5)	4.0	1.7	5.7
		H-AMG-GM2	– / 0.01	**15** (2.5)	2.9	2.2	**5.1**
128	2,567,202	U-AMG	– / –	26 (2.3)	5.3	1.1	6.4
		H-AMG	– / –	23 (2.3)	4.4	1.7	6.1
		H-AMG-GM2	– / 0.01	**15** (2.4)	3.0	2.3	**5.3**
512	10,254,402	U-AMG	– / –	29 (2.2)	6.0	1.3	7.3
		H-AMG	– / –	25 (2.2)	4.8	1.9	6.7
		H-AMG-GM2	– / 0.01	**16** (2.3)	3.2	2.3	**5.5**
2048	40,988,802	U-AMG	– / –	48 (2.2)	10.2	1.4	11.6
		H-AMG	– / –	26 (2.2)	5.1	1.9	7.0
		H-AMG-GM2	– / 0.01	**18** (2.2)	3.6	2.4	**6.0**
8192	163,897,602	U-AMG	– / –	51 (2.2)	11.0	1.6	12.6
		H-AMG	– / –	26 (2.2)	5.1	2.0	7.1
		H-AMG-GM2	– / 0.01	**18** (2.2)	3.6	2.5	**6.1**
32,768	655,475,202	U-AMG	– / –	54 (2.2)	11.9	1.8	13.7
		H-AMG	– / –	30 (2.2)	5.9	2.0	7.9
		H-AMG-GM2	– / 0.01	**19** (2.2)	3.8	2.5	**6.3**
131,072	2,621,670,402	U-AMG	– / –	59 (2.2)	13.4	2.0	15.4
		H-AMG	– / –	29 (2.2)	5.8	2.1	7.9
		H-AMG-GM2	– / 0.01	**20** (2.2)	4.1	2.7	**6.8**

Table 2 Same problem setup and notation as in Table 1, but larger problem sizes

#MPI ranks (=#Cores)	Problem size	Preconditioner	Pmax/Q-th	It. (Cop)	Time GMRES (s)	Time BoomerSetup (s)	Time Setup + Solve (s)
32	1,644,162	U-AMG	-/-	24 (2.4)	17.5	3.1	20.6
		H-AMG	-/-	23 (2.5)	18.5	4.3	22.8
		H-AMG-GM2	-/0.01	**14 (2.5)**	12.5	5.4	**17.9**
128	6,565,122	U-AMG	-/-	28 (2.3)	20.4	3.1	23.5
		H-AMG	-/-	24 (2.3)	19.9	4.4	24.3
		H-AMG-GM2	-/0.01	**16 (2.3)**	14.0	5.5	**19.5**
512	26,237,442	U-AMG	-/-	44 (2.2)	32.8	3.1	35.9
		H-AMG	-/-	26 (2.2)	21.8	4.5	26.3
		H-AMG-GM2	-/0.01	**17 (2.3)**	15.2	5.6	**20.8**
2048	104,903,682	U-AMG	-/-	51 (2.2)	38.0	3.3	41.3
		H-AMG	-/-	26 (2.2)	21.9	4.6	26.5
		H-AMG-GM2	-/0.01	**18 (2.3)**	16.2	5.7	**21.9**
8192	419,522,562	U-AMG	-/-	54 (2.2)	40.8	3.5	44.3
		H-AMG	-/-	27 (2.2)	23.1	4.6	27.7
		H-AMG-GM2	-/0.01	**18 (2.2)**	16.4	5.8	**22.2**
32,768	1,677,905,922	U-AMG	-/-	58 (2.2)	43.9	3.7	47.6
		H-AMG	-/-	30 (2.2)	25.4	4.8	30.2
		H-AMG-GM2	-/0.01	**19 (2.2)**	17.2	6.0	**23.3**
131,072	6,711,255,042	U-AMG	-/-	83 (2.2)	63.6	3.9	67.5
		H-AMG	-/-	52 (2.2)	44.7	4.9	49.6
		H-AMG-GM2	-/0.01	**21 (2.2)**	19.1	6.2	**25.3**

approach *U-AMG*, the hybrid approach *H-AMG*, and, representing the interpolation approaches, the GM2 approach. The GM1 and LN approaches performed similarly to or worse than GM2 here, but are included in a more detailed discussion on the results in three dimensions, where differences between the approaches are more interesting.

In the weak scaling results in Table 1, the number of GMRES iterations for the unknown approach increases from 23 to 59 iterations, resulting in a noticeable increase in the iteration time as well. However, both the hybrid and GM2 approaches achieve good weak scalability. Comparing the hybrid and the GM2 approaches, the AMG setup times are slightly higher with the GM2 approach. This increased computational cost is expected due to the exact interpolation of the rotation. Since iteration counts and thus the iteration times are lower, the GM2 approach is always the fastest approach in this comparison.

Table 2 also contains weak scaling results for the 2D beam, but the problem sizes are approximately 2.6 times larger per core. Results are similar to the results in Table 1, but here, for the largest problem with 6.7 billion degrees of freedom, the hybrid approach needs 52 compared to only 21 GMRES iterations for the GM2 approach. This improvement leads to a much faster convergence for GM2; see also Fig. 3 for a visualization.

We can conclude that with our settings, all three approaches work well for smaller problems. For larger problems (and larger numbers of cores), the GM2 approach remains scalable whereas U-AMG and H-AMG experience an increase in the number of iterations. The setup cost for the GM2 approach is slightly higher,

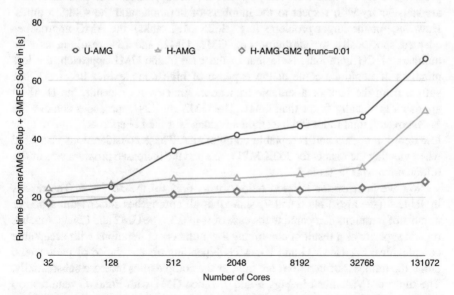

Fig. 3 Weak scalability of total solution time for the two-dimensional beam with $\nu = 0.3$ and $E = 210$; cf. Table 2

compared to the other two approaches, but the setup time is scalable and amortized in the iteration phase; see also Fig. 3.

6.2   Results in Three Dimensions

Now we present results for several three-dimensional domains. In particular, we first investigate weak scalability for a *3D beam* problem. We also investigate the effect of a higher Poisson ratio v on the *3D beam*, showing scalability results and presenting a small study that increases v. Second, we examine doubling the beam length. And for a third model problem, we consider a heterogeneous material with different boundary condition, called the *3D cuboid*.

6.2.1   3D Beam Problem

Similar to the *2D beam*, the *3D beam* problem is defined on the domain $[0, 8] \times [0, 1] \times [0, 1]$ for $v = 0.3$, $v = 0.45$ and $v = 0.49$. First, we present weak scalability results in Table 3 for the *3D beam* with $v = 0.3$ for all approaches. For the 262K MPI ranks case, we also include a larger problem to show the effect of increasing problem size on performance at large scale.

From the results in Table 3 (see also Figs. 4 and 5), we conclude that for smaller problems, a set of parameters can be found for all approaches such that the results are satisfactory with respect to the numbers of iterations and the solution times. However, for the larger problems (e.g., 262K MPI ranks), the AMG approaches adapted specifically for elasticity, i.e., GM1, GM2, and LN, result in smaller numbers of CG iterations. Note that in the case of the GM1 approach, the low numbers of iterations come at the expense of high complexities because GM1 suffers from the lack of a suitable truncation strategy. As a result, the H-AMG approach is actually faster than GM1. The GM2 and LN approaches achieve the fastest overall total times (with a slight advantage for the LN approach) due to their low iteration counts and acceptable complexities. These considerations also hold when viewing the results for 262K MPI ranks and the increased problem size of 6.3 billion unknowns in Table 3.

Now we increase the Poisson ratio to $v = 0.45$ for the *3D beam*. The results in Table 4 (see also Figs. 6 and 7) show that all approaches suffer from a higher number of iterations compared to the case of $v = 0.3$. The GM2 and LN approaches remain superior as a result of combining low numbers of iterations with acceptable complexities. For U-AMG and H-AMG, depending on the choice of parameters, either the numbers of iterations are high or the complexities increase substantially. The times are visualized in Figs. 6 and 7. Since GM1 with Pmax=3 requires too much memory, we use it here with Pmax=2. Note that GM1 fails for the largest problem considered.

Table 3 Weak scalability of the 3D beam problem with $E = 210$ and $\nu = 0.3$; iterative solver: preconditioned CG; stopping tolerance for the relative residual: 1e-6; linear tetrahedral finite elements; 2 MPI ranks per Blue Gene/Q core are used; *Preconditioner* denotes the AMG approach (one V-cycle); *Pmax/Q-th* denotes the truncation of the interpolation operators for P (max non-zeros per row) and Q (absolute threshold); *It.* denotes the number of CG iterations and *(Cop)* the operator complexity; *Time CG* denotes the runtime of the AMG-CG solve phase; *Time BoomerSetup* denotes the time spent in the BoomerAMG setup; *Setup + Solve* denotes the total solution time spent in the AMG setup and the AMG-CG solve

#MPI ranks (= 2×#Cores)	Problem size	Preconditioner	Pmax/Q-th	It. (Cop)	Time CG (s)	Time BoomerSetup (s)	Time Setup + Solve (s)
64	839,619	U-AMG	2 / –	88 (2.76)	23.50	2.42	25.92
		U-AMG	3 / –	58 (2.94)	15.30	3.19	28.49
		U-AMG	4 / –	44 (3.14)	12.25	4.64	16.89
		H-AMG	3 / –	58 (2.42)	12.11	3.44	15.55
		H-AMG	4 / –	50 (2.83)	11.64	5.31	16.95
		H-AMG-GM1	2 / 0.05	52 (2.82)	12.19	5.25	17.44
		H-AMG-GM1	3 / 0.05	**37** (3.61)	10.34	9.18	19.52
		H-AMG-GM2	3 / 0.05	47 (2.45)	10.06	4.54	**14.60**
		H-AMG-LN	3 / 0.05	48 (2.44)	10.26	4.75	15.01
512	6 502,275	U-AMG	2 / –	118 (2.81)	36.27	3.77	40.04
		U-AMG	3 / –	73 (3.02)	23.85	5.21	29.06
		U-AMG	4 / –	54 (3.23)	17.48	6.52	24.00
		H-AMG	3 / –	70 (2.45)	15.22	4.35	19.57
		H-AMG	4 / –	59 (2.87)	14.34	6.39	20.73
		H-AMG-GM1	2 / 0.05	71 (2.84)	18.24	7.17	25.41
		H-AMG-GM1	3 / 0.05	**44** (3.66)	12.81	12.37	25.18
		H-AMG-GM2	3 / 0.05	55 (2.47)	12.35	5.09	**17.44**
		H-AMG-LN	3 / 0.05	57 (2.46)	12.77	5.29	18.06

(continued)

Table 3 (continued)

#MPI ranks (= 2×#Cores)	Problem size	Preconditioner	Pmax/Q-th	It. (Cop)	Time CG (s)	Time BoomerSetup (s)	Time Setup + Solve (s)
4096	51,171,075	U-AMG	2 / –	149 (2.86)	50.64	5.12	55.76
		U-AMG	3 / –	89 (3.09)	33.16	7.14	40.30
		U-AMG	4 / –	67 (3.32)	25.21	8.76	33.97
		H-AMG	3 / –	86 (2.47)	19.34	5.08	24.42
		H-AMG	4 / –	67 (2.89)	16.98	7.39	24.37
		H-AMG-GM1	2 / 0.05	78 (2.84)	20.48	8.25	28.73
		H-AMG-GM1	3 / 0.05	**47** (3.69)	14.05	14.50	28.55
		H-AMG-GM2	3 / 0.05	68 (2.48)	15.83	6.18	22.01
		H-AMG-LN	3 / 0.05	67 (2.48)	15.48	6.38	**21.86**
32,786	406,003,203	U-AMG	2 / –	189 (2.89)	70.94	8.73	79.67
		U-AMG	3 / –	112 (3.13)	49.73	12.90	62.63
		U-AMG	4 / –	86 (3.36)	40.69	15.36	56.05
		H-AMG	3 / –	95 (2.47)	21.97	6.72	28.69
		H-AMG	4 / –	87 (2.90)	22.95	8.98	31.93
		H-AMG-GM1	2 / 0.05	100 (2.84)	26.82	8.89	35.71
		H-AMG-GM1	3 / 0.05	**64** (3.70)	19.54	15.86	35.40
		H-AMG-GM2	3 / 0.05	81 (2.48)	19.53	7.36	26.89
		H-AMG-LN	3 / 0.05	74 (2.48)	17.78	7.60	**25.38**

262,144	3,234,610,179	U-AMG	2 / –	232 (2.90)	95.95	15.36	111.31
		U-AMG	3 / –	135 (3.15)	73.26	27.78	101.04
		U-AMG	4 / –	101 (3.49)	67.16	38.35	105.51
		H-AMG	3 / –	124 (2.48)	29.64	8.53	38.17
		H-AMG	4 / –	106 (2.90)	29.33	9.68	39.01
		H-AMG-GM1	2 / 0.05	138 (2.84)	37.81	8.70	46.51
		H-AMG-GM1	3 / 0.05	**73** (3.70)	22.99	21.39	44.38
		H-AMG-GM2	3 / 0.05	94 (2.48)	23.84	11.01	34.85
		H-AMG-LN	3 / 0.05	84 (2.48)	21.22	11.21	**32.43**
Increased problem size							
262,144	6,312,364,803	U-AMG	3 / –	143 (3.10)	118.72 s	36.94 s	155.66 s
		H-AMG	3 / –	134 (2.52)	62.48 s	13.76 s	76.24 s
		H-AMG-GM2	3 / 0.05	102 (2.53)	48.64 s	16.89 s	65.53 s
		H-AMG-LN	3 / 0.05	**88** (2.53)	41.76 s	16.89 s	**58.65 s**

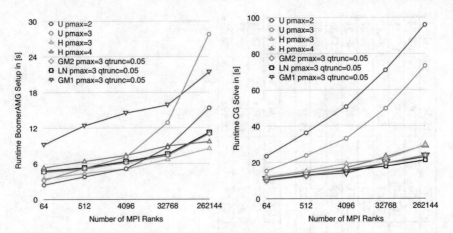

Fig. 4 Weak scalability of the BoomerAMG Setup (*left*) and the time spent in the AMG-CG solve phase (*right*) for the three-dimensional beam with $v = 0.3$ and $E = 210$; cf. Table 3

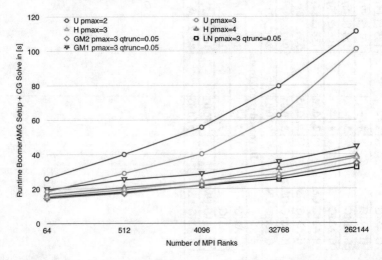

Fig. 5 Weak scalability of total solution time for the three-dimensional beam with $v = 0.3$ and $E = 210$; cf. Table 3

Next, in Table 5, the effect of the Poisson ratio on the different AMG approaches is studied. We see that H-AMG does not converge within the limit of 1000 iterations for $v = 0.49$. For the other approaches, the convergence rate suffers from an increasing value of v towards almost incompressibility. This deterioration is also the case for the AMG approaches which are especially adapted for (compressible) elasticity problems, i.e., GM1, GM2, and LN, but which are based on H-AMG. For $v = 0.49$, U-AMG, while exhibiting the highest Cop, is the fastest variant in terms of total time.

Table 4 Same problem setup and notation as in Table 3, but larger problem sizes, $\nu = 0.45$. On 32 K MPI ranks H-AMG-GM1 hits the maximal iteration number of 1000 (marked with *max It.*)

#MPI ranks (=2×#Cores)	Problem size	Preconditioner	Pmax/Q-th	It. (Cop)	Time CG (s)	Time BoomerSetup (s)	Time Setup + Solve (s)
64	1,618,803	U-AMG	2 / –	98 (3.18)	56.75	6.10	62.85
		U-AMG	3 / –	69 (3.52)	43.12	9.09	52.21
		U-AMG	4 / –	**54** (3.84)	35.86	12.15	48.01
		H-AMG	3 / –	151 (2.55)	65.27	8.52	73.79
		H-AMG	4 / –	80 (2.97)	38.67	11.86	50.53
		H-AMG-GM1	2 / 0.01	77 (3.07)	38.01	11.08	49.09
		H-AMG-GM2	3 / 0.01	64 (3.07)	31.94	13.52	45.46
		H-AMG-GM2	3 / 0.05	72 (2.56)	31.78	9.52	41.30
		H-AMG-LN	3 / 0.05	69 (2.56)	30.45	10.07	**40.52**
512	12,616,803	U-AMG	2 / –	128 (3.26)	78.80	9.11	87.91
		U-AMG	3 / –	85 (3.59)	56.54	15.10	71.64
		U-AMG	4 / –	**65** (3.94)	46.60	20.58	67.18
		H-AMG	3 / –	232 (2.57)	103.70	9.80	113.50
		H-AMG	4 / –	104 (2.99)	51.60	14.16	65.76
		H-AMG-GM1	2 / 0.01	96 (3.10)	48.72	15.38	64.10
		H-AMG-GM2	3 / 0.01	71 (2.66)	33.29	14.62	**47.91**
		H-AMG-GM2	3 / 0.05	96 (2.58)	43.76	11.45	55.21
		H-AMG-LN	3 / 0.05	89 (2.58)	40.52	11.99	52.51

(continued)

Table 4 (continued)

#MPI ranks (=2×#Cores)	Problem size	Preconditioner	Pmax/Q-th	It. (Cop)	Time CG (s)	Time BoomerSetup (s)	Time Setup + Solve (s)
4096	99,614,403	U-AMG	2 / –	141 (3.30)	95.04	13.13	108.17
		U-AMG	3 / –	106 (3.64)	79.87	21.70	101.57
		U-AMG	4 / –	**85** (4.00)	68.90	27.33	96.23
		H-AMG	3 / –	375 (2.58)	174.54	11.57	186.11
		H-AMG	4 / –	184 (3.01)	95.46	16.32	111.78
		H-AMG-GM1	2 / 0.01	115 (3.12)	59.89	17.58	77.47
		H-AMG-GM2	3 / 0.01	90 (2.60)	42.97	15.81	**58.78**
		H-AMG-GM2	3 / 0.05	125 (2.59)	58.93	13.58	72.52
		H-AMG-LN	3 / 0.05	109 (2.58)	51.15	13.80	64.95
32,768	791,664,003	U-AMG	2 / –	202 (3.31)	146.31	23.41	169.72
		U-AMG	3 / –	128 (3.65)	124.37	48.18	172.55
		U-AMG	4 / –	**102** (4.02)	114.02	54.98	169.00
		H-AMG	3 / –	692 (2.58)	340.75	14.44	355.19
		H-AMG	4 / –	320 (3.01)	174.94	19.26	195.22
		H-AMG-GM1	2 / 0.01	max It.	–	–	–
		H-AMG-GM2	3 / 0.01	124 (2.59)	61.91	19.15	81.06
		H-AMG-GM2	3 / 0.05	146 (2.59)	72.67	17.08	89.75
		H-AMG-LN	3 / 0.05	118 (2.58)	57.24	16.19	**73.43**

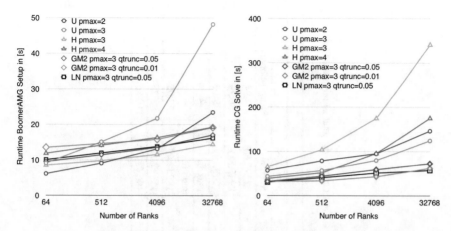

Fig. 6 Weak scalability of the BoomerAMG Setup (*left*) and the time spent in the AMG-CG solve phase (*right*) for the three-dimensional beam with $\nu = 0.45$ and $E = 210$; cf. Table 4

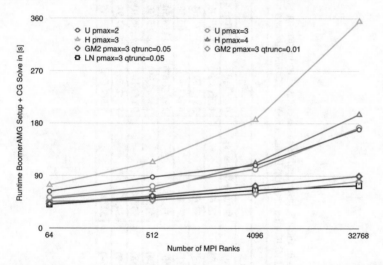

Fig. 7 Weak scalability of total solution time for the three-dimensional beam with $\nu = 0.45$ and $E = 210$; cf. Table 4

6.2.2 3D Beam Problem with Double Length

For $\nu = 0.3$, we examine the effect of doubling the length of the *3D beam* such that the domain is $[0, 16] \times [0, 1] \times [0, 1]$. Table 6 lists the results obtained for the *3D beam* with double the length, using up to 16 of the total 28 racks of the JUQUEEN supercomputer. Again, these experiments show the clear advantage of the GM2 and LN approaches for this problem over the standard methods. The largest three dimensional problem with approximately 13 billion unknowns is solved in less than 81 s using the LN approach. Here, the solve phase time of LN is twice as fast as that of the fastest standard approach H-AMG.

Table 5 Same problem setup and notation as in Table 3. Investigation of the effect of an increasing ν; *Setup + Solve* denotes the total solution time spent in the AMG setup and the AMG-CG solve; H-AMG hits the maximal iteration number of 1000 (marked with *max It.*)

512 MPI ranks, 12,616,803 dofs

Preconditioner	Pmax/Q-th	$\nu = 0.3$		$\nu = 0.45$		$\nu = 0.49$	
		It. (Cop)	Setup + Solve (s)	It. (Cop)	Setup + Solve (s)	It. (Cop)	Setup + Solve (s)
U-AMG	2 / –	128 (2.79)	76.61	128 (3.26)	87.91	125 (3.60)	102.94
U-AMG	3 / –	79 (2.98)	53.69	85 (3.59)	71.64	89 (3.81)	85.39
U-AMG	4 / –	57 (3.21)	56.16	**65** (3.94)	67.18	**72** (3.89)	**79.10**
H-AMG	3 / –	76 (2.50)	41.47	232 (2.57)	113.50	max It.	–
H-AMG	4 / –	59 (2.94)	41.24	104 (2.99)	65.76	max It.	–
H-AMG-GM1	2 / 0.01	56 (2.88)	39.74	96 (3.10)	64.10	189 (3.49)	127.05
H-AMG-GM2	3 / 0.01	49 (2.59)	34.07	71 (2.66)	**47.91**	159 (2.79)	99.25
H-AMG-LN	3 / 0.01	**47** (2.54)	**32.34**	79 (2.62)	50.06	196 (2.71)	110.85

Table 6 Weak scalability of the larger $[0, 16] \times [0, 1] \times [0, 1]$ *3D beam* problem with $E = 210$ and $\nu = 0.3$. Same notation as in Table 3. Computations carried out on JUQUEEN BlueGene/Q at Jülich Supercomputing Centre (JSC)

#MPI ranks (=2×#Cores)	Problem size	Preconditioner	Pmax/Q-th	It. (Cop)	Time CG (s)	Time BoomerSetup (s)	Time Setup + Solve (s)
16	424,683	U-AMG	3 / –	96 (2.83)	35.18	**3.83**	39.01
		H-AMG	3 / –	92 (2.49)	27.39	4.73	32.12
		H-AMG-GM2	3 / 0.01	**55** (3.50)	21.63	11.38	33.01
		H-AMG-LN	3 / 0.01	56 (3.14)	**20.19**	9.22	**29.41**
128	3,232,563	U-AMG	3 / –	118 (2.93)	64.49	**6.05**	70.54
		H-AMG	3 / –	117 (2.48)	49.88	6.71	56.59
		H-AMG-GM2	3 / 0.01	84 (3.04)	41.66	12.04	43.70
		H-AMG-LN	3 / 0.01	**65** (2.68)	**29.68**	10.23	**39.91**
1024	25,213,923	U-AMG	3 / –	145 (2.99)	82.55	8.79	91.34
		H-AMG	3 / –	138 (2.50)	59.94	**8.57**	68.51
		H-AMG-GM2	3 / 0.01	86 (2.59)	38.88	12.12	51.00
		H-AMG-LN	3 / 0.01	**80** (2.54)	**35.77**	11.41	**47.18**
8192	199,151,043	U-AMG	3 / –	188 (3.04)	113.42	10.61	124.03
		H-AMG	3 / –	160 (2.51)	70.76	**9.01**	79.77
		H-AMG-GM2	3 / 0.01	108 (2.53)	48.94	11.82	60.76
		H-AMG-LN	3 / 0.01	**103** (2.52)	**46.64**	11.94	**58.58**
65,536	1,583,018,883	U-AMG	3 / –	225 (3.11)	157.34	19.84	177.18
		H-AMG	3 / –	195 (2.52)	89.01	**11.32**	100.33
		H-AMG-GM2	3 / 0.01	120 (2.53)	55.63	14.80	70.43
		H-AMG-LN	3 / 0.01	**115** (2.53)	**53.29**	14.86	**68.15**
524,288	12,623,496,963	U-AMG	3 / –	270 (3.06)	229.40	39.18	268.58
		H-AMG	3 / –	253 (2.52)	118.38	**13.34**	131.72
		H-AMG-GM2	3 / 0.01	144 (2.53)	68.56	18.17	86.73
		H-AMG-LN	3 / 0.01	**131** (2.53)	**62.52**	18.11	**80.63**

6.2.3 3D Cuboid Problem

Finally, we consider a *3D cuboid* problem. The cuboid has the same form and size as the original *3D beam*, but is fixed on the two opposite sides with $x = 0$ and $x = 8$. We then compress the cuboid to 95 % of its length. Note that for the *3D cuboid*, we have a core material with $E = 210$ and $\nu = 0.45$ in the part of the cuboid where $0.25 < y < 0.75$ and $0.25 < z < 0.75$. Here (x, y, z) denotes the coordinates in the undeformed reference configuration of the cuboid. In the remaining hull, we have $E = 210$ and $\nu = 0.3$.

The results for the *3D cuboid* problem in Table 7 show that the AMG approaches benefit from the larger Dirichlet boundary as compared to the *3D beam*. However, the GM2 and LN approaches show the best numerical scalability, i.e., the numbers of iterations only increase from 29 to 44 for GM2 and from 24 to 39 for LN when scaling weakly from 64 to 262 K MPI ranks. For this problem, the H-AMG approach remains competitive as well for the largest number of ranks with regard to total times as a result of its low setup time.

6.3   Parallel Problem Assembly and Reordering Process

Although the focus of this paper is on the parallel performance of AMG for linear elasticity problems, we also comment on the parallel problem assembly and setup, presenting timing results in Table 8. In order to assemble the global elasticity problems in two and three dimensions, we first decompose the domain into nonoverlapping parts of equal size, one for each MPI rank. We then assemble local stiffness matrices corresponding to these local parts. These computations are completely local to the ranks and thus perfectly scalable. The local assembly process is denoted as **Local Asm.** in Table 8. To assemble the local stiffness matrices to one global and parallel stiffness matrix, some global communication is necessary. This global assembly process is denoted as **Global Asm.** in Table 8. This process scales fine up to 32 K ranks. Scaling further, the amount of communication and synchronization slows the global assembly down. A classical lexicographical ordering of the global indices is often not optimal for the convergence, especially using hybrid approaches, and we therefore reorder the indices. After the **reordering** process, each rank holds a portion of the global stiffness matrix which has a shape close to a square in two dimensions and a cube in three dimensions. The implementation of the index reordering step is very fast (see Table 8) but makes use of the same communication patterns as the global assembly process leading to the same deterioration on more than 32 K cores.

Table 7 Weak scalability results for the *3D cuboid* problem; notation as in Table 3

#MPI ranks (=2×#Cores)	Problem size	Preconditioner	Pmax/Q-th	It. (Cop)	Time CG (s)	Time BoomerSetup (s)	Time Setup + Solve (s)
64	1,618,803	U-AMG	2 / –	49 (2.73)	25.50	4.53	30.03
		U-AMG	3 / –	34 (2.95)	18.76	6.93	25.69
		H-AMG	3 / –	34 (2.48)	14.57	7.67	22.24
		H-AMG-GM2	3 / 0.01	29 (3.04)	14.51	12.81	27.32
		H-AMG-LN	3 / 0.01	**24 (2.68)**	11.09	10.86	**21.95**
512	12,616,803	U-AMG	2 / –	65 (2.79)	35.83	6.24	42.07
		U-AMG	3 / –	43 (2.98)	24.77	8.57	33.34
		H-AMG	3 / –	41 (2.50)	18.06	8.33	26.39
		H-AMG-GM2	3 / 0.01	31 (2.59)	14.25	11.83	26.08
		H-AMG-LN	3 / 0.01	**23 (2.55)**	10.57	11.25	**21.83**
4096	99,614,403	U-AMG	2 / –	84 (2.84)	49.76	8.06	57.82
		U-AMG	3 / –	53 (3.04)	32.02	10.75	42.77
		H-AMG	3 / –	45 (2.51)	20.18	9.23	29.41
		H-AMG-GM2	3 / 0.01	33 (2.53)	15.32	12.28	27.60
		H-AMG-LN	3 / 0.01	**31 (2.53)**	14.34	12.30	**26.64**
32,768	791,664,003	U-AMG	2 / –	102 (2.87)	64.11	11.83	75.94
		U-AMG	3 / –	63 (3.09)	43.83	18.31	62.14
		H-AMG	3 / –	49 (2.52)	22.58	10.01	32.59
		H-AMG-GM2	3 / 0.01	38 (2.53)	17.90	13.58	31.48
		H-AMG-LN	3 / 0.01	**34 (2.53)**	16.03	14.05	**30.08**
262,144	6,312,364,803	U-AMG	2 / –	126 (2.86)	81.26	17.51	98.77
		U-AMG	3 / –	73 (3.10)	60.98	37.99	98.97
		H-AMG	3 / –	64 (2.52)	30.15	12.60	42.75
		H-AMG-GM2	3 / 0.01	44 (2.53)	21.20	17.85	39.05
		H-AMG-LN	3 / 0.01	**39 (2.53)**	18.84	17.83	**36.67**

Table 8 Presentation of problem assembly and setup timings, which are independent of the chosen AMG preconditioner. Values are averages over the measured values in all runs presented in Table 3. The total runtime of the complete *3D beam* application can be obtained by adding these three times to the time *Setup + Solve* from Table 3

#MPI ranks	Problem size	Local Asm. (s)	Global Asm. (s)	Reorder (s)
64	839,619	19.10	0.81	0.67
512	6,502,275	19.14	0.86	0.84
4096	51,171,075	19.14	0.93	0.77
32,786	406,003,203	19.05	1.44	1.57
262,144	3,234,610,179	19.03	8.82	9.35

7 Conclusions

We investigated the performance of hypre's AMG variants for elasticity for several 2D and 3D linear elasticity problems with varying Poisson ratios v. We compared the unknown and hybrid approaches, which use prolongation operators that only interpolate the translations, with three approaches, GM1, GM2 and LN, that are based on the hybrid approach and also incorporate the rotations. In all cases, GM1, GM2 and LN showed improved convergence over the hybrid approach when using the same truncation for P. For $v = 0.3$, all hybrid approaches scaled better than the unknown approach, and the GM2 and LN approaches were overall faster for very large problems. For the largest problem in three dimensions with 14 billion unknowns and using the largest number of processes considered, i.e., 524,288 processes, the LN approach was 40 % faster than the standard approaches. For $v = 0.45$, GM2 and LN clearly scale better than the other approaches and are more than twice as fast on 32,768 processes with better complexities and five times as fast as the hybrid approach with the same operator complexity.

We also found that the unknown approach was more robust with regard to an increase in v than the other approaches, solving the problem with $v = 0.49$ faster than any of the other approaches, but generally needed larger complexities. While the hybrid approach did not converge within 1000 iterations for $v = 0.49$, GM1, GM2 and LN were able to solve the problem in less than 200 iterations.

Overall, our study shows that the inclusion of the rigid body modes into AMG interpolation operators is generally beneficial, especially at large scale. We conclude that, for elasticity problems, using enhancements of the interpolation, parallel AMG methods are able to scale to the largest supercomputers currently available.

Acknowledgements This work was supported in part by the German Research Foundation (DFG) through the Priority Program 1648 "Software for Exascale Computing" (**SPPEXA**) under **KL 2094/4-1** and **RH 122/2-1**. The authors also gratefully acknowledge the use of the **Vulcan** supercomputer at Lawrence Livermore National Laboratory. Partial support for this work was provided through Scientific Discovery through Advanced Computing (**SciDAC**) program funded by U.S. Department of Energy, Office of Science, Advanced Scientific Computing Research (and Basic Energy Sciences/Biological and Environmental Research/High Energy Physics/Fusion

Energy Sciences/Nuclear Physics). This work was performed under the auspices of the U.S. Department of Energy by Lawrence Livermore National Laboratory under Contract **DE-AC52-07NA27344**. The authors gratefully acknowledge the Gauss Centre for Supercomputing (GCS) for providing computing time through the John von Neumann Institute for Computing (NIC) on the GCS share of the supercomputer **JUQUEEN** [24] at Jülich Supercomputing Centre (JSC). GCS is the alliance of the three national supercomputing centres HLRS (Universität Stuttgart), JSC (Forschungszentrum Jülich), and LRZ (Bayerische Akademie der Wissenschaften), funded by the German Federal Ministry of Education and Research (BMBF) and the German State Ministries for Research of Baden-Württemberg (MWK), Bayern (StMWFK) and Nordrhein-Westfalen (MIWF).

References

1. Augustin, C.M., Neic, A., Liebmann, M., Prassl, A.J., Niederer, S.A., Haase, G., Plank, G.: Anatomically accurate high resolution modeling of human whole heart electromechanics: a strongly scalable algebraic multigrid solver method for nonlinear deformation. J. Comput. Phys. **305**, 622–646 (2016)
2. Baker, A.H., Kolev, T.V., Yang, U.M.: Improving algebraic multigrid interpolation operators for linear elasticity problems. Numer. Linear Algebra Appl. **17**(2–3), 495–517 (2010). http://dx.doi.org/10.1002/nla.688
3. Blatt, M., Ippisch, O., Bastian, P.: A massively parallel algebraic multigrid preconditioner based on aggregation for elliptic problems with heterogeneous coefficients. arXiv preprint arXiv:1209.0960 (2013)
4. Braess, D.: Towards algebraic multigrid for elliptic problems of second order. Computing **55**(4), 379–393 (1995). http://dx.doi.org/10.1007/BF02238488
5. Braess, D.: Finite Elemente, vol. 4. Springer, Berlin (2007)
6. Brezina, M., Cleary, A.J., Falgout, R.D., Jones, J.E., Manteufel, T.A., McCormick, S.F., Ruge, J.W.: Algebraic multigrid based on element interpolation (AMGe). SIAM J. Sci. Comput. **22**, 1570–1592 (2000). Also LLNL technical report UCRL-JC-131752
7. Brezina, M., Tong, C., Becker, R.: Parallel algebraic multigrid methods for structural mechanics. SIAM J. Sci. Comput. **27**(5), 1534–1554 (2006)
8. Bulgakov, V.E.: Multi-level iterative technique and aggregation concept with semi-analytical preconditioning for solving boundary value problems. Commun. Numer. Methods Eng. **9**(8), 649–657 (1993). http://dx.doi.org/10.1002/cnm.1640090804
9. Cleary, A.J., Falgout, R.D., Henson, V.E., Jones, J.E., Manteuffel, T.A., McCormick, S.F., Miranda, G.N., Ruge, J.W.: Robustness and scalability of algebraic multigrid. SIAM J. Sci. Comput. **21**, 1886–1908 (2000)
10. Clees, T.: AMG Strategies for ODE Systems with Applications in Industrial Semiconductor Simulation. Shaker Verlag GmbH, Germany (2005)
11. De Sterck, H., Yang, U.M., Heys, J.J.: Reducing complexity in parallel algebraic multigrid preconditioners. SIAM J. Matrix Anal. Appl. **27**(4), 1019–1039 (2006). http://dx.doi.org/10.1137/040615729
12. De Sterck, H., Falgout, R.D., Nolting, J.W., Yang, U.M.: Distance-two interpolation for parallel algebraic multigrid. Numer. Linear Algebra Appl. **15**, 115–139 (2008)
13. Dohrmann, C.R.: Interpolation operators for algebraic multigrid by local optimization. SIAM J. Sci. Comput. **29**(5), 2045–2058 (electronic) (2007). http://dx.doi.org/10.1137/06066103X
14. Griebel, M., Oeltz, D., Schweitzer, A.: An algebraic multigrid for linear elasticity. J. Sci. Comput. **25**(2), 385–407 (2003)
15. Henson, V.E., Vassilevski, P.S.: Element-free AMGe: general algorithms for computing interpolation weights in AMG. SIAM J. Sci. Comput. **23**(2), 629–650 (electronic) (2001). http://dx.doi.org/10.1137/S1064827500372997. copper Mountain Conference (2000)

16. Henson, V.E., Yang, U.M.: BoomerAMG: a parallel algebraic multigrid solver and preconditioner. Appl. Numer. Math. **41**, 155–177 (2002)
17. hypre: High performance preconditioners. http://www.llnl.gov/CASC/hypre/
18. Lanser, M.: Nonlinear FETI-DP and BDDC Methods. Ph.D. thesis, Universität zu Köln (2015)
19. Muresan, A.C., Notay, Y.: Analysis of aggregation-based multigrid. SIAM J. Sci. Comput. **30**, 1082–1103 (2008)
20. Notay, Y.: An aggregation-based algebraic multigrid method. Electron. Trans. Numer. Anal. **37**, 123–146 (2010)
21. Notay, Y., Napov, A.: Algebraic analysis of aggregation-based multigrid. Numer. Linear Algebra Appl. **18**, 539–564 (2011)
22. Ruge, J.W.: AMG for problems of elasticity. Appl. Math. Comput. **19**, 293–309 (1986)
23. Ruge, J.W., Stüben, K.: Algebraic multigrid (AMG). In: McCormick, S.F. (ed.) Multigrid Methods. Frontiers in Applied Mathematics, vol. 3, pp. 73–130. SIAM, Philadelphia (1987)
24. Stephan, M., Docter, J.: JUQUEEN: IBM blue gene/Q®supercomputer system at the Jülich Supercomputing Centre. JLSRF **1**, A1 (2015). http://dx.doi.org/10.17815/jlsrf-1-18
25. Stüben, K.: An introduction to algebraic multigrid. In: Multigrid, pp. 413–532. Academic Press, London/San Diego (2001). also available as GMD Report 70, November 1999
26. Trottenberg, U., Oosterlee, C.W., Schüller, A.: Multigrid. Academic Press, London/San Diego (2001)
27. Vaněk, P., Mandel, J., Brezina, M.: Algebraic multigrid by smooth aggregation for second and fourth order elliptic problems. Computing **56**, 179–196 (1996)
28. Yang, U.M.: Parallel algebraic multigrid methods – high performance preconditioners. In: Bruaset, A., Tveito, A. (eds.) Numerical Solutions of Partial Differential Equations on Parallel Computers. Lecture Notes in Computational Science and Engineering, pp. 209–236. Springer, Berlin (2006)
29. Yang, U.M.: On long-range interpolation operators for aggressive coarsening. Numer. Linear Algebra Appl. **17**, 453–472 (2010)

**Part IV
EXAHD: An Exa-Scalable Two-Level
Sparse Grid Approach
for Higher-Dimensional Problems
in Plasma Physics and Beyond**

Recent Developments in the Theory and Application of the Sparse Grid Combination Technique

Markus Hegland, Brendan Harding, Christoph Kowitz, Dirk Pflüger, and Peter Strazdins

Abstract Substantial modifications of both the choice of the grids, the combination coefficients, the parallel data structures and the algorithms used for the combination technique lead to numerical methods which are scalable. This is demonstrated by the provision of error and complexity bounds and in performance studies based on a state of the art code for the solution of the gyrokinetic equations of plasma physics. The key ideas for a new fault-tolerant combination technique are mentioned. New algorithms for both initial- and eigenvalue problems have been developed and are shown to have good performance.

1 Introduction

The solution of moderate- to high-dimensional PDEs (larger than four dimensions) comes with a high demand for computational power. This is due to the curse of dimensionality, which manifests itself by the fact that very large computational grids are required even for moderate accuracy. In fact, the grid sizes are an exponential function of the dimension of the problem. Regular grids are thus not feasible even when future exascale systems are to be utilized. Fortunately, hierarchical

M. Hegland (✉) • B. Harding
Mathematical Sciences Institute, The Australian National University, Canberra, Australia
e-mail: markus.hegland@anu.edu.au;brendan.harding@anu.edu.au

C. Kowitz
Chair of Scientific Computing, Technische Universität München, Munich, Germany
e-mail: kowitz@in.tum.de

D. Pflüger
Institute for Parallel and Distributed Systems, University of Stuttgart, Stuttgart, Germany
e-mail: dirk.pflueger@ipvs.uni-stuttgart.de

P. Strazdins
Engineering and Computer Science, The Australian National University, Canberra, Australia
e-mail: peter.strazdins@anu.edu.au

© Springer International Publishing Switzerland 2016

143

H.-J. Bungartz et al. (eds.), *Software for Exascale Computing – SPPEXA 2013-2015*, Lecture Notes in Computational Science and Engineering 113, DOI 10.1007/978-3-319-40528-5_7

discretization schemes come to the rescue. So-called sparse grids [53] mitigate the curse of dimensionality to a large extent.

Nonetheless, the need for HPC resources remains. The aim of two recent projects, one (EXAHD) within the German priority program "Software for exascale computing" and one supported through an Australian Linkage grant and Fujitsu Laboratories of Europe, has been to study the sparse grid combination technique for the solution of moderate-dimensional PDEs which arise in plasma physics for the simulation of hot fusion plasmas. The combination technique is well-suited for such large-scale simulations on future exascale systems, as it adds a second level of parallelism which admits scalability. Furthermore, its hierarchical principle can be used to support algorithm-based fault tolerance [38, 46]. In this work, we focus on recent developments with respect to the theory and application of the underlying methodology, the *sparse grid combination technique*.

The sparse grid combination technique utilizes numerical solutions $u(\gamma)$ of partial differential equations computed for selected values of the parameter vector γ which controls the underlying grids. As the name suggests, the method then proceeds by computing a linear combination of the component solutions $u(\gamma)$:

$$u_I = \sum_{\gamma \in I} c_\gamma \, u(\gamma) \, . \tag{1}$$

Computationally, the combination technique thus consists of a reduction operation which evaluates the linear combination of the computationally independent components $u(\gamma)$. A similar structure is commonly found in data analytic problems and is exploited by the Map Reduce method. Since the inception of the combination technique, parallel algorithms were studied which made use of the computational structure [15, 18, 19]. The current work is based on the same principles as these earlier works, see [2, 24, 25, 28, 31, 32, 38–40, 48].

The combination technique computes a sparse grid approximation without having to implement complex sparse grid data structures. The result is a proper sparse grid function. In the case of the interpolation problem one typically obtains the exact sparse grid interpolant but for other problems (like finite element solutions) one obtains an approximating sparse grid function. Mathematically, the combination technique is an extrapolation method, and the accuracy is established using error expansions, see [5, 44, 45]. However, specific error expansions are only known for simple cases. Some recent work on errors of the sparse grid combination technique can be found in [16, 20, 21, 47]. The scarcity of theoretical results, however, did not stop its popularity in applications. Examples include partial differential equations in fluid dynamics, the advection and advection-diffusion equation, the Schrödinger equation, financial mathematics, and machine learning, see, e.g., [8–13, 17, 41, 51]. However, as the combination technique is an extrapolation method, it is inherently unstable and large errors may occur if the error expansions do not hold. This is further discussed in [30] where also a stabilized approach, the so-called Opticom method, is analyzed. Several new applications based on this stabilized approach are discussed in [1, 7, 23, 26, 35, 51, 52]. Other non-standard combination approximations are considered in [4, 35, 37, 43].

The main application considered in the following deals with the solution of the gyrokinetic equations by the software code GENE [14]. These equations are an approximation for the case of a small Larmor-radius of the Vlasov equations for densities f_s of plasmas,

$$\frac{\partial f_s}{\partial t} + \mathbf{v} \cdot \frac{\partial f_s}{\partial \mathbf{x}} + \frac{q_s}{m_s} (\mathbf{E} + \mathbf{v} \times \mathbf{B}) \cdot \frac{\partial f_s}{\partial \mathbf{v}} = 0 . \tag{2}$$

The densities are distribution functions over the state space and E and B are the electrostatic and electromagnetic fields (both external and induced by the plasma), $\mathbf{v}$ is the velocity and $\mathbf{x}$ the location. The fields E and B are then the solution of the Maxwell equations for the charge and current densities defined by

$$\rho(\mathbf{x}, t) = \sum_s q_s \int f_s(\mathbf{x}, \mathbf{v}, t) \, dv, \quad \text{and} \quad \mathbf{j}(\mathbf{x}, t) = \sum_s q_s \int f_s(\mathbf{x}, \mathbf{v}, t) \mathbf{v} dv . \tag{3}$$

While the state space has 6 dimensions (3 space and 3 velocity), the gyrokinetic equations reduce this to 5 dimensions. The index s numbers the different species (ions and electrons). The numerical scheme uses both finite differences and spectral approximations. As complex Fourier transforms are used, the densities f_s are complex.

In Sect. 2 a general combination technique suitable for our application is discussed. In this section the set I occurring in the combination formula (1) uniquely determines the combination coefficients c_γ in that formula. Some parallel algorithms and data structures supporting the sparse grid combination technique are presented in Sect. 3. In order to stabilize the combination technique, the combination coefficients need to be modified and even chosen dependent on the solution. This is covered in Sect. 4. An important application area relates to eigenvalue problems in Sect. 5, where we cover challenges and algorithms for this problem.

2   A Class of Combination Techniques

Here we call a *combination technique* a method which is obtained by substituting some of the hierarchical surpluses by zero. This includes the traditional sparse grid combination technique [19], the truncated combination technique [4], dimension adaptive variants [10, 29] and even some of the fault tolerant methods [24]. The motivation for this larger class is that often the basic *error splitting assumption*— which can be viewed as an assumption about the surplus—does not hold in these cases. We will now formally define this combination technique.

We assume that we have at our disposition a computer code which is able to produce approximations of some real or complex number, some vector or some function. We denote the *quantity of interest* by u and assume that the space of all possible u is a Euclidean vector space (including the numbers) or a Hilbert space of functions. The computer codes are assumed to compute a very special class of approximations $u(\gamma)$ which in some way are associated with regular d-dimensional grids with step size $h_i = 2^{-\gamma_i}$ in the i-th coordinate. For simplicity we will assume that in principle our code can compute $u(\gamma)$ for any $\gamma \in \mathbb{N}_0^d$. Furthermore, $u(\gamma) \in V(\gamma)$ where the spaces $V(\gamma) \subset V$ are hierarchical, such that $V(\alpha) \subset V(\beta)$ when $\alpha \leq \beta$ (i.e. where $\alpha_i \leq \beta_i$ for all $i = 1, \ldots d$). For example, if $V = \mathbb{R}$ then so are all $V(\gamma) = \mathbb{R}$. Another example is the space of functions with bounded (in L_2) mixed derivatives $V = H^1_{\text{mix}}([0,1]^d)$. In this case one may choose $V(\gamma)$ to be appropriate spaces of multilinear functions.

The quantities of interest include solutions of partial differential equations, minima of convex functionals and eigenvalues and eigenfunctions of differential operators. They may also be functions or functionals of solutions of partial differential equations. They may be moments of some particle densities which themselves are solutions to some Kolmogorov, Vlasov, or Boltzmann equations. The computer codes may be based on finite difference and finite element solvers, least squares and Ritz solvers but could also just be interpolants or projections. In all these cases, the combination technique is a method which combines multiple approximations $u(\gamma)$ to get more accurate approximations. Of course the way how the underlying $u(\gamma)$ are computed will have some impact on the final combination approximation.

The combination technique is fundamentally tied to the concept of the hierarchical surplus [53] which was used to introduce the sparse grids. However, there is a subtle difference between the surplus used to define the sparse grids and the one at the foundation of the combination technique. The surplus used for sparse grids is based on the representation of functions as a series of multiples of hierarchical basis functions. In contrast, the combination technique is based on a more general decomposition. It is obtained from the following result which follows from two lemmas in chapter 4 of [22].

Proposition 1 (Hierarchical surplus) *Let $V(\gamma)$ be linear spaces with $\gamma \in \mathbb{N}_0^d$ such that $V(\alpha) \subset V(\beta)$ if $\alpha \leq \beta$ and let $u(\gamma) \in V(\gamma)$. Then there exist $w(\alpha) \in V(\alpha)$ such that*

$$\sum_{\alpha \leq \gamma} w(\alpha) = u(\gamma) . \tag{4}$$

Moreover, the $w(\gamma)$ are uniquely determined and one has

$$w(\alpha) = \sum_{\gamma \in B(\alpha)} (-1)^{|\alpha - \gamma|} u(\gamma) \tag{5}$$

where $B(\alpha) = \{\gamma \geq 0 \mid \alpha - 1 \leq \gamma \leq \alpha\}$ and $1 = (1, \ldots, 1) \in \mathbb{N}^d$.

The set of γ is countable and the proposition is proved by induction over this set. Note that the equations are cumulative sums and the solution is given in the form of a finite difference. For the case of $d = 2$ and $\gamma \leq (2,2)$ one gets the following system of equations:

$$
\begin{bmatrix} u(2,2) \\ u(1,2) \\ u(2,1) \\ u(0,2) \\ u(1,1) \\ u(2,0) \\ u(0,1) \\ u(1,0) \\ u(0,0) \end{bmatrix}
=
\begin{bmatrix}
1 & 1 & 1 & 1 & 1 & 1 & 1 & 1 & 1 \\
  & 1 &   & 1 & 1 &   & 1 & 1 & 1 \\
  &   & 1 &   & 1 & 1 & 1 & 1 & 1 \\
  &   &   & 1 &   &   & 1 &   & 1 \\
  &   &   &   & 1 &   & 1 & 1 & 1 \\
  &   &   &   &   & 1 &   & 1 & 1 \\
  &   &   &   &   &   & 1 &   & 1 \\
  &   &   &   &   &   &   & 1 & 1 \\
  &   &   &   &   &   &   &   & 1
\end{bmatrix}
\begin{bmatrix} w(2,2) \\ w(1,2) \\ w(2,1) \\ w(0,2) \\ w(1,1) \\ w(2,0) \\ w(0,1) \\ w(1,0) \\ w(0,0) \end{bmatrix} .
\tag{6}
$$

Note that all the components of the right hand side and the solution are elements of linear spaces. The vector of $w(\alpha)$ is for the example:

$$
\begin{bmatrix} w(2,2) \\ w(1,2) \\ w(2,1) \\ w(0,2) \\ w(1,1) \\ w(2,0) \\ w(0,1) \\ w(1,0) \\ w(0,0) \end{bmatrix}
=
\begin{bmatrix}
+1 & -1 & -1 &   & +1 &   &   &   &   \\
   & +1 &   & -1 & -1 &   & +1 &   &   \\
   &   & +1 &   & -1 & -1 &   & +1 &   \\
   &   &   & +1 &   &   & -1 &   &   \\
   &   &   &   & +1 &   & -1 & -1 & +1 \\
   &   &   &   &   & +1 &   & -1 &   \\
   &   &   &   &   &   & +1 &   & -1 \\
   &   &   &   &   &   &   & +1 & -1 \\
   &   &   &   &   &   &   &   & +1
\end{bmatrix}
\begin{bmatrix} u(2,2) \\ u(1,2) \\ u(2,1) \\ u(0,2) \\ u(1,1) \\ u(2,0) \\ u(0,1) \\ u(1,0) \\ u(0,0) \end{bmatrix} .
\tag{7}
$$

For any set of indices $I \subset \mathbb{N}_0^d$ we now define the combination technique as any method delivering the approximation

$$
u_I = \sum_{\alpha \in I} w(\alpha) .
\tag{8}
$$

In practice, the approximation u_I is computed directly from the $u(\gamma)$. The combination formula is directly obtained from Proposition 1 and one has

Proposition 2 *Let $u_I = \sum_{\alpha \in I} w(\alpha)$ where $w(\alpha)$ is the hierarchical surplus for the approximations $u(\gamma)$. Then there exists a subset I' of the smallest downset which contains the set I and some coefficients $c_\gamma \in \mathbb{Z}$ for $\gamma \in I'$ such that*

$$
u_I = \sum_{\gamma \in I'} c_\gamma u(\gamma) .
\tag{9}
$$

Furthermore, one has

$$c_\gamma = \sum_{\alpha \in C(\gamma)} (-1)^{|\gamma - \alpha|} \chi_I(\alpha) \tag{10}$$

where $C(\gamma) = \{\alpha \mid \gamma \leq \alpha \leq \gamma + 1\}$ and where $\chi_I(\alpha)$ is the characteristic function of I.

The proof of this result is a direct application of Proposition 1, see also [22]. For the example $d = 2$ and $n = 2$ one gets

$$u_n^C = u(0, 2) + u(1, 1) + u(2, 0) - u(0, 1) - u(1, 0) . \tag{11}$$

Note the coefficients $c_\gamma = 1$ for the finest grids, $c_\gamma = -1$ for some grids which are slightly coarser and $c_\gamma = 0$ for all the other grids. There are both positive and negative coefficients. Indeed, the results above can also be shown to be a consequence of the *inclusion-exclusion principle*. One can show that if $0 \in I$ then $\sum_{\gamma \in I} c_\gamma = 1$.

An implementation of the combination technique will thus compute a linear combination of a potentially large number of component solutions $u(\gamma)$. One thus requires two steps, first the independent computation of the components $u(\gamma)$ and then the reduction to the combination u_I. Thus the computations require a collection of computational clusters which are loosely connected. This is a great advantage on HPC systems as the need for global communication is significantly reduced to a loose coupling.

Many variants of the combination technique are obtained using the technique introduced above. They differ by their choice of the summation set I. The classical combination technique utilizes

$$I = \{\alpha \mid |\alpha| \leq n + d - 1\} . \tag{12}$$

Many variants are subsets of this set. This includes the *truncated sparse grids* [3, 4] defined by

$$I = \downarrow \{\alpha \mid |\alpha| \leq n + d - 1, \, \alpha \geq \beta\} \tag{13}$$

where $\downarrow$ is the operator producing the smallest downset containing the operand. Basically the same class is considered in [49] (there called partial sparse grids):

$$I = \downarrow \{\alpha \mid |\alpha| \leq n + |\beta| - 1, \, \alpha \geq \beta\} \tag{14}$$

for some $\beta \geq 1$. Sparse grids with faults [24] include sets of the form

$$I = \{\alpha \mid |\alpha| \leq n + d - 1, \alpha \neq \beta\} \tag{15}$$

for some β with $|\beta| = n$. Finally, one may consider the *two-scale* combination with

$$I = \bigcup_{k=1}^{d} \{\alpha \mid \alpha \leq n_0 1 + n_k e_k\} \tag{16}$$

where e_k is the standard k-th basis vector in $\mathbb{R}^d$. This has been considered in [3] for the case of $n_0 = n_k = n$. Another popular choice is

$$I = \{\alpha \mid |\operatorname{supp}\alpha| \leq k\} . \tag{17}$$

This corresponds to a truncated ANOVA-type decomposition. An alternative ANOVA decomposition is obtained by choosing $\beta^{(k)}$ with $|\operatorname{supp}\beta^{(k)}| = k$ and setting

$$I = \bigcup_{k=1}^{d} \{\alpha \mid \alpha \leq \beta^{(k)}\} . \tag{18}$$

The sets I are usually *downsets*, i.e., such that $\beta \in I$ if there exists an $\alpha \in I$ such that $\beta \leq \alpha$. Note that any downset I especially contains the zero vector. The corresponding vector space $V(0)$ typically contains the set of constant functions.

We will now consider errors. First we reconsider the error of the $u(\gamma)$. In terms of the surpluses, one has from the surplus decomposition of $u(\gamma)$ that

$$e(\gamma) = u - u(\gamma) = \sum_{\alpha \nleq \gamma} w(\alpha) . \tag{19}$$

Let $I_s(\gamma) = \{\alpha \mid \alpha_s > \gamma_s\}$. Then one has

$$\{\alpha \nleq \gamma\} = \bigcup_{s=1}^{d} I_s(\gamma) \tag{20}$$

as any α which is not less or equal to γ contains at least one element $\alpha_s > \gamma_s$. We now define

$$I(\gamma; \sigma) = \bigcap_{s \in \sigma} I_s(\gamma) \tag{21}$$

for any non-empty subset $\sigma \subseteq \{1, \ldots, d\}$. A direct application of the inclusion-exclusion principle then leads to the *error splitting*

$$e(\gamma) = \sum_{\emptyset \neq \sigma \subseteq \{1, \ldots, d\}} (-1)^{|\sigma|-1} z(\gamma, \sigma) \tag{22}$$

where

$$z(\gamma, \sigma) = \sum_{\alpha \in I(\gamma;\sigma)} w(\alpha) . \tag{23}$$

This is an ANOVA decomposition of the approximation error of $u(\gamma)$. From this one gets the result

Proposition 3 *Let $u_I = \sum_{\gamma \in I'} c_\gamma u(\gamma)$ and the combination coefficients c_γ be such that $\sum_{\gamma \in I} c_\gamma = 1$. Then*

$$u - u_I = \sum_{\emptyset \neq \sigma \subseteq \{1,\ldots,d\}} (-1)^{|\sigma|-1} \sum_{\gamma \in I'} c_\gamma \, z(\gamma, \sigma) . \tag{24}$$

Proof This follows from the discussion above and because $0 \in I$ one has

$$u - u_I = \sum_{\gamma \in I'} c_\gamma \, e(\gamma) . \tag{25}$$

□

An important point to note here is that this error formula does hold for any coefficients c_γ, not just the ones defined by the general combination technique. This thus leads to a different way to choose the combination coefficients which results in a small error. We will further discuss such choices in the next section. Note that for the general combination technique the coefficients are uniquely determined by the set I. In this case one has a complete description of the error using the hierarchical surplus

$$e_I = \sum_{\alpha \notin I} w(\alpha) . \tag{26}$$

In summary, we have now two strategies to design a combination approximation: one may choose either

- the set I which contains all the $w(\alpha)$ which are larger than some threshold
- or the combination coefficients such that the sums $\sum_{\alpha \in I(\gamma,\sigma)} c_\gamma \, z(\gamma, \sigma)$ are small.

One approach is to select the $w(\alpha)$ adaptively, based on their size so that

$$I = \{\alpha \mid \|w(\alpha)\| \geq \epsilon\} . \tag{27}$$

Such an approach is sometimes called dimension adaptive to distinguish it from the spatially adaptive approach where grids are refined locally. One may be interested in finding an approximation for some $u(\gamma)$, for example, for $\gamma = (n, \ldots, n)$. In this case one considers

$$I = \{\alpha \leq \gamma \mid \|w(\alpha)\| \geq \epsilon\} \tag{28}$$

and one has the following error bound:

Proposition 4 *Let* $I = \{\alpha \leq \gamma \mid \|w(\alpha)\| \geq \epsilon\}$ *and* $u(\gamma) - u_I$ *be the error of the combination approximation based on the set* I *relative to* $u(\gamma)$. *Then one has the bound*

$$\|u(\gamma) - u_I\| \leq \prod_{i=1}^{d}(\gamma_i + 1)\,\epsilon\;. \tag{29}$$

The result is a simple application of the triangle inequality and the fact that

$$|I| = \prod_{i=1}^{d}(\gamma_i + 1)\;. \tag{30}$$

In particular, one has if all $\gamma_i = n$:

$$\|u(\gamma) - u_I\| \leq (n+1)^d\epsilon\;. \tag{31}$$

While this bound is very simple, it is asymptotically (in n and d) tight due to the concentration of measure. Note also, that a similar bound for the spatially adaptive method is not available. An important point to note is that this error bound holds always, independently of how good the surplus is at approximating the exact result. For $\gamma = (n, \ldots, n)$ one can combine the estimate of Proposition 4 with a bound on $u - u(\gamma)$ to obtain

$$\|u - u_I\| \leq \|u - u(\gamma)\| + \|u(\gamma) - u_I\| \leq K\,4^{-n} + (n+1)^d\epsilon\;. \tag{32}$$

One can then choose n which minimizes this for a given ϵ by balancing the two terms. Conversely, for a given n the corresponding ϵ is given by $\epsilon_n = (n+1)^{-d}K4^{-n}$. In Fig. 1 we plot ϵ_n/K against $\|u-u_I\|/K$ for several different d to demonstrate how the error changes with the threshold.

While the combination approximation is the sum of the surpluses $w(\alpha)$ over all $\alpha \in I$, the result only depends on a small number of $u(\gamma)$ close to the maximal elements of I. In particular, any errors of the values $u(\alpha)$ for small α have no effect for approximations based on larger α. Thus when doing an adaptive approximation, the earlier errors are forgotten.

Finally, if one has a model for the hierarchical surplus, for example, if it is of the form

$$\|w(\alpha)\| \leq 4^{-|\alpha|}y(\alpha) \tag{33}$$

Fig. 1 Scaled error against
threshold

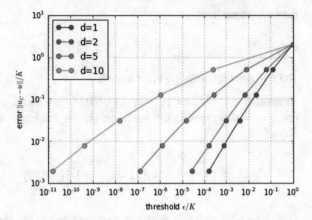

for some bounded $y(\alpha)$ then one can get specific error bounds for the combination technique, in particular the well-known bounds for the classical sparse grid technique. In this case one gets $\|w(\alpha)\| \leq K4^{-|\alpha|}$ if one chooses $|\alpha| \geq n$ as for the classical combination technique. One can show that the terms in the error formula for the components $u(\gamma)$ satisfy

$$\|z(\gamma, \sigma)\| \leq \left(\frac{4}{3}\right)^d 4^{-\sum_{s=1}^{|\sigma|}(\gamma_{\sigma_s}+1)} K \ . \tag{34}$$

3   Algorithms and Data Structures

In this section we consider the parallel implementation of the combination technique for partial differential equation solvers. For large-scale simulations, for example as being the final target for the EXAHD project in the second phase, even a single component grid (together with the data structures to solve the underlying PDE on it) will not fit into the memory of a single node any more. Furthermore, the storage of a full grid representation of a sparse grid will exceed the predicted RAM of a whole exascale machine. Furthermore, the communication overhead across a whole HPC systems' network cannot be neglected. In this section we will assume that the component grids $u(\gamma)$ are implemented as distributed regular grids. In a first stage we consider the case where the combined solution u_I is also a distributed regular grid. Later we will then discuss distributed sparse grid data structures.

The combination technique is a reduction operation combining the components according to Eq. (1). This reduction is based on the sum $u' \leftarrow u' + cu$ of a component u (we omit the parameters γ for simplicity) to the resulting combination u' (or u_I). Assume that the u and u' are distributed over P and P' processors, respectively.

The *direct SGCT algorithm* involves for each of the component processes sending all its points of u to the respective combination process. This is denoted as the

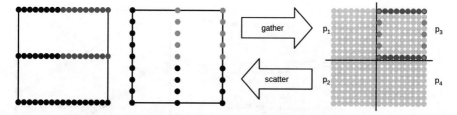

Fig. 2 Gather and scatter steps

gather stage. In a second stage, the combination processes then first *interpolates* the gathered points to the combination grid u before adding them. In a third stage, the *scatter stage*, the data on each combination process is sampled and the samples sent to the corresponding component processes, see Fig. 2.

In the direct SGCT algorithm, the components and combination are represented by the function values on the grid points or coefficients of the nodal basis. We have also considered a *hierarchical SGCT algorithm* which is based on the coefficients of the hierarchical basis which leads to a hierarchical surplus representation. When the direct SGCT algorithm is applied to these hierarchical surpluses there is no need for interpolation, and the sizes of the corresponding surplus vectors are exactly the same for both the components and the combination. However, for performance, it is necessary to *coalesce* the combination of surpluses as described in [49]. As the largest surpluses only occur for one component they do not need to be communicated. Despite the savings in the hierarchical algorithm, we found that the direct algorithm is always faster than the hierarchical, and it scales better with both n, d and the number of processes (cores). This does however require that the representation of the combined grid u' is sparse, as is described below. We also found that the formation of the hierarchical surpluses (and its inverse) took a relatively small amount of time, and concluded that, even when the data is originally stored in hierarchical form, it is faster to dehierarchize it, apply the direct algorithm and hierarchize it again [49].

New adapted algorithms and implementations have been developed with optimal communication overhead, see Fig. 3 (left) and the corresponding paper in this proceedings [27]. The gather–scatter steps described above have to be invoked multiple times for the solution of time-dependent PDEs. (We found that for eigenvalue problems it is often sufficient to call the gather–scatter only once, see the Sect. 5.) In any case, the gather–scatter step is the only remaining global communication of the combination technique and thus has to be examined well. In previous work [31] we have thus analyzed communication schemes required for the combination step in the framework of BSP-models and developed new algorithmic variants with communication that is optimal up to constant factors. This way, the overall makespan volume, the maximal communicated volume, can be drastically reduced with a slightly increased number of messages that have to be sent.

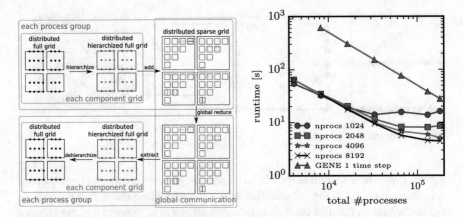

Fig. 3 Distributed hierarchical combination with optimal communication overhead *(left)* and runtime results on Hazel Hen *(right)* for different sizes of process groups (local parallelism with nprocs processors). The results measure only the communication (local + global) and distributed hierarchization, not the computation. The saturation for large numbers of component grids is due to the logarithmic scaling of the global reduce step for large numbers of process groups and up to 180,224 processors in total. In comparison, the time for a single time step with GENE for a process group size of 4096 is shown, see [27] in this proceedings for further details

A distributed sparse grid data structure is described in [49]. The index set I for this case is a variant of a truncated sparse grid set, see Eq. (14). Recall that the sparse grid points are obtained by taking the union of all the component grid points. As the number of sparse grid points is much less than the number of full grid points it makes sense to compute only the combinations for the sparse grid points. A sparse grid data structure has been developed which is similar to the CSR data structure used for sparse matrices. In this case one stores both information about the value u at the grid point and the location of the grid point. Due to the regularity of the sparse grid this can be done efficiently.

With optimal communication, distributed data structures and corresponding algorithms, excellent scaling can be obtained for large numbers of process groups as shown in Fig. 3 (right) on Hazel Hen, which includes local algorithmic work to hierarchize, local communication and global communication. See the corresponding paper in this proceedings [27].

4 Modified Combination Coefficients

Here we consider approximations which are based on a vector $(u(\gamma))_{\gamma \in I}$ of numerical results. It has been seen, however, that the standard way to choose the combination coefficients is not optimal and may lead to large errors. In fact one may interpret the truncated combination technique as a variant where some of the coefficients have been chosen to be zero and the rest adapted. In the following we provide a more radical approach to choosing the coefficients c_γ. An advantage of this approach is that it does not depend so much on properties of the index set I, in fact, this set does not even need to be a downset.

A first method was considered in [30, 52] for convex optimization problems. Here, let the component approximations be

$$u(\gamma) = \mathrm{argmin}\{J(v) \mid v \in V(\gamma)\} . \tag{35}$$

Then the Opticom method, a Ritz approximation over the span of given $u(\gamma)$ computes

$$u^O = \mathrm{argmin}\left\{ J(v) \mid v = \sum_{\gamma \in I} c_\gamma u(\gamma) \right\} . \tag{36}$$

Computationally the Opticom method consists of the determination of minimization of a convex function $P(c)$ of $|I|$ variables of the form

$$\Phi(c) = J\left(\sum_{\gamma \in I} c_\gamma u(\gamma) \right) \tag{37}$$

to get the combination coefficients. Once they have been determined, the approximation u^O is then computed as in the Sects. 2 and 3. By design, one has $J(u^O) \le J(u(\gamma))$ for all $\gamma \in I$. If I gives rise to a combination approximation u^C then one also has $J(u^O) \le J(u^C)$. A whole family of other convex functions $\Phi(c)$ for the combination coefficients were considered in [30]. Using properties of the Bregman divergence, one can derive error bounds and quasi-optimality criteria for the Opticom method, see [52].

A similar approach was suggested for the determination of combination coefficients for faulty sets I. Let I be any set and I' be the smallest downset which contains I. Then let the $w(\alpha)$ be the surpluses computed from the set of all $u(\gamma)$ for $\gamma \in I$ and $\alpha \in I'$. Finally, let the regular combination technique be defined as

$$u^R = \sum_{\alpha \in I'} w(\alpha) \tag{38}$$

and let for any c_γ a combination technique be

$$u^C = \sum_{\gamma \in I} c_\gamma u(\gamma) . \tag{39}$$

Then the difference between the new combination technique and the regular combination technique is

$$u^C - u^R = \sum_{\gamma \in I} c_\gamma \, u(\gamma) - \sum_{\alpha \in I'} w(\alpha)$$

$$= \sum_{\gamma \in I} c_\gamma \sum_{\alpha \leq \gamma} w(\alpha) - \sum_{\alpha \in I'} w(\alpha) \tag{40}$$

$$= \sum_{\alpha \in I'} w(\alpha) \left(\sum_{\gamma \in I(\alpha)} c_\gamma - 1 \right)$$

where $I(\alpha) = \{\gamma \in I \mid \gamma \geq \alpha\}$. Using the triangle inequality one obtains

$$\|u^C - u^R\| \leq \Phi(c) \tag{41}$$

with

$$\Phi(c) = \sum_{\alpha \in I'} \theta(\alpha) \left| \sum_{\gamma \in I(\alpha)} c_\gamma - 1 \right| , \tag{42}$$

where θ is such that $\|w(\alpha)\| \leq \theta(\alpha)$. Minimizing the $\Phi(c)$ thus seems to lead to a good choice of combination coefficients, and this is confirmed by experiments as well [22]. The resulting combination technique forms the basis for a new fault-tolerant approach which has been discussed in [24].

5 Computing Eigenvalues and Eigenvectors

Here we consider the eigenvalue problem in V where one would like to compute complex eigenvalues λ and the corresponding eigenvectors u such that

$$Lu = \lambda u , \tag{43}$$

where L is a given linear operator defined on V. We assume we have a code which computes approximations $\lambda(\gamma) \in \mathbb{C}$ and $u_\lambda(\gamma) \in V(\gamma)$ of the eigenvalues λ and the corresponding eigenvectors u. We have chosen to discuss the eigenvalue problem separately as it does exhibit particular challenges which do not appear for initial and boundary value problems.

Consider now the determination of the *eigenvalues* λ. Note that one typically has a large number of eigenvalues for any given operator L. First one needs to decide which eigenvalue to compute. For example, if one is interested in stability of a system, one would like to determine the eigenvalue with the largest real part. It is possible to use the general combination technique, however, one needs to make sure that the (non-zero) combination coefficients c_γ used are such that the eigenvectors of $L(\gamma)$ contain approximations of the eigenvector u which is of interest. However, computing the surplus $v(\alpha)$ for the eigenvalues $\lambda(\gamma)$ and including all the ones which satisfy $|v(\alpha)| \geq \epsilon$ for some ϵ would be a good way to make sure that we get a good result. Furthermore, the error bound given in Sect. 2 does hold here. As any surplus $v(\alpha)$ does only depend on the values $\lambda(\gamma)$ for γ close to α any earlier $\lambda(\gamma)$ with a large error will not influence the final result. Practical computations confirmed the effectiveness of this approach, see [34]. If one knows which spaces $V(\gamma)$ produce reasonable approximations for the eigenvector corresponding to some eigenvalue λ then one can define a set $I(\lambda)$ containing only those γ. Combinations over $I(\gamma)$ will then provide good approximations of λ. (However, as stated above, the combination technique is asymptotically stable against wrong or non-existing eigenvectors.)

Computing the eigenvectors faces the same problem one has for computing the eigenvalues. In addition, however, one has an extra challenge as the eigenvectors are only determined up to some complex factor. In particular, if one uses the eigenvectors $u(\gamma)$ to compute the surplus functions $w(\alpha)$ one may get very wrong results. One way to deal with this is to first normalize the eigenvectors. For this one needs a functional $s \in V^*$. One then replaces the $u(\gamma)$ by $u(\gamma)/\langle s, u(\gamma)\rangle$ when computing the surplus, i.e., one solves the surplus equations

$$\sum_{\alpha \leq \gamma} w(\alpha) = \frac{u(\gamma)}{\langle s, u(\gamma)\rangle} \tag{44}$$

and computes the combination approximation as

$$u_I = \sum_{\gamma \in I} c_\gamma \frac{u(\gamma)}{\langle s, u(\gamma)\rangle}. \tag{45}$$

In practice, this did give good results and it appears reasonable that bounds on the so computed surplus provide a foundation for the error analysis. In any case the error bound of the adaptive method holds. Actually, this bound even holds when the eigenvectors are not normalized. The advantage of the normalization is really that the number of surpluses to include are much smaller—i.e. a computational advantage. Practical experiments also confirmed this. It remains to be shown that error splitting assumptions are typically invariant under the scaling done above.

5.1   An Opticom Approach for Solving the Eigenvalue Problem

An approach to solving the eigenvalue problem which does not require scaling has been proposed and investigated by Kowitz and collaborators [34, 36]. The approach is based on a minimization problem which determines combination coefficients in a similar manner as the opticom method in Sect. 3. It is assumed that I is given and the $u(\gamma)$ for $\gamma \in I$ have been computed and solve $L(\gamma)u(\gamma) = \lambda(\gamma)u(\gamma)$. Let the matrix $G = [u(\gamma)]_{\gamma \in I}$ and the vector $c = [c_\gamma]_{\gamma \in I}^T$, then the combination approximation for the eigenvector can be written as the matrix-vector product

$$Gc = \sum_{\gamma \in I} c_\gamma u(\gamma) . \tag{46}$$

This eigenvalue problem can be solved by computing

$$(c, \lambda) = \mathrm{argmin}_{c,\lambda} \|LGc - \lambda Gc\| \tag{47}$$

with the normal equations

$$(LG - \lambda G)^*(LG - \lambda G)c = 0 \tag{48}$$

for the solution of c. Osborne et al. [33, 42] solved this by considering the problem

$$\begin{pmatrix} K(\lambda) & t \\ s^* & 0 \end{pmatrix} \begin{pmatrix} c \\ \beta \end{pmatrix} = \begin{pmatrix} 0 \\ 1 \end{pmatrix} \tag{49}$$

with $K(\lambda) = (LG - \lambda G)^*(LG - \lambda G)$. Here λ is a parameter. One obtains the solution

$$\beta(\lambda) = -\langle s^*, K(\lambda)^{-1}t \rangle^{-1} \tag{50}$$

for which one then uses Newton's method to solve $\beta(\lambda) = 0$ with respect to λ. With $\beta(\lambda) = 0$ it follows that $K(\lambda)c = 0$ and $\langle s^*, c \rangle = 1$. Thus one obtains a normalized solution of the nonlinear eigenvalue problem (i.e., where λ occurs in a nonlinear way in $K(\lambda)$).

Another approach for obtaining the least squares solution is its interpretation as an overdetermined eigenvalue problem. Das et al. [6] developed an algorithm based on the QZ decomposition which allows the computation of the eigenvalue and the eigenvector in $\mathcal{O}(mn)$ complexity, where $n = |I|$ and $m = |V|$.

The approaches have both been investigated for a simple test problem (see left of Fig. 4) and for large eigenvalue computations with GENE (see right of Fig. 4). The combination approximations (though computed serially here) can be usually obtained faster than the full grid approximations. Note that the run-times here have been obtained in a prototypical implementation before the development of the scalable algorithms described in Sect. 3. For large problems, the combination approximation can be expected to be even significantly faster as the combination technique exhibits a better parallel scalability than the full grid solution. For further details, see [34, 36].

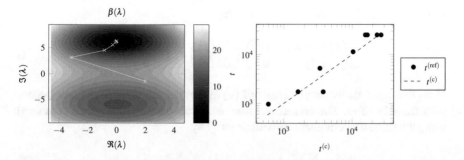

Fig. 4 The convergence of the Newton iteration towards the root of $\beta(\lambda)$ for the simple test problem *(left)* and the time for obtaining the combination approximation $t^{(c)}$ compared to the time to compute an eigenpair on a full grid of similar accuracy $t^{(\mathrm{ref})}$ for linear GENE computations *(right)*

5.2   Iterative Refinement and Iterative Methods

Besides the adaptation of the combination coefficients, the combination technique for eigenvalue problems can also be improved by refining the $u(\gamma)$ iteratively. Based on the iterative refinement procedure introduced by Wilkinson [50], the approximation of the eigenvalue λ_I and the corresponding eigenvector u_I can be improved towards λ and u with corrections $\Delta\lambda$ and Δu by

$$u = u_I + \Delta u \qquad \lambda = \lambda_I + \Delta\lambda . \tag{51}$$

Putting this into $0 = Lu - \lambda u$, the corrections can be obtained by solving

$$0 = Lu_I - \lambda_I u_I - \Delta\lambda u_I + L\Delta u - \lambda_I \Delta u, \tag{52}$$

where the quadratic term $\Delta\lambda\Delta u$ is neglected. This system is underdetermined. An additional scaling condition $\langle s^*, \Delta u \rangle = 0$ with $s \in V$ ensures that the correction Δu does not change the magnitude of u_I. Solving the linear system

$$\begin{pmatrix} L - \lambda_I I \; u_I \\ s^* \qquad 0 \end{pmatrix} \begin{pmatrix} \Delta u \\ \Delta\lambda \end{pmatrix} = \begin{pmatrix} \lambda_I u_I - Lu_I \\ 0 \end{pmatrix} , \tag{53}$$

we obtain the corrections $\Delta\lambda$ and Δu. The linear operator L has a large rank and its inversion is generally infeasible for high-dimensional settings. Nevertheless computing a single matrix vector product Lu_I is feasible, so that the right-hand side is easily computed. In the framework of the combination technique the corrections Δu and $\Delta\lambda$ are computed on each subspace $V(\gamma)$. Therefore, the residual $r = Lu - \lambda u$ and the initial combination approximation u_I are projected on $V(\gamma)$ using suitable prolongation operators [18]. The corrections $\Delta u(\gamma)$ and $\Delta\lambda(\gamma)$ are computed on each subspace $V(\gamma)$ by solving

$$\begin{pmatrix} L(\gamma) - \lambda_I I & u_I(\gamma) \\ s^*(\gamma) & 0 \end{pmatrix} \begin{pmatrix} \Delta u(\gamma) \\ \Delta \lambda(\gamma) \end{pmatrix} = \begin{pmatrix} -r(\gamma) \\ 0 \end{pmatrix} . \tag{54}$$

Here, the significantly smaller rank of $L(\gamma)$ allows the solution of the linear system with feasible effort. The corrections from each subspace $V(\gamma)$ are then combined using the standard combination coefficients c_γ by

$$\Delta u_I = \sum_{\gamma \in I} c_\gamma \Delta u_I(\gamma) \qquad \Delta \lambda_I = \sum_{\gamma \in I} c_\gamma \Delta \lambda_I(\gamma) . \tag{55}$$

After adding the correction to u_I and λ_I, the process can be repeated up to marginal $\Delta \lambda_I$ and Δu_I.

Instead of using the standard combination coefficients c_γ, we can also adapt the combination coefficients in order to minimize the residual r. The minimizer

$$(\Delta u, \Delta \lambda) = \text{argmin}_c \| r - \lambda_I u_I + L\Delta u_I - \lambda_I \Delta u_I - \Delta \lambda_I u_I \| \tag{56}$$

is then the best combination of the corrections. Both approaches have been tested for the Poisson problem as well as GENE simulations. For details see [34].

6 Conclusions

Early work on the combination technique revealed that it leads to a suitable method for the solution of simple boundary value problems on computing clusters. The work presented here demonstrated, that if combined with strongly scalable solvers for the components, one can develop an approach which is suitable for exascale architectures. This was investigated for the plasma physics code GENE which was used to solve initial and eigenvalue problems and stationary solutions. In addition to the 2 levels of parallelism exhibited by the combination technique, the flexibility of the choice of the combination coefficients led to a totally new approach to algorithm-based fault tolerance which further enhanced the scalability of the approach.

Acknowledgements The work presented here reviews some results of a German-Australian collaboration going over several years which was supported by grants from the German DFG (SPP-1648 SPPEXA: EXAHD) and the Australian ARC (LP110200410), contributions by Fujitsu Laboratories of Europe (FLE) and involved researchers from the ANU, FLE, TUM, and the Universities of Stuttgart and Bonn. Contributors to this research included Stephen Roberts, Jay Larson, Moshin Ali, Ross Nobes, James Southern, Nick Wilson, Hans-Joachim Bungartz, Valeriy Khakhutskyy, Alfredo Hinojosa, Mario Heene, Michael Griebel, Jochen Garcke, Rico Jacob, Philip Hupp, Yuan Fang, Matthias Wong, Vivien Challis and several others.

References

1. Ali, M.M., Southern, J., Strazdins, P.E., Harding, B.: Application level fault recovery: Using fault-tolerant open MPI in a PDE solver. In: 2014 IEEE International Parallel & Distributed Processing Symposium Workshops, Phoenix, 19–23 May 2014, pp. 1169–1178. IEEE (2014)
2. Ali, M.M., Strazdins, P.E., Harding, B., Hegland, M., Larson, J.W.: A fault-tolerant gyrokinetic plasma application using the sparse grid combination technique. In: Proceedings of the 2015 International Conference on High Performance Computing & Simulation (HPCS 2015), pp. 499–507. IEEE, Amsterdam (2015). Outstanding paper award
3. Benk, J., Bungartz, H.J., Nagy, A.E., Schraufstetter, S.: Variants of the combination technique for multi-dimensional option pricing. In: Günther, M., Bartel, A., Brunk, M., Schöps, S., Striebel, M. (eds.) Progress in Industrial Mathematics at ECMI 2010, pp. 231–237. Springer, Berlin/Heidelberg (2010)
4. Benk, J., Pflüger, D.: Hybrid parallel solutions of the Black-Scholes PDE with the truncated combination technique. In: Smari, W.W., Zeljkovic, V. (eds.) 2012 International Conference on High Performance Computing & Simulation, HPCS 2012, Madrid, 2–6 July 2012, pp. 678–683. IEEE (2012)
5. Bungartz, H.J., Griebel, M., Rüde, U.: Extrapolation, combination, and sparse grid techniques for elliptic boundary value problems. Comput. Method. Appl. M. 116(1–4), 243–252 (1994)
6. Das, S., Neumaier, A.: Solving overdetermined eigenvalue problems. SIAM J. Sci. Comput. 35(2), 541–560 (2013)
7. Fang, Y.: One dimensional combination technique and its implementation. ANZIAM J. Electron. Suppl. 52(C), C644–C660 (2010)
8. Franz, S., Liu, F., Roos, H.G., Stynes, M., Zhou, A.: The combination technique for a two-dimensional convection-diffusion problem with exponential layers. Appl. Math. 54(3), 203–223 (2009)
9. Garcke, J.: Regression with the optimised combination technique. In: Proceedings of the 23rd International Conference on Machine Learning (ICML 2006), vol. 2006, pp. 321–328. ACM, New York (2006)
10. Garcke, J.: A dimension adaptive sparse grid combination technique for machine learning. ANZIAM J. 48(C), C725–C740 (2007)
11. Garcke, J., Griebel, M.: On the computation of the eigenproblems of hydrogen and helium in strong magnetic and electric fields with the sparse grid combination technique. J. Comput. Phys. 165(2), 694–716 (2000)
12. Garcke, J., Hegland, M.: Fitting multidimensional data using gradient penalties and combination techniques. In: Modeling, Simulation and Optimization of Complex Processes, pp. 235–248. Springer, Berlin (2008)
13. Garcke, J., Hegland, M.: Fitting multidimensional data using gradient penalties and the sparse grid combination technique. Computing 84(1–2), 1–25 (2009)
14. Gene Development Team: GENE. http://www.genecode.org/
15. Griebel, M.: The combination technique for the sparse grid solution of PDEs on multiprocessor machines. Parallel Process. Lett. 2, 61–70 (1992)
16. Griebel, M., Harbrecht, H.: On the convergence of the combination technique. Lect. Notes Comput. Sci. 97, 55–74 (2014)
17. Griebel, M., Thurner, V.: The efficient solution of fluid dynamics problems by the combination technique. Int. J. Numer. Method. H. 5(3), 251–269 (1995)
18. Griebel, M., Huber, W., Rüde, U., Störtkuhl, T.: The combination technique for parallel sparse-grid-preconditioning or -solution of PDEs on workstation networks. In: Bougé, L., Cosnard, M., Robert, Y., Trystram, D.(eds.) Parallel Processing: CONPAR 92 – VAPP V, Lecture Notes in Computer Science. vol. 634, pp. 217–228. Springer, Berlin/Heidelberg/London (1992). Proceedings of the Second Joint International Conference on Vector and Parallel Processing, Lyon, 1–4 Sept 1992

19. Griebel, M., Schneider, M., Zenger, C.: A combination technique for the solution of sparse grid problems. In: Iterative Methods in Linear Algebra (Brussels, 1991), pp. 263–281. North-Holland, Amsterdam (1992)
20. Harding, B.: Adaptive sparse grids and extrapolation techniques. In: Proceedings of Sparse Grids and Applications 2014. Lecture Notes in Computational Science and Engineering, vol. 109, pp. 79–102. Springer, New York (2015)
21. Harding, B.: Combination technique coefficients via error splittings. ANZIAM J. **56**, C355–C368 (2016). (Online)
22. Harding, B.: Fault tolerant computation of hyperbolic PDEs with the sparse grid combination technique. Ph.D. thesis, The Australian National University (2016)
23. Harding, B., Hegland, M.: A robust combination technique. ANZIAM J. Electron. Suppl. **54**(C), C394–C411 (2012)
24. Harding, B., Hegland, M.: A parallel fault tolerant combination technique. Adv. Parallel Comput. **25**, 584–592 (2014)
25. Harding, B., Hegland, M.: Robust solutions to PDEs with multiple grids. In: Garcke, J., Pflüger, D. (eds.) Sparse Grids and Applications, Munich 2012. Lecture Notes in Computer Science. vol. 97, pp. 171–193. Springer, Cham (2014)
26. Harding, B., Hegland, M., Larson, J.W., Southern, J.: Fault tolerant computation with the sparse grid combination technique. SIAM J. Sci. Comput. **37**(3), C331–C353 (2015)
27. Heene, M., Pflüger, D.: Scalable algorithms for the solution of higher-dimensional PDEs. In: Proceedings of SPPEXA Symposium 2016. Lecture Notes in Computational Science and Engineering. Springer, Berlin/Heidelberg (2016)
28. Heene, M., Kowitz, C., Pflüger, D.: Load balancing for massively parallel computations with the sparse grid combination technique. In: Bader, M., Bungartz, H.J., Bode, A., Gerndt, M., Joubert, G.R. (eds.) Parallel Computing: Accelerating Computational Science and Engineering (CSE). pp. 574–583. IOS Press, Amsterdam (2014)
29. Hegland, M.: Adaptive sparse grids. In: Burrage, K., Sidje, R.B. (eds.) Proceedings of 10th Computational Techniques and Applications Conference CTAC-2001, vol. 44, pp. C335–C353 (2003)
30. Hegland, M., Garcke, J., Challis, V.: The combination technique and some generalisations. Linear Algebra Appl. **420**(2–3), 249–275 (2007)
31. Hupp, P., Jacob, R., Heene, M., Pflüger, D., Hegland, M.: Global communication schemes for the sparse grid combination technique. Adv. Parallel Comput. **25**, 564–573 (2014)
32. Hupp, P., Heene, M., Jacob, R., Pflüger, D.: Global communication schemes for the numerical solution of high-dimensional PDEs. Parallel Comput. **52**, 78–105 (2016)
33. Jennings, L.S., Osborne, M.: Generalized eigenvalue problems for rectangular matrices. IMA J. Appl. Math. **20**(4), 443–458 (1977)
34. Kowitz, C.: Applying the sparse grid combination technique in Linear Gyrokinetics. Ph.D. thesis, Technische Universität München (2016)
35. Kowitz, C., Hegland, M.: The sparse grid combination technique for computing eigenvalues in linear gyrokinetics. Procedia Comput. Sci. **18**, 449–458 (2013). 2013 International Conference on Computational Science
36. Kowitz, C., Hegland, M.: An opticom method for computing eigenpairs. In: Garcke, J., Pflüger, D. (eds.) Sparse Grids and Applications, Munich 2012 SE – 10. Lecture Notes in Computer Science. vol. 97, pp. 239–253. Springer, Cham (2014)
37. Kowitz, C., Pflüger, D., Jenko, F., Hegland, M.: The combination technique for the initial value problem in linear gyrokinetics. Lecture Notes in Computer Science, vol. 88, pp. 205–222. Springer, Heidelberg (2013)
38. Larson, J.W., Hegland, M., Harding, B., Roberts, S., Stals, L., Rendell, A., Strazdins, P., Ali, M.M., Kowitz, C., Nobes, R., Southern, J., Wilson, N., Li, M., Oishi, Y.: Fault-tolerant grid-based solvers: Combining concepts from sparse grids and mapreduce. Procedia Comput. Sci. **18**, 130–139 (2013). 2013 International Conference on Computational Science
39. Larson, J.W., Strazdins, P.E., Hegland, M., Harding, B., Roberts, S.G., Stals, L., Rendell, A.P., Ali, M.M., Southern, J.: Managing complexity in the parallel sparse grid combination

technique. In: Bader, M., Bode, A., Bungartz, H.J., Gerndt, M., Joubert, G.R., Peters, F.J. (eds.) PARCO. Advances in Parallel Computing, vol. 25, pp. 593–602. IOS Press, Amsterdam (2013)

40. Larson, J., Strazdins, P., Hegland, M., Harding, B., Roberts, S., Stals, L., Rendell, A., Ali, M., Southern, J.: Managing complexity in the parallel sparse grid combination technique. Adv. Parallel Comput. **25**, 593–602 (2014)

41. Lastdrager, B., Koren, B., Verwer, J.: The sparse-grid combination technique applied to time-dependent advection problems. In: Multigrid Methods, VI (Gent, 1999). Lecture Notes of Computer Science & Engineering, vol. 14, pp. 143–149. Springer, Berlin (2000)

42. Osborne, M.R.: A new method for the solution of eigenvalue problems. Comput. J. **7**(3), 228–232 (1964)

43. Parra Hinojosa, A., Kowitz, C., Heene, M., Pflüger, D., Bungartz, H.J.: Towards a fault-tolerant, scalable implementation of GENE. In: Recent Trends in Computation Engineering – CE2014. Lecture Notes in Computer Science, vol. 105, pp. 47–65. Springer, Cham (2015)

44. Pflaum, C.: Convergence of the combination technique for second-order elliptic differential equations. SIAM J. Numer. Anal. **34**(6), 2431–2455 (1997)

45. Pflaum, C., Zhou, A.: Error analysis of the combination technique. Numer. Math. **84**(2), 327–350 (1999)

46. Pflüger, D., Bungartz, H.J., Griebel, M., Jenko, F., Dannert, T., Heene, M., Kowitz, C., Parra Hinojosa, A., Zaspel, P.: EXAHD: An exa-scalable two-level sparse grid approach for higher-dimensional problems in plasma physics and beyond. In: Euro-Par 2014: Parallel Processing Workshops, Porto. Lecture Notes in Computer Science, vol. 8806, pp. 565–576. Springer, Cham (2014)

47. Reisinger, C.: Analysis of linear difference schemes in the sparse grid combination technique. IMA J. Numer. Anal. **33**(2), 544–581 (2013)

48. Strazdins, P.E., Ali, M.M., Harding, B.: Highly scalable algorithms for the sparse grid combination technique. In: IPDPS Workshops, Hyderabad, pp. 941–950. IEEE (2015)

49. Strazdins, P.E., Ali, M.M., Harding, B.: The design and analysis of two highly scalable sparse grid combination algorithms (2015, under review)

50. Wilkinson, J.H.: Rounding Errors in Algebraic Processes. Her Majesty's Stationery Office, London (1963)

51. Wong, M., Hegland, M.: Maximum a posteriori density estimation and the sparse grid combination technique. ANZIAM J. Electron. Suppl. **54**(C), C508–C522 (2012)

52. Wong, M., Hegland, M.: Opticom and the iterative combination technique for convex minimisation. Lect. Notes Comput. Sci. **97**, 317–336 (2014)

53. Zenger, C.: Sparse grids. In: Hackbusch, W. (ed.) Parallel Algorithms for Partial Differential Equations. Notes on Numerical Fluid Mechanics, vol. 31, pp. 241–251. Vieweg, Braunschweig (1991)

Scalable Algorithms for the Solution of Higher-Dimensional PDEs

Mario Heene and Dirk Pflüger

Abstract The solution of higher-dimensional problems, such as the simulation of plasma turbulence in a fusion device as described by the five-dimensional gyrokinetic equations, is a grand challenge for current and future high-performance computing. The sparse grid combination technique is a promising approach to the solution of these problems on large-scale distributed memory systems. The combination technique numerically decomposes a single large problem into multiple moderately-sized partial problems that can be computed in parallel, independently and asynchronously of each other. The ability to efficiently combine the individual partial solutions to a common sparse grid solution is a key to the overall performance of such large-scale computations. In this work, we present new algorithms for the recombination of distributed component grids and demonstrate their scalability to 180,225 cores on the supercomputer *Hazel Hen*.

1 Introduction

The solution of higher-dimensional problems, especially higher-dimensional partial differential equations (PDEs) that require the joint discretization of more than the usual three spatial dimensions plus time, is one of the grand challenges in current and future high-performance computing (HPC). Resolving the simulation domain as fine as required by the physical problem is in many cases not feasible due to the exponential growth of the number of unknowns—the so-called curse of dimensionality. A resolution of, for example, 1000 grid points in each dimension would result in 10^{15} grid points in five dimensions. Note that this is not an exaggerated example: simulations of physical phenomena as e.g. a turbulent flow require the resolution of length scales that can span many orders of magnitude, from meters to less than millimeters.

This can be observed in the simulation of plasma turbulence in a fusion reactor with the code GENE [9], which solves the five-dimensional gyrokinetic equations.

M. Heene (✉) • D. Pflüger
Institute for Parallel and Distributed Systems, University of Stuttgart, Universitätsstr. 38, 70569 Stuttgart, Germany
e-mail: mario.heene@ipvs.uni-stuttgart.de; dirk.pflueger@ipvs.uni-stuttgart.de

© Springer International Publishing Switzerland 2016   165
H.-J. Bungartz et al. (eds.), *Software for Exascale Computing – SPPEXA 2013-2015*, Lecture Notes in Computational Science and Engineering 113, DOI 10.1007/978-3-319-40528-5_8

With classical discretization techniques the resolutions necessary for physically accurate turbulence simulations of a large fusion device, such as the world's flagship fusion experiment ITER, quickly hit the computational borders even on today's largest HPC systems.

Sparse grids are a hierarchical approach to mitigate the curse of dimensionality to a large extent by drastically reducing the number of unknowns, while preserving a similar accuracy as classical discretization techniques that work on regular grids [4]. However, due to their recursive and hierarchical structure and the resulting global coupling of basis functions, the direct sparse grid approach is not feasible for large-scale distributed-memory parallelization.

A scalable approach to solve higher-dimensional problems is the sparse grid combination technique [11]. It is based on an extrapolation scheme and decomposes a single large problem (i.e. discretized with a fine resolution) into multiple moderately-sized problems that have coarse and anisotropic resolutions. This introduces a second level of parallelism, enabling one to compute the partial problems in parallel, independently and asynchronously of each other. This breaks the demand for full global communication and synchronization, which is expected to be one of the limiting factors with classical discretization techniques to achieve scalability on future exascale systems. Furthermore, by mitigating the curse of dimensionality, it offers the means to tackle problem sizes that would be out of scope for the classical discretization approaches. This allows us to significantly push the computational limits of plasma turbulence simulations and other higher-dimensional problems and is the driving motivation for our project EXAHD [21]. Additionally, we investigate novel approaches to enable fault-tolerance on the algorithmic level based on the combination technique [19, 20], which is a key issue in ongoing exascale research [5].

We are currently preparing our combination technique software framework to handle large-scale global computations with GENE. The key part that had been missing until now were efficient and scalable algorithms for the recombination of distributed component grids, which we present in this work. We present experiments that demonstrate the scalability of our algorithms on up to 180,225 cores on Germany's Tier-0/1 supercomputer *Hazel Hen*. In order to get meaningful results, we used the problem sizes for our scaling experiments that we aim to use for our future large-scale production runs. The experiments not only demonstrate the scalability of our recombination algorithm, but also that it would even be fast enough to be applied after each GENE time step, if this was necessary.

1.1 Sparse Grid Combination Technique

The sparse grid combination technique [4, 11] computes the sparse grid approximation of a function f by a linear combination of component solutions f_l. Each f_l is an approximation of f that has been computed on a coarse and anisotropic Cartesian component grid Ω_l. In our case f is the solution of the gyrokinetic equations, a higher-dimensional PDE, and the corresponding approximation f_l is the result of a simulation with the application code GENE (see Sect. 1.2) computed on the grid Ω_l. In general this can be any kind of function which fulfills certain smoothness conditions.

The discretization of each d-dimensional component grid Ω_l is defined by the level vector $\mathbf{l} = (l_1, \cdots, l_d)^T$, which determines the uniform mesh width 2^{-l_i} in dimension i. The number of grid points of a component grid is $|\Omega_l| = \prod_{i=1}^{d}(2^{l_i} \pm 1)$ ($+1$ if the grid has boundary points in dimension i and -1 if not).

In order to retrieve a sparse grid approximation $f_\mathbf{n}^{(c)} \approx f$ one can combine the partial solutions $f_l(\mathbf{x})$ as

$$f_\mathbf{n}^{(c)}(\mathbf{x}) = \sum_{l \in \mathscr{I}} c_l f_l(\mathbf{x}), \tag{1}$$

where c_l are the combination coefficients and $\mathscr{I}$ is the set of level vectors used for the combination. $\mathbf{n}$ denotes the maximum discretization level in each dimension. It also defines the discretization of the corresponding full grid solution $f_\mathbf{n}$ on $\Omega_\mathbf{n}$. Figure 1 shows a two-dimensional example.

There exist different approaches to determine the combination coefficients c_l and the index set $\mathscr{I}$ [14, 15]. Usually,

$$f_\mathbf{n}^{(c)}(\mathbf{x}) = \sum_{q=0}^{d-1}(-1)^q \binom{d-1}{q} \sum_{l \in \mathscr{I}_{\mathbf{n},q}} f_l(\mathbf{x}) \tag{2}$$

is referred to as the classical combination technique with the index set [11]

$$\mathscr{I}_{\mathbf{n},q} = \{\mathbf{l} \in \mathbb{N}^d : |\mathbf{l}|_1 = |\mathbf{l}_{\min}|_1 + c - q : \mathbf{n} \geq \mathbf{l} \geq \mathbf{l}_{\min}\}, \tag{3}$$

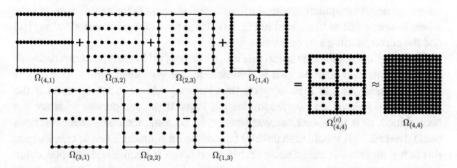

$\Omega_{(4,1)}$ $\Omega_{(3,2)}$ $\Omega_{(2,3)}$ $\Omega_{(1,4)}$ $\Omega_{(4,4)}^{(c)}$ $\Omega_{(4,4)}$

$\Omega_{(3,1)}$ $\Omega_{(2,2)}$ $\Omega_{(1,3)}$

Fig. 1 The classical combination technique with $\mathbf{n} = (4, 4)$ and $\mathbf{l}_{\min} = (1, 1)$. Seven component grids are combined to obtain a sparse grid approximation (on the grid $\Omega_{(4,4)}^{(c)}$) to the full grid solution on the grid $\Omega_{(4,4)}$

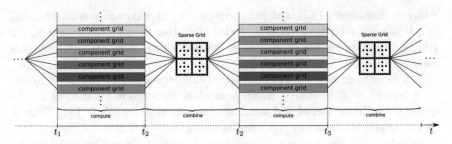

Fig. 2 Concept of recombination. On each component grid, the PDE is solved for a certain time interval Δt. Then the component grids are combined in the sparse grid space, and the combined solution is set as the new initial value for each component grid for the computation of the next time interval

where $\mathbf{l}_{\min} = \mathbf{n} - c \cdot \mathbf{e}$, $c \in \mathbb{N}_0$ s.th. $\mathbf{l}_{\min} \geq \mathbf{e}$ specifies a minimal resolution level in each direction, $\mathbf{e} = (1, \dots, 1)^T$ and $\mathbf{l} \geq \mathbf{j}$ if $\mathbf{l}_i \geq \mathbf{j}_i \ \forall i$. The computational effort (with respect to the number of unknowns) decreases from $\mathcal{O}(2^{nd})$ for the full grid solution $f_{\mathbf{n}}$ on $\Omega_{\mathbf{n}}$ to $\mathcal{O}(dn - 1)$ partial solutions of size $\mathcal{O}(2^n)$. If f fulfills certain smoothness conditions, the approximation quality is only deteriorated from $\mathcal{O}(2^{-2n})$ for $f_{\mathbf{n}}$ to $\mathcal{O}(2^{-2n}n^{d-1})$ for $f_{\mathbf{n}}^{(c)}$. The minimum level $l_{\min}$ has been introduced in order to exclude component grids from the combination. In some cases, if the resolution of a component grid is too coarse this could lead to numerically unstable or even physically meaningless results.

In case of time-dependent initial value computations, as we have them in GENE, advancing the combined solution $f_{\mathbf{n}}^{(c)}(t)$ in time requires to recombine the component solutions every few time steps. This is necessary to guarantee convergence and stability of the combined solution. Recombination means to combine the component solutions $f_{\mathbf{l}}(t_i)$ according to Eq. (1) and to set the corresponding sparse grid solution $f_{\mathbf{n}}^{(c)}(t_i)$ as the new initial value for each component grid. After that, the independent computations are continued until the next recombination point t_{i+1}, where $t_i = t_0 + i\Delta t$ as illustrated in Fig. 2 (t_0 is the time of a given initial value $f(t_0)$ and the corresponding approximation $f_{\mathbf{l}}(t_0)$).

This does not necessarily mean that all component grids must be computed with the same time step size Δt. If it is desirable to use individual time step sizes for the component grids, or even adaptive time stepping, Δt can be understood as the time interval in-between two recombination steps. If the component grids are not recombined, or if the recombination interval is too long, the component solutions could diverge. This would then destroy the combined solution. Even in cases where this is not an issue, it might be desirable to compute the combined solution every few time steps in order to trace certain physical properties of the solution field over time. Being the only remaining step that involves global communication, an efficient implementation is crucial for the overall performance of computations with the combination technique.

The function space $V_{\mathbf{l}}$ of a component grid is spanned by classical nodal basis functions (e.g. piece-wise linear). $V_{\mathbf{l}}$ can be uniquely decomposed into hierarchical

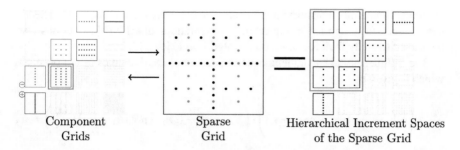

Fig. 3 Component grids of the combination technique *(left)* with $\mathbf{l}_{min} = (1, 1)$ and $\mathbf{n} = (4, 4)$, the corresponding sparse grid *(middle)* and its hierarchical increment spaces *(right)*. The hierarchical increment spaces, as well as the corresponding grid points in the sparse grid, of the component grid with $\mathbf{l} = (2, 3)$ are marked in *green*

increment spaces $W_{\mathbf{l}'}$ (also called hierarchical subspaces in the following) where $V_{\mathbf{l}} = \oplus_{\mathbf{l}' \leq \mathbf{l}} W_{\mathbf{l}'}$. We refer to the operation of decomposing a component grid into its hierarchical subspaces as *hierarchization* and the inverse operation as *dehierarchization*. The sparse grid which corresponds to a particular combination is a union of the hierarchical subspaces of the component grids that contribute to the combination. Figure 3 shows the component grids of the combination technique with $\mathbf{n} = (4, 4)$ and the corresponding sparse grid. The component grids can be combined according to Eq. (1) in the function space of the sparse grid by adding up the hierarchical subspaces of the component grids. This is explained in detail in [17].

The naive way to perform the recombination would be to interpolate each partial solution onto the full grid $\Omega_{\mathbf{n}}$ and to obtain the combined solution by adding up grid point by grid point weighted with the corresponding combination coefficient. However, for the large-scale computations with GENE that we aim for, this is not feasible. The size of the full grid would be so large that even storing it would be out of scope. The only efficient (or even feasible) way is to recombine the component grids in the corresponding sparse grid space. For large-scale setups, where the component grids are distributed onto several thousand processes, an efficient and scalable implementation of the recombination is crucial for the overall performance, because this is the only remaining step that requires global communication. In [16, 17] we already have presented and analyzed different algorithms for the recombination when a component grid is stored on a single node. In this work we present new, scalable algorithms for the recombination step with component grids that are distributed over a large number of nodes on an HPC system.

1.2 Large Scale Plasma Turbulence Simulations with GENE

A limiting factor for the generation of clean sustainable energy from plasma fusion reactors are microinstabilities that arise during the fusion process [8]. Simulation codes like GENE play an important role in understanding the mechanisms of the resulting anomalous transport phenomena. The combination technique has already been successfully applied to eigenvalue computations in GENE [18]. It has also been used to study fault tolerance on the algorithmic level [2, 19].

The 5-dimensional gyrokinetic equation describes the dynamics of hot plasmas, which is given by

$$
\frac{\partial f_s}{\partial t} + \left(v_\| \mathbf{b}_0 + \frac{B_0}{B_{0\|}^*} (\mathbf{v}_{E_\chi} + \mathbf{v}_{\nabla B_0} + \mathbf{v}_c) \right) \cdot \left(\nabla f_s + \frac{1}{m_s v_\|} \left(q\bar{\mathbf{E}}_1 - \mu \nabla \left(B_0 + \bar{B}_{1\|} \right) \right) \frac{\partial f_s}{\partial v_\|} \right) = 0 ,
\tag{4}
$$

where $f_s \equiv f_s(\mathbf{x}, v_\|, \mu; t)$ is (5+1)-dimensional due to the gyrokinetic approximation (see [3, 6] for a thorough description of the model and an explanation of all identifiers) and s denotes the species (electron, ion, etc.). If we consider g_s (the perturbation of f_s with respect to the Maxwellian background distribution) instead of f_s, Eq. (4) can be written as a sum of a (nonsymmetric) linear and a nonlinear operator, namely

$$
\frac{\partial g}{\partial t} = \mathcal{L}[g] + \mathcal{N}[g] ,
\tag{5}
$$

where g is a vector including all species in g_s.

There are different simulation modes in GENE to solve (4). *Local* (or flux-tube) simulations treat the x and y coordinates in a pseudo-spectral way, and background quantities like density or temperature (and their gradients) are kept constant in the simulation domain. The simulation domain is essentially only parallelized in four dimensions, because no domain decomposition is used in the radial direction. In *global* runs, only the y direction is treated in a spectral way, and all 5 dimensions are parallelized, with background quantities varying radially according to given profiles.

Additionally, GENE handles three main physical scenarios: multiscale problems (local mode, typical grids of size $1024 \times 512 \times 24 \times 48 \times 16$ $(x, y, z, v_\|, \mu)$ with two species), stellarator problems (local mode, typical grids of size $128 \times 64 \times 512 \times 64 \times 16$ with two species), and global simulations (expected grid size $8192 \times 64 \times 32 \times 128 \times 64$ or higher, with up to 4 species). Especially for the global simulations, there exist certain scenarios that require such high resolutions that they are too expensive to compute on current machines.

GENE makes use of highly sophisticated domain decomposition schemes that allow it to scale very well on massively parallel systems. To name an extreme example: the largest run performed so far was carried out on the JUGENE supercomputer in Jülich using up to 262k cores [7]. Thus, GENE is able to efficiently exploit the first fine level of parallelism. By adding a second level of parallelization with the combination technique we can ensure the scalability of GENE on future machines and make it possible to compute problem sizes that would otherwise be out of scope on current machines.

2 Software Framework for Large-Scale Computations with the Combination Technique

We are developing a general software framework for large-scale computations with the combination technique as a part of our sparse grid library SG++ [1]. In order to distribute the component grids over the available compute resources, the framework implements the manager–worker pattern. A similar concept has already been successfully used for the combination technique in [10]. The available compute resources, in form of MPI processes, are arranged in process groups, except one process, which is the dedicated manager process. In the framework the computation of and access to the component grids is abstracted by so-called *compute tasks*. The manager distributes these tasks to the process groups. This is illustrated in Fig. 4. The actual application code is then executed by all processes of a process group in order to compute the task. Note that this concept implements the two-level parallelism of the combination technique. Apart from the recombination step, there is no communication between the process groups and the process groups can compute the tasks independently and asynchronously of each other. The communication effort between the manager and the dedicated master process of each process group to orchestrate the compute tasks can be neglected. It only consists of very few small messages that are necessary to coordinate the process groups.

Our framework uses a *Task* class (see Fig. 5) as an abstract and flexible interface that can be conveniently adapted to the application code. This class hides the application's implementation details from the framework. A task is specified by the level vector l and the combination coefficient c_l. The *Task* base class already contains all the necessary functionalities which are independent of the actual application code. Only a few absolutely necessary functions have to be provided by the user in a derived class.

The *run* method takes care of everything that is necessary to start the application in the process group with a discretization compatible with the combination technique and specified by l. Furthermore, the specialization of the Task class

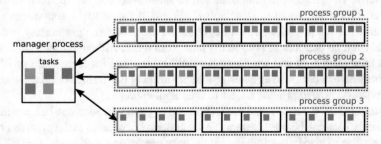

Fig. 4 The manager–worker pattern used by the software framework. The manager distributes compute tasks over the process groups. In each process group a dedicated master process is responsible for the communication with the manager and the coordination of the group. In the example, each process group consists of three nodes with four processes each.

```
 1 class Task {
 2
 3 public:
 4   inline const LevelVector& getLevelVector() const;
 5
 6   inline real getCoeff() const;
 7
 8   virtual void run(CommunicatorType& lcomm) = 0;
 9
10   virtual
11   DistributedFullGrid<DataType>& getDistributedFullGrid() = 0;
12 };
```

Fig. 5 The essential functions of the task interface (the actual file contains more functions). Only *run* and *getDistributedFullGrid* have to be implemented in the derived class by the user

has to provide access to the application data, i.e., the actual component grid corresponding to the task by the method *getDistributedFullGrid*. Many parallel applications in science and engineering, like GENE, typically use MPI parallelization and computational grids that are distributed onto a large numbers of processes of an HPC system. Such a distributed component grid is represented by the class *DistributedFullGrid*. This is a parallel version of the combination-technique component grids as introduced in Sect. 1.1. Since our component grids have a regular Cartesian structure, they are parallelized by arranging the processes on a Cartesian grid as well. It is defined by the d-dimensional level-vector $\mathbf{l}$ and a d-dimensional parallelization vector $\mathbf{p}$ which specifies the number of processes in each dimension. Furthermore, information about the actual domain decomposition must be provided.

The actual way in which the underlying component grid of the application is accessed depends on the application. In the best case, the application code can directly work with our data structure. This is the most efficient way, because no additional data transformations are necessary for the recombination step. However, this is not possible for most application codes, because they use their own data structures. In this case, the interface has to make sure that the application data is converted into the format of *DistributedFullGrid*. If access to the application's data at runtime is not possible (which might be the case for some proprietary codes) it would even be possible to read in the data from a file. Although this would be the most inefficient way and we doubt it would be feasible for large-scale problems, we mention this option to emphasize that the users of our framework have maximal flexibility on how to provide access to their application.

Furthermore, a custom load model tailored to the application can be specified by the user, which can be used to improve the assignment of tasks to the process groups with respect to load balancing. In [13] we have presented a load model for GENE. If such a model is not available, the standard model would estimate the cost of a task based on the number of grid points of the corresponding component grid.

3 Scalable Algorithms for the Combination Step with Distributed Component Grids

The recombination step is the only step of the combination technique which requires global communication between the process groups. Especially for time-dependent simulations, such as the initial value problems in GENE, which require a high number of recombination steps (in the worst case after each time step), an efficient and scalable implementation of this step is crucial for the overall performance and the feasibility of the combination technique. Efficient and scalable in this context means that the time for the recombination must not consume more than a reasonable fraction of the overall run time and that this fraction must not scale worse than the run time of the application. In our context we refer to global communication as the communication between processes of different process groups, whereas local communication happens between processes which belong to the same process group. As indicated in Sect. 2, we neglect communication between the process groups and the manager process, because no data is transferred apart from very few small messages that are necessary for the coordination of the process groups.

In [16, 17] we presented a detailed analysis of different communication strategies for the recombination step with component grids that exist on a single node. Note, however, that any approach which is based on gathering a distributed component grid on a single node cannot scale for large problem sizes. On the one hand, such approaches limit the problem size to the main memory of a single node. On the other hand, the network bandwidth of a node is limited and the time to gather a distributed grid on a single node does not decrease with increasing number of processes.

Figure 6 shows the substeps of the distributed recombination. The starting point are multiple process groups which hold one or more component grids each. Each component grid is distributed over all processes in a group. First, each process group transfers all its component grids from the nodal basis representation into the hierarchical basis of the sparse grid by distributed hierarchization. Then, the hierarchical coefficients of each component grid are multiplied by the combination coefficient and added to a temporary distributed sparse grid data structure. Note that each component grid only contributes to a subset of the hierarchical subspaces in the sparse grid. The component grid visualized in Fig. 6 has discretization level $l = (3, 2)$. Hence, it only contributes to the hierarchical subspaces $W_{l'}$ with $l' \leq (3, 2)$. In the following we will refer to the operation of adding up (and multiplying with the combination coefficient) all the component grids in the sparse grid data structure as *reduction*. The *reduce* operation is done in a local and a global substep. First, each process group has to locally reduce its component grids. Then, the individual distributed sparse grids of the process groups are globally reduced to the combined solution, which now exists on each process group in the distributed sparse grid. As a next step, for each component grid the relevant hierarchical coefficients are extracted from the combined solution. This operation happens inside each process group, and it is the inverse of the local reduction. We refer to it as *scatter* operation. Afterwards, the combined solution is available on

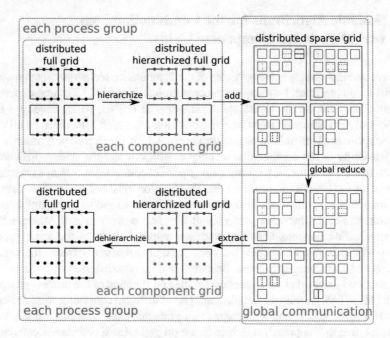

Fig. 6 The recombination step. Each component grid is hierarchized and then added to a temporary distributed sparse grid. After the distributed sparse grids in each process group have been globally reduced to the combined solution, the hierarchical coefficients of each component grid are extracted and the component grid is brought back into the nodal basis representation by dehierarchization

each component grid in the hierarchical basis and is transferred back into the nodal basis by dehierarchization. In the following sections we will discuss the substeps of the distributed recombination step in detail.

3.1 Distributed Hierarchization/Dehierarchization

In [12], we presented a new, scalable distributed hierarchization algorithm to efficiently hierarchize large, distributed component grids. In the following, we summarize this work. Figure 7 shows a component grid distributed onto eight different MPI processes. The corresponding dependency graph for the hierarchization in x_2-direction illustrates that exchange of data between the processes is necessary. Our algorithm follows the uni-directional principle which means that we hierarchize the d dimensions of the component grid one after the other. The hierarchization in each dimension consists of two steps:

1. Each process determines from the dependency graph the values it has to exchange with other processes. The dependency graph can be deduced from the discretization and the domain partitioning and is known to each process without additional communication.

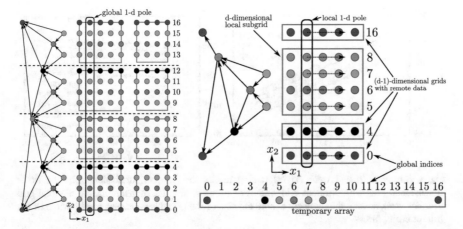

Fig. 7 *Left:* Component grid distributed over eight processes and the corresponding dependency graph for the hierarchization of the x_2 direction. *Right:* Distributed hierarchization with remote data from the local view of one process

2. The data is exchanged using non-blocking MPI calls. Each process stores the received remote data in the form of $(d - 1)$-dimensional grids.

After the communication step, each process locally performs the hierarchization. The local subgrid is traversed with one-dimensional poles (see Fig. 7). If during the update of each point we would need to determine whether its dependencies are contained in the local or in the remote data, we would obtain bad performance due to branching and address calculations. In order to avoid this, we copy the local and the remote data to a temporary 1-d array which has the full global size of the current dimension. We then perform the hierarchization of the current pole in the temporary array using an efficient 1-d hierarchization kernel. Afterwards, the updated values are copied back to the subgrid. For reasonable problem sizes the maximum global size per dimension is not much more than a few thousand grid points. So, the temporary array will easily fit into the cache, even for very large grids. The poles traverse the local subgrid and the remote data in a cache-optimal order. Thus, no additional main memory transfers are required. With this approach we can achieve high single core performance due to cache optimal access patterns. Furthermore, only the absolutely necessary data is exchanged in the communication step.

Figure 8 shows strong and weak scaling results for 5-dimensional component grids of different sizes. The experiments were performed from 32 to 32,768 cores on the supercomputer *Hornet*. The grid sizes used for the strong scaling experiments would correspond to the sizes of component grids of very large GENE simulations. The largest grid used for the weak scaling experiments had a total size of 36 TByte distributed over 32,768 cores. The experiments show that our distributed hierarchization algorithm scales very well. More detailed information on the distributed hierarchization experiments can be found in [12].

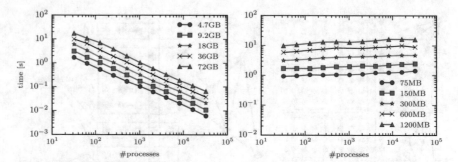

Fig. 8 Strong *(left)* and weak *(right)* scaling results for the distributed hierarchization of 5-dimensional grids with different grid sizes. For strong scaling the numbers denote the total grid size. For weak scaling the numbers denote the size of the (roughly) constant local portion of the grid for each process

There, also further issues such as different dimensionalities and anisotropic discretizations or anisotropic domain decompositions of the component grids are investigated.

3.2 Local Reduction/Scatter of Component Grids Inside the Process Group

For the local reduction step inside the process group we use two variants of a distributed sparse grid. In the first variant, the hierarchical subspaces are distributed over the processes of the group, so that each subspace is assigned to exactly one process. In the second variant, each hierarchical subspace is geometrically decomposed in the same way as the distributed component grids. Each process stores its part of the domain of each hierarchical subspace. Unlike the first variant, no communication between the processes is required for the local reduction step. However, this variant can only be applied if all component grids on the different process groups have exactly the same geometrical domain partitioning.

In the following we will present the two variants of the distributed sparse grid and discuss their advantages and disadvantages for the local reduction step. Note that the scatter step, where the hierarchical coefficients of the component grids are extracted from the combined solution, is just the inverse operation of the reduction step. Thus, we will only discuss the reduction here.

3.2.1 Variant 1: General Reduction of Distributed Component Grids

In the first variant of the distributed sparse grid, each hierarchical subspace is stored on exactly one process. A sensible rule to assign the hierarchical subspaces to the processes would be to distribute them such that a balanced number of grid points is achieved. This is important for the global reduction step, which has minimal cost when the number of grid points on each process in a process group is equal. We use a straightforward way to assign the subspaces to the processes: first we sort the subspaces by size in descending order. Then we assign the subspaces to the computed nodes in a round-robin fashion. As a last step, we distribute the subspaces to the processes of each node in the same way. A balanced distribution over the compute nodes is more important than over the processes, because the processes on each node share main memory and network bandwidth. In this way, if the number of hierarchical subspaces is large enough (several thousands or ten thousands in higher dimensions), a balanced distribution can easily be achieved.

The advantage of this sparse grid variant is that it is general. It can be conveniently used for the local reduction step to add up distributed component grids that have different parallelization. Futhermore, with this variant the global reduction step can easily be extended to process groups of different sizes. Adding a distributed component grid to a distributed sparse grid means adding up the coefficients of each hierarchical subspace common to both the component grid and the sparse grid. The grid points of each hierarchical subspace in the distributed full grid are geometrically distributed over the processes of the process group. Hence, to add them to the corresponding subspace in the distributed sparse grid, they have to be gathered on the process to which the subspace is assigned. This is illustrated in Fig. 9 (left). The communication overhead that comes with these additional gather operations is the major disadvantage of this sparse grid variant.

In the following, we will analyze the communication costs that incur when a distributed component grid is added to a distributed sparse grid. These costs resemble the costs to (almost) completely redistribute the grid points of the distributed full grid: we start with a distributed full grid which has N/m grid points per process (N grid points distributed over m processes) and we end up with a distributed sparse grid which has N/m grid points per process. In a simple

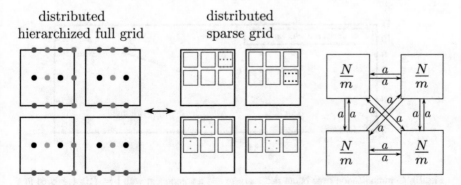

Fig. 9 Each hierarchical subspace of the distributed sparse grid is stored on exactly one process. Gathering the hierarchical subspaces yields the shown communication pattern *(right)*. Each of the m processes contains N/m grid points of the distributed full grid (of total size N) and of the distributed sparse grid. Each process sends $a = N/m^2$ grid points to $m - 1$ other processes

communication model, which only considers the number of grid points to be sent, this means each process sends N/m^2 grid points to $(m-1)$ other processes (see Fig. 9). Thus, in total each process has to send (and receive)

$$\frac{N}{m^2}(m-1) = \frac{N}{m} - \frac{N}{m^2} \approx \frac{N}{m} \tag{6}$$

grid points, which is approximately N/m for large m. Hence, the amount of data to be transferred by each process (or node) scales with the number of processes. However, on an actual system the time to send a message to another process does not only depend on the size, but is bounded from below by a constant latency. A common model for the message time is $t = t_{lat} + K/B$, with latency t_{lat}, message size K and bandwidth B. The message latency impedes the scalability of the redistribution step: with increasing m the size of the messages becomes so small that the time to send the message is dominated by the latency. But the number of messages increases with m.

We have implemented this redistribution operation: first, each process locally collects all the data that it has to send to the other processes in a send buffer for each process. Likewise, a receive buffer is created for each of the processes. Then, non-blocking *MPI_Isend* and *MPI_Irecv* calls are used to transfer the data. In theory, this allows the system to optimize the order of the messages, so that an optimal bandwidth usage can be achieved. Afterwards, the received values have to be copied (or added) to the right place in the underlying data structures. Figure 10 shows the communication time (neglecting any local operations) for a large component grid. We can observe a small decrease in run time from 512 to 1024 processes. But after this point, the run time increases almost linearly. We tried different measures to reduce the number of messages at the price of higher communicated data volume by using collective operations like *MPI_Reduce*, but could not observe a significant reduction in communication time.

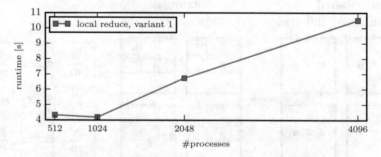

Fig. 10 Communication time being necessary to add a component with $\mathbf{l} = (10, 4, 6, 6, 6)$ to a distributed sparse grid of variant 1 for different numbers of processes

3.2.2 Variant 2: Communication-Free Local Reduction of Uniformly Parallelized Component Grids

For the special case where all the component grids share the same domain decomposition, this variant of the distributed sparse grid can be used to add up the distributed full grids in the local reduction operation without any communication. Here, each hierarchical subspace of the distributed sparse grid is geometrically decomposed and distributed over the processes in the same way as the component grids (see Fig. 11 (left)). The assignment of a grid point to a process depends on its coordinate in the simulation domain. This assignment is equal for the distributed sparse grid and for the distributed full grid. This is visualized for a one-dimensional example in Fig. 11 (right): although the component grids have different numbers of grid points, grid points with the same coordinates are always assigned to the same process. This is due to the nested structure of the component grids.

Compared to variant 1 this has the advantage that no redistribution of the hierarchical coefficients is necessary and therefore no data has to be exchanged between processes. When a distributed full grid is added to the distributed sparse grid, each process adds the hierarchical coefficients of its local part of the distributed full grid to its local part of the distributed sparse grid. With a balanced number of grid points per process N/m in the distributed full grid, the time for the local reduction scales with m. The major drawback of this approach is that it works only for the special case of uniformly parallelized component grids. However, uniformly parallelized component grids might not be possible for all applications. In this case only variant 1 remains. Also, the chosen parallelization might not be the optimal one for all component grids. For applications where both variants are possible, we face a trade-off between the computation time for the component grids, and the time for the local reduction step. This will heavily depend on the frequency of the

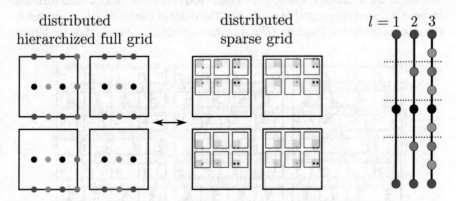

Fig. 11 *Left*: Each process stores for each hierarchical subspace only its local part of the domain. *Right*: One-dimensional component grids of different levels distributed over four processes. Grid points with the same coordinates are always on the same process due to the geometrical domain decomposition (and due to the nested structure of the component grids)

recombination step. If the recombination has to be done after each time step, variant 2 probably is the better choice. If the combination has to be done only once in the end or at least rather rarely, variant 1 might be more desirable.

3.3 Global Reduction of the Combination Solution

In this work we will discuss the global reduction step for process groups with equal numbers of processes. For a well-scaling application like GENE this is a sensible assumption, because it reduces the complexity of the global reduction. However, for applications with worse scaling behavior, in terms of overall efficiency it might be advantageous to use process groups of different sizes, especially when the recombination step is performed only rarely. However, a detailed discussion of the distributed recombination step for process groups with different sizes is out of scope for this work.

The global reduction step is basically identical for both variants of the distributed sparse grid: An *MPI_Allreduce* operation is performed by all the processes which have the same position (the same local rank) in the process group (see Fig. 12). These are the processes which store the same coefficients of the distributed sparse grid in each process group. The algorithm to do this is rather simple: first, an MPI communicator is created, which contains all processes which have the same local rank in all process groups. This communicator only has to be created once and can then be reused for subsequent recombination steps. Second, each process copies its local partition of the distributed sparse grid into a buffer. The order of coefficients in the buffer is equal on all processes. It can happen that a particular hierarchical subspace does not exist in a process group. In this case corresponding entries in the buffer are filled up with zeros. This procedure is identical for both variants of the distributed sparse grid. Then, each process copies its local partition of the distributed sparse grid into a buffer. The order of coefficients in the buffer is equal on all processes. Next, an *MPI_Allreduce* operation is executed on the buffer.

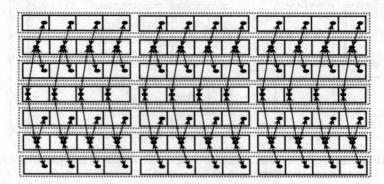

Fig. 12 Communication graph of the global reduction step. The example shows six process groups. Each process group consists of three nodes with 4 processes each

In a last step, each process extracts the coefficients from the buffer and copies them back to its local part of the distributed sparse grid. Now, the combined solution is available in the distributed sparse grid of each process group.

In the following we will analyze the cost of the global reduction step. For m processes per process group, the global reduction step is m-way parallel. It consists of m *MPI_Allreduce* operations that can be performed in parallel and individually of each other. We can estimate the time for each *MPI_Allreduce* by

$$t_{ar} = 2t_l \log(ngroups) + 2\frac{N_{SG}/m}{B/M} \log(ngroups) . \tag{7}$$

Here, t_l is the message latency, *ngroups* is the number of process groups and N_{SG}/m the size of the buffer (the size of the local part of the distributed sparse grid per process). Since M processes per node share the node's network bandwidth B, the effective bandwidth per process is B/M. Thus, the global reduction scales with the process group size, but increases logarithmically with the number of process groups. However, for our large scale experiments with GENE, there will be several thousand processes per process group, but one or two orders of magnitude less process groups. We have already used the same model in [17] to analyze the costs of the recombination step for non-distributed component grids. It is based on the assumption that *MPI_Allreduce* is performed in a reduce and a broadcast step using binomial trees. Although this does not necessarily hold for actual implementations [22], it would only change the communication model by a constant factor.

4   Results

We investigated the scalability of our distributed combination algorithm on up to 180,225 (out of 185,088) cores of Germany's Tier-0/1 supercomputer Hazel Hen (rank 8 on the Top500 list of November 2015). It consists of 7712 dual-socket nodes, which have 24 cores of Intel Xeon E5-2680 v3 (Haswell) and 128 GByte of main memory.

We chose the problem size in the range that we aim to use for future large-scale experiments with GENE, $\mathbf{n} = (14, 6, 6, 8, 7)$ and $\mathbf{l}_{min} = (9, 4, 4, 6, 4)$. This results in 182 component grids. The 76 largest component grids have between 74 and 83 GByte (complex-valued numbers). Our experiments correspond to the recombination of one species. The time for the recombination, as well as the computation time of GENE, multiplies with the number of species. We used a distributed sparse grid of variant 2 and equal parallelization for each component grid. Directly computing a global GENE simulation with the chosen $\mathbf{n}$ with GENE would result in a resolution of $16384 \times 64 \times 64 \times 256 \times 128$. This problem would be 32 times larger than what GENE users already consider to be infeasible (compare the numbers in Sect. 1.2). We used process groups of size *nprocs* = 1024, 2048, 4096 and 8192. The number of process groups *ngroups* ranges between 1 and 128 (only powers of two were used). Furthermore, 22, 44, 88, and 176 process groups were used for the largest experiments with 180,225 processes (one process for the manager). We were not able to perform the experiments with less than 4096 cores, because the necessary memory to store the component grids exceeded the nodes' main memory.

Figure 13 shows the individual times for the hierarchization, the local reduction and the global reduction. Furthermore, it shows the total time of hierarchization, local reduction and global reduction. We do not present results for the scatter

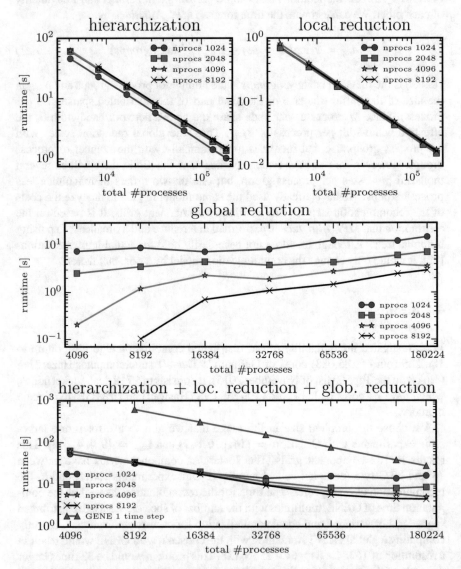

Fig. 13 Timings for the individual steps of the distributed recombination step for different sizes of the process groups. The last plot *(bottom)* shows the total time of hierarchization, local reduction and global reduction. We also included a rough estimate of the computation time for one time step of GENE with process groups of size 4096

step and dehierarchization here, because they behave so similar to their inverse counterparts, reduction and hierarchization, that this would not bring any further insight. The run times presented are the average times per process group. For hierachization and local reduction they are the accumulated times of all component grids assigned to the process group. The abscissa of the plots presents the total number of processes used: *ngroups* × *nprocs* (the manager process is neglected).

The hierarchization scales well with the total number of processes. However, we can observe a small increase in run time when larger process groups are used. The distributed hierarchization consists of two steps, a communication step and a computation step. The time for the latter one does only depend on the number of grid points per process. Thus it scales perfectly with the total number of processes. The time for the commmunication step scales as well (see Fig. 7), but not perfectly. This explains the small increase in run time when larger process groups are used.

The local reduction perfectly scales with the total number of processes. Since it does not require any communication between the processes, the run time only depends on the number of grid points per process. The number of grid points per process halves when the size of the process group doubles. Likewise, when twice as much process groups are used, the number of component grids per group will halve, and thus the number of grid points per process.

The time for the global reduction increases with the number of process groups. Even though we cannot see a strict logarithmic increase of the run time with the number of process groups here, as predicted by Eq. (7), this formula is still sensible as a rough guide. As already discussed in [17], actual implementations of *MPI_Allreduce* might use other algorithms than binomial trees. Also, the actual network bandwidth between two distinct compute nodes depends on many factors, like their position in the network topology and the current overall load on the network. Furthermore, we did not observe a worse than logarithmic increase in run time. For example, with *nprocs* 1024 the run time increased from 4 process groups (4096 processes) to 176 process groups (180,224 processes) by a factor of 2.2, whereas $\log(176)/\log(4) \approx 3.7$. The run time decreases for larger process groups, because with less grid points per process the message size for the *MPI_Allreduce* decreases. The strong increase from one to two process groups for *nprocs* 4096 and *nprocs* 8192 comes from the fact that if only one process group is used, *MPI_Allreduce* essentially is a no-op and so no actual communication happens.

The total time for hierarchization, local reduction and global reduction scales until it is dominated by the time for the global reduction. For small total process counts, the time for the hierarchization dominates. The time for the local reduction is so small in comparison to the time for the hierarchization that it can be neglected. Strong scalability by adding more and more process groups can only be achieved until the point when the global reduction time dominates. This point, and thus the strong scalability limit of the recombination step, can be shifted to higher process counts by using larger process groups. However, enlarging the process groups is only sensible until the application's scalability limit is reached.

In Fig. 13 we included a rough estimate of the average time per process group to advance the GENE simulation on each component grid by one time step.

It corresponds to the computation time per species on process groups of size 4096. For the largest component grids it was not possible to use less than 4096 processes per species, because of main memory limitations. Note that GENE computations use a lot more main memory than what is necessary to store the computational grid. Therefore, for our future experiments we will rather use 8192 or 16384 processes per species for such large component grids. However, a meaningful investigation of the scalability of the global reduction step requires a large enough number of process groups. Thus, we also chose smaller process groups in our experiments. We did not measure all the individual times for all 182 component grids, but instead we measured one component grid of each of the five different grid sizes that we used for the combination as specified by $|\mathbf{l}|_1$, cf. Eq. (1). We estimated the total time under the assumption that all component grids with the same level-sum have the same run time. In general this is not true, because the run time does not only depend on the size of the component grids, but it is also significantly influenced by the actual discretization and parallelization [13]. However, for the purpose of presenting a rough guide how the run time of the actual GENE computations relates to the time of the recombination step, these numbers are accurate enough.

So far it is not yet clear whether it is necessary to recombine the component grids after each time step, or if it would be sufficient to recombine only every 10 or even every 100 time steps. This depends on the application and problem at hand, and investigating it for the different GENE operation modes will be part of our future research. However, the results show that the run time for the new and optimized recombination algorithms is now below the computation time for one time step. So even recombining after each time step would be feasible. Furthermore, this is just a first implementation of the recombination step and there is still optimization potential for all substeps. For the distributed hierarchization it is possible to further reduce the communication effort and to speed up the local computations (see [12]). For the global communication an improved algorithm based on the communication patterns presented in [17] can be used.

5   Conclusion and Future Work

In this work we presented new, scalable algorithms to recombine the distributed component grids. We provide a detailed analysis of the individual steps of the recombination, hierarchization, local reduction and global reduction. Additionally, we present experimental results that show the scalability of a first implementation of the distributed recombination on up to 180,225 cores on the supercomputer *Hazel Hen*. The experiments demonstrate that our recombination algorithms lead to highly parallel and scalable implementations of the remaining global communication for the solution of higher-dimensional PDEs with the sparse grid combination technique. It would even be possible to recombine the partial solutions very frequently if required.

As a next step we plan to perform actual GENE simulations in order to investigate the effect of the recombination (length of the recombination interval) on the error of the combined solution (compared to a the full grid solution or to experimental results, if available). At the time of writing this was not possible, because a parallel algorithm for the adaption of GENE's boundary conditions (which has to be done after each recombination) was not yet realized in our software framework.

We did not perform the large scaling experiments with distributed sparse grids of variant 1, because a scalable implementation for the local reduction/scatter of component grids with non-uniform domain decomposition does not exist yet. The reasons are discussed in Sect. 3.2. Although for many applications this is not a severe restriction, in order to improve the generality of our method, finding a better algorithm for the reduction of component grids with non-uniform domain decomposition will be a topic of future work.

Acknowledgements This work was supported by the German Research Foundation (DFG) through the Priority Program 1648 "Software for Exascale Computing" (SPPEXA).

References

1. SG++ library, http://sgpp.sparsegrids.org/
2. Ali, M.M., Strazdins, P.E., Harding, B., Hegland, M., Larson, J.W.: A fault-tolerant gyrokinetic plasma application using the sparse grid combination technique. In: International Conference on High Performance Computing & Simulation (HPCS), Amsterdam, pp. 499–507. IEEE (2015)
3. Brizard, A., Hahm, T.: Foundations of nonlinear gyrokinetic theory. Rev. Mod. Phys. **79**, 421–468 (2007)
4. Bungartz, H.J., Griebel, M.: Sparse grids. Acta Numer. **13**, 147–269 (2004)
5. Cappello, F., Geist, A., Gropp, W., Kale, S., Kramer, B., Snir, M.: Toward exascale resilience: 2014 update. Supercomput. Front. Innov. **1**(1), 5–28 (2014)
6. Dannert, T.: Gyrokinetische Simulation von Plasmaturbulenz mit gefangenen Teilchen und elektromagnetischen Effekten. Ph.D. thesis, Technische Universität München (2004)
7. Dannert, T., Görler, T., Jenko, F., Merz, F.: Jülich blue gene/p extreme scaling workshop 2009. Technical report, Jülich Supercomputing Center (2010)
8. Doyle, E.J., Kamada, Y., Osborne, T.H., et al.: Chapter 2: plasma confinement and transport. Nucl. Fusion **47**(6), S18 (2007)
9. Görler, T., Lapillonne, X., Brunner, S., Dannert, T., Jenko, F., Merz, F., Told, D.: The global version of the gyrokinetic turbulence code GENE. J. Comput. Phys. **230**(18), 7053–7071 (2011)
10. Griebel, M., Huber, W., Rüde, U., Störtkuhl, T.: The combination technique for parallel sparse-grid-preconditioning or -solution of PDEs on workstation networks. In: Parallel Processing: CONPAR 92 VAPP V. LNCS, vol. 634. Springer, Berlin/New York (1992)
11. Griebel, M., Schneider, M., Zenger, C.: A combination technique for the solution of sparse grid problems. In: de Groen, P., Beauwens, R. (eds.) Iterative Methods in Linear Algebra. IMACS, pp. 263–281. Elsevier/North Holland (1992)
12. Heene, M., Pflüger, D.: Efficient and scalable distributed-memory hierarchization algorithms for the sparse grid combination technique. In: Parallel Computing: On the Road to Exascale. Advances in Parallel Computing, vol. 27. IOS Press, Amsterdam (2016)

13. Heene, M., Kowitz, C., Pflüger, D.: Load balancing for massively parallel computations with the sparse grid combination technique. In: Parallel Computing: Accelerating Computational Science and Engineering (CSE). Advances in Parallel Computing, vol. 25, pp. 574–583. IOS Press, Amsterdam (2014)
14. Hegland, M., Garcke, J., Challis, V.: The combination technique and some generalisations. Linear Algebra Appl. **420**(2–3), 249–275 (2007)
15. Hegland, M., Harding, B., Kowitz, C., Pflüger, D., Strazdins, P.: Recent developments in the theory and application of the sparse grid combination technique. In: Proceedings of the SPPEXA Symposium 2016, Garching. Lecture Notes in Computational Science and Engineering. Springer (2016)
16. Hupp, P., Jacob, R., Heene, M., et al.: Global communication schemes for the sparse grid combination technique. Par. Comput.: Accel. Comput. Sci. Eng. **25**, pp. 564–573 (2014)
17. Hupp, P., Heene, M., Jacob, R., Pflüger, D.: Global communication schemes for the numerical solution of high-dimensional {PDEs}. Parallel Comput. **52**, 78–105 (2016)
18. Kowitz, C., Hegland, M.: The sparse grid combination technique for computing eigenvalues in linear gyrokinetics. Procedia Comput. Sci. **18**(0), 449–458 (2013). 2013 International Conference on Computational Science
19. Parra Hinojosa, A., Kowitz, C., Heene, M., Pflüger, D., Bungartz, H.J.: Towards a fault-tolerant, scalable implementation of GENE. In: Proceedings of ICCE 2014, Nara. Lecture Notes in Computational Science and Engineering. Springer (2015)
20. Parra Hinojosa, A., Harding, B., Hegland, M., Bungartz, H.J.: Handling silent data corruption with the sparse grid combination technique. In: Proceedings of the SPPEXA Symposium 2016, Garching. Lecture Notes in Computational Science and Engineering. Springer (2016)
21. Pflüger, D., Bungartz, H.J., Griebel, M., Jenko, F., et al.: EXAHD: an exa-scalable two-level sparse grid approach for higher-dimensional problems in plasma physics and beyond. In: Euro-Par 2014: parallel processing workshops, Porto. Lecture Notes in Computer Science, vol. 8806, pp. 565–576. Springer International Publishing (2014)
22. Thakur, R., Rabenseifner, R., Gropp, W.: Optimization of collective communication operations in MPICH. Int. J. High Perform. C. **19**, 49–66 (2005)

Handling Silent Data Corruption
with the Sparse Grid Combination Technique

Alfredo Parra Hinojosa, Brendan Harding, Markus Hegland,
and Hans-Joachim Bungartz

Abstract We describe two algorithms to detect and filter silent data corruption (SDC) when solving time-dependent PDEs with the Sparse Grid Combination Technique (SGCT). The SGCT solves a PDE on many regular full grids of different resolutions, which are then combined to obtain a high quality solution. The algorithm can be parallelized and run on large HPC systems. We investigate silent data corruption and show that the SGCT can be used with minor modifications to filter corrupted data and obtain good results. We apply sanity checks before combining the solution fields to make sure that the data is not corrupted. These sanity checks are derived from well-known error bounds of the classical theory of the SGCT and do not rely on checksums or data replication. We apply our algorithms on a 2D advection equation and discuss the main advantages and drawbacks.

1 Introduction

Faults in high-end computing systems are now considered the norm rather than the exception [13]. The more complex these systems become, and the larger the number of components they have, the higher the frequency at which faults occur. Following the terminology in [33], a fault is simply the cause of an error. Errors, in turn, are categorized into three groups: (1) detected and corrected by hardware (DCE), (2) detected but uncorrectable errors (DUE), and (3) silent errors (SE). If an error leads to system failure, it is called *masked*; otherwise it is *unmasked*. We say that a system failed if there is a deviation from the correct service of a system function [2].

The field of fault tolerance explores ways to avoid system failures when faults occur. Different strategies can be followed depending on the type of fault. For example, one might be interested in tolerating the failure of single MPI processes,

A. Parra Hinojosa (✉) • H.-J. Bungartz
Technische Universität München, München, Germany
e-mail: hinojosa@in.tum.de; bungartz@in.tum.de

B. Harding • M. Hegland
Mathematical Sciences Institute, The Australian National University, Canberra, ACT, Australia
e-mail: brendan.harding@anu.edu.au; markus.hegland@anu.edu.au

© Springer International Publishing Switzerland 2016   187
H.-J. Bungartz et al. (eds.), *Software for Exascale Computing – SPPEXA*
2013-2015, Lecture Notes in Computational Science and Engineering 113,
DOI 10.1007/978-3-319-40528-5_9

since one process failure can cause the whole application to crash, and new parallel libraries have been developed to handle these issues [5]. Another option is to use Checkpoint/Restart (C/R) algorithms, where the state of the simulation is stored to memory and retrieved in case of failure. The simulation is then restarted from the last complete checkpoint. Alternatively, developers could make replicas of certain critical processes as backups in case one of them fails [12].

These algorithms are usually applied when errors trigger a signal and thus can be easily detected. But we might run into problems if the errors don't trigger any signal. This is the case for *silent data corruption* (SDC), a common type of unmasked silent errors. SDC arises mainly in the form of undetected errors in arithmetic operations (most prominently as bit flips) and memory corruption [33]. Although SDC is expected to occur less often than detectable errors (such as hardware failure), one single occurrence of SDC could lead to entirely incorrect results [10]. The frequency at which silent errors occur has not been quantified rigorously, but evidence suggests that they occur frequently enough to be taken seriously [33].

Previous research has focused on algorithms that deal with detectable errors when using the SGCT [20]. We now want to understand the effect of SDC on the SGCT when solving PDEs. Elliott et al. [11] have outlined a methodology to model and simulate SDC, and they have described guidelines to design SDC-resilient algorithms. We adopt their recommendations in this paper, and we now briefly describe their main ideas.

1.1 Understanding Silent Data Corruption

Many algorithm designers start by assuming that SDC will occur exclusively in the form of bit flips. For this reason, they have chosen to simulate SDC by randomly injecting bit flips into an existing application, and then attempting to detect and correct wrong data. But this can only tell us how the application behaves in average, and one might fail to simulate the worst-case scenarios. Additionally, the exact causes of SDC in existing and future parallel systems are still poorly understood. For this reason we should avoid making assumptions about the exact causes of SDC. This lack of certainty does not mean that we should not attempt to simulate SDC (and, if possible, overcome it). On the contrary, by making no assumptions about the exact origins and types of SDC, we can focus on what really matters in terms of algorithm design: numerical errors. A robust algorithm should be able to handle numerical errors of arbitrary magnitude in the data without any knowledge of the specific sources of the error. In this way, the problem can be posed purely in terms of numerical analysis and error bounds.

But how does one actually design robust algorithms? There are several things to keep in mind. For instance, it is important to determine in which parts of the algorithm we cannot afford faults to occur, and in which parts we can relax this condition. This is called *selective reliability* [6]. It is also useful to identify *invariants* in a numerical algorithm. Energy conservation is a typical example, as well as requiring a set of vectors to remain orthogonal. These invariants can be good places to start when searching for anomalous data. Furthermore, algorithm experts should try to develop cheap sanity checks to bound or exclude wrong results. Our research is largely based on this last recommendation. Finally, SDC can cause the control flow of the algorithm to deviate from its normal behavior. Although it is difficult to predict what this would mean for a specific application code, one can still turn to selective reliability to make sure that vulnerable sections of the code are dealt with properly by specifying conditions of correctness. We do not address problems in control flow explicitly in this work, but we do mention briefly how our algorithms could encompass this type of faults.

Many authors opt for a much more elaborate (and expensive) methodology based on checksums. (See [33], Sect. 5.4.2 for an extensive list of examples.) Implementing checksums even for a simple algorithm can prove a very difficult task. A good example is the self-healing, fault-tolerant preconditioned CG algorithm described in [8]. To perform the checksums, the authors require local diskless checkpointing, additional checkpoint processes, and a fault-tolerant MPI implementation. The programming effort and computational costs are substantial. In many cases checksums cannot be applied at all. Data replication can be sometimes useful (see [9]), but it also comes at a cost. These experiences motivate the search for new algorithmic, numerics-based solutions.

1.2 Statement of the Problem

We now want to translate these ideas into an SDC-resilient version of the Sparse Grid Combination Technique (SGCT) algorithm, which we describe in detail in the next section. The SGCT is a powerful algorithm that has been used to solve a wide variety of problems, from option pricing [31] and machine learning [14] to plasma physics [28] and quantum mechanics [16]. Our focus will be high-dimensional, time-dependent PDEs. Full grids with high discretization resolution are usually too computationally expensive, especially in higher dimensions. The SGCT solves the original PDE on different coarse, anisotropic full grids. Their coarseness makes them computationally cheap. The solutions on these coarse grids are then combined properly to approximate the full grid solution in an extrapolation-like manner. (We will see in Sect. 2.1 what it means to combine grids of different resolutions.) Our main concern is the following: the solution on one (or more) of the coarse grids might be wrong due to SDC, which can cause the final combined solution to be wrong as well. We therefore want to implement cheap sanity checks to make sure that wrong solutions are filtered and not considered for the combination. In a sense, the fact that the SGCT solves the same PDE on different grids means that it inherently shows data replication, and it is precisely this fact that we will exploit. But before continuing our discussion of SDC we take a small detour to recall the theory of sparse grids, and we describe the SGCT in detail.

2   Basics of Sparse Grids

Let us first introduce some notation. To discretize the unit interval $[0, 1]$ we use a one-dimensional grid Ω_l with $2^l - 1$ inner points and one point on each boundary ($2^l + 1$ points in total). This grid has mesh size $h_l := 2^{-l}$ and grid points $x_{l,j} := j \cdot h_l$ for $0 \leq j \leq 2^l$, with $l \in \mathbb{N} = \{1, 2, \ldots\}$.

In d dimensions we use underlined letters to denote multi-indices, $\mathbf{l} = (l_1, \ldots, l_d) \in \mathbb{N}^d$, and we discretize the d-unit cube using a d-dimensional full grid, $\Omega_{\mathbf{l}} := \Omega_{l_1} \times \cdots \times \Omega_{l_d}$. This grid has mesh sizes

$$h_{\mathbf{l}} := (h_{l_1}, \ldots, h_{l_d}) := 2^{-\mathbf{l}} \tag{1}$$

and grid points

$$x_{\mathbf{l},\mathbf{j}} := (x_{l_1 j_1}, \ldots, x_{l_d j_d}) := \mathbf{j} \cdot h_{\mathbf{l}} \quad \text{for} \quad \mathbf{0} \leq \mathbf{j} \leq 2^{\mathbf{l}} . \tag{2}$$

Comparisons between multi-indices are done componentwise: two multi-indices $\mathbf{i}$ and $\mathbf{j}$ satisfy $\mathbf{i} \leq \mathbf{j}$ if $i_k \leq j_k$ for all $k \in \{1, \ldots, d\}$. (The same applies for similar operators.) We will also use discrete l_p-norms $|\cdot|_p$ for multi-indices. For example, $|\mathbf{l}|_1 := l_1 + \cdots + l_d$. Additionally, the operation $\mathbf{i} \wedge \mathbf{j}$ denotes the componentwise minimum of $\mathbf{i}$ and $\mathbf{j}$, i.e., $\mathbf{i} \wedge \mathbf{j} := (\min\{i_1, j_1\}, \ldots, \min\{i_d, j_d\})$. Finally, if $\mathscr{I}$ is a set of multi-indices, we define the *downset* of $\mathscr{I}$ as $\mathscr{I}_\downarrow := \{\mathbf{l} \in \mathbb{N}^d : \exists \mathbf{k} \in \mathscr{I} \text{ s.t. } \mathbf{l} \leq \mathbf{k}\}$. The downset $\mathscr{I}_\downarrow$ includes all multi-indices smaller or equal to all multi-indices in $\mathscr{I}$.

Suppose $u(\mathbf{x}) \in V \subset C([0, 1]^d)$ is the exact solution of a d-dimensional PDE. A numerical approximation of u will be denoted $u_{\mathbf{i}}(\mathbf{x}) \in V_{\mathbf{i}} \subset V$, where $V_{\mathbf{i}} = \bigotimes_{k=1}^{d} V_{i_k}$ is the space of piecewise d-linear functions defined on a grid $\Omega_{\mathbf{i}}$ [15],

$$V_{\mathbf{i}} := \text{span}\{\phi_{\mathbf{i},\mathbf{j}} : \mathbf{0} \leq \mathbf{j} \leq 2^{\mathbf{i}}\} . \tag{3}$$

The d-dimensional hat functions $\phi_{\mathbf{i},\mathbf{j}}$ are the tensor product of one-dimensional hat functions,

$$\phi_{\mathbf{i},\mathbf{j}}(\mathbf{x}) := \prod_{k=1}^{d} \phi_{i_k j_k}(x_k) , \tag{4}$$

with

$$\phi_{i,j}(x) := \max(1 - |2^i x - j|, 0) . \tag{5}$$

As a result, the interpolation of $u_{\mathbf{i}}(\mathbf{x})$ on grid $\Omega_{\mathbf{i}}$ can be written as

$$u_{\mathbf{i}}(\mathbf{x}) = \sum_{\mathbf{0} \leq \mathbf{j} \leq 2^{\mathbf{i}}} u_{\mathbf{i},\mathbf{j}} \phi_{\mathbf{i},\mathbf{j}}(\mathbf{x}) . \tag{6}$$

The coefficients $u_{\mathbf{i},\mathbf{j}} \in \mathbb{R}$ are simply the height of the hat functions $\phi_{\mathbf{i},\mathbf{j}}$ (see Fig. 1, left). We call (6) the *nodal representation* of $u_{\mathbf{i}}(\mathbf{x})$, and $u_{\mathbf{i},\mathbf{j}}$ are the *nodal coefficients*.

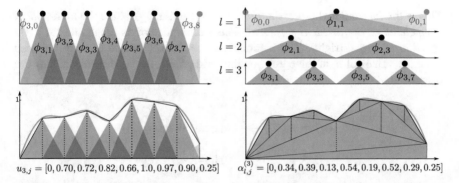

$u_{3,j} = [0, 0.70, 0.72, 0.82, 0.66, 1.0, 0.97, 0.90, 0.25]$ $\alpha_{l,j}^{(3)} = [0, 0.34, 0.39, 0.13, 0.54, 0.19, 0.52, 0.29, 0.25]$

Fig. 1 Two different bases to interpolate a one-dimensional function using discretization level $i = 3$. *Left*: nodal representation. We store the values of $u_{i,j}$, which correspond to the height of the nodal hat functions. *Right*: hierarchical basis. We store the hierarchical coefficients $\alpha_{l,j}^{(i)}$, which represent the increments w.r.t. the previous level $l - 1$. Their magnitude decreases as the level l increases

Apart from V_i we will also consider *hierarchical spaces* W_l defined as

$$W_l := \operatorname{span}\left\{\phi_{l,j}(\mathbf{x}) : \mathbf{j} \in \mathscr{I}_l\right\} , \tag{7}$$

where the index set $\mathscr{I}_l$ is given by

$$\mathscr{I}_l := \left\{\mathbf{j} : 1 \leq j_k \leq 2^{l_k} - 1, j_k \text{ odd}, 1 \leq k \leq d\right\} . \tag{8}$$

A hierarchical space W_i can be defined as the space of all functions $u_i \in V_i$ such that u_i is zero on all grid points in the set $\bigcup_{l < i} \Omega_l$ [20]. The space W_1 is treated separately, since it is endowed with two additional basis functions $\phi_{0,0}$ and $\phi_{0,1}$ to include the boundary conditions, as illustrated in Fig. 1.[1] Using hierarchical spaces allows us to decompose a space V_i as

$$V_i = \bigoplus_{l \leq i} W_l . \tag{9}$$

Equations (7), (8), and (9) tell us that each $u_i \in V_i$ can be written alternatively as

$$u_i(\mathbf{x}) = \sum_{l \leq i} h_l(\mathbf{x}), \quad h_l(\mathbf{x}) \in W_l \tag{10}$$

$$= \sum_{l \leq i} \sum_{\mathbf{j} \in \mathscr{I}_l} \alpha_{l,j}^{(i)} \phi_{l,j}(\mathbf{x}) . \tag{11}$$

---

[1]For a detailed discussion on the boundary treatment, see [30].

This is the *hierarchical representation* of $u_i(\mathbf{x})$, and $\alpha_{\mathbf{l},\mathbf{j}}^{(i)} \in \mathbb{R}$ are the *hierarchical coefficients* or *hierarchical surpluses*. This decomposition is illustrated for a 1D function in Fig. 1 (right). The hierarchical coefficients can be directly obtained from the values of u_i at the corresponding grid points. In one dimension, they are calculated as

$$\alpha_{l,j}^{(i)} = u_i(x_{l,j}) - \frac{1}{2}(u_i(x_{l,j-1}) + u_i(x_{l,j+1}))$$
$$= \left[-\frac{1}{2} \quad 1 \quad -\frac{1}{2}\right]_{l,j} u_i(x_{l,j}) . \tag{12}$$

This operation is called *hierarchization*, and can be extended to d dimensions using the one-dimensional stencil above,

$$\alpha_{\mathbf{l},\mathbf{j}}^{(i)} = \left(\prod_{k=1}^{d} \left[-\frac{1}{2} \quad 1 \quad -\frac{1}{2}\right]_{l_k j_k}\right) u_i(x_{\mathbf{l},\mathbf{j}}) . \tag{13}$$

This is simply a transformation from the nodal to the hierarchical basis. The inverse operation (calculating the nodal coefficients from the hierarchical coefficients) is called *dehierarchization*.

If we discretize a problem on a uniform d-dimensional full grid $\Omega_\mathbf{n}$ (with mesh size $h_n = 2^{-n}$ in every dimension), this grid will have $\mathcal{O}(h_n^{-d}) = \mathcal{O}(2^{nd})$ grid points. This exponential dependence on n and d makes running algorithms on such grids infeasible, a fact commonly referred to as the *curse of dimensionality*. *Sparse grids* aim to alleviate the curse of dimensionality via a hierarchical approach [7, 15]. The classical sparse grid space $V_n^s \subset V_n$ is defined as

$$V_n^s := \bigoplus_{|\mathbf{l}|_1 \leq n+d-1} W_\mathbf{l} , \tag{14}$$

which we have illustrated in Fig. 2 for $d = 2$ and $n = 4$. A sparse grid has $\mathcal{O}(h_n^{-1}(\log h_n^{-1})^{d-1})$ points, which represents a dramatic reduction from the $\mathcal{O}(h_n^{-d})$

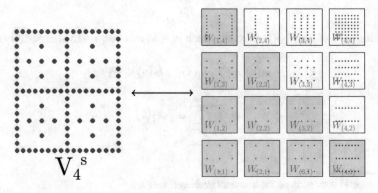

Fig. 2 A sparse grid of level 4 and the hierarchical subspaces that compose it

discretization points required by a full grid of the same level n. However, for functions whose mixed second derivatives are bounded, the interpolation error on a sparse grid is in $\mathcal{O}(h_n^2(\log h_n^{-1})^{d-1})$, only slightly larger than on a full grid, which is in $\mathcal{O}(h_n^2)$ [30].

2.1 The Sparse Grid Combination Technique

Sparse grids allow us to reduce the number of degrees of freedom in a discrete problem without sacrificing much in terms of accuracy. However, it is in general difficult to discretize a problem on a sparse grid. Luckily, there exists a variant of sparse grids, the *Sparse Grid Combination Technique* (SGCT)[17, 18], which can overcome this problem. We illustrate the SGCT using a simple time-dependent PDE, the linear advection equation in two dimensions plus time given by

$$\frac{\partial u}{\partial t} + c_x\frac{\partial u}{\partial x} + c_y\frac{\partial u}{\partial y} = 0 , \tag{15}$$

in the unit square $(x, y) \in [0, 1]^2$ with initial condition $u(x, y, t = 0) = \sin(2\pi x)\sin(2\pi y)$ and periodic boundary conditions in x and y. The velocities c_x and c_y are real positive constants. The analytical solution of (15) is $u(x, y, t) = \sin(2\pi(x - c_x t))\sin(2\pi(y - c_y t))$.

Suppose we use an explicit discretization scheme in time, such as Lax-Wendroff [34]. In Fig. 3 (left) we have plotted the solution of (15) at time $t = 0.5$ with $c_x = c_y = 0.5$ using the Lax-Wendroff scheme on a grid $\Omega_{\mathbf{n}}$ of discretization level $\mathbf{n} = (4, 4)$ (i.e. with $(2^4 + 1) \times (2^4 + 1)$ points). On the right, we have done the same but on five coarser grids of different discretization level, each of which has four times fewer discretization points than the grid $\Omega_{(4,4)}$. The idea behind the SGCT is to *combine* those five grids with weights $+1$ and -1 (as indicated in the

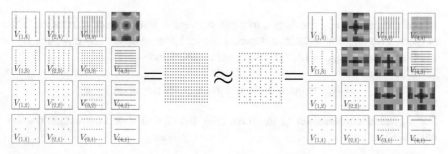

Fig. 3 A classical SGCT to solve the 2D advection equation using $\mathbf{n} = (4, 4)$ and $\tau = 2$ ($\mathbf{i}_{\min} = (2, 2)$). The combination of the five resulting grids results in a sparse grid approximation of the full grid $\Omega_{(4,4)}$

figure) to obtain an approximation of the solution on grid $\Omega_{(4,4)}$. The union of these grids results in a sparse grid, depicted on the middle right.

The grids are usually combined either using interpolation or hierarchization. The former means that each combination grid is interpolated to the full grid space (in our example, $\Omega_{(4,4)}$) and the grids are combined together in this space. The latter requires us to transform each solution into the hierarchical basis (using Eq. (13)), after which the hierarchical coefficients $\alpha_{\mathbf{l},\mathbf{j}}^{(i)}$ can be added directly in the sparse grid space [23, 25].

We can now write down the classical definition of the SGCT. We approximate the full grid solution $u_{\mathbf{n}}$ by a function $u_{\mathbf{n}}^{(c)}$ as follows:

$$u_{\mathbf{n}} \approx u_{\mathbf{n}}^{(c)} = \sum_{\mathbf{i} \in \mathscr{I}} c_{\mathbf{i}} u_{\mathbf{i}} . \tag{16}$$

The weights $c_{\mathbf{i}} \in \mathbb{R}$ are called *combination coefficients*, and each $u_{\mathbf{i}}$ is a numerical approximation of u on a coarse, anisotropic full grid $\Omega_{\mathbf{i}}$. We call the $u_{\mathbf{i}}$ *combination solutions*, the grids $\Omega_{\mathbf{i}}$ *combination grids*, $u_{\mathbf{n}}^{(c)}$ the *combined solution*, and its corresponding grid $\Omega_{\mathbf{n}}^{(c)}$ the *combined grid* (which is a sparse grid). $\mathscr{I}$ is a set of multi-indices.

The approximation quality of the combined solution (16) strongly depends on the coefficients $c_{\mathbf{i}}$ and the set $\mathscr{I}$, since only certain choices yield reasonable results. One such choice is the *classical combination technique* given by

$$u_{\mathbf{n}}^{(c)} = \underbrace{\sum_{q=0}^{d-1} (-1)^q \binom{d-1}{q}}_{=c_{\mathbf{i}}} \sum_{\mathbf{i} \in \mathscr{I}_{n,q,\tau}} u_{\mathbf{i}} . \tag{17}$$

Here, the index set $\mathscr{I}_{\mathbf{n},q,\tau}$ is defined as

$$\mathscr{I}_{\mathbf{n},q,\tau} := \{\mathbf{i} \in \mathbb{N}^d : |\mathbf{i}|_1 = |\mathbf{i}_{\min}|_1 + \tau - q \quad \text{and} \quad \mathbf{i}_{\min} = \mathbf{n} - \tau \cdot \mathbf{1} > \mathbf{0}\} , \tag{18}$$

where $\tau \in \mathbb{N}$, and $\mathbf{i}_{\min}$ specifies a minimal resolution level in each dimension. The classical combination technique depicted in Fig. 3 was generated by choosing $\mathbf{n} = (4, 4)$ and $\tau = 2$ ($\mathbf{i}_{\min} = (2, 2)$), giving the combination

$$u_{\mathbf{n}}^{(c)} = u_{(2,4)} + u_{(3,3)} + u_{(4,2)} - u_{(2,3)} - u_{(3,2)} .$$

For a general combination of the form (16), the combination coefficients can be calculated as

$$c_{\mathbf{i}} = \sum_{\mathbf{i} \leq \mathbf{j} \leq \mathbf{i}+\mathbf{1}} (-1)^{|\mathbf{j}-\mathbf{i}|} \chi_{\mathscr{I}}(\mathbf{j}) , \tag{19}$$

where $\chi_{\mathscr{I}}$ is the indicator function of set $\mathscr{I}$ [19].

The main advantage of the SGCT is that it approximates a full grid solution very well by using a combination of solutions on coarse anisotropic grids. The combination of these grids results in a sparse grid, but we avoid discretizing our problem directly on the sparse grid, which is cumbersome and requires complex data structures. However, by using the combination technique we have an extra storage overhead compared to a single sparse grid, since there is some data redundancy among the combination grids. The storage requirements of the combination technique are of order $\mathcal{O}(d(\log h_n^{-1})^{d-1}) \times \mathcal{O}(h_n^{-1})$ [30]. This data redundancy is the key feature of the SGCT that will allow us to deal with data corruption.

Finally, it is important to mention that the combination coefficients c_i and the index set $\mathscr{I}$ can be chosen in various different ways, and the resulting combinations vary in approximation quality. This is the underlying idea behind dimension-adaptive sparse grids [24], and their construction is inherently fault tolerant.

3 The SGCT in Parallel and Fault Tolerance with the Combination Technique

The SGCT offers two levels of parallelism. First, since we solve the same PDE on different grids, each solver call is independent of the rest. Second, on each grid one can use domain decomposition (or other parallelization techniques). But if the PDE is time-dependent we have to combine the solutions every certain number of time steps to avoid divergence, which requires communication. The fraction of time spent in the solver and in the communication steps depends on how often one combines the grids. If the PDE is not time-dependent we combine only once at the end. Algorithm 2 summarizes the main components of a parallel implementation of the classical SGCT. It can be implemented using a master/slave scheme. The master distributes the work and coordinates the combination of the grids. The slaves solve the PDE on the different grids and communicate the results to the master. Most of

---

Algorithm 2 Classical SGCT in Parallel

1: **input:** A function SOLVER; maximum resolution $\mathbf{n}$; parameter τ; time steps per combination N_t
2: **output:** Combined solution $u_n^{(c)}$
3: Generate index set $\mathscr{I}_{n,q,\tau}$ ▷ Eq. (18)
4: Calculate combination coefficients c_i ▷ Eq. (19)
5: **for** $i \in \mathscr{I}_{n,q,\tau}$ **do**
6:      $u_i \leftarrow u(\mathbf{x}, t = 0)$ ▷ Set initial conditions by sampling
7: **while** not converged **do**
8:      **for** $i \in \mathscr{I}_{n,q,\tau}$ **do in parallel**
9:          $u_i \leftarrow$ SOLVER(u_i, N_t) ▷ Solve the PDE on grid Ω_i (N_t time steps)
10:          $u_i \leftarrow$ HIERARCHIZE(u_i) ▷ Transform to hier. basis, Eq. (13)
11:      $u_n^{(c)} \leftarrow$ REDUCE$(c_i \, u_i)$ ▷ Combined solution (in the hier. basis)
12:      $u_n^{(c)} \leftarrow$ DEHIERARCHIZE$(u_n^{(c)})$ ▷ Transform back to nodal basis
13: **for** $i \in \mathscr{I}_{n,q,\tau}$ **do**
14:      $u_i \leftarrow$ SCATTER$(u_n^{(c)})$ ▷ Sample each u_i from new $u_n^{(c)}$

---

the time and computational resources are spent on the calls to the actual PDE solver, line 9.

We are currently working on a massively parallel implementation of the SGCT [22]. It is a $C++$ framework that can call existing PDE solvers (e.g. *DUNE* [3]) and apply the SGCT around them. To develop such an environment, three major issues have to be carefully taken into consideration:

1. *Load balancing.* The work load (calculating all combination solutions u_i, line 9) has to be distributed properly among the computing nodes. The time to solution of each u_i depends on the number of unknowns and the anisotropy of each grid Ω_i. This problem has been studied in [21].
2. *Communication.* After performing N_t time steps, the different u_i have to be combined, which requires communication (line 11). The combined solution $u_n^{(c)}$ is used as initial condition for the next N_t time steps for all combination solutions u_i (line 14), and this also requires communication. Efficient communication patterns for the SGCT have been studied in detail in [26]. The problem of determining how often the grids should be combined is still under investigation, since the frequency depends on the specific PDE.
3. *Fault tolerance.* In light of increasing hardware and software faults, the *Fault Tolerant Combination Technique* (FTCT) has been developed [20]. It has been applied to plasma physics simulations and proved to scale well when *hard faults* occur [1, 29]. This is the area where our group contributes to the $C++$ framework.

All three points raise interesting algorithmic questions, but since the third plays a central role in our discussion of SDC, we should add a few words about it. In Fig. 4 (left) we depict a classical SGCT with $\mathbf{n} = (5, 5)$ and $\tau = 3$ ($\mathbf{i}_{min} = (2, 2)$). Suppose the system encounters a fault during the call to solver. As a result, one or more combination solutions u_i will be lost. In Fig. 4 (left) we have assumed that solution $u_{(4,3)}$ has been lost due to a fault. Instead of recomputing this lost solution, the FTCT attempts to find alternative ways of combining the successfully calculated solutions, excluding the solutions lost due to faults. The possible alternative combinations are almost as good as the original combination. In Fig. 4 (right) we see an alternative combination that excludes solution $u_{(4,3)}$, using instead solution $u_{(3,2)}$. But notice that $u_{(3,2)}$ was not part of the original set of solutions. The main idea of the FTCT is to compute some extra solutions beforehand (such as $u_{(3,2)}$) and use them only in case of faults. This results in a small extra overhead, but it has been shown to scale [20]. In the original SGCT we solve the PDE on grids Ω_i with index $\mathbf{i} \in \mathscr{I}_{\mathbf{n},q,\tau}$ for $q = 0, \ldots, d-1$, Eq. (17). The FTCT extends this set to include the indices that result from setting $q = d, d+1$. We call this set $\mathscr{I}_{\mathbf{n},q,\tau}^{\text{ext}}$. The combination coefficients c_i that correspond to these extra solutions are set to zero if no faults occur and can become nonzero if faults occur.

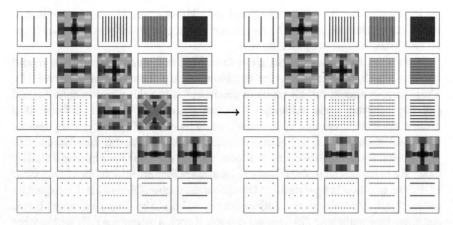

Fig. 4 Example of FTCT with $\mathbf{n} = (5, 5)$ and $\tau = 3$. The original index set $\mathscr{I}_{(5,5),q,3}$ is extended with the additional sets $\mathscr{I}_{(5,5),2,3} = \{(2, 3), (3, 2)\}$ and $\mathscr{I}_{(5,5),3,3} = \{(2, 2)\}$, and the additional solutions are used only in case of faults

3.1 SDC and the Combination Technique

If SDC occurs at some stage of Algorithm 2, it is most likely to happen during the call to the solver, line 9, which is where most time is spent. This means that the solver could return a wrong answer, and this would taint the combined solution $u_{\mathbf{n}}^{(c)}$ during the combination (reduce) step in line 11. Since the combination step is a linear operation, the error introduced in the combined solution would be of the same magnitude of the affected combination solution. Additionally, if the convergence criterion in line 7 has not been met, the scatter step would propagate the spurious data to all other combination solutions, potentially ruining the whole simulation.

To simulate an occurrence of SDC we follow an approach similar to [10]. Let us first assume that SDC only affects one combination solution $u_{\mathbf{i}}$, and that only one of its values $u_{\mathbf{i}}(x_{\mathbf{i},\mathbf{j}})$ is altered in one of the following ways:

1. $\tilde{u}_{\mathbf{i}}(x_{\mathbf{i},\mathbf{j}}) = u_{\mathbf{i}}(x_{\mathbf{i},\mathbf{j}}) \times 10^{+5}$ (very large)[2]
2. $\tilde{u}_{\mathbf{i}}(x_{\mathbf{i},\mathbf{j}}) = u_{\mathbf{i}}(x_{\mathbf{i},\mathbf{j}}) \times 10^{-0.5}$ (slightly smaller)
3. $\tilde{u}_{\mathbf{i}}(x_{\mathbf{i},\mathbf{j}}) = u_{\mathbf{i}}(x_{\mathbf{i},\mathbf{j}}) \times 10^{-300}$ (very small)

This fault injection is performed only once throughout the simulation,[3] at a given time step. Altering only one value of one combination solution is a worst-case scenario, since a solution with many wrong values should be easier to detect. If the wrong solution is not detected, the wrong data would propagate to other grids during

---

[2] The authors in [10] use a factor of 10^{+150} to cover all possible orders of magnitude, but we choose 10^{+5} simply to keep the axes of our error plots visible. The results are equally valid for 10^{+150}.
[3] The assumption that SDC occurs only once in the simulation is explained in [10].

the scatter step. In our tests we also simulated additive errors of various magnitudes, as well as random noise, and the results did not offer new insights. We thus focus on multiplicative errors in what follows.

If we want to make sure that all $u_\mathbf{i}$ have been computed correctly, we should introduce *sanity checks* before the combination step (between lines 10 and 11). These checks should not be problem-dependent, since the function *solver* could call any arbitrary code. Although it is not possible in general to know if a given combination solution $u_\mathbf{i}$ is correct, *we have many of them* (typically tens or hundreds), each with a different discretization resolution. We can therefore use this redundant information to determine if one or more solutions have been affected by SDC. We know that all $u_\mathbf{i}$ should look similar, since they are all solutions of the same PDE. The question is how similar? Or equivalently, how *different* can we expect an arbitrary pair of combination solutions (say u_s and u_t) to be from each other? If two solutions look somehow different we can ask if such a difference falls within what is theoretically expected or not. The theory of the SGCT provides a possible answer to this question.

Early studies of the SGCT show that convergence can be guaranteed if each $u_\mathbf{i}$ satisfies the *error splitting assumption* (ESA) [18], which for arbitrary dimensions can be written as [19]

$$u - u_\mathbf{i} = \sum_{k=1}^{d} \sum_{\substack{\{e_1,\dots,e_k\} \\ \subset \{1,\dots,d\}}} C_{e_1,\dots,e_k}(\mathbf{x}, h_{i_{e_1}}, \dots, h_{i_{e_k}}) h_{i_{e_1}}^p \cdots h_{i_{e_k}}^p , \tag{20}$$

where $p \in \mathbb{N}$ and each function $C_{e_1,\dots,e_k}(\mathbf{x}, h_{i_{e_1}}, \dots, h_{i_{e_k}})$ depends on the coordinates $\mathbf{x}$ and on the different mesh sizes $h_\mathbf{i}$. Additionally, for each $\{e_1, \dots, e_k\} \subset \{1, \dots, d\}$ one has $|C_{e_1,\dots,e_k}(\mathbf{x}, h_{i_{e_1}}, \dots, h_{i_{e_k}})| \leq \kappa_{e_1,\dots,e_k}(\mathbf{x})$, and all $\kappa_{e_1,\dots,e_k}$ are bounded by $\kappa_{e_1,\dots,e_k}(\mathbf{x}) \leq \kappa(\mathbf{x})$. Equation (20) is a *pointwise* relation, i.e. it must hold for all points $\mathbf{x}$ independently, which can be seen by the explicit dependence of each function $C_{e_1,\dots,e_k}$ on $\mathbf{x}$.

In one dimension ($d = 1$), the ESA is simply

$$u - u_i = C_1(x_1, h_i) h_i^p, \quad |C_1(x_1, h_i)| \leq \kappa_1(x_1) . \tag{21}$$

From (12) it follows that the hierarchical coefficients also satisfy the ESA

$$\alpha_{l,j} - \alpha_{l,j}^{(i)} = D_1(x_{l,j}, h_i) h_i^p, \quad |D_1(x_{l,j}, h_i)| \leq 2\kappa_1(x_{l,j}) , \tag{22}$$

where $\alpha_{l,j}$ is the exact hierarchical coefficient at point $x_{l,j}$.

Similarly, in two dimensions we have

$$u - u_\mathbf{i} = C_1(x_1, x_2, h_{i_1}) h_{i_1}^p + C_2(x_1, x_2, h_{i_2}) h_{i_2}^p + C_{1,2}(x_1, x_2, h_{i_1}, h_{i_2}) h_{i_1}^p h_{i_2}^p . \tag{23}$$

There are univariate contributions from each dimension and a cross term that depends on both dimensions. Analogously, for the hierarchical coefficients we have

$$\alpha_{1,j} - \alpha_{1,j}^{(i)} = D_1(x_{1,j}, h_{i_1})h_{i_1}^p + D_2(x_{1,j}, h_{i_2})h_{i_2}^p + D_{1,2}(x_{1,j}, h_{i_1}, h_{i_2})h_{i_1}^p h_{i_2}^p , \qquad (24)$$

with $|D_1| \leq 4\kappa_1(x_{1,j})$, $|D_2| \leq 4\kappa_2(x_{1,j})$, and $|D_{1,2}| \leq 4\kappa_{1,2}(x_{1,j})$. This follows from (13).

Now suppose we take two arbitrary combination solutions u_s and u_t in two dimensions. If these two solutions satisfy the ESA it is straightforward to show that the difference of their corresponding hierarchical coefficients satisfies

$$\begin{aligned} \alpha_{1,j}^{(t)} - \alpha_{1,j}^{(s)} = &D_1(x_{1,j}, h_{t_1})h_{t_1}^p + D_2(x_{1,j}, h_{t_2})h_{t_2}^p + D_{1,2}(x_{1,j}, h_{t_1}, h_{t_2})h_{t_1}^p h_{t_2}^p \\ &- D_1(x_{1,j}, h_{s_1})h_{s_1}^p - D_2(x_{1,j}, h_{s_2})h_{s_2}^p - D_{1,2}(x_{1,j}, h_{s_1}, h_{s_2})h_{s_1}^p h_{s_2}^p . \end{aligned} \qquad (25)$$

Clearly, this equation holds only for the hierarchical spaces common to both grids Ω_s and Ω_t, i.e. for all W_l with $(1,1) \leq l \leq s \wedge t$. Equation (25) tells us that the difference between the hierarchical coefficients of two combination solutions depends mainly on two things: (1) how coarse or fine the grids are, and (2) the distance $|t - s|_1$, which tells us whether grids Ω_s and Ω_t have similar discretization resolutions. The former can be observed by the explicit dependence on h_t and h_s, dominated by the univariate terms. The latter means that if two grids have similar discretizations, the terms in (25) will tend to cancel each other out. Equation (25) is (pointwise) bounded by

$$\beta_{1,j}^{(s,t)} := |\alpha_{1,j}^{(t)} - \alpha_{1,j}^{(s)}| \leq 4 \cdot \kappa(x_{1,j}) \cdot (h_{t_1}^p + h_{s_1}^p + h_{t_2}^p + h_{s_2}^p + h_{t_1}^p h_{t_2}^p + h_{s_1}^p h_{s_2}^p), \quad 1 \leq s \wedge t . \qquad (26)$$

This result can help us to detect SDC, as we soon show.

Our goal is to implement Algorithm 3. A sanity check is done before the combination step (line 11). This is where we attempt to detect and filter wrong

---

Algorithm 3 FTCT with sanity checks for SDC

1: **input:** A function SOLVER; maximum resolution n; parameter τ; time steps per combination N_t
2: **output:** Combined solution $u_n^{(c)}$
3: Generate extended index set $\mathscr{I}_{n,q,\tau}^{ext}$
4: Calculate combination coefficients c_i ▷ Eq. (19)
5: **for** $i \in \mathscr{I}_{n,q,\tau}^{ext}$ **do in parallel**
6:      $u_i \leftarrow u(\mathbf{x}, t = 0)$ ▷ Set initial conditions by sampling
7: **while** not converged **do**
8:      **for** $i \in \mathscr{I}_{n,q,\tau}^{ext}$ **do in parallel**
9:          $u_i \leftarrow$ SOLVER(u_i, N_t)   ▷ Solve the PDE on grid Ω_i (N_t time steps)
10:          $u_i \leftarrow$ HIERARCHIZE(u_i)   ▷ Transform to hier. basis, Eq. (13)
11:      $\{i_{sdc}\} \leftarrow$ SDCSANITYCHECK($\{u_i\}$)   ▷ Check for SDC in all u_i
12:      **if** $\{i_{sdc}\}$ not empty **then**   ▷ Did SDC affect any u_i?
13:          $\{c_i\} \leftarrow$ COMPUTENEWCOEFFS($\{i_{sdc}\}$)   ▷ Update combination coeffs.
14:      $u_n^{(c)} \leftarrow$ REDUCE($c_i u_i$)   ▷ Combined solution (in the hier. basis)
15:      $u_n^{(c)} \leftarrow$ DEHIERARCHIZE($u_n^{(c)}$)   ▷ Transform back to nodal basis
16:      **for** $i \in \mathscr{I}_{n,q,\tau}^{ext}$ **do**
17:          $u_i \leftarrow$ SCATTER($u_n^{(c)}$)   ▷ Sample each u_i from new $u_n^{(c)}$

combination solutions, based on (26). If we are able to detect whether one or more combination solutions are wrong, we can apply the FTCT, treating wrong solutions in the same way as when hard faults occur, finding a new combination of solutions that excludes them, see Fig. 4. We now describe two possible implementations of the function sdcSanityCheck.

3.2 Sanity Check 1: Filtering SDC via Comparison of Pairs of Solutions

The first possible implementation of a simple sanity check is to apply (26) directly: we compare pairs of solutions u_s and u_t in their hierarchical basis and make sure that the bound (26) is fulfilled. If one of the solutions is wrong due to SDC, the quantity $\beta_{\mathbf{l},\mathbf{j}}^{(s,t)}$ will be large and the bound might not be fulfilled, indicating that something is wrong. Unfortunately, the constant $\kappa(x_{\mathbf{l},\mathbf{j}})$ in the bound is in fact a function of space and is problem-dependent. This means that it has to be approximated somehow at all points $x_{\mathbf{l},\mathbf{j}}$, which is not trivial. It is only possible to estimate it once the solutions u_i have been calculated, but this is done assuming that all u_i have been computed correctly. Of course, this assumption does not hold if SDC can occur.

Despite these disadvantages, it is still possible to use bound (26) to detect and filter SDC. First, note that the function $\kappa(x_{\mathbf{l},\mathbf{j}})$ decays exponentially with increasing level $\mathbf{l}$. This is due to the fact that the hierarchical coefficients $\alpha_{\mathbf{l},\mathbf{j}}$ themselves decay exponentially with $\mathbf{l}$ (see Fig. 1, right). For a simple interpolation problem they behave as [7]

$$|\alpha_{\mathbf{l},\mathbf{j}}| \leq 2^{-d} \cdot \left(\frac{2}{3}\right)^{d/2} \cdot 2^{-(3/2)\cdot|\mathbf{l}|_1} \cdot \left\| D^2(u|_{\mathrm{supp}\,\phi_{\mathbf{l},\mathbf{j}}}) \right\|_{L_2}, \tag{27}$$

with $D^2(u) := \frac{\partial^4 u}{\partial x_1^2 \partial x_2^2}$. We can account for this exponential decay by normalizing the quantity $\beta_{\mathbf{l},\mathbf{j}}^{(s,t)}$ as follows:

$$\hat{\beta}_{\mathbf{l},\mathbf{j}}^{(s,t)} := \frac{|\alpha_{\mathbf{l},\mathbf{j}}^{(t)} - \alpha_{\mathbf{l},\mathbf{j}}^{(s)}|}{\min\left\{|\alpha_{\mathbf{l},\mathbf{j}}^{(t)}|, |\alpha_{\mathbf{l},\mathbf{j}}^{(s)}|\right\}} \quad \text{for all } \mathbf{l} \leq \mathbf{s} \wedge \mathbf{t}, \quad \mathbf{0} \leq \mathbf{j} \leq 2^{\mathbf{l}}. \tag{28}$$

If no SDC occurs, $|\alpha_{\mathbf{l},\mathbf{j}}^{(s)}|$ and $|\alpha_{\mathbf{l},\mathbf{j}}^{(t)}|$ should be very similar, so it does not matter which of the two we use for the normalization. But if SDC occurs and their difference is large, dividing by the smaller one will amplify this difference. We can then take the largest $\hat{\beta}_{\mathbf{l},\mathbf{j}}^{(s,t)}$ over all grid points $x_{\mathbf{l},\mathbf{j}}$,

$$\hat{\beta}^{(s,t)} := \max_{\mathbf{l} \leq \mathbf{s} \wedge \mathbf{t}} \max_{\mathbf{j} \in \mathscr{I}_{\mathbf{l}}} \hat{\beta}_{\mathbf{l},\mathbf{j}}^{(s,t)}. \tag{29}$$

Pair	$\hat{\beta}^{(s,t)}$
(7, 9) (7, 8)	**8.71e+04**
(7, 8) (9, 7)	**4.95e+04**
(8, 7) (7, 7)	2.50e-02
(7, 9) (8, 8)	3.67e-02
(7, 7) (8, 8)	4.91e-02
(8, 7) (8, 8)	2.50e-02
(7, 9) (8, 7)	6.03e-02
(7, 9) (7, 7)	3.76e-02
(9, 7) (7, 7)	3.76e-02
(9, 7) (8, 8)	3.67e-02
(7, 8) (7, 7)	**5.01e+04**
(8, 7) (7, 8)	**4.97e+04**
(8, 7) (9, 7)	1.24e-02
(7, 9) (9, 7)	7.24e-02
(7, 8) (8, 8)	**8.68e+04**

Algorithm 4 Sanity check via comparison of solutions

1: **input:** The set of all combination solutions $\{u_i\}$ (in the hierarchical basis)
2: **output:** The set of indices corresponding to the solutions affected by SDC,
3:          $\{i_{sdc}\}$
4: **function** SDCSANITYCHECK($\{u_i\}$)
5:     **for** all pairs (u_s, u_t) with $s, t \in \mathscr{I}_{n,q,\tau}^{ext}$ **do**
6:         Compute $\hat{\beta}^{(s,t)}$ ▷ Eq. (29)
7:         **if** $\hat{\beta}^{(s,t)}$ too large **then**
8:             Mark pair (s, t) as corrupted
9:     From list of corrupted pairs (s, t), determine corrupted grids $\{i_{sdc}\}$
10:
11:     Return$\{i_{sdc}\}$

This quantity simply gives us the largest (normalized) difference between two combination solutions in the hierarchical basis, and it does not decay exponentially in l. Our goal is to keep track of this quantity, expecting it to be small for all pairs of combination solutions. If this is not the case for a specific pair (u_s, u_t) we can conclude that one solution (or both) was not computed correctly during the call to the function `solver`.

Algorithm 4 summarizes this possible implementation of the function `sdc-SanityCheck`. The table on the right illustrates what the function generates for a simple implementation of a 2D FTCT. We solved once again the advection equation (15) for $t = 0.25$ and 129 time steps, with $c_x = c_y = 0.5$. The FTCT parameters used where $\mathbf{n} = (9, 9)$ and $\tau = 2$, which results in six combination grids. We injected SDC of small magnitude (case 2 from the previous section) into one of the combination grids at the very last time step. The table shows a list of all pairs (s, t) and the calculated value of $\hat{\beta}^{(s,t)}$. Some pairs have unusually large values of $\hat{\beta}^{(s,t)}$, shown in boldface. It should be evident that solution $u_{(7,8)}$ has been affected by SDC, being the only one appearing in all five pairs marked as corrupted.

There remains one unanswered question: which values of $\hat{\beta}^{(s,t)}$ should be considered and marked as "too large" (lines 7 and 8)? We discussed that it is difficult to calculate a specific value for the upper bound (26), since $\kappa(x_{l,j})$ is problem-dependent. But we might not need to. We simply need to recognize that some values of $\hat{\beta}^{(s,t)}$ are disproportionately large *compared to the rest*. The values highlighted in the table are indeed clear outliers, and the wrong solution can be identified. The idea of detecting outliers leads us to our second implementation of `sdcSanityCheck`.

3.3   Sanity Check 2: Filtering SDC via Outlier Detection

So far we have used two facts about the combination technique to be able to deal with SDC. First, although we cannot tell in general if one single combination solution has been computed correctly, we know that all combination solutions should look somewhat similar. And second, this similarity can be measured, and the difference between two solutions cannot be arbitrarily large, since it is bounded.

Consider the value of the combined solution $u_n^{(c)}$ at an arbitrary grid point $x_{1,j}$ of the combined grid $\Omega_n^{(c)}$. This value, $u_n^{(c)}(x_{1,j})$, is obtained from the combination of the different solutions u_i that include that grid point (with the appropriate combination coefficients). For every grid point $x_{1,j}$ there is always at least one combination solution u_i that includes it, and at most $|\mathscr{I}|$ such grids. For example, all combination grids u_i include the grid points with $l = 1$ ($x_{1,j}$, corresponding to subspace W_1). In other words, we have $|\mathscr{I}|$ solutions of the PDE at the grid points $x_{1,j}$. Let's call $N_l = 1, \ldots, |\mathscr{I}|$ the number of combination solutions u_i that contain the grid points $x_{1,j}$. We expect the different versions of a point $u_n^{(c)}(x_{1,j})$ to be similar, but with slight variations. This variance is given by

$$\mathrm{Var}[u_n^{(c)}(x_{1,j})] = \frac{1}{N_l} \sum_{l' \geq l} \left(u_{l'}(x_{1,j}) - \mathrm{E}[u_n^{(c)}(x_{1,j})] \right)^2, \quad l, l' \in \mathscr{I}, \tag{30}$$

since a grid point $x_{1,j}$ can be found in all combination solutions $u_{l'}$ with $l' \geq l$ (see Eq. (9)). The mean value of $u_n^{(c)}(x_{1,j})$ over all combination solutions is defined as

$$\mathrm{E}[u_n^{(c)}(x_{1,j})] = \frac{1}{N_l} \sum_{l' \geq l} u_{l'}(x_{1,j}) . \tag{31}$$

Since we have been working in the hierarchical basis, the variance of the value at point $x_{1,j}$ in this basis is given by

$$\mathrm{Var}[\alpha_n^{(c)}(x_{1,j})] = \frac{1}{N_l} \sum_{l' \geq l} \left(\alpha_{1,j}^{(l')} - \mathrm{E}[\alpha_n^{(c)}(x_{1,j})] \right)^2 . \tag{32}$$

This quantity is in fact bounded, due to (26), by

$$\mathrm{Var}[\alpha_n^{(c)}(x_{1,j})] = \frac{1}{2N_l^2} \sum_{s \geq l} \sum_{t \geq l} \left(\alpha_{1,j}^{(s)} - \alpha_{1,j}^{(t)} \right)^2$$

$$\leq \frac{8 \cdot \kappa^2(x_{1,j})}{N_l^2} \sum_{s \geq l} \sum_{\substack{t \geq l \\ t \neq s}} g^2(h_s^p, h_t^p) , \tag{33}$$

with $g(h_s^p, h_t^p) := h_{t_1}^p + h_{s_1}^p + h_{t_2}^p + h_{s_2}^p + h_{t_1}^p h_{t_2}^p + h_{s_1}^p h_{s_2}^p$.

Equation (33) tells us that if we observe how the solution of our PDE at point $x_{l,j}$ varies among the different combination solutions, the variance will not be arbitrarily large. This gives us a second way to perform a sanity check to filter SDC, summarized in Algorithm 5. Using the fact that the variance of each point is bounded, we can apply existing algorithms from robust statistics to find outliers among the different versions of each point. The algorithms used are described in Sect. 4.1. This allows us to filter solutions with unusually large variation and we can be certain that the rest of the solutions has been computed correctly. Just as we did in the first implementation of sdcSanityCheck, we do not need to find a value for the upper bound of the variance (33), but simply to detect unusually large variations.

There is one special case to consider. What happens for subspaces W_l for which we only have one version of the solution ($N_l = 1$)? These are the grid points found in the highest hierarchical subspaces (largest l). In the very unfortunate case where one of these values is wrong *and the fault does not propagate to neighboring points*, we have no other values with which to compare it and thus it cannot be filtered with this approach (nor with the previous). A possible way to detect such errors can be deduced from the fact that the hierarchical coefficients should decrease exponentially in magnitude with increasing level l (Eq. (27)). This means that the coefficients on the highest hierarchical subspace should be very small compared to the rest. We verified this exponential decay for our advection problem as well as for the more complex plasma simulation code *GENE* [27]. One can try to verify that the hierarchical coefficients at the highest level are smaller than those at a lower level, say, m levels lower,

$$|\alpha_{l,j}^{(l)}| < |\alpha_{l-m\cdot e_k,j}^{(l)}| . \tag{34}$$

The direction e_k should be chosen preferably to be the most finely discretized one (i.e. that for which l_k is largest, which will be large in exascale simulations). For our experiments, $m = 3$ worked well. This check could return false positives if lower coefficients are small, so more robust checks could be useful. Note, however, that this check is not even necessary for the combination solutions with the finest discretization, since the combination step would not propagate the fault to other combination solutions. This further reduces the significance of this special case.

Algorithm 5 Sanity check via outlier detection

1: **input**: The set of all combination solutions $\{u_i\}$ (in the hierarchical basis)
2: **output**: The set of indices corresponding to the solutions affected by SDC, $\{i_{sdc}\}$
3: **function** SDCSANITYCHECK($\{u_i\}$)
4:     **for** all grid points $x_{l,j}$ in $\Omega_n^{(c)}$ **do in parallel**
5:         $\alpha[l'] \leftarrow$ GATHER($\alpha_{l,j}^{(l')}$) for all $l' \geq l$
6:         **if** any OUTLIER_TEST($\alpha[l']$) **then**
7:             Add outlier l' to set of corrupted indices $\{i_{sdc}\}$
8:     **return** $\{i_{sdc}\}$

---

4   Numerical Tests

4.1   Experimental Setup

We implemented Algorithm 3 with both types of sanity checks (Algorithms 4 and 5) in Python for our 2D advection equation (15) with $c_x = c_y = 1$. Despite this being a toy problem, the algorithms presented are general enough to be applied to any PDE solver for which the SGCT converges. In other words, if a PDE can be solved using the SGCT, our sanity checks are guaranteed to work, thanks to the error splitting assumption (20). The function `solver` is a Lax-Wendroff solver, which has order two in space and time. For the FTCT we use a maximum resolution $\mathbf{n} = (9, 9)$ and $\tau = 3$ (giving $\mathbf{i}_{\min} = (6, 6)$). This results in a classical index set $\mathscr{I}_{\mathbf{n},q,\tau}$ with 7 elements and an extended set $\mathscr{I}_{\mathbf{n},q,\tau}^{\text{ext}}$ with 10 elements. We calculate the solution at time $t = 0.5$ using 512 total time steps for all combination solutions, which ensures that the CFL condition is met. We combine the solutions twice during the simulation (`reduce` step), once at the middle (after 256 time steps) and at the end (after 512 time steps).

For this discussion we use only the second version of the function `sdcSanity-Check` (detecting outliers), since we found it to be more robust and to have more potential for parallelization. In particular, the `gather` step can be combined with the `reduce` step, so we would only need to communicate once instead of twice. And second, this `gather+reduce` step can be performed efficiently using the algorithm *Subspace Reduce* [26] with small modifications.

All simulations presented here were carried out serially. To detect if any of the hierarchical coefficients $\alpha_{\mathbf{l},\mathbf{j}}^{(\mathbf{l}')}$ is an outlier, we used the Python library *statsmodels* [32], specifically, the function `outlier_test` from the module `linear_model`. As of version 0.7.0 the function implements seven outlier detection methods, all of which performed very similarly. For our tests we chose `method='fdr_by'` which is based on a false discovery rate (FDR) method described in [4]. We consider a grid to be affected by SDC if at least one of its values is detected as an outlier.

4.2   Results

In Fig. 5 we have plotted six sets of simulation results. Each set has an iteration number (from 0 to 511) on the x-axis, which represents the time step in the function `solver` at which SDC was injected. This means that each of the six plots shows 512 different simulations. On the y-axis we have plotted the L_2 relative error of the combined solution $u_{\mathbf{n}}^{(c)}$ with respect to the exact solution at the end of each simulation. The three rows of plots show the different magnitudes of the SDC (10^5, $10^{-0.5}$, and 10^{-300}). For all simulations, the wrong value was injected into the same combination solution, $u_{(7,8)}$. (Choosing different solutions made no difference in the

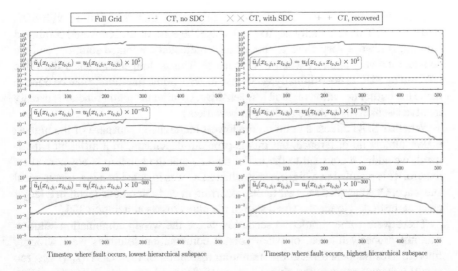

Fig. 5 L_2 relative error of the FTCT when SDC of various magnitudes are injected into one combination solution

results.) Finally, the plots on the left differ from those on the right by the choice of the grid point where SDC was injected. Recall that each combination solution u_i is transformed to the hierarchical basis after the call to the solver (line 10). If the solver returns a u_i with wrong values, the hierarchization step can propagate these wrong values to various degrees depending on which grid point(s) were affected. For the three plots on the left we injected SDC on the lowest hierarchical level (more precisely, in the middle of the domain), whereas for the plots on the right, the wrong value was inserted on one point of the highest hierarchical level (right next to the middle of the domain).

The blue line on each plot is the error of the full grid solution $u_{(9,9)}$; the green dotted line is the error of the combined solution $u_n^{(c)}$ when no SDC is injected; each red dot represents the error of the combined solution when SDC has been injected and not filtered; and the blue crosses are the error of the combined solution after detecting and filtering the wrong solution ($u_{(7,8)}$) and combining the rest of the grids with different coefficients.

As discussed earlier, the error of the combined solution is proportional to the error of the wrong combination solution. In all but one of the $6 \times 512 = 3072$ simulations the wrong solution was detected and filtered. This was the case when SDC of magnitude $10^{-0.5}$ was injected on the lowest hierarchical subspace during the last iteration, because the value of the solution of the PDE at that point is very close to zero. (Recall that the exact solution is a product of sine functions, so it is equal to zero at various points.) This is also true during (roughly) the first and last ten iterations, and from the plots we can see that the outlier detection method as we applied it is *too* sensitive. Even when SDC is barely noticeable (thus not affecting the quality of the combined solution), it is still detected, and the recovered solution (blue crosses) can actually be slightly *worse* than the solution with SDC (red circles). This is actually not too bad, since the error of the recovered combined solution is always very close to that of the unaffected combined solution, and we consider it a very unlikely worst-case scenario. This no longer happens if the value affected by SDC is not originally very close to zero. Some fine-tuning can be done to make the outlier detection method less sensitive. (Outlier detection functions usually involve a sensitivity parameter that can be varied.)

Whether SDC affects a point on the lowest hierarchical subspace or the highest makes almost no difference, but this is problem-dependent. In a different experiment, we added a constant to the initial field so that the solution is nowhere close to zero. This resulted in a higher error when SDC was injected in a low hierarchical subspace. One can also see that faults occurring in early iterations result in a larger error at the end. (Notice the small step in the blue crosses at iteration 256.)

There were some simulation scenarios where the wrong combination solution was not properly filtered, or when correct combination solutions were wrongly filtered. This happened when the minimum resolution of the SGCT (i_{min}) was too small. For our problem, the choice $i_{min} \geq (5, 5)$ was large enough for the outlier detection algorithms to work properly. As long i_{min} is chosen large enough, both the SGCT and the sanity checks work as expected.

Finally, if SDC causes alterations in the control flow of the program, two scenarios are possible (assuming once again that the fault occurs when calling the solver): either the solver returns a wrong solution or does not return at all. The first case can be treated by our algorithms. The second case can either translate into a hard fault (i.e., an error signal is produced—and this can be dealt with) or cause the solver to hang indefinitely. We plan to investigate this last scenario in the future.

Despite these fine-tuning issues, our approach offers several advantages over existing techniques. We do not implement any complicated checksum schemes; there is no checkpointing involved at any memory level; and there is no need to replicate MPI processes nor data. We simply make use of the existing redundancy in the SGCT to either calculate a norm or to look for outliers. These two algorithms are inexpensive and should not be difficult to implement in parallel. We plan to investigate robust, parallel implementations in future work, as well as to carry out further experiments.

5 Conclusions

The SGCT and its fault tolerant version, the FTCT, offer an inherent type of data redundancy that can be exploited to detect SDC. Assuming that one or more combination solutions can be affected by SDC of arbitrary magnitude, one can perform sanity checks before combining the results. The sanity checks work even in the worst-case scenario where only one value in one field is affected by a factor of arbitrary magnitude. These recovery algorithms do not require checkpointing. Existing outlier detection techniques from robust statistics can be directly incorporated into the FTCT, which requires only minimal modifications. Only some fine-tuning is required to minimize false positives or negatives, but this algorithmic approach avoids the drawbacks of the alternative techniques. We plan to demonstrate the applicability of these algorithms in massively large parallel simulations in the near future.

Acknowledgements This work was supported in part by the German Research Foundation (DFG) through the Priority Programme 1648 "Software for Exascale Computing" (SPPEXA). We thank the reviewers for their valuable comments. A. Parra Hinojosa thanks the TUM Graduate School for financing his stay at ANU Canberra, and acknowledges the additional support of CONACYT, Mexico.

References

1. Ali, M.M., Strazdins, P.E., Harding, B., Hegland, M., Larson, J.W.: A fault-tolerant gyrokinetic plasma application using the sparse grid combination technique. In: Proceedings of the 2015 International Conference on High Performance Computing & Simulation (HPCS 2015), pp. 499–507. IEEE, Amsterdam (2015)
2. Avižienis, A., Laprie, J.C., Randell, B., Landwehr, C.: Basic concepts and taxonomy of dependable and secure computing. IEEE Trans. Dependable Secure Comput. **1**(1), 11–33 (2004)
3. Bastian, P., Blatt, M., Dedner, A., Engwer, C., Klöfkorn, R., Ohlberger, M., Sander, O.: A generic grid interface for parallel and adaptive scientific computing. Part I: abstract framework. Computing **82**(2–3), 103–119 (2008)
4. Benjamini, Y., Yekutieli, D.: The control of the false discovery rate in multiple testing under dependency. Ann. Stat. **29**(4), 1165–1188 (2001)
5. Bland, W., Bouteiller, A., Herault, T., Bosilca, G., Dongarra, J.J.: Post-failure recovery of MPI communication capability: design and rationale. Int. J. High Perform. Comput. Appl. **27**(3), 244–254 (2013)
6. Bridges, P.G., Ferreira, K.B., Heroux, M.A., Hoemmen, M.: Fault-tolerant linear solvers via selective reliability. Preprint arXiv:1206.1390 (2012)
7. Bungartz, H.J., Griebel, M.: Sparse grids. Acta Numer. **13**, 147–269 (2004)
8. Chen, Z., Dongarra, J.: Highly scalable self-healing algorithms for high performance scientific computing. IEEE Trans. Comput. **58**(11), 1512–1524 (2009)
9. van Dam, H.J.J., Vishnu, A., De Jong, W.A.: A case for soft error detection and correction in computational chemistry. J. Chem. Theory Comput. **9**(9), 3995–4005 (2013)
10. Elliott, J., Hoemmen, M., Mueller, F.: Evaluating the impact of SDC on the GMRES iterative solver. In: 2014 IEEE 28th International Parallel and Distributed Processing Symposium, pp. 1193–1202. IEEE (2014)
11. Elliott, J., Hoemmen, M., Mueller, F.: Resilience in numerical methods: a position on fault models and methodologies. Preprint arXiv:1401.3013 (2014)
12. Ferreira, K., Stearley, J., Laros III, J.H., Oldfield, R., Pedretti, K., Brightwell, R., Riesen, R., Bridges, P.G., Arnold, D.: Evaluating the viability of process replication reliability for exascale systems. In: Proceedings of 2011 International Conference for High Performance Computing, Networking, Storage and Analysis, p. 44. ACM (2011)
13. Fiala, D., Mueller, F., Engelmann, C., Riesen, R., Ferreira, K., Brightwell, R.: Detection and correction of silent data corruption for large-scale high-performance computing. In: Proceedings of the International Conference on High Performance Computing, Networking, Storage and Analysis, p. 78. IEEE Computer Society Press (2012)

14. Garcke, J.: A dimension adaptive sparse grid combination technique for machine learning. ANZIAM J. **48**, 725–740 (2007)
15. Garcke, J.: Sparse grids in a nutshell. In: Garcke, J., Griebel, M. (eds.) Sparse Grids and Applications. Lecture Notes in Computational Science and Engineering, pp. 57–80. Springer, Berlin/Heidelberg (2013)
16. Garcke, J., Griebel, M.: On the computation of the eigenproblems of hydrogen and helium in strong magnetic and electric fields with the sparse grid combination technique. J. Comput. Phys. **165**(2), 694–716 (2000)
17. Griebel, M.: The combination technique for the sparse grid solution of PDE's on multiprocessor machines. Parallel Process. Lett. **2**, 61–70 (1992)
18. Griebel, M., Schneider, M., Zenger, C.: A combination technique for the solution of sparse grid problems. In: Iterative Methods in Linear Algebra, pp. 263–281. IMACS, Elsevier, North Holland (1992)
19. Harding, B.: Adaptive sparse grids and extrapolation techniques. In: Sparse Grids and Applications. Lecture Notes in Computational Science and Engineering, pp. 79–102. Springer, Cham (2015)
20. Harding, B., Hegland, M., Larson, J., Southern, J.: Fault tolerant computation with the sparse grid combination technique. SIAM J. Sci. Comput. 37(3), C331–C353 (2015)
21. Heene, M., Kowitz, C., Pflüger, D.: Load balancing for massively parallel computations with the sparse grid combination technique. In: PARCO, pp. 574–583. IOS Press, Garching (2013)
22. Heene, M., Pflüger, D.: Scalable algorithms for the solution of higher-dimensional PDEs. In: Proceedings of the SPPEXA Symposium. Lecture Notes in Computational Science and Engineering. Springer, Garching (2016)
23. Heene, M., Pflüger, D.: Efficient and scalable distributed-memory hierarchization algorithms for the sparse grid combination technique. In: Parallel Computing: On the Road to Exascale, Advances in Parallel Computing, vol. 27, pp. 339–348. IOS Press, Garching (2016)
24. Hegland, M.: Adaptive sparse grids. ANZIAM J. **44**, C335–C353 (2003)
25. Hupp, P.: Performance of unidirectional hierarchization for component grids virtually maximized. Procedia Comput. Sci. **29**, 2272–2283 (2014)
26. Hupp, P., Jacob, R., Heene, M., Pflüger, D., Hegland, M.: Global communication schemes for the sparse grid combination technique. Adv. Parallel Comput. **25**, 564–573 (2013). IOS Press
27. Jenko, F., Dorland, W., Kotschenreuther, M., Rogers, B.N.: Electron temperature gradient driven turbulence. Phys. Plasmas **7**(5), 1904–1910 (2000). http://www.genecode.org/
28. Kowitz, C., Hegland, M.: The sparse grid combination technique for computing eigenvalues in linear gyrokinetics. Procedia Comput. Sci. **18**, 449–458 (2013)
29. Parra Hinojosa, A., Kowitz, C., Heene, M., Pflüger, D., Bungartz, H.J.: Towards a fault-tolerant, scalable implementation of gene. In: Recent Trends in Computational Engineering – CE2014. Lecture Notes in Computational Science and Engineering, vol. 105, pp. 47–65. Springer, Cham (2015)
30. Pflüger, D.: Spatially Adaptive Sparse Grids for High-Dimensional Problems. Verlag Dr. Hut, München (2010)
31. Reisinger, C., Wittum, G.: Efficient hierarchical approximation of high-dimensional option pricing problems. SIAM J. Sci. Comput. **29**(1), 440–458 (2007)
32. Seabold, S., Perktold, J.: Statsmodels: econometric and statistical modeling with python. In: Proceedings of the 9th Python in Science Conference, pp. 57–61 (2010). http://statsmodels.sourceforge.net/
33. Snir, M., Wisniewski, R.W., Abraham, J.A., Adve, S.V., Bagchi, S., Balaji, P., Belak, J., Bose, P., Cappello, F., Carlson, B., et al.: Addressing failures in exascale computing. Int. J. High Perform. Comput. Appl. **28**, 129–173 (2014)
34. Winter, H.: Numerical advection schemes in two dimensions (2011). www.lancs.ac.uk/~winterh/advectionCS.pdf

Part V
TERRA-NEO: Integrated Co-Design of an Exascale Earth Mantle Modeling Framework

Hybrid Parallel Multigrid Methods
for Geodynamical Simulations

Simon Bauer, Hans-Peter Bunge, Daniel Drzisga, Björn Gmeiner,
Markus Huber, Lorenz John, Marcus Mohr, Ulrich Rüde, Holger Stengel,
Christian Waluga, Jens Weismüller, Gerhard Wellein, Markus Wittmann,
and Barbara Wohlmuth

Abstract Even on modern supercomputer architectures, Earth mantle simulations
are so compute intensive that they are considered grand challenge applications. The
dominating roadblocks in this branch of Geophysics are model complexity and
uncertainty in parameters and data, e.g., rheology and seismically imaged mantle
heterogeneity, as well as the enormous space and time scales that must be resolved
in the computational models. This article reports on a massively parallel all-at-once
multigrid solver for the Stokes system as it arises in mantle convection models.
The solver employs the hierarchical hybrid grids framework and demonstrates
that a system with coupled velocity components and with more than a trillion
$(1.7 \cdot 10^{12})$ degrees of freedom can be solved in about 1,000 s using 40,960 compute
cores of JUQUEEN. The simulation framework is used to investigate the influence
of asthenosphere thickness and viscosity on upper mantle velocities in a static
scenario. Additionally, results for a time-dependent simulation with a time-variable
temperature-dependent viscosity model are presented.

S. Bauer • H.-P. Bunge • M. Mohr • J. Weismüller
Department of Earth and Environmental Sciences, Ludwig-Maximilians-Universität München,
München, Germany
e-mail: simon.bauer@lmu.de

D. Drzisga • M. Huber • L. John (✉) • C. Waluga • B. Wohlmuth
Institute for Numerical Mathematics, Technische Universität München, München, Germany
e-mail: john@ma.tum.de

B. Gmeiner • U. Rüde
Department of Computer Science 10, Friedrich-Alexander-University Erlangen-Nuremberg,
Erlangen, Germany

H. Stengel • G. Wellein • M. Wittmann
Erlangen Regional Computing Center (RRZE), Friedrich-Alexander-University
Erlangen-Nuremberg, Erlangen, Germany

© Springer International Publishing Switzerland 2016   211
H.-J. Bungartz et al. (eds.), *Software for Exascale Computing – SPPEXA*
2013-2015, Lecture Notes in Computational Science and Engineering 113,
DOI 10.1007/978-3-319-40528-5_10

1 Introduction

The surface of our planet is shaped by processes deep beneath our feet. For a better understanding of these physical phenomena, simulations with high resolution are essential. The solid Earth's mantle extends from some tens of kilometers below the surface down to the core-mantle boundary at about 3,490 km depth. On geologic time scales, the mantle behaves like a highly viscous convecting fluid with one complete overturn taking about 100 million years [14]. It is this motion that is finally responsible for plate tectonics, mountain and ocean building, volcanism, and the accumulation of stresses leading to earthquakes. Convection itself is driven by internal heating, resulting from the decay of radioactive rocks in the mantle, by heat flux from the Earth's core, and by secular cooling. Due to the enormous length scales advection predominates heat transport through the planet over heat conduction.

The governing equations for mantle convection are formulations for the balance of forces and the conservation of mass and energy. While the general structure of convection within the mantle is relatively well understood, see e.g. [13, 44, 53, 61], a rich spectrum of physics is compatible with these equations and many details of the physical processes are poorly known [36]. Major unresolved questions include the potential thermo-chemical nature of the convection currents (essentially a statement on the buoyancy forces) [20, 52], appropriate rheological parameters and the importance of lateral viscosity variation [51]. In fact, a better understanding of the processes of convection inside the Earth belongs to the *10 Grand Research Questions in the Earth Sciences* identified in [18].

Due to the extreme conditions of the deep Earth and the large time scales involved, answering these questions is mostly outside the scope of laboratory experiments. Instead, further progress in geodynamics relies on extracting answers from the geologic record through a careful assimilation of observations into models by means of fluid dynamics inverse simulations [12]. There are three aspects making the inversion feasible: the strongly advective nature of the heat transport, mentioned above, the availability of terminal conditions from seismic tomography [24, 45], which provides present day temperatures and densities inside the mantle, and the availability of boundary conditions, i.e. surface velocity fields for the past 130 million years, from paleomagnetic reconstructions [40, 48].

Clearly, the geological and geophysical observations used for mantle convection simulations are subject to numerous sources of uncertainties, as e.g. in the context of the aforementioned buoyancy term, and we expect stochastic models to become of significant importance also in this area [43]. Both extensions require multiple evaluations of a forward problem, be it within an outer loop for an inverse problem, or a stochastic solution algorithm, e.g., a multi-level Monte Carlo method in the case of uncertainty quantification. Consequently, fast time stepping is essential. This in turn requires very fast and robust solution techniques for the stationary problem.

However, Earth's parameter regime dictates a resolution for real-world simulations of the mantle on the order of at least 10 km. Therefore, high resolution simulations and parallel solver performance have traditionally been a focus of research in this area and remain to be, see e.g. [3, 11, 19, 41, 46, 54, 62]. On the discretization side, while alternative approaches are sometimes employed, the finite element method is typically the method of choice, due to its flexibility with respect to geometry and parameter representation. Adaptive techniques and high-order finite elements are being investigated, see e.g. [1, 15, 16, 46, 50]. Note, however, that these techniques come at a price, e.g., in case of adaptivity dynamical data structures are required, which can lead to significant additional computational costs, parallel overhead, and extensive additional memory traffic.

Our objective in this paper is to present an approach, based on low-order elements, the regular refinement of a coarse (potentially) unstructured base mesh and a matrix-free implementation, discuss its performance and scalability and demonstrate the applicability for geodynamical simulations. In doing so, we mostly focus on the stationary (static) problem, which is a Stokes type problem, see Sect. 2. It constitutes the computationally most demanding part of any mantle convection model. The article summarizes selected results obtained in [28–30, 58] and extends these by scaling results for the massive parallel case, fault tolerant algorithms and geophysical simulations with real data. Additionally, results for a time-dependent geodynamical simulation will also be given.

Our approach is based on the hierarchical hybrid grids framework (HHG) [6, 28, 29, 59]. HHG presents a compromise between structured and unstructured grids. The former is essential, as it is well known that harvesting the full potential of current generation supercomputers is not feasible with completely unstructured grids. HHG exploits the flexibility of finite elements and capitalizes on the parallel efficiency of geometric multigrid methods. The HHG package was originally designed as a multigrid library for elliptic problems, and its excellent efficiency was demonstrated for scalar equations in [5, 6]. Meanwhile it was extended to the Stokes systems, see [27–30, 38] for details. HHG provides excellent scalability on current state-of-the-art supercomputers. By using all-at-once Uzawa-type multigrid methods, it is possible to solve more than a trillion (10^{12}) unknowns in a few minutes compute time. This corresponds to a spatial resolution of approximately 1 km.

This paper is structured as follows: In Sect. 2, we discuss geodynamical modeling aspects, state the governing equations and discuss suitable non-dimensionalization. In Sect. 3, we introduce a stabilized finite element discretization, present an all-at-once multigrid method based on a Uzawa smoother and recall the idea of the hierarchical hybrid grids framework. Scalability and parallel performance will be addressed in Sect. 4. We complete the paper by presenting selected geodynamical simulations. In Sect. 5, the question of flow-velocities in the asthenosphere is addressed using our code for real-world data of plate velocities and mantle temperatures, and in Sect. 6 we give results for a time-dependent simulation.

2   Geodynamical Modeling

A general mathematical description of mantle convection can be derived from the principles of conservation of momentum, mass and energy. It is common practice to neglect inertial terms and the Coriolis force due to their relative insignificance. For a detailed explanation and justification see, e.g., [44]. Doing so one arrives at

$$- \operatorname{div} \boldsymbol{\sigma} = \rho \mathbf{g} \qquad \text{in } \Omega \,, \tag{1}$$

$$\partial_t \rho + \operatorname{div}(\rho \mathbf{u}) = 0 \qquad \text{in } \Omega \,, \tag{2}$$

$$\partial_t(\rho e) + \operatorname{div}(\rho e \mathbf{u}) + \operatorname{div} \mathbf{q} - H - \boldsymbol{\sigma} : \dot{\boldsymbol{\varepsilon}} = 0 \qquad \text{in } \Omega \,. \tag{3}$$

The meaning of the different symbols is detailed in Table 1.

One can reformulate (1)–(2) in terms of the primary quantities velocity $\mathbf{u}$ and pressure p employing the coupling of stress and strain rate tensor and their relation to pressure and velocity

$$\boldsymbol{\sigma} = 2\nu \dot{\boldsymbol{\varepsilon}}(\mathbf{u}) - p\mathbf{I} \,, \quad \dot{\boldsymbol{\varepsilon}}(\mathbf{u}) = \frac{1}{2}\left(\nabla \mathbf{u} + (\nabla \mathbf{u})^{\top}\right) . \tag{4}$$

We point out that the dynamic viscosity ν depends on temperature, pressure/depth and velocity. However, the precise variations of ν with these quantities belongs to the big open questions of geodynamics. The energy equation (3) is typically re-cast in terms of the temperature τ

$$c_p \rho \partial_t \tau - \alpha \tau \partial_t p + c_p \rho \mathbf{u} \cdot \nabla \tau - \alpha \tau \mathbf{u} \cdot \nabla p + \operatorname{div}(\kappa \nabla \tau) - H - \boldsymbol{\sigma} : \dot{\boldsymbol{\varepsilon}} = 0 \tag{5}$$

and the system is closed by adding an equation of state relating density to pressure and temperature $\rho = \rho(p, T)$. The precise details of the latter vary between models as the composition and mineralogical behavior of the mantle is also an open research problem.

Density can be split into two contributions $\rho = \rho_{\text{ref}} + \rho_{\text{var}}$. Here ρ_{ref} is a so called *background density* that is derived from the time-constant hydrostatic pressure. The term ρ_{var} then represents the density variations resulting from thermal expansion that drives the convection. The latter, however, are very small compared to ρ_{ref} and allows to consider their effects only in the buoyancy term of the momentum

Table 1 Physical quantities and their representing symbols

ρ	Density	$\mathbf{u}$	Velocity	p	Pressure
$\mathbf{g}$	Gravitational acceleration	ν	Dynamic viscosity	$\boldsymbol{\sigma}$	Cauchy stress tensor
$\dot{\boldsymbol{\varepsilon}}$	Rate of strain tensor	τ	Temperature	e	Internal energy
$\mathbf{q}$	Heat flux per unit area	H	Heat production rate	α	Thermal expansivity
c_p	Specific heat capacity	κ	Heat conductivity		

equation (1) and to neglect them in the continuity equation (2). This results in the so called *anelastic approximation* $\mathrm{div}(\rho_{\mathrm{ref}}\mathbf{u}) = 0$, which is similar to the *Boussinesq approximation*, but takes into account that the background density increases by a factor of about two from the top to the bottom of the mantle, see, e.g., [21].

Treatment of $\mathrm{div}(\rho_{\mathrm{ref}}\mathbf{u}) = 0$ introduces an additional complexity that we will not consider in the current paper. Note, however, that it is possible to reformulate the problem, based on geophysical arguments as $\mathrm{div}\,\mathbf{u} = -\beta\rho\mathbf{g}\cdot\mathbf{u}$ with compressibility β, for details see, e.g., [17]. In both cases the anelastic approximation adds a grad-div term to the stress tensor (4). For purpose of this article we will always assume an incompressible flow field, which results in the Stokes system

$$
\begin{aligned}
-\,\mathrm{div}\,\sigma &= \rho\mathbf{g} &&\text{in } \Omega\,, \\
\mathrm{div}\,\mathbf{u} &= 0 &&\text{in } \Omega\,.
\end{aligned}
\tag{6}
$$

Firstly, we will consider the static problem (6) and in Sect. 6 we additionally couple it with a simplified version of the temperature equation (3).

The problem is posed on a domain $\Omega \subset \mathbb{R}^3$ representing the Earth's mantle. It is common practice to employ a thick spherical shell centered around the origin. Its boundary $\Gamma = \partial\Omega$ consists of two mutually separate parts, namely the surface, denoted here by Γ_{srf}, and the core-mantle boundary, denoted by Γ_{cmb}. On these two surfaces appropriate boundary conditions for $\mathbf{u}$ and τ must be specified to obtain a well-posed problem. By default temperature boundary conditions are of Dirichlet type, while velocity boundary conditions always include a no-outflow constraint, i.e. a vanishing radial component.

The tangential velocity components can then either be given, e.g., in the form of (measured) plate velocities, or the no-outflow condition is combined with the constraint of vanishing tangential shear stress $\sigma\,\mathbf{n}\cdot\mathbf{t} = 0$ resulting in a free-slip condition. Here $\mathbf{n}$ denotes the outward normal vector on the boundary and $\mathbf{t}$ the tangential vector. The free-slip condition is commonly associated with Γ_{cmb}, as the Earth's outer core is liquid, consisting mostly of molten iron.

For our simulations, we prefer to non-dimensionalize the governing equations (6). We describe the procedure for a simple equation of state model, which links density and temperature via a linear thermal expansion

$$
\rho - \rho_0 = -\alpha\,\rho_0\,\tau\,,
$$

where α denotes the coefficient of thermal diffusion and $\rho_0 = \rho_{\mathrm{srf}} = \text{const}$, with typical values of $\alpha = 2\cdot 10^{-5}\,\mathrm{K}^{-1}$ and $\rho_0 = 3.3\cdot 10^3\,\mathrm{kg/m}^3$. We then introduce the new variables

$$
\mathbf{U} = \frac{\mathbf{u}}{U^*}, \qquad \widetilde{P} = \frac{p}{P^*}, \qquad \mathbf{X} = \frac{\mathbf{x}}{X^*}, \qquad T = \frac{\tau}{T^*}\,.
$$

Choosing $X^* = 6.371\cdot 10^6\,\mathrm{m}$ (Earth radius) results in a spherical shell Ω with unit outer radius $r_{\max} = 1$ and an inner radius of $r_{\min} = 0.55$ for the scaled core-mantle

boundary. $U^* = 1$ cm/a is a value suitable for Earth mantle simulations. For the
temperature we select $T^* = T_{max}$ such that $T \in (0, 1)$.

Finally we set $M^* = \nu_{max} = 10^{22}$ Pa s, which results from our viscosity model detailed in Sect. 5. With this, we introduce then the scaled pressure
$P = P^* X^* / (U^* M^*) \widetilde{P}$. Multiplying the momentum equation with $(M^*)^{-1}$, which
constitutes a scaling for the viscosity and using a gravity constant of 10 m/s^2, we
obtain

$$- \mathrm{div} \left(2 \frac{\nu}{M^*} \dot{\varepsilon}(\mathbf{U}) \right) + \nabla P = \frac{(X^*)^2 \alpha \rho_0 T^* c_g}{M^* U^*} \overline{T} \frac{\mathbf{x}}{\|\mathbf{x}\|} = R T \frac{\mathbf{x}}{\|\mathbf{x}\|}$$

with a value of $R = 34{,}965$.

For ease of notation, we revert to lowercase letters again, which now represent
the non-dimensional quantities. The generalized Stokes equations (6) then read

$$- \mathrm{div} \left(2\nu(\tau, \mathbf{x}) \dot{\varepsilon}(\mathbf{u}) \right) + \nabla p = \mathbf{f} \qquad \text{in } \Omega \ ,$$
$$\mathrm{div}\, \mathbf{u} = 0 \qquad \text{in } \Omega \ , \tag{7}$$

with right-hand side $\mathbf{f} = R \tau \mathbf{x} \|\mathbf{x}\|^{-1}$ and additional boundary conditions.

The analysis of the Stokes system is well understood. Thus, we do not dwell on
it here, but refer the reader for instance to [10, 26].

3 Discretization and Hybrid Parallel Multigrid Methods

In the following we present the finite element discretization and fast solution
techniques based on hybrid parallel multigrid methods.

3.1 Finite Element Discretization

The basic idea of hierarchical hybrid grids (HHG) is to subdivide the computational
domain into a conforming tetrahedral initial mesh $\mathcal{T}_{-2}$ with possibly unstructured
and anisotropic elements. This initial grid is decomposed into different primitive
types (vertices, edges, faces and volumes). A hierarchy of grids $\mathcal{T} := \{\mathcal{T}_\ell, \ell = 0, 1, \ldots, L\}$ is constructed by successive uniform refinement, conforming to the
array-based data structures used in the HHG implementation, cf. [6, 28] for
details. The coarse grid $\mathcal{T}_0$ is a two times refined initial grid, which guarantees
that each primitive type owns at least one interior node. We point out that this
uniform refinement strategy guarantees that all our meshes satisfy a uniform shape-
regularity.

For simplicity, we describe the discretization for the case of homogeneous Dirichlet boundary conditions. We use linear, conforming finite elements, i.e., for a given mesh $\mathscr{T}_\ell \in \mathscr{T}$, $\ell \in \mathbb{N}_0$, we define the function space of piecewise linear and globally continuous functions by

$$S_\ell^1(\Omega) := \{v \in \mathscr{C}(\overline{\Omega}) : v|_T \in P_1(T), \ \forall T \in \mathscr{T}_\ell\} .$$

The conforming finite element spaces for velocity and pressure are then given by

$$\mathbf{V}_\ell = [S_\ell^1(\Omega) \cap H_0^1(\Omega)]^3, \qquad Q_\ell = S_\ell^1(\Omega) \cap L_0^2(\Omega) , \tag{8}$$

where $L_0^2(\Omega) := \{q \in L^2(\Omega) : \langle q, 1\rangle_\Omega = 0\}$ and $\langle \cdot, \cdot\rangle_\Omega$ denotes the inner product in $L^2(\Omega)$. Since the above finite element pair (8) does not satisfy a uniform inf–sup condition, we need to stabilize the method. Here, we consider the standard PSPG stabilization, see, e.g., [9]. This leads to the following discrete variational formulation of the Stokes system (6). Find $(\mathbf{u}_\ell, p_\ell) \in \mathbf{V}_\ell \times Q_\ell$ such that

$$\begin{aligned} a(\mathbf{u}_\ell, \mathbf{v}_\ell) + b(\mathbf{v}_\ell, p_\ell) &= f(\mathbf{v}_\ell) && \forall \mathbf{v}_\ell \in \mathbf{V}_\ell , \\ b(\mathbf{u}_\ell, q_\ell) - c_\ell(q_\ell, p_\ell) &= g_\ell(q_\ell) && \forall q_\ell \in Q_\ell , \end{aligned} \tag{9}$$

where the bilinear and linear forms are given by

$$a(\mathbf{u}, \mathbf{v}) := 2\langle \nu D(\mathbf{u}), D(\mathbf{v})\rangle_\Omega , \qquad b(\mathbf{u}, q) := -\langle \operatorname{div}\mathbf{u}, q\rangle_\Omega , \qquad f(\mathbf{v}) := \langle \mathbf{f}, \mathbf{v}\rangle_\Omega ,$$

for all $\mathbf{u}, \mathbf{v} \in H_0^1(\Omega)^3$, $q \in L_0^2(\Omega)$. Furthermore, the level-dependent stabilization terms $c_\ell(\cdot, \cdot)$ and $g_\ell(\cdot)$ are given by

$$c_\ell(q_\ell, p_\ell) := \sum_{T\in\mathscr{T}_\ell} \delta_T h_T^2 \langle \nabla p_\ell, \nabla q_\ell\rangle_T \quad \text{and} \quad g_\ell(q_\ell) := -\sum_{T\in\mathscr{T}_\ell} \delta_T h_T^2 \langle \mathbf{f}, \nabla q_\ell\rangle_T ,$$

with $h_T = (\int_T dx)^{1/3}$. The stabilization parameter $\delta_T > 0$ has to be chosen carefully to avoid unwanted effects due to over-stabilization, we fix $\delta_T = 1/12$ which is a good choice in practice, see, e.g., [22].

Let us denote by $n_{\ell,u} = \dim\mathbf{V}_\ell$ and $n_{\ell,p} = \dim Q_\ell$ for $\ell = 0, \ldots, L$ the number of degrees of freedom for velocity and pressure, respectively. Then, the isomorphisms $\mathbf{u}_\ell \leftrightarrow \mathbf{u} \in \mathbb{R}^{n_{\ell,u}}$ and $p_\ell \leftrightarrow \mathbf{p} \in \mathbb{R}^{n_{\ell,p}}$ are satisfied, and the algebraic form of the variational formulation (9) reads as

$$\mathscr{K}\begin{pmatrix}\mathbf{u}\\\mathbf{p}\end{pmatrix} := \begin{pmatrix}A & B^{\top}\\B & -C\end{pmatrix}\begin{pmatrix}\mathbf{u}\\\mathbf{p}\end{pmatrix} = \begin{pmatrix}\mathbf{f}\\\mathbf{g}\end{pmatrix} , \tag{10}$$

with the system matrix $\mathscr{K} \in \mathbb{R}^{(n_{\ell,u}+n_{\ell,p})\times(n_{\ell,u}+n_{\ell,p})}$. Note, the matrix $A \in \mathbb{R}^{n_{\ell,u}\times n_{\ell,u}}$ consists of a 3×3 block structure.

Remark 1 The correct treatment of free-slip boundary in the context of curved boundaries is more involved than for boundaries aligned to an axis, see, e.g., [23, 56] and more recently [55]. We implement a vanishing tangential stress by the condition $(I - \mathbf{n}\mathbf{n}^\top)\,\sigma(\mathbf{u}, p)\,\mathbf{n} = \mathbf{0}$ in a point-wise fashion. This approach requires the normal vector in a node on the free-slip boundary, where it is particularly important to construct the normals in such a way that they are not interfering with the mass conservation, see, e.g., [23]. Such a global mass conservative definition of the normal vector is given for the node $\mathbf{x}_i$ by

$$\mathbf{n}_i := \|\langle \nabla \varphi_i, 1 \rangle_\Omega \|_2^{-1} \langle \nabla \varphi_i, 1 \rangle_\Omega , \tag{11}$$

where φ_i denotes the scalar basis function which corresponds to the i-th node on the boundary. Note, the integral has to be understood component wise and can be easily computed by a local stencil application of $B^\top$ onto the vector $\mathbf{1}$.

3.2 Multigrid Solvers and the HHG Framework

For the numerical solution of the saddle point problem (10) several possibilities are available, such as a pressure Schur complement CG algorithm, a preconditioned MINRES method or an all-at-once multigrid method using Uzawa-type smoothers, see [28] for details and comparisons. In the given setting, the all-at-once multigrid method using Uzawa-type smoothers for saddle point system is more efficient with respect to time-to-solution and memory consumption. Furthermore, the inexact Uzawa smoother has the advantage that it can be implemented with the existing communication routines in HHG. This is different from the distributed smoothers proposed in [7, 8] that require a more complex parallel communication pattern. Uzawa-type multigrid methods for the Stokes system have been successfully applied, see, e.g., [2, 25, 47, 63]. Thus we shall choose this solution approach.

Let us briefly recall the idea of the Uzawa smoother, which is based on a preconditioned Richardson iteration. In the $(k + 1)$th-iteration, we solve the system

$$\begin{pmatrix} \hat{A} & 0 \\ B & -\hat{S} \end{pmatrix} \begin{pmatrix} \mathbf{u}_{k+1} - \mathbf{u}_k \\ \mathbf{p}_{k+1} - \mathbf{p}_k \end{pmatrix} = \begin{pmatrix} \mathbf{f} \\ \mathbf{g} \end{pmatrix} - \mathcal{K} \begin{pmatrix} \mathbf{u}_k \\ \mathbf{p}_k \end{pmatrix} ,$$

for $(\mathbf{u}_{k+1}, \mathbf{p}_{k+1})^\top$, where $\hat{A}$ and $\hat{S}$ denote preconditioners for A and the Schur-complement $S := BA^{-1}B^\top + C$, respectively. The application of the preconditioner results in the algorithm of the *inexact Uzawa method*, where we smooth the velocity part in a first step and in a second step the pressure, i.e.

$$\begin{aligned} \mathbf{u}_{k+1} &= \mathbf{u}_k + \hat{A}_i^{-1}(\mathbf{f} - A_i\mathbf{u}_k - B^\top\mathbf{p}_k) , \\ \mathbf{p}_{k+1} &= \mathbf{p}_k + \hat{S}_i^{-1}(B\mathbf{u}_{k+1} - C\mathbf{p}_k - \mathbf{g}) . \end{aligned} \tag{12}$$

For the convergence analysis of these methods, see, e.g., [25, 47, 63].

The HHG framework is a carefully designed and implemented high performance finite element geometric multigrid software framework [5, 6]. It combines the flexibility of unstructured finite element meshes with the performance advantage of structured grids in a block-structured approach. The grid is organized into primitive types according to the original input grid structure: vertices, edges, faces, and volumes. These primitives become container data structures for the nodal values of the refined mesh. We exploit this data structure in all multigrid operations such as smoothing, prolongation, restriction, and residual calculation. Each of these subroutines operates locally on the primitive itself. The dependencies between primitives are updated by ghost layer exchanges in an ordering from the vertex primitives via the edges and faces to the volume primitives. Communication is applied only in one way, i.e., copying of data is always handled by the primitive of higher geometrical order, see Fig. 1 for an illustration in 2D. This design decision imposes a natural ordering of a block Gauss-Seidel iteration based on the primitive classes. To facilitate the parallelization and reduce the communication, the primitives within each class are decoupled resulting in a block Jacobi structure. Finally, we are free to specify the smoother acting on the nodes of each primitive.

For the Uzawa multigrid method (12), we consider for the velocity part $\hat{A}$ a pseudo-symmetric hybrid block Gauss-Seidel smoother, which consists of the sequential execution of a so-called forward hybrid Gauss-Seidel and backward hybrid Gauss-Seidel, which reverses the ordering within each primitive but not the ordering in the primitive hierarchy. For the pressure, we consider a forward hybrid Gauss-Seidel smoother, applied to the stabilization matrix C, with under-relaxation $\omega = 0.3$. These smoothers are then applied within a variable $V(3, 3)$ cycle, where two additional smoothing steps are applied on each coarser level. As a stopping criterion we consider the relative reduction $\varepsilon = 10^{-8}$ of the error of the residual with respect to the Euclidean norm. Further, as a coarse grid solver simple

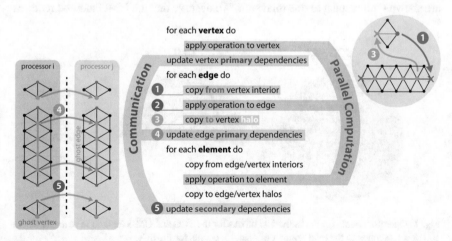

Fig. 1 Illustration of the communication structure in HHG

preconditioned MINRES iterations are employed. For a more detailed description
of the HHG framework and the smoother, see [5, 6].

4   Scalability and Performance of the Multigrid Method

In the following we present results for the Uzawa-type multigrid method using
the HHG framework and briefly discuss the concept of parallel textbook multigrid
efficiency (parTME) and node performance.

4.1   Operator Counts

To demonstrate the efficiency of the Uzawa-type multigrid method (UMG), we
study the number of operator evaluations $n_{op}(\cdot)$ for the solver of the individual
blocks (A, B and C) of the linear system (10). Here, we sum the number of
operator evaluations of all levels of the multigrid hierarchy and take into account
that coarsening reduces the evaluation cost by a factor $1/8$. We compare the UMG
to the Schur complement CG algorithm (SCG), where we use as a preconditioner for
the Schur complement $C + BA^{-1}B^{\mathsf{T}}$ the lumped mass matrix M. As a test problem
we consider the unit cube with 8 uniform refinement levels, which correspond to
$6.6 \cdot 10^7$ degrees of freedom and consider a relative residual of 10^{-8} as a stopping
criterion. We refer to [28] for a more detailed description and further results.

In Fig. 2, we present the number of operator counts in percent for the SCG and
the UMG method. The SCG solver case is set as reference case, where the sum of
the evaluations of all blocks, A, B, C and M corresponds to 100 %. The percentages
are shown with respect to this total sum. We observe that the UMG method requires

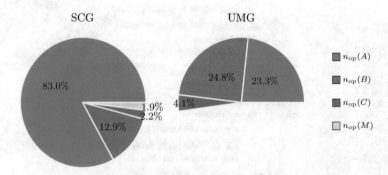

Fig. 2 Operator counts for the individual blocks for the SCG and UMG method in percents where
100 % corresponds to the total number of operator counts for the SCG

only half of the total number of operator evaluations in comparison to the SCG algorithm. This is also reflected within the time-to-solution of the UMG.

4.2 Scalability

In the following, we present numerical examples, illustrating the robustness and scalability of the Uzawa-type multigrid solver. For simplicity we consider the homogeneous Dirichlet boundary value problem. As a computational domain, we consider the spherical shell

$$\Omega = \{\mathbf{x} \in \mathbb{R}^3 : 0.55 < \|\mathbf{x}\|_2 < 1\}.$$

We present weak scaling results, where the initial mesh $\mathcal{T}_{-2}$ consists of 240 tetrahedra for the case of 5 nodes and 80 threads on 80 compute cores, cf. Table 3. This mesh is constructed using an icosahedron to provide an initial approximation of the sphere, see e.g. [19, 41], where we subdivide each resulting spherical prismatoid into three tetrahedra. The same approach is used for all other experiments below that employ a spherical shell. We consider 16 compute cores per node (one thread per core) and assign 3 tetrahedra of the initial mesh per compute core. The coarse grid $\mathcal{T}_0$ is a twice refined initial grid, where in one uniform refinement step each tetrahedron is decomposed into 8 new ones. Thus, the number of degrees of freedoms grows approximately from 10^3 to 10^6, when we perform a weak scaling experiment starting on the coarse grid $\mathcal{T}_0$.

The right-hand side $\mathbf{f} = \mathbf{0}$ and a constant viscosity $\nu = 1$ are considered. Due to the constant viscosity it is also possible to use the simplification of the Stokes equations using Laplace-operators. In the following numeral experiments, we shall compare this formulation with the generalized form that employs the $\dot{\boldsymbol{\varepsilon}}$-operator. For the initial guess $\mathbf{x}_0 = (\mathbf{u}_0, \mathbf{p}_0)^\top$, we distinguish between velocity and pressure part. Here, $\mathbf{u}_0$ is a random vector with values in $[0, 1]$, while the initial random pressure $\mathbf{p}_0$ is scaled by h^{-1}, i.e., with values in $[0, h^{-1}]$. This is important in order to represent a less regular pressure, since the velocity is generally considered as a $H^1(\Omega)$ and the pressure as a $L^2(\Omega)$ function. Numerical examples are performed on JUQUEEN[1] (Jülich Supercomputing Center, Germany). We present weak scaling results for up to 40,960 compute cores (threads) and more than 10^{12} degrees of freedom (DoFs). The finest computational level is obtained by an 8 times refined initial mesh. The fragment of JUQUEEN employed in this experiment is less than a tenth of the full machine and provides about 40 TByte of memory. Note here that storing the solution vector with $1.7 \cdot 10^{12}$ unknowns in double precision floating point values requires alone already 13.6 TByte of memory. We point out that any conventional

---

[1] listed top 9 of the TOP500 list, Nov. 2015.

Table 2 Weak scaling results for the UMG method with the Laplace-operator formulation

Nodes	Threads	DoFs	Iter	Time	Time w.c.g.	Time c.g. in %
5	80	$2.7 \cdot 10^9$	10	685.88	678.77	1.04
40	640	$2.1 \cdot 10^{10}$	10	703.69	686.24	2.48
320	5,120	$1.2 \cdot 10^{11}$	10	741.86	709.88	4.31
2,560	40,960	$1.7 \cdot 10^{12}$	9	720.24	671.63	6.75

Table 3 Weak scaling results for the UMG method with the $\dot{\varepsilon}$-operator formulation

Nodes	Threads	DoFs	Iter	Time	Time w.c.g.	Time c.g. in %
5	80	$2.7 \cdot 10^9$	10	1,134.28	1,120.85	1.18
40	640	$2.1 \cdot 10^{10}$	9	1,054.24	1,028.69	2.42
320	5,120	$1.2 \cdot 10^{11}$	8	974.95	936.89	3.90
2,560	40,960	$1.7 \cdot 10^{12}$	8	1,037.00	958.59	7.56

FE technique that is based on assembling and storing the stiffness matrix would inevitably consume at least one order of magnitude more memory.

In Table 2, we present scaling results for the Laplace-operator formulation, see also [28], in form of iteration numbers (iter), time-to-solution (time) in seconds, time-to-solution without the time on the coarse grid (time w.c.g.) and the time in % spent on the coarse grid (time c.g. in %). We observe excellent scalability of the multigrid method. We point out that the coarse grid solver in these examples considered as a stand-alone solver lacks optimal complexity. However, even for the largest example with more than 10^{12} DoFs it still requires less than 10 % of the overall time.

In Table 3, we extend our previous study by considering the same set-up for the $\dot{\varepsilon}$-operator formulation. Again, excellent scalability of the algorithm is achieved and also in this case the time spent on the coarse grid is less than 10 %.

Compared by the overall time the Laplace-operator formulation is approximately a factor of 1.4 faster than the $\dot{\varepsilon}$-operator formulation.

4.3 Fault Tolerance

Let us now turn towards another aspect of algorithm design important in the context of HPC which is fault-tolerance. On current supercomputing systems and also in the future era of exa-scale computing, highly scalable implementations will execute up to billions of parallel threads on millions of compute nodes. In this scenario, it may become essential that fault tolerance is also supported algorithmically, i.e. the algorithms themselves are made tolerant against errors and are augmented with intelligent and adaptive strategies to detect and to compensate for faults. Commonly used redundancy approaches, such as global check-pointing, may here become too costly, due to the high memory and energy consumption. In [37] an algorithmic

Table 4 CA results for an elliptic model equation using $\eta_s = 5$

n_F	1	2	3	4	5	6
(DD)	0.33	0.00	0.00	0.08	0.17	0.17
(DN)	0.33	0.00	0.00	0.00	0.08	0.17

alternative is suggested. In particular, it is found that data lost due to a hard fault of a core or node can be compensated for by solving local recovery problems using local multigrid cycles. This local recovery can benefit from a special acceleration. The recovery from one or multiple faults becomes especially efficient, when the global solution process is suitably decoupled from the recovery, but is itself continued asynchronously in parallel. Several variants of fault tolerant multigrid strategies are proposed and compared in [37]. To quantify a possible performance loss, we introduce a *relative Cycle Advantage* (CA) parameter κ defined as

$$\kappa := \frac{k_{\text{faulty}} - k_{\text{free}}}{k_F}, \tag{13}$$

where k_{faulty} is the number of multigrid iterations required in the case of a faulty execution with a fault occurring after k_F cycles, and k_{free} stands for the number of multigrid iterations required in the case of a fault-free execution. Intuitively, we expect that $\kappa \in [0, 1]$, and the smaller κ is the more effective is the recovery. Both our strategies, a Dirichlet–Dirichlet (DD) and a Dirichlet–Neumann (DN) approach, are based on a tearing and interconnecting idea. During n_F multigrid cycles the faulty and the healthy domain are separated and asynchronous sub-jobs take place. For (DD), we perform n_F multigrid cycles on the healthy domain with Dirichlet data at the interface to the faulty domain while for (DN), Neumann data are specified. In both cases, we run $n_F \eta_s$ local multigrid cycles with Dirichlet data on the faulty domain. After n_F global steps both sub-jobs are interconnected.

In Table 4 we present the CA results for an elliptic model equation. A standard V(2,1) multigrid scheme with a hybrid Gauss-Seidel smoother is applied, and the injected fault occurs after $k_F = 12$ iterations. As we can see, the (DN) approach is more robust with respect to the choice of n_F compared to the (DD) scheme. But both schemes can fully compensate the fault and achieve $\kappa = 0$.

4.4 Performance

Scalability is a necessary but no sufficient condition for a fast solver. On modern architectures, the real-life performance is increasingly determined by the intra-node and intra-core performance. Therefore it is essential that the algorithms and their implementation also exploit the available instruction level parallelism and the respective CPU micro-architecture in the best way possible. Here, we develop a concept to quantify the efficiency for parallel multigrid algorithms based on Achi

Brandt's notion of textbook multigrid efficiency (TME) as in [30]. We recall that the classical definition of TME requires that a PDE is solved with an effort that is at most the cost of ten operator evaluations. Such an operator evaluation is defined to be the elementary work unit (WU). Thus, TME means that computing the solution of a linear system (resulting from the discretization of the PDE) should cost no more than performing ten sparse matrix vector multiplications with the corresponding system matrix. In [30] this is elaborated in terms of scalar equations with constant and varying coefficients as well as linear systems with saddle-point structure.

The new step of our analysis consists in an extension of this idea to a parallel, architecture-aware setting. To this end we have developed a new characterization of a work unit (WU) that uses systematic performance modeling techniques. The goal of this WU characterization is to provide a simple but realistic criterion for which performance can at best be delivered for the problem class and the architecture that are under consideration. As a first step, the smoother kernel is executed on a single socket and is benchmarked. As in classical system analysis, we assume that the performance is limited by a critical resource and that identifying the critical resource is an essential aspect in understanding and analyzing the performance. In many cases, a good starting point will be the roofline model [60] that states that either the floating point throughput or the main memory bandwidth constitute the critical resource for numerical kernels. Since the roofline model does not take the data transfer through the cache hierarchy into account, the more sophisticated execution-cache-memory (ECM) model [33] is applied in [30].

Let us illustrate the essence of this analysis on the basis of HHG, where the smoother kernel must apply a 15-point stencil on a structured mesh. Since this has a low floating point intensity, one may expect that the critical resource is not the floating point throughput, but rather memory bandwidth. Against conventional wisdom, however, the ECM analysis in [30] and summarized here in Fig. 3 shows that the critical resource is more complex. The red-black stride-2 access to memory by the special Gauss-Seidel smoother and the tetrahedral (though structured) memory layout require complex memory operations before each core can employ SIMD vectorization. These memory access- and arrangement operations are found to be the critical limit to the smoother intra-core performance. The evaluation of measured against analytically predicted performance on a socket of the architecture, as displayed in Fig. 3 shows that the code has in fact been optimized to reach the performance limit that is dictated by the given micro-architecture. The measured performance values lie consistently between the upper and lower limits as they are predicted by the ECM model.

Having such a quantitative characterization of the critical resource is the first essential ingredient of the parallel TME analysis. For the new parallel TME analysis it is fundamental that an optimal implementation of the application specific smoother kernel has been found, since this becomes the basis to assess the algorithmic performance together with the parallel overhead in the second step of the parallel TME analysis.

We note here that in practice identifying the optimal smoother kernel is rarely reached in one single step. A given kernel will often still permit optimizations. It

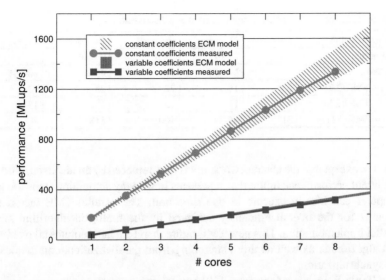

Fig. 3 Sandy Bridge (SNB) single-chip performance scaling of the stencil smoothers on a tetrahedral grid with 257 grid points along each edge of a volume primitive. Measured data and ECM prediction ranges are shown

is quite common that the performance of an initial smoother kernel is hampered by unfavorable data structures and memory layouts (such as a structure of arrays versus an array of structures), or by an unsatisfactory vectorization, or by other suboptimal code structures, including suboptimal compiler optimizations. Many yet unoptimized codes will initially underperform by orders of magnitude. We emphasize here that the novel parallel TME paradigm requires that all these node- and core-level inefficiencies are carefully accounted for, documented, and wherever possible removed, until the point is reached when the fundamental architectural limitations with respect to the given application have been identified. However, note here that this detailed and labor intensive work is in our analysis focused only on the smoother of the multigrid algorithm (and thus a simple single algorithmic component) and that here yet only intra-node parallelism must be accounted for.

For the next step of this analysis, the ideal aggregate smoother performance for the full parallel system is computed. This figure then forms the basis against which the measured run time of the solver can be put in relation. To be more precise we define a *parallel textbook efficiency factor*

$$E_{\text{ParTME}}(N, U) = \frac{T(N, U)}{T_{\text{WU}}(N, U)} ,$$

which relates $T(N, U)$, the time to solve a problem with N unknowns on a computer with U processor cores, to the idealized time $T_{\text{WU}}(N, U)$ required for one work unit. The rationale behind this is the same line of thought as in a classical TME

Table 5 TME factors for different problem sizes and discretizations

Setting/Measure	E_{TME}	E_{SerTME}	$E_{NodeTME}$	$E_{ParTME1}$	$E_{ParTME2}$
Grid points	–	$2 \cdot 10^6$	$3 \cdot 10^7$	$9 \cdot 10^9$	$2 \cdot 10^{11}$
Processor cores U	–	1	16	4,096	16,384
(CC) – FMG(2,2)	6.5	15	22	26	22
(VC) – FMG(2,2)	6.5	11	13	15	13
(SF) – FMG(2,1)	31	64	100	118	–

analysis, except that the abstract work unit is now replaced by an idealized, hardware dependent performance figure that represents how many smoothing steps the given computer can ideally execute in one time unit. The parallel TME factor, thus, accounts for the overall efficiency achieved by the multigrid algorithm and its parallel implementation. This new TME figure therefore incorporates all overheads, both algorithmic as well as those stemming from parallelization, communication, and synchronization.

In Table 5 we reproduce from [30] typical values as they are obtained in the parallel TME analysis for HHG when using full multigrid. The row denoted by (CC) refers to the case of a constant coefficient scalar problem, (VC) to the situation with variable coefficients when the local stencils are assembled on the fly. (SF) finally refers to the pressure correction Stokes solver using 4 CG cycles with a V(2,1) multigrid solver. The column E_{TME} presents the classical TME factor as defined by Brandt. E_{SerTME} and $E_{NodeTME}$ refer to the single core and node (i.e. using 16 cores) performance, respectively. $E_{ParTME1}$ then presents the new efficiency factor in the case of a parallel execution with 256 nodes ($= 4096$ cores) and $E_{ParTME2}$ drives this up to using 16,384 cores which is sufficient to handle a discretization with $2 \cdot 10^{11}$ grid points.

5 Application to the Earth's Upper Mantle

As an example for an application of HHG to an actual problem of mantle convection, we turn our attention to the question of the thickness of the asthenosphere. The latter is a mechanically weak layer in the uppermost mantle, but its depth and viscosity remain debated. From observations of the Earth's response to the melting of ice sheets, we know, however, that depth and viscosity are closely coupled via the relation

$$v_a \propto d_a^3 , \tag{14}$$

with a viscosity of $v_a = 10^{21}$ Pa s in the asthenosphere, ranging down to a depth of $d_a = 1,000$ km. This is the so called Haskell constraint, see [35, 39].

At the same time, recent geological observations hint at high velocities in the Earth's upper mantle [34, 42, 57], giving rise to the thought that the common assumption of an asthenosphere with a thickness of 660 km might not necessarily be true. To assess the implications of a thinner and less viscous asthenosphere, we consider the model (7). For the viscosity, we set up four scenarios with asthenospheric depths of $d_a \in \{1000, 660, 410, 200\}$ km according to

$$
\nu = \begin{cases} 10^{22} \text{ Pa s} & \text{for } |\mathbf{x}| < r_{\text{srf}} - d_a \\ \nu_a = 10^{21} \left(\frac{d_a}{d_{a,\text{ref}}}\right)^3 \text{ Pa s} & \text{for } |\mathbf{x}| \geq r_{\text{srf}} - d_a, \end{cases} \tag{15}
$$

as suggested by [59] and motivated from the Haskell constraint. Here, $d_{a,\text{ref}} = 1000$ km is the reference radius for the viscosity jump at the upper-lower mantle transition, r_{srf} the radius of the Earth and $d_a \in [r_{a,\text{ref}}, r_{\text{srf}})$ the key parameter we will vary for our different simulation runs. The scenario with 1000 km will serve as our reference case, whereas the 660 and 410 km scenarios are chosen because they correspond to the main seismic reflector depths. In addition, the 200 km case serves to demonstrate the possibility of a very thin and fast layer. This corresponds to the setup by [59], where the same viscosity profiles were used. However, we replace their synthetic buoyancy term and boundary conditions with indirect observations from the Earth as follows: we impose inhomogeneous Dirichlet conditions at the outer boundary to represent the movement of the tectonic plates, with velocities as given by [40]. Tangential stresses are set to zero at the inner boundary, since the outer core below the Earth's mantle is liquid. In addition, we prohibit in- and outflow at all boundaries. This yields

$$
\mathbf{u} = \mathbf{u}_{\text{plt}} \qquad \text{on } \Gamma_{\text{srf}}, \tag{16}
$$

$$
\sigma \, \mathbf{n} \cdot \mathbf{t} = \mathbf{0} \qquad \text{on } \Gamma_{\text{cmb}}, \tag{17}
$$

$$
\mathbf{u} \cdot \mathbf{n} = 0 \qquad \text{on } \Gamma_{\text{cmb}} \cup \Gamma_{\text{srf}}, \tag{18}
$$

where $\mathbf{u}_{\text{plt}}$ m/s represents the plate velocities, and $\mathbf{n}$, $\mathbf{t}$ the boundaries' normal, tangential vectors. In Fig. 4 we show the corresponding velocity field $\mathbf{u}_{\text{plt}}$.

The forcing term $\mathbf{f}$ is derived directly from the buoyancy ($-\rho \mathbf{g}$), i.e. the product of a density field ρ kg/m^3 and the gravitational acceleration within the mantle $\mathbf{g}$ m/s^2, which we prescribe as a vector of magnitude 10 m/s^2 pointing towards the origin (the center of the Earth). The density ρ is obtained from a tomographic model of seismic wave speeds within the Earth [32], converted to densities with the mineralogic model of [49].

As an example, the resulting velocity field for the 660 km and the 410 km scenarios are visualized in Fig. 5. Well visible are the increased velocities in the asthenosphere compared to the lower mantle, especially below the Eastern Pacific. Also clearly visible are strong upwellings below the Icelandic hotspot.

We now compare our simulation results to a theoretical prediction. If we approximate the asthenosphere as a two-dimensional, plane channel, driven by

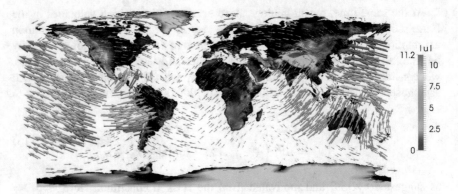

Fig. 4 Velocity field $\mathbf{u}_{\text{plt}}$ cm/a, derived from the movement of the tectonic plates, and imposed as upper boundary velocity

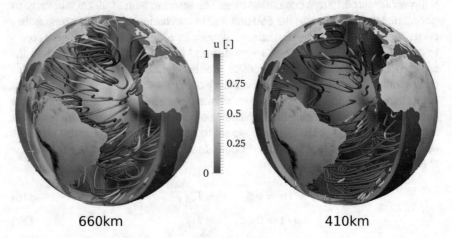

Fig. 5 Velocity magnitudes for an assumed asthenospheric depth of 660 km (*left*) and 410 km (*right*), normalized to the corresponding largest value. Visualized is a slice through the Earth's mantle together with the corresponding flow lines

pressure gradients alone, we can apply a Poiseuille flow model, which states that the average velocity scales as

$$\hat{u} \propto \frac{d_a^2}{v_a} \nabla p \ . \tag{19}$$

Note that the pressure p is linear for this particular case and thus its gradient is constant. This gives for the average velocity $\hat{u}_a$ in the upper mantle compared to the reference upper mantle

$$\frac{\hat{u}_a}{\hat{u}_{a,\text{ref}}} = \frac{v_{a,\text{ref}} \, d_a^2}{v_a \, d_{a,\text{ref}}^2} \ . \tag{20}$$

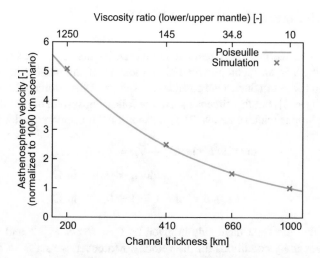

Fig. 6 Averaged flow velocities in the asthenosphere relative to the 1000 km reference scenario. Shown are simulation results *(red)* and the theoretical prediction by a Poiseuille flow law *(blue)*. Upper and lower axis are tied together via (14)

Using (14), this can then be simplified to

$$\frac{\hat{u}_a}{\hat{u}_{a,\text{ref}}} = \frac{d_{a,\text{ref}}}{d_a} . \tag{21}$$

We point out that this model is free of any fitting parameters. The corresponding curve is plotted in Fig. 6 together with the resulting average velocities from the simulations.

We observe a very good correspondence between the Poiseuille flow model and the simulations, with deviations that are consistently smaller than 2.3 %. In the synthetic setting with a buoyancy based on a spherical harmonic heterogeneity of degree two and without a additional driving force at the boundary in [59], the Poiseuille model underestimated the resulting velocities by up to 25 %. Since the moving plates add a Couette-component to the flow, it is quite surprising that we obtain even better agreement with the Poiseuille model in the realistic setup.

In Fig. 5, it can clearly be seen that even in the 660 km scenario, the velocities in the asthenosphere can exceed plate velocities even under the fast plates, i.e. in the Eastern Pacific under the Nazca plate. This is a clear indication that motion in the asthenosphere is not solely driven by the tectonic plates (Couette flow), but that at least some Poiseuille component is present.

Since the scaling of the buoyancy, as derived from tomography, is uncertain to about a factor of two [4], we might overestimate the velocities by a factor of two. However, since the average velocities in the asthenosphere in the different scenarios vary by a factor of five, a factor of two in the buoyancy should not significantly alter our interpretation, at least not for the very thin channels.

6   Simulations of the Coupled Problem

In this section, we present numerical simulation results for a thermal convection problem which is an incompressible simplification of the non-isothermal system (1), (2), and (3). As in Sect. 5 we consider the spherical shell domain $\Omega = \{x \in \mathbb{R}^3 : 0.55 < \|x\|_2 < 1\}$, but this time we solve the following system of non-dimensional balance equations which extends (7) by an additional transport equation:

$$-\operatorname{div}\left(2\nu(\tau, \mathbf{x})\dot{\varepsilon}(\mathbf{u})\right) + \nabla p = \mathbf{f} \qquad \text{in } \Omega ,$$

$$\operatorname{div}\mathbf{u} = 0 \qquad \text{in } \Omega , \qquad (22)$$

$$\partial_t \tau + \operatorname{div}(\tau\mathbf{u} - \kappa\nabla\tau) = 0, \qquad \text{in } \Omega .$$

We recall that the right hand side is given by $\mathbf{f} = R\,\tau\,\mathbf{x}\,\|\mathbf{x}\|^{-1}$ and we impose Dirichlet boundary conditions for the velocity. Moreover, we set $R = 10^7$ and the viscosity model is temperature-dependent and given by $\nu = e^{1/2-\tau}$, where $\tau \in [0, 1]$ denotes the scaled temperature which is initialized with an interpolation between the surface temperature ($\tau = 0$) and core-mantle boundary temperature ($\tau = 1$). To trigger the formation of thermal plume patterns we add small perturbations using spherical harmonic functions.

The system of non-linear equations is solved in an explicit time stepping fashion, which has the advantage that the problem decouples into a Stokes problem of the form (7), and a scalar transport equation, that in this case is integrated by a strongly stability preserving Runge-Kutta scheme of second order. To reduce the amount of implicit solves of the instantaneous Stokes subproblem, we employ a splitting approach which is detailed in [58], leaving us with one implicit saddle-point solve and two applications of the convection–diffusion operator per time step. Thus, the most time-consuming part of the coupled solver is again the Stokes problem for which we discussed efficient solver strategies in Sect. 3.

However, besides the performance aspect, the above type of equations comes with an additional challenge: For such non-linear bidirectional couplings between incompressible flow- and transport-type equations it is crucial to avoid spurious compressibility effects in the energy equation which can lead to unphysical temperature solutions, which in turn have an effect on the buoyant forcing terms and the viscosity [58]. Since these unphysical effects typically amplify over time it is of uttermost importance to preserve mass-conservation, especially in long-term simulations. We thus employ a recently developed technique to enforce a mass-conservative coupling, using equal-order linear elements for the Stokes part and a finite volume method with local flux-corrections for the transport part [31, 58]. This combination has the advantage that it can be implemented in a matrix-free fashion, using the collocated nodal data structures in the HHG framework.

In Fig. 7 we depict the characteristic plume-structures of the temperature iso-surfaces for two different time steps of a typical test run on what can today be considered a mid-sized department cluster. The machine, which is operated by the

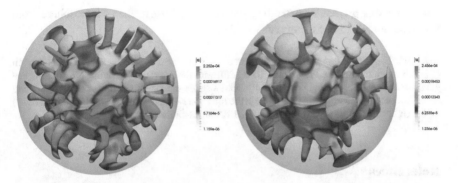

Fig. 7 Isosurfaces for $\tau = 0.4$ colored by the velocity magnitude after 7000 time steps (*left*) and 10,500 time steps (*right*)

Institute for System Simulation at FAU Erlangen, consists of 8 compute nodes connected by a QDR Infiniband network. Each node is equipped with 4 Intel Xeon E7-4830 CPUs (8 cores each) and 256 GB RAM. The computational mesh is again an icosahedral sphere mesh consisting of roughly a hundred million (94,371,840) tetrahedra, and the discretization of the above model problem consists of 77,824,950 degrees of freedom per time step. In this experiment, we solve the coupled problem for 11,500 time steps, where we employ the SCG algorithm for the Stokes part. For the initial solve we conduct 20 SCG iterations and in every subsequent solve, we apply 8 SCG steps, where we approximate the inverse of the viscous operator by a $V(3, 3)$ cycle. The SCG method is preconditioned by a lumped mass matrix which is scaled by the reciprocal viscosity, and we restart the method every 4 iterations due to the inexactness of the Krylov space which is built up with respect to the approximate Schur-complement.

7 Conclusion

In this article we presented an approach, currently under development, for a new framework for mantle convection simulation models suitable for the upcoming exa-scale era. Our approach is based on the hierarchical hybrid grids (HHG) idea and uses low-order finite elements. We demonstrated that models with global resolutions of about 1 km, corresponding to more than a trillion ($1.7 \cdot 10^{12}$) degrees of freedom, can be solved in about 1000 s, even on today's top architectures. This is accomplished by combining the lightweightedness of the HHG structure, its sophisticated matrix-free, highly-optimized implementation, as demonstrated by our ECM analysis, with modern multigrid algorithms. Using our mass-conservative coupling shows that stable transient simulations for low-order elements are possible. We also reported on our work on fault-tolerant algorithms which addresses another crucial aspect of future systems. The suitability of the approach for mantle convection modelling was proven by extending our study of flow in the asthenosphere from synthetic models to one using real-world data.

Acknowledgements This work was supported (in part) by the German Research Foundation (DFG) through the Priority Programme 1648 "Software for Exascale Computing" (SPPEXA) and grant WO 671/11-1. The authors gratefully acknowledge the Gauss Centre for Supercomputing (GCS) for providing computing time through the John von Neumann Institute for Computing (NIC) on the GCS share of the supercomputer JUQUEEN at Jülich Supercomputing Centre (JSC).

References

1. Bangerth, W., Burstedde, C., Heister, T., Kronbichler, M.: Algorithms and data structures for massively parallel generic adaptive finite element codes. ACM Trans. Math. Soft. **38**(2), 14:1–14:28 (2011)
2. Bank, R.E., Welfert, B.D., Yserentant, H.: A class of iterative methods for solving saddle point problems. Numer. Math. **56**(7), 645–666 (1990)
3. Baumgardner, J.R.: Three-dimensional treatment of convective flow in the Earth's mantle. J. Stat. Phys. **39**(5/6), 501–511 (1985)
4. Becker, T.W., Boschi, L.: A comparison of tomographic and geodynamic mantle models. Geochem. Geophy. Geosy. **3**, 1525–2027 (2002)
5. Bergen, B., Gradl, T., Rüde, U., Hülsemann, F.: A massively parallel multigrid method for finite elements. Comput. Sci. Eng. **8**(6), 56–62 (2006)
6. Bergen, B., Wellein, G., Hülsemann, F., Rüde, U.: Hierarchical hybrid grids: achieving TERAFLOP performance on large scale finite element simulations. Int. J. Parallel Emergent Distrib. Syst. **22**(4), 311–329 (2007)
7. Brandt, A.: Guide to multigrid development. In: Multigrid methods, pp. 220–312. Springer, Berlin/Heidelberg (1982). Republished as: *Multigrid Techniques: 1984 guide with applications to fluid dynamics*, revised edition, SIAM, 2011
8. Brandt, A., Dinar, N.: Multigrid solutions to elliptic flow problems. In: Numerical methods for partial differential equations (Proc. Adv. Sem., Math. Res. Center, Univ. Wisconsin, Madison, Wis., 1978), Publ. Math. Res. Center Univ. Wisconsin, vol. 42, pp. 53–147. Academic Press, New York/London (1979)
9. Brezzi, F., Douglas, Jr., J.: Stabilized mixed methods for the Stokes problem. Numer. Math. **53**(1–2), 225–235 (1988)
10. Brezzi, F., Fortin, M.: Mixed and Hybrid Finite Element Methods. Springer, New York (1991)
11. Bunge, H.P., Baumgardner, J.R.: Mantle convection modeling on parallel virtual machines. Comput. Phys. **9**(2), 207–215 (1995)
12. Bunge, H.P., Hagelberg, C.R., Travis, B.J.: Mantle circulation models with variational data assimilation: inferring past mantle flow and structure from plate motion histories and seismic tomography. Geophys. J. Int. **152**(2), 280–301 (2003). http://www.geophysik.uni-muenchen. de/Members/bunge/download/adjoint-paper.pdf
13. Bunge, H.P., Richards, M.A., Baumgardner, J.R.: A sensitivity study of three-dimensional spherical mantle convection at 10^8 Rayleigh number: effects of depth-dependent viscosity, heating mode, and an endothermic phase change. J. Geophys. Res. **102**, 11991–12007 (1997)
14. Bunge, H.P., Richards, M., Lithgow-Bertelloni, C., Baumgardner, J.R., Grand, S., Romanow-icz, B.: Time scales and heterogeneous structure in geodynamic earth models. Science **280**, 91–95 (1998). http://www.geophysik.uni-muenchen.de/~bunge/downloads/gemlab.pdf
15. Burstedde, C., Stadler, G., Alisic, L., Wilcox, L.C., Tan, E., Gurnis, M., Ghattas, O.: Large-scale adaptive mantle convection simulation. Geophys. J. Internat. **192**(3), 889–906 (2013)

16. Burstedde, C., Wilcox, L.C., Ghattas, O.: `p4est`: Scalable algorithms for parallel adaptive mesh refinement on forests of octrees. SIAM J. Sci. Comp. **33**(3), 1103–1133 (2011)
17. CIG – Computational Infrastructure for Geodynamics: ASPECT: Advanced Solver for Problems in Earth's ConvecTion, User Manual (2015), version 1.3
18. Council, N.R.: Origin and Evolution of Earth: Research Questions for a Changing Planet. The National Academies Press, Washington, DC (2008). http://www.nap.edu/catalog/12161/origin-and-evolution-of-earth-research-questions-for-a-changing
19. Davies, D.R., Davies, J.H., Bollada, P.C., Hassan, O., Morgan, K., Nithiarasu, P.: A hierarchical mesh refinement technique for global 3-D spherical mantle convection modelling. Geosci. Model Dev. **6**(4), 1095–1107 (2013)
20. Davies, D.R., Goes, S., Davies, J.H., Schuberth, B.S.A., Bunge, H.P., Ritsema, J.: Reconciling dynamic and seismic models of Earth's lower mantle: the dominant role of thermal heterogeneity. Earth Planet. Sci. Lett. **353–354**(1), 253–269 (2012)
21. Dziewonski, A.M., Anderson, D.L.: Preliminary reference Earth model. Phys. Earth Plan. Int. **25**, 297–356 (1981)
22. Elman, H.C., Silvester, D.J., Wathen, A.J.: Finite Elements and Fast Iterative Solvers: With Applications in Incompressible Fluid Dynamics. Oxford University Press, New York (2005)
23. Engelman, M.S., Sani, R.L., Gresho, P.M.: The implementation of normal and/or tangential boundary conditions in finite element codes for incompressible fluid flow. Int. J. Numer. Methods Fluids **2**(3), 225–238 (1982)
24. Fichtner, A., Kennett, B.L.N., Igel, H., Bunge, H.P.: Full seismic waveform tomography for upper-mantle structure in the Australasian region using adjoint methods. Geophys. J. Int. **179**(3), 1703–1725 (2009)
25. Gaspar, F.J., Notay, Y., Oosterlee, C.W., Rodrigo, C.: A simple and efficient segregated smoother for the discrete Stokes equations. SIAM J. Sci. Comput. **36**(3), A1187–A1206 (2014)
26. Girault, V., Raviart, P.A.: Finite Element Methods for Navier-Stokes Equations. Springer, New York (1986)
27. Gmeiner, B., Huber, M., John, L., Rüde, U., Waluga, C., Wohlmuth, B.: Massively parallel large scale stokes flow simulation. In: Binder, K., Müller, M., Kremer, M., Schnurpfeil, A. (eds.) NIC Symposium 2016. Schriften des Forschungszentrums Jülich, NIC Series, vol. 48, pp. 333–341. ISBN:978-3-95806-109-5
28. Gmeiner, B., Huber, M., John, L., Rüde, U., Wohlmuth, B.: A quantitative performance analysis for Stokes solvers at the extreme scale (submitted, arXiv:1511.02134)
29. Gmeiner, B., Rüde, U., Stengel, H., Waluga, C., Wohlmuth, B.: Performance and scalability of hierarchical hybrid multigrid solvers for stokes systems. SIAM J. Sci. Comput. **37**(2), C143–C168 (2015)
30. Gmeiner, B., Rüde, U., Stengel, H., Waluga, C., Wohlmuth, B.: Towards textbook efficiency for parallel multigrid. Numer. Math. Theory Methods Appl. **8**, 22–46 (2015)
31. Gmeiner, B., Waluga, C., Wohlmuth, B.: Local mass-corrections for continuous pressure approximations of incompressible flow. SIAM J. Numer. Anal. **52**(6), 2931–2956 (2014)
32. Grand, S.P., van der Hilst, R.D., Widiyantoro, S.: Global seismic tomography: a snapshot of convection in the earth. GSA Today **7**, 1–7 (1997)
33. Hager, G., Treibig, J., Habich, J., Wellein, G.: Exploring performance and power properties of modern multi-core chips via simple machine models. Concurr. Comput. **28**, 1–2 (2014)
34. Hartley, R.A., Roberts, G.G., White, N., Richardson, C.: Transient convective uplift of an ancient buried landscape. Nat. Geosci. **4**, 562–565 (2011)
35. Haskell, N.A.: The motion of a fluid under a surface load. Physics **6**, 265–269 (1935)
36. Höink, T., Lenardic, A.: Three-dimensional mantle convection simulations with a low-viscosity asthenosphere and the relationship between heat flow and the horizontal length scale of convection. Geophys. Res. Lett. **35**, L10304 (2008)
37. Huber, M., Gmeiner, B., Rüde, U., Wohlmuth, B.: Resilience for multigrid software at the extreme scale (preprint, arXiv:1506.06185)
38. Huber, M., John, L., Pustejovska, P., Rüde, U., Waluga, C., Wohlmuth, B.: Solution Techniques for the Stokes System: a priori and a posteriori modifications, resilient algorithms. In Proceedings of the ICIAM, Beijing (2015). arXiv:151105759

39. Mitrovica, J.X.: Haskell [1935] revisited. J. Geophys. Res. **101**, 555–569 (1996)
40. Müller, R.D., Sdrolias, M., Gaina, C., Roest, W.R.: Age, spreading rates, and spreading asymmetry of the world's ocean crust. Geochem. Geophy. Geosy. **9**, 1525–2027 (2008)
41. Oeser, J., Bunge, H.P., Mohr, M.: Cluster Design in the Earth Sciences: TETHYS. In: Gerndt, M., Kranzlmüller, D. (eds.) High Performance Computing and Communications – Second International Conference, HPCC 2006, Munich. Lecture Notes in Computer Science, vol. 4208, pp. 31–40. Springer (2006). http://www.springerlink.com/content/l18628n708k11127
42. Parnell-Turner, R., White, N., Henstock, T., Murton, B., Maclennan, J., Jones, S.M.: A continuous 55 million year record of transient mantle plume activity beneath Iceland. Nat. Geosci. **7**, 914–919 (2014)
43. Resovsky, J., Trampert, J.: Using probabilistic seismic tomography to test mantle velocity–density relationships. Earth Planet. Sci. Lett. **215**(1), 121–134 (2003)
44. Ricard, Y.: Physics of mantle convection. In: Schubert, G. (ed.) Treatise on Geophysics, vol. 7. Elsevier, Amsterdam (2007)
45. Ritsema, J., von Heijst, H.J., Woodhouse, J.H.: Global transition zone tomography. J. Geophys. Res. **109**, B02302 (2004)
46. Rudi, J., Malossi, A.C.I., Isaac, T., Stadler, G., Gurnis, M., Staar, P.W.J., Ineichen, Y., Bekas, C., Curioni, A., Ghattas, O.: An Extreme-scale Implicit Solver for Complex PDEs: Highly Heterogeneous Flow in Earth's Mantle. In: Proceedings of the International Conference for High Performance Computing, Networking, Storage and Analysis (SC '15), pp. 5:1–5:12. ACM, New York (2015). http://doi.acm.org/10.1145/2807591.2807675
47. Schöberl, J., Zulehner, W.: On Schwarz-type smoothers for saddle point problems. Numer. Math. **95**(2), 377–399 (2003)
48. Seton, M., Müller, R.D., Zahirovic, S., Gaina, C., Torsvik, T.H., Shephard, G., Talsma, A., Gurnis, M., Turner, M., Maus, S., Chandler, M.: Global continental and ocean basin reconstructions since 200 ma. Earth-Sci. Rev. **113**, 212–270 (2012)
49. Stixrude, L., Lithgow-Bertelloni, C.: Thermodynamics of mantle minerals – I. Physical properties. Geophys. J. Int. **162**, 610–632 (2005)
50. Sundar, H., Stadler, G., Biros, G.: Comparison of multigrid algorithms for high-order continuous finite element discretizations. Numer. Linear Algebra Appl. **22**(4), 664–680 (2015)
51. Tackley, P.J.: Effects of strongly variable viscosity on three-dimensional compressible convection in planetary mantles. J. Geophys. Res. **101**, 3311–3332 (1996)
52. Tackley, P.J.: Mantle convection and plate tectonics: toward an integrated physical and chemical theory. Science **16**, 2002–2007 (2000)
53. Tackley, P.J., Stevenson, D.J., Glatzmaier, G.A., Schubert, G.: Effects of multiple phase transitions in a three-dimensional spherical model of convection in earth's mantle. J. Geophys. Res. **99**(B8), 15877–15901 (1994)
54. Tan, E., Choi, E., Thoutireddy, P., Gurnis, M., Aivazis, M.: GeoFramework: coupling multiple models of mantle convection within a computational framework. Geochem. Geophy. Geosy. **7**(6), Q06001 (2006)
55. Urquiza, J.M., Garon, A., Farinas, M.I.: Weak imposition of the slip boundary condition on curved boundaries for Stokes flow. J. Comput. Phys. **256**, 748–767 (2014)
56. Verfürth, R.: Finite element approximation of incompressible Navier-Stokes equations with slip boundary condition. Numer. Math. **50**(6), 697–721 (1987)
57. Vogt, P.R.: Asthenosphere motion recorded by the ocean floor south of Iceland. Earth Planet. Sci. Lett. **13**, 153–160 (1971), http://www.sciencedirect.com/science/article/pii/0012821X7190118X
58. Waluga, C., Wohlmuth, B., Rüde, U.: Mass-corrections for the conservative coupling of flow and transport on collocated meshes. J. Comp. Phys. 305, 319–332 (2016)
59. Weismüller, J., Gmeiner, B., Ghelichkhan, S., Huber, M., John, L., Wohlmuth, B., Rüde, U., Bunge, H.P.: Fast asthenosphere motion in high-resolution global mantle flow models. Geophys. Res. Lett. **42**(18), 7429–7435 (2015)

60. Williams, S.W., Waterman, A., Patterson, D.A.: Roofline: an insightful visual performance model for floating-point programs and multicore architectures. Tech. Rep. UCB/EECS-2008-134, EECS Department, University of California, Berkeley (Oct 2008)

61. Zhong, S., McNamara, A., Tan, E., Moresi, L., Gurnis, M.: A benchmark study on mantle convection in a 3-D spherical shell using CitcomS. Geochem. Geophy. Geosy. **9**, Q10017 (2008)

62. Zhong, S., Zuber, M.T., Moresi, L., Gurnis, M.: The role of temperature-dependent viscosity and surface plates in spherical shell models of mantle convection. J. Geophys. Res. **105**(B5), 11063–11082 (2000)

63. Zulehner, W.: Analysis of iterative methods for saddle point problems: a unified approach. Math. Comput. **71**(238), 479–505 (2002)

Part VI
ExaFSA: Exascale Simulation
of Fluid–Structure–Acoustics Interactions

Partitioned Fluid–Structure–Acoustics Interaction on Distributed Data: Coupling via preCICE

Hans-Joachim Bungartz, Florian Lindner, Miriam Mehl, Klaudius Scheufele, Alexander Shukaev, and Benjamin Uekermann

Abstract One of the great prospects of exascale computing is to simulate challenging highly complex multi-physics scenarios with different length and time scales. A modular approach re-using existing software for the single-physics model parts has great advantages regarding flexibility and software development costs. At the same time, it poses challenges in terms of numerical stability and parallel scalability. The coupling library preCICE provides communication, data mapping, and coupling numerics for surface-coupled multi-physics applications in a highly modular way. We recapitulate the numerical methods but focus particularly on their parallel implementation. The numerical results for an artificial coupling interface show a very small runtime of the coupling compared to typical solver runtimes and a good parallel scalability on a number of cores corresponding to a massively parallel simulation for an actual, coupled simulation. Further results for actual application scenarios from the field of fluid–structure–acoustic interactions are presented in the next chapter.

1 Introduction

The upcoming exascale era will allow the simulation of a new range of multi-physics simulations that are unfeasible with current compute resources. These simulations promise breakthrough insights in climate simulation, human body simulation, and many engineering problems [10]. To succeed in setting up a multi-physics simulation environment in an acceptable time, we have to face the challenge of reducing the complexity of the involved software and the involved numerical

H.-J. Bungartz • A. Shukaev • B. Uekermann (✉)
Scientific Computing in Computer Science, Technical University of Munich, München, Germany
e-mail: bungartz@in.tum.de; uekerman@in.tum.de

F. Lindner • M. Mehl • K. Scheufele
Institute for Parallel and Distributed Systems, University of Stuttgart, Stuttgart, Germany
e-mail: florian.lindner@ipvs.uni-stuttgart.de; miriam.mehl@ipvs.uni-stuttgart.de; klaudius.scheufele@ipvs.uni-stuttgart.de

© Springer International Publishing Switzerland 2016   239
H.-J. Bungartz et al. (eds.), *Software for Exascale Computing – SPPEXA 2013-2015*, Lecture Notes in Computational Science and Engineering 113, DOI 10.1007/978-3-319-40528-5_11

methods of such simulations. At the same time, we have to exploit the potential of massively parallel exascale computers fully to achieve the required accuracy by a combination of an accurate multi-physics model with the correspondingly high grid resolutions. In addition, the computational requirements are further increased due to multi-scale effects, instabilities, and complex and changing computational domain geometries.

By breaking up multi-physics simulation into single-physics problems and their coupling, we can benefit from decades of experience in every single-physics discipline. This approach is called the partitioned approach in contrast to the monolithic approach that considers the complete multi-physics model as a single large system and, thus, discretizes and solves it as such. The monolithic approach allows the developers to fully adapt the solver software to the underlying numerical discretization, the layout of data structures, and the linear and nonlinear system solvers in an optimal way for the given multi-physics model. However, this requires a costly and time-consuming new implementation of a solver for the specific combination of physical phenomena. Therefore, many simulation tools use a partitioning of the coupled model into single-physics parts at various levels starting from the preconditioning of the (non-)linear solvers, over the matrix assembly and time stepping up to the simulation software itself. The advantages of the partitioned approach come at the price of possible instabilities due to a too loose numerical coupling of the model parts. Also, the partitioned simulation environment is not trivial to parallelize in an efficient way. We focus on the case where we bring the partitioning to its limits by using black-box solvers for the involved single-physics model parts. This means that our solvers are accessible only via their interfaces for input and output values. This allows to include legacy codes and to study the coupling independent from the details of the numerical approach of each single-physics field.

Still, sub-iterating over all single-physics fields allows to recover the monolithic solution. A coupling software has to cover three building blocks to cope with this situation. Each of them has to work efficiently with distributed data: communication of data between separate executables, interpolation methods between non-matching meshes, and efficient solvers for fixed-point equations derived from coupling conditions. The open-source[1] coupling library preCICE[2] offers methods for all three building blocks while allowing for a minimally invasive integration into existing single-physics codes [5, 9]. The library focuses on surface-coupled problems, which differ from volume-coupling in terms of parallel communication layouts and memory requirements. For a comparison of preCICE to other coupling software, the reader is referred to [5]. In this chapter, we describe how we realized the three building blocks on distributed data and analyze their efficiency. In the next chapter (Partitioned Fluid-Structure-Acoustics Interaction on Distributed Data: Numerical Results and Visualization), we study fluid–structure–acoustic interactions as an

---

[1] preCICE is licensed under LGPL3.

[2] www.precice.org

example of a complex surface-coupled multi-physics problem. In this chapter, Sect. 2 describes the algorithmic approach and its realization on distributed data for each of the three building blocks. Sect. 3 presents scalability results for each part as well as the overall coupling using an artificial test case. This chapter concludes in Sect. 4.

2 Coupling Building Blocks on Distributed Data

In this section, we present the methodological and algorithmic realization of the three main building blocks of partitioned black-box coupling that have to be available to glue together different solvers at the coupling surface: (1) communication between processes of several separate parallel solvers, (2) data mapping between non-matching meshes, and (3) iterative solvers for the interface fixed-point equations. We briefly recapitulate the theoretical background for each of them, but mainly focus on the efficient parallelization of the work by time step. The initialization phase is not the current focus of optimization. Longer setup times can be tolerated for a high number of time steps. Future work will, however, focus on the initialization as well. Please also compare the remarks in Sect. 4.

We restrict the description to two coupled solvers A and B, which we refer to as bi-coupling. However, all components can be generalized to coupled simulations with more than two solvers (cf. [6]). Both solvers are parallel solvers using their own domain/mesh partitioning. This partitioning of both participants implies also a partitioning of their surface meshes at the coupling interface, cf. Fig. 1. Without loss of generality, we assume that the ranks of the respective processes are $0, \ldots, p_A - 1$ for solver A and $0, \ldots, p_B - 1$ for solver B. We have

$$\bigcup_{i=0}^{p_A-1} \Gamma_A^i = \Gamma_A, \quad \bigcup_{i=0}^{p_B-1} \Gamma_B^i = \Gamma_B,$$

Fig. 1 Schematic example for two two-dimensional solvers A and B both using domain partitioning with some, but not all of their partitions involved in the coupling at the common surface (coupling surface)

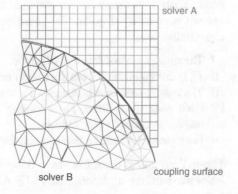

for the respective coupling surface partitions Γ_A^i and Γ_B^i of the solvers and the realizations Γ_A and Γ_B of the complete coupling surface in the solvers. In addition, we assume that rank 0 of each solver runs in master mode, all others in slave mode. At the current state, we assume static partitioning of the coupling surface in both solvers. This is true if both solvers use Lagrangian or Arbitrary Lagrangian Eulerian (ALE) mesh concepts and no re-meshing or dynamical load balancing. In Eulerian meshes, the coupling surface, however, can move relative to the computational mesh. Thus, we get a dynamical partitioning of the coupling surface in this case even for an overall static partitioning.

We present the details of the communication between the surface partitions of solver A and B in Sect. 2.1, the data mapping methods between the non-matching meshes at the coupling surface in Sect. 2.2, and the iterative methods solving the interface equation in Sect. 2.3.

2.1 Communication of Distributed Data

In order to define an overall communication between participants A and B, communication relations between the processes of the participants have to be established. The communication requirements are closely connected to the data mapping (see Sect. 2.2). The data mapping is executed by the processes of one of the solvers. Without loss of generality, we assume, that this is solver A. Accordingly, the surface mesh Γ_B is copied from B to A and re-partitioned at A. This copy operation and the re-partitioning are the first part of the communication initialization and described in Sect. 2.1.1, whereas the second part comprises the setup of the point-to-point communication relations between solver processes (see Sect. 2.1.2).

2.1.1 Surface Mesh Re-Partitioning

We treat surface meshes as unstructured point clouds, which can optionally also hold connectivity information as for example edges or triangles. Partitions of meshes are not necessarily disjoint. The mesh copy and re-partitioning consists of five algorithmic steps:

  I The master rank of B gathers Γ_B from all ranks of B.
 II Γ_B is communicated from the master rank of B to the master rank of A.
III The master rank of A broadcasts Γ_B to all ranks of A.
IV Each rank of A filters Γ_B according to its partition of Γ_A and the defined mappings between both meshes.
 V Each rank of A sends a feedback on the filtering to the master rank of A.

Figure 2 visualizes these five steps. After the re-partitioning, the master rank of A holds two partitioning descriptions of Γ_B. As the re-partitioning builds on the storage

I: Gather Mesh
II: Communicate Mesh
III: Broadcast Mesh
IV: Filter Mesh
V: Feedback

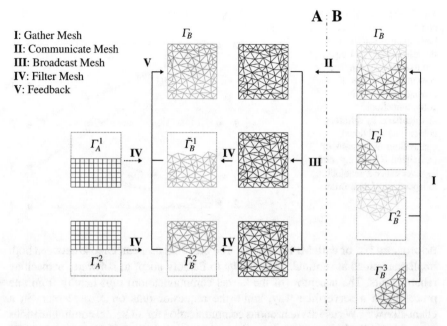

Fig. 2 Step wise description of the copy and mesh re-partitioning in the initialization phase of the inter-solver communication: coupling mesh Γ_B is send from participant A to B at initialization and re-partitioned there. Participant A and B run on three respectively four processors. The user defines the local partitioning of Γ_A on A, marked in *black*, and the local partitioning of Γ_B on B, marked in *blue*. The re-partitioning constructs the partitioning of Γ_B on A, marked in *green*

of the complete coupling mesh at a single rank, we rely on the minimal storage requirements of a surface coupling. This, still, contradicts exascale computing paradigms, and can, therefore, be substituted at a later software development step in a modular way without interfering with the implementation of the data mapping or the equation coupling.

2.1.2 Point-to-Point Communication

After the mesh re-partitioning described above, participants A and B play the role of the acceptor and the requestor of the communication, respectively. This is consistent with the classic nomenclature of, for example, MPI ports: the acceptor publishes connection information while the requestor pulls for such information. In our implementation in preCICE, the M:N communication between A and B builds upon multiple 1:N communications, which mark the kernel of our communication (cf. Fig. 3). The 1:N communications are implemented either with TCP/IP based on

Fig. 3 Example of an M:N communication between participant A (on three processors) and participant B (on five processors). The communication layout consists of three 1:N communications, distinguished by different colors. The individual connections at one rank of participant B (here e.g. rank 2) never cause a deadlock, independent of their order

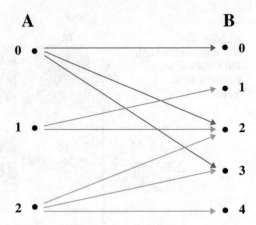

Boost.Asio[3] or with MPI-2.0 ports. The preCICE user can choose between both implementations at runtime. This allows to flexibly adapt to software or machine requirements. The acceptor (of the kernel communication) runs hereby from one processor in a server-like way, while the requestor runs on N processors in a client-like way. We use asynchronous communication for all kernel communications to avoid deadlocks resulting from the ordering of multiple 1:N communication channels at one processor. We tested an alternative realization based on OpenMP threads, but asynchronous communication appeared to be more reliable over a broad range of MPI implementations due to thread-safety guarantee inconsistencies, though more involved to establish.

The initialization of the communication works as follows: after the mesh re-partitioning, the master ranks of both participants hold a description of the partitioning of Γ_B, i.e., all coupling surface mesh points and the corresponding processes in solver A or solver B, respectively. Both descriptions are exchanged and broadcast to all slaves afterwards. Finally, every rank of both solvers extracts which vertices it has to communicate with whom. Every rank of A accepts one 1:N communication as a server, where the N clients correspond to its communication partners among the ranks of solver B. Every rank of B requests multiple 1:N communications as a client (cf. Fig. 3). Note that this approach avoids any deadlocks though the accept/request of every kernel communication is blocking and every rank of B requests its communication channels sequentially and in an arbitrary order. A 1:1 kernel communication would not allow for such a simple, but elegant initialization. While the complete initialization still relies on global operations, the communication itself, once established, is completely local and needs no global synchronization. Furthermore, asynchronous communication results in a substantial efficiency boost. The point-to-point communication was realized as part of the master thesis of Alexander Shukaev [14].

---

[3] www.boost.org

2.2 Interpolation Methods on Distributed Data

The mapping of data between the two non-matching surface meshes of solver A and solver B is a prerequisite for any coupling scheme. The type of mapping has to be chosen carefully in order to ensure the required accuracy and to fulfill physical conservation laws. Currently, preCICE provides two standard interpolation methods: projection-based mapping and radial basis function mapping [4]. Both types work in a pure black-box setting, meaning that no discretization details of the underlying single-physics solvers are needed. Let $U_A \in \mathbb{R}^{N_A}$ and $U_B \in \mathbb{R}^{N_B}$ denote the values that we want to map, located at the nodes of Γ_A and Γ_B. An interpolation from Γ_B to Γ_A is a linear mapping and, thus, can be written as

$$U_A = H^{AB} U_B \,,$$

with the mapping matrix $H^{AB} \in \mathbb{R}^{N_A \times N_B}$. A mapping is called consistent if the entries of every row sum up to one [3]:

$$\sum_{j=1}^{N_B} H_{ij}^{AB} = 1 \quad \forall i = 1 \dots N_A \,.$$

This guarantees the exact mapping of constant functions. Consistent mappings are usually applied to values such as positions, fluxes, or densities [3]. If the entries of every column sum up to one, a mapping is called conservative:

$$\sum_{i=1}^{N_A} H_{ij}^{AB} = 1 \quad \forall j = 1 \dots N_B \,.$$

This guarantees the conservation of the sum of values and is usually applied to integral values such as forces. Obviously, every consistent mapping from Γ_A to Γ_B implies a conservative mapping in the reverse direction by simply transposing the coupling matrix $H^{BA} = (H^{AB})^T$. Thus, we restrict our considerations in this work to consistent mappings without loss of generality. In preCICE, consistent mappings are always realized from Γ_B to Γ_A, i.e., from the re-partitioned mesh to the local mesh. This simplifies the parallelization drastically. This often does not impose a constraint as, typically, one of the two mappings between the solvers is required to be consistent whereas the other one has to be conservative. Thus, we can always postulate that the consistent mapping is from Γ_B to Γ_A whereas the conservative mapping transports surface data from Γ_A to Γ_B if we choose the roles of solver A and solver B accordingly. Figure 4 visualizes this concept. In case that we need two consistent or two conservative mappings, both meshes have to be communicated to the partner solver and re-partitioned.

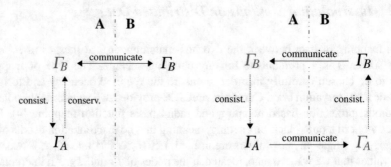

Fig. 4 Valid mapping combinations. preCICE only allows consistent mappings from a re-partitioned mesh to a local mesh and conservative mappings in the reverse direction (*left*). Changing such a conservative mapping to a consistent one, requires the computation of the mapping on the other participant (*right*). Now, both meshes need to be re-partitioned. Re-partitioned meshes are marked in *blue*

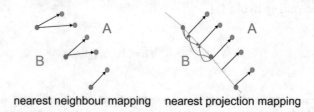

Fig. 5 Schematic view of the two projection-based mapping methods provided by preCICE in a simple two-dimensional case: the nearest neighbor mapping and the nearest projection mapping. Both mappings are displayed in their consistent variant. *Arrows* point in data transfer direction

2.2.1 Projection-Based Interpolation

Depending on whether participant B provides connectivity information for the mesh at its surface Γ_B, simple projection-based interpolation can be realized as a nearest neighbor (NN) or a nearest projection (NP) mapping.

Algorithmic Building Blocks of Projection-Based Interpolation The NN mapping simply looks for the closest neighbor of a Γ_A vertex among the vertices on Γ_B and copies the respective value. This is only a first order accurate method, but can be very useful to handle matching meshes (with non-matching partitions). The NP mapping looks for the closest neighboring element of a Γ_A vertex among the surface elements in Γ_B and computes the projection point of the Γ_A vertex on the Γ_B element (for example a triangle or a quad element). This is followed by a (bi-)linear interpolation from the element's vertices to the projection point. This interpolated value is then copied to the Γ_A vertex. If the mesh element size of Γ_B is significantly larger than the orthogonal distance between both meshes, this results in a second order accurate method. Figure 5 schematically shows the principles of both NN and NP in a very simple two-dimensional setting.

Realization of Projection-Based Interpolation on Distributed Data The realization of such projection methods on distributed data is almost trivial. The filter step of the re-partitioning of Γ_B has to be realized such that each rank of A chooses all vertices or elements from Γ_B that are the result of a projection or nearest neighbor search for a vertex from Γ_A. We realize this by a bounding box filter, followed by the computation of a preliminary mapping. There can be vertices or elements of Γ_B associated to several vertices and, thus, potentially several ranks of Γ_A (compare Fig. 5). This is not an issue, however, for our consistent mapping since mapping data from one vertex or element of Γ_B to several vertices of Γ_A does not destroy the consistency. For the backward mapping from Γ_A to Γ_B, using the transpose of the consistent mapping to achieve a conservative mapping, we have to accumulate mapping results on vertices of Γ_B from all ranks of Γ_A that contribute. During the actual computation of the mapping, no communication between ranks of A is needed.

2.2.2 Radial Basis Function (RBF) Interpolation

To map a variable from Γ_B to Γ_A, radial basis function interpolation builds up a global interpolant on Γ_B, which is then evaluated on Γ_A. As a basis, we use radially symmetric basis functions centered at the vertices of Γ_B. To ensure the exact interpolation of constant and linear functions, this basis is enriched with a global first order polynomial. In literature (compare e.g. [16]), various types of radial basis functions, with local or global support, are studied. Different approaches to parallelization of the algorithm have been discussed by [7] and [20]. Table 1 lists the classical choices, which are also implemented in preCICE.

The interpolant $s : \mathbb{R}^3 \to \mathbb{R}$, thus, reads

$$s(x) = \sum_{i=1}^{N_B} \gamma_i \cdot \varphi(\|x - x_i\|) + q(x) \,,$$

with the radial basis functions φ centered at the vertices x_i of Γ_B and the global linear function $q(x) = \beta_0 + \beta_1 x_1 + \beta_2 x_2 + \beta_3 x_3$ (three-dimensional case).

Algorithmic Building Blocks of the RBF Interpolation We aim for coefficients $\gamma_i \in \mathbb{R}$ and $\beta_i \in \mathbb{R}$ that fulfill the interpolation condition

$$s(x_i) = v_i^B \ \forall i = 1 \ldots N_B \,, \tag{1}$$

with v_i^B denoting the respective variable value at vertex x_i on Γ_B. In addition, we introduce the polynomial condition

$$\sum_{i=1}^{N_B} \gamma_i \cdot p(x^i) = 0 \,,$$

for every polynomial p of degree less or equal than q. This regularizes the underdetermined system (1). In matrix notation, we get:

$$
\underbrace{\begin{pmatrix} 0 & Q^T \\ Q & P \end{pmatrix}}_{C} \underbrace{\begin{pmatrix} \beta \\ \gamma \end{pmatrix}}_{p} = \begin{pmatrix} 0 \\ \omega \end{pmatrix}, \tag{2}
$$

where $P \in \mathbb{R}^{N_B \times N_B}$, $P_{i,j} = \varphi\left(\|x_i - x_j\|_2\right)$ and where the ith row of $Q \in \mathbb{R}^{N_B \times 4}$ looks like $(1\ x_{i,1}\ x_{i,2}\ x_{i,3})$. Once, we have resolved this system, we can evaluate the interpolant at the vertices of Γ_A:

$$
v_j^A = s(y_j) = \sum_{i=1}^{N_B} \gamma_i \varphi(\|y_j - x_i\|_2) + q(y_j) \quad \forall j = 1 \ldots N_A, \tag{3}
$$

where y_j denotes vertex j on Γ_A.

Realization of the RBF Interpolation on Distributed Data To solve the system (2), we use PETSc [2]. We decompose C row-wise among the ranks of A to distribute the compute effort. Since the basis functions often decay quickly to zero or are restricted to a local carrier, we use a sparse format to save C. The row-wise decomposition gives the master rank of solver A the first block of rows of C. Those rows contain the matrix Q^T. These master rows are dense independent

Table 1 Types of radial basis functions implemented in preCICE for data mapping between non-matching meshes. a is the so-called shape-parameter, $\|x\|$ the Euclidian distance of the evaluation point from the origin/center of the basis function. For basis functions with local support, the support radius is given by r, i.e., $\phi(\|x\|) = 0$ for $\|x\| > r$ and $\xi = \|x\|/r$ denotes the normalized distance from the origin/center

	φ	Support
Gaussian	$\exp\left(-(a\|x\|)^2\right)$	Global
Multiquadrics	$\sqrt{a^2 + \|x\|^2}$	Global
Inverse multiquadrics	$1/\sqrt{a + \|x\|^2}$	Global
Thin plate splines	$\|x\|^2 \log(\|x\|)$	Global
Volume splines	$\|x\|$	Global
Compact thin plate splines C2	$1 - 30\xi^2 - 10\xi^3 + 45\xi^4 - 6\xi^5 - 60\xi^3 \log \xi$	Local
Compact polynomial C0	$(1 - \xi)^2$	Local
Compact polynomial C6	$(1 - \xi)^8(32\xi^3 + 25\xi^2 + 8\xi + 1)$	Local

on whether we use radial basis functions with local or global support. During each mapping step, C is solved in a distributed manner using the PETSc GMRES solver. Afterwards, the complete solution vector p is broadcast, such that every rank can compute (3) for its part of Γ_A.

Matrix C has to be assembled only once at initialization if the solvers use Lagrangian or Arbitrarily Lagrangian Eulerian (ALE) meshes and we perform the mapping based on the positions of vertices in the reference domain instead of the physical domain. In this case, only the right-hand side of (2), i.e., the variable values at the vertices of Γ_B change after each mapping step. This leaves room for improvement as C could be factorized at initialization, only, and a potentially expensive preconditioner would pay off. Also, we currently start every GMRES iteration from zero. Restarting from the last solution, should result in a further speed-up. A last non-optimal part of the current preCICE implementation of the RBF mapping stems from the fact that we do not filter Γ_B on each rank of A during the re-partitioning of Γ_B in case of an RBF mapping, such that every rank of A has access to the complete mesh. For RBFs with global support, this is optimal, but for a local support, each rank of solver A only needs to store those vertices on Γ_B whose basis functions are non-zero at its own vertices of Γ_A. Exploiting this should reduce the memory requirements drastically.

2.3 Fixed-Point Acceleration Methods on Distributed Data

Interface coupling is numerically realized via fixed-point equations at the coupling surface. In case the respective fixed-point equation is fulfilled, the coupling conditions are fulfilled as well and thus, the partitioned solution for the respective time step is equivalent to the solution of the monolithic system. We illustrate this using the example of fluid–structure interactions. Here, we have two coupling conditions at the interface between fluid and structure: the kinematic conditions (equality of velocities and displacements) and the dynamics conditions (equality of forces). This allows us to formulate various fixed-point equations depending on, e.g., the execution order of the two solvers. The standard approach is the so-called sequential or staggered coupling where interface velocities/displacements are used as an input for the flow solver which computes the respective forces exerted by the fluid on the structure. These forces are used in a second step as an input to the subsequent structure solver instance that calculates a new iterate for the velocities/displacements. If this iterative process reaches a fixed-point, i.e., velocities/displacements reach a converged state, we fulfill both coupling conditions. An alternative, parallel, fixed-point equation is based on using both velocities/displacements and forces as an input vector for the parallel execution of flow and structure solver yielding new forces and velocities/displacements as an output. Figure 6 schematically shows the iteration concept for these two approaches in a general setting with two arbitrary solvers A and B. Whereas the serial coupling

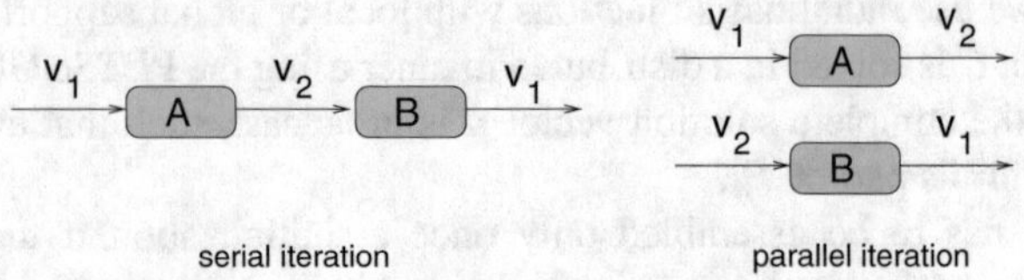

Fig. 6 Schematic illustration of two coupling iteration types with two involved solvers A and B and the respective input and output variables v_1 and v_2

is the standard approach, we favor the parallel coupling as only this allows for an efficient hardware usage in parallel simulations (see [17]).

In practice, the pure fixed-point iteration often needs to be accelerated in order to achieve convergence or a reasonable convergence speed for physically strongly coupled fields. The range of such acceleration methods yields from constant and dynamic under-relaxation to a more sophisticated and very powerful class of acceleration methods, the so-called quasi-Newton methods. As all these can be applied to any type of fixed-point interface equation, we use the generic fixed-point equation

$$H(x) = x \iff R(x) := H(x) - x = 0 \tag{4}$$

in the following description of the methods implemented in preCICE. We start with a short recapitulation of the serial algorithms before presenting the respective parallel algorithms in more detail.

2.3.1 Theory of Robust Quasi-Newton Fixed-Point Acceleration

The simple fixed-point iteration or its variant with under-relaxation

$$x^{k+1} = x^k + \omega \cdot H(x^k) \text{ with } \omega \in [0; 1]$$

are trivial to implement both in the serial and in the parallel case if the required communication and data mapping functionalities are given as described in Sects. 2.1 and 2.2. We, thus, do not present algorithmical details for these methods but restrict our description to the more sophisticated quasi-Newton methods.

The quasi-Newton methods we consider accelerate the fixed-point iteration by a subsequent Newton step:

$$x^{k+1} = \underbrace{H(x^k)}_{=: \tilde{x}^k} - J_{\tilde{R}}^{-1} (\underbrace{H(x^k) - x^k}_{= \tilde{x}^k - H^{-1}(\tilde{x}^k)}) \tag{5}$$

with $\tilde{R} := I - H^{-1}$ mapping the Picard iterate $\tilde{x}$ to the residual $r^k := R(x^k) = \tilde{R}(\tilde{x}^k) = H(x^k) - x^k$. As the Jacobian of R is not accessible for a black box approach, the task is to find a good approximation $\widehat{J}_{\tilde{R},k}^{-1}$ of the inverse Jacobian $J_{\tilde{R},k}^{-1}$. This can be done in several ways that are all based on the secant equation

$$\widehat{J}_{\tilde{R},k}^{-1} V_k = W_k \tag{6}$$

with the matrices V_k and W_k being composed of differences of input-output data throughout the coupling iterations within the current time step:

$$W_k = \left[\Delta \tilde{x}^k, \Delta \tilde{x}^{k-1}, \cdots, \Delta \tilde{x}^1 \right], \text{ with } \Delta \tilde{x}^i = \tilde{x}^i - \tilde{x}^{i-1},$$
$$V_k = \left[\Delta r^k, \Delta r^{k-1}, \cdots, \Delta r^1 \right], \text{ with } \Delta r^i = R(x^i) - R(x^{i-1}).$$

preCICE provides quasi-Newton methods based on the following two approaches to regularize the under-determined system (6) based on different norm-minimization conditions for $\widehat{J}_{\tilde{R},k}^{-1}(\tilde{x}^k)$: the **classical interface quasi-Newton (ILS) approach** (see [1, 11, 18]) uses the approximate with the minimal Frobenius norm, i.e.,

$$\| \widehat{J}_{\tilde{R},k}^{-1} \|_F \to \min . \tag{7}$$

The resulting Jacobian estimate $\widehat{J}_{\tilde{R},k}^{-1} = W_k (V_k^T V_k)^{-1} V_k^T$ is however not computed explicitly but instead, the update formula

$$x^{k+1} = H(x^k) - W_k \underbrace{(V_k^T V_k)^{-1} V_k^T r^k}_{=:\alpha_k}$$

is used, where the vector α is computed solving the least-squares optimization problem $\| V_k \alpha_k + r^k \|_2 \to \min$. The convergence properties can in some cases be greatly improved by incorporating information from previous time steps in the matrices V_k and W_k.

The **multi-vector quasi-Newton (IMVJ) approach** (see [11]) can be seen as generalized Broyden method as it minimizes the distance between $\widehat{J}_{\tilde{R},k}^{-1}$ and the approximate $\widehat{J}_{\tilde{R}}^{-1,(N)}$ from the previous time step:

$$\| \widehat{J}_{\tilde{R},k}^{-1} - \widehat{J}_{\tilde{R}}^{-1,(N)} \|_F \to \min . \tag{8}$$

This results in the Jacobian estimate

$$\widehat{J}_{\tilde{R},k}^{-1} = \widehat{J}_{\tilde{R}}^{-1,(N)} + \left(W_k - \widehat{J}_{\tilde{R}}^{-1,(N)} V_k \right) (V_k^T V_k)^{-1} V_k^T . \tag{9}$$

Using this method, we have to compute and store the full Jacobian $\widehat{J}_{\tilde{R},k}^{-1}$ which increases the computational cost and the storage requirements remarkably. Though, due to the implicit use of information from past time steps via the norm minimization condition, there is no need to retain old time steps explicitly in the secant equation (6).

2.3.2 Implementational Aspects of Quasi-Newton Coupling Iterations

Algorithmic Building Blocks The primary kernel of our quasi-Newton variants is the computation of the pseudo-inverse $(V_k^T V_k)^{-1} V_k^T$ of V_k (IMVJ) or the vector $\alpha = (V_k^T V_k)^{-1} V_k^T r^k$ (ILS). It is easy to show that finding $(V_k^T V_k)^{-1} V_k^T y$ for a given vector y is equivalent to solving the unconstrained least-squares minimization

$$\text{find } z \in \mathbb{R}^n \text{ with } z = \text{argmin}_{\bar{z} \in \mathbb{R}^n} \| V \bar{z} - y \|_2 \, . \tag{10}$$

As we use a suitable implementation of the **QR-decomposition** of V_k, this is not only better conditioned than the calculation of $(V_k^T V_k)^{-1}$, but also particularly efficient as we solve successive similar least-squares problems where V_k only grows by one column in each iteration. More specifically, economy sized QR-factorization turned out to be a good choice due to a good trade off between efficiency and accuracy (see [8, 19]). From the decomposition $V_k = QR$, we get z from solving the quadratic system

$$\widehat{R} z = \widehat{Q}^T y \tag{11}$$

via **backward substitution**. $\widehat{R} \in \mathbb{R}^{m \times m}$ denotes the first m rows of R and $\widehat{Q}$ contains the first m columns of Q (if $V_k \in \mathbb{R}^{n \times m}$). For the ILS approach, we calculate $z = \alpha = (V_k^T V_k)^{-1} V_k^T r^k$ using the right-hand side $y := r^k$, while, for the IMVJ method, we have to compute the columns of the matrix $Z_k = (V_k^T V_k)^{-1} V_k^T$ via solving (10) for all unit vectors $y = e_i \in \mathbb{R}^n, i = 1, \ldots, n$.

The matrix $V_k = [\Delta r^k, V_{k-1}(1 : m - 1)]$ is updated in each iteration by adding a column with the most recent information at the beginning and possibly dropping one column at the end to maintain the size m of a given sliding window or to avoid linear dependencies between the columns of V_k. In case of insertColumn, we correspondingly update the QR-factorization executing first a modified Gram-Schmidt orthogonalization of the new column v against the previous columns in $\widehat{Q}$ followed by Given rotations eliminating the newly created sub-diagonal entries in $\widehat{R}$. For deleteColumn, the matrix $\widehat{R}$ after deletion of column v is still upper triangular. See Algorithm 6 for both operations.

The backward substitution based on the QR-factorization delivers the vector $\alpha \in \mathbb{R}^m$ (for ILS) or the pseudo-inverse $Z_k = (V_k^T V_k)^{-1} V_k^T$ of V_k (for IMVJ), respectively. For the update of x^k to x^{k+1}, we additionally need a matrix–vector multiplication $W_k \alpha$ for the ILS method, several matrix–matrix products for the computation of $(W_k - J_{\widehat{R}}^{-1,(N)} V_k) Z_k$ as well as a larger matrix–vector product $\widehat{J}_{\widehat{R},k}^{-1} r^k$.

Algorithm 6 Pseudo code for the routines `insertColumn` and `deleteColumn` that represent the updated QR-factorization

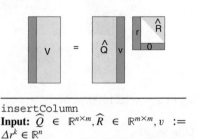 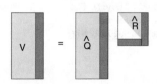

`insertColumn`
Input: $\widehat{Q} \in \mathbb{R}^{n \times m}, \widehat{R} \in \mathbb{R}^{m \times m}, v :=$
$\Delta r^k \in \mathbb{R}^n$
Output: $\widehat{Q} \in \mathbb{R}^{n \times m+1}, \widehat{R} \in \mathbb{R}^{m+1 \times m+1}$

for $j = 1$ to m **do**
$\quad r(j) \leftarrow \langle \widehat{Q}(:,j), v \rangle$
$\quad v \leftarrow v - r(j) \cdot \widehat{Q}(:,j)$
$r(m+1) \leftarrow \|v\|_2$
$\widehat{Q}(:, m+1) \leftarrow v/r(m+1)$
$\widehat{R} = \left[r, \begin{pmatrix} \widehat{R} \\ 0 \end{pmatrix} \right]$
choose Given rotations $G_{i,j}$ s. t.
$\quad \widehat{R} \leftarrow G_{1,2} \cdots G_{m,m+1}\widehat{R}$ is up. triangular
$\quad \widehat{Q} \leftarrow \widehat{Q}G_{m,m+1} \cdots G_{1,2}$ orthonormal

`deleteColumn`
input: $\widehat{Q} \in \mathbb{R}^{n \times m}, \widehat{R} \in \mathbb{R}^{m \times m}$
output: $\widehat{Q} \in \mathbb{R}^{n \times m-1}, \widehat{R} \in \mathbb{R}^{m-1 \times m-1}$

$\widehat{R} \leftarrow \widehat{R}(1 : m-1, 1 : m-1)$

Updated QR-Factorization on Distributed Data In order to make the QR-factorization as our first main algorithmic building block feasible for the execution on parallel systems with distributed data, a communication avoiding QR-factorization based on the `deleteColumn` and `insertColumn` operations described above is used. For p processors, $\widehat{Q}$ is decomposed into row-blocks corresponding to the interface data belonging to the respective surface partition. Thus, each process holds n/p rows in the ideally balanced case (see Fig. 7, left). As the number of columns m is very small (typically $m \ll n$), a copy of the matrix $\widehat{R}$ is held on each processor. With this distribution of data, the only communication between processors required for inserting a column is incurred by the dot-products $\langle \widehat{Q}(:,j), \Delta r^k \rangle$ and $\|\Delta r^k\|_2$. The local contributions of the dot-product are summed up in a reduce step and afterwards, the results are broadcast to all processors. The transformation using Givens rotations to restore a proper QR-factorization is fully local, i.e., no interprocess communication is involved. This is due to the fact that $\widehat{R}$ exists redundantly on each processor and every update operation on $\widehat{Q}$ solely requires information from other elements in $\widehat{Q}$ in the respective row.

If we consider the runtime on a parallel machine with p processors, we get the following estimates: the orthogonalization of the new column for `insertColumn`

Fig. 7 Decomposition and storage distribution of data of the matrices $\widehat{Q} \in \mathbb{R}^{n \times m}$ (*left*) and $Z_k \in \mathbb{R}^{m \times n}$ (*right*). The matrices V and W are distributed analogously to $\widehat{Q}$

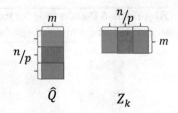

requires m dot-products of length n resulting in a parallel runtime of

$$O(mn/p) + O(m \log p) .$$

The $m^2/2$ Givens Rotations required in `insertColumn` each include the calculation of two dot-products of length m for the coefficients of the rotation (locally at each processor without communication), adding two columns of Q of length n, and adding two rows of $\widehat{R}$ of length m. The columns of Q are added in parallel according to the row-blocks stored in the processors. Thus, we get a total runtime for the Givens rotations of

$$O(m^2 n/p) + O(m^3) .$$

Backward Substitution on Distributed Data The next step after computing the QR-factorization is solving the quadratic system (11). This comprises computing $\widetilde{Q}^T y$ and the actual backward substitution. Since $\widehat{Q}$ is available in a distributed form in row-blocks, each processor computes an additive contribution to $\widehat{Q}^T y$ (runtime $O(mn/p)$). All contributions are summed up in a reduce step and redistributed to all processors via broadcast (runtime $O(m \log p)$) that each solve the small system (11) based on their copy of $\widehat{R}$ (runtime $O(m^2)$). The theoretical runtime, thus, accumulates to

$$O(mn/p) + O(m \log p) + O(m^2)$$

for the ILS method. For the IMVJ method, we have to do this for all unit vectors, i.e., $y = e_j, j = 1, \ldots, n$. Here, each processor can handle those n/p unit vectors that are associated to its row-block of $\widehat{Q}$. The matrix–vector products $\widehat{Q}^T e_j$ correspond to simply choosing column j from $\widehat{Q}^T$, for which no communication is required. Accordingly, each processor also only solves its n/p equations $\widehat{R} z_j = \widehat{Q}^T e_j$. Thus, the matrix Z_k is distributed among processors in column-blocks as depicted in Fig. 7 (right) and we get a total parallel runtime of

$$O(nm^2/p) .$$

Matrix–Vector Product $W_k \alpha$ on Distributed Data Analogue to $\widehat{Q}$, the matrix W_k is distributed to the processors in row-blocks. Thus, each processor can calculate its n/p entries of $W_k\alpha$ in $O(nm/p)$ runtime without communicating with the other processes.

Matrix–Matrix Products in $\widehat{J}_{\widetilde{R}}^{-1,(N)} + \left(W_k - \widehat{J}_{\widetilde{R}}^{-1,(N)} V_k\right) Z_k$ on Distributed Data

The realization of the IMVJ on distributed data is more complex due to its explicit computation of the Jacobian matrix and the resulting matrix multiplications. In particular, we are given two full matrix multiplications that need to be realized efficiently on distributed data, namely $\widehat{J}_{\widetilde{R}}^{-1,(N)} V_k$ and $\widetilde{W}_k Z_k$ with $\widetilde{W}_k = W_k - \widehat{J}_{\widetilde{R}}^{-1,(N)} V_k$.

On a machine with distributed data, the Jacobian matrix is distributed as column-blocks of size $n \times n/p$, V_k as line-blocks of size $n/p \times m$. The first multiplication $\widehat{J}_{\widetilde{R}}^{-1,(N)} V_k$ can thus be done by multiplying the local matrix-blocks $\widehat{J}_{\widetilde{R}}^{-1,(N)}|_p \cdot V|_p$ and sum up the p sub-results via a reduce operation (see Fig. 8). The overall result needs to be scattered to the sub-processes, i.e., each processor receives its corresponding block of rows. The costs for this product are, thus,

$$O(n^2 m/p) + O(nm \log p) .$$

The second multiplication $\widetilde{W}_k \cdot Z_k$ is more involved as each processor needs information, i.e., the corresponding sub-block of $\widetilde{W}_k$ from all other processors. This is solved by using a cyclic communication principle among the involved processors. In each cycle, each processor computes locally the product of the currently available block of $\widetilde{W}_k$ with Z_k and, afterwards, hands over its block of $\widetilde{W}_k$ to the next processor. After p cycles the matrix–matrix product is readily available and the sub-blocks of the result are completely available at the processors where they are supposed to be stored. Hence, no costly re-distribution has to be done. A schematic representation of the matrix-blocks and cyclic communication and computation is given in Fig. 8 on the right and in Fig. 9 for three processors p_A–p_C. The left part illustrates the communication and computation of certain matrix blocks with respect to time, while the right part shows the storage location of the respecting blocks in the distributed matrix. It is clearly visible that all the matrix entries within one processor are also computed locally. In order to interlace computation and communication we use asynchronous send and receive operations for the matrix blocks $\widetilde{W}_k$. Each matrix-block has a complexity of $O(m \cdot n/p \cdot n/p)$. The total runtime

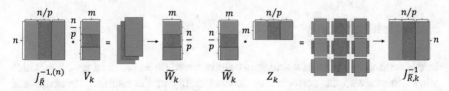

Fig. 8 Schematic view of the block-partitioned matrices and the parallel implementation of their products for the IMVJ quasi-Newton method

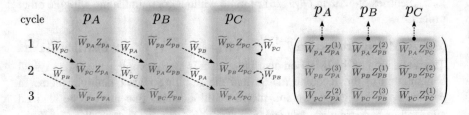

Fig. 9 Schematic representation of the cyclic communication and computation paradigm for the realization of the matrix multiplication $\widetilde{W}_k Z_k$ with $\widetilde{W}_k \in \mathbb{R}^{n \times m}, Z_k \in \mathbb{R}^{m \times n}$ on distributed data. The *left side* shows the aspects of communication of needed matrix blocks and computation of sub-results with respect to time. The *right side* gives the storage location of the computed blocks within the distributed $n \times n$ matrix

of this matrix–matrix product, meaning for all p blocks, thus, is

$$O(n^2 m/p) .$$

Matrix–Vector Product $\widehat{J}_{\widetilde{R},k}^{-1} r^k$ As displayed in Fig. 8, $\widehat{J}_{\widetilde{R},k}^{-1}$ is available on the processors in column-blocks that correspond exactly to the row-blocks of r^k such that each processor can calculate its additive contribution to all entries of the matrix–vector product without communication to other processors in $O(n^2/p)$ time. Afterwards, the complete result has to be accumulated in a reduce step and its row-blocks have to be scattered to the responsible processors. This sums to a parallel runtime of

$$O(n^2/p) + O(n \log p) .$$

Table 2 summarizes the parallel runtimes of all algorithmic building blocks of the ILS and the IMVJ quasi-Newton method and gives the complexities of the total parallel runtimes of both methods, as well.

Remark: The matrix $\widehat{J}_{\widetilde{R},k}^{-1}$ has a rank that is substantially smaller than n. This can be exploited to reduce the storage requirements and the computational costs of the IMVJ method. This is subject of ongoing research.

3 Scalability Study

In this section, we present a scalability study based on an artificial testcase, which allows to evaluate the performance and scalability of the interface numerics and communication isolated from solver runs. In an actual multi-physics simulation, the runtime of the interface numerics should be insignificant compared to the runtime of the involved solvers as the interface has a lower dimensionality than

Table 2 Overview of the parallel runtime complexities of the algorithmic building blocks of one iteration of the ILS and the IMVJ quasi-Newton approach

	ILS	IMVJ
QR-factorization	$O\left(\frac{m^2n}{p}\right) + O\left(m^3\right) + O\left(m\log p\right)$	$O\left(\frac{m^2n}{p}\right) + O\left(m^3\right) + O\left(m\log p\right)$
$\alpha = (V_k^T V_k)^{-1}V_k^T r^k$	$O\left(\frac{mn}{p}\right) + O(m\log p) + O(m^2)$	—
$Z_k = (V_k^T V_k)^{-1}V_k^T$	—	$O(nm^2/p)$
$W_k\alpha$	$O\left(\frac{mn}{p}\right)$	—
$\widehat{J}_{\bar{R}}^{-1,(N)} + (W_k - \widehat{J}_{\bar{R}}^{-1,(N)}V_k)Z_k$	—	$O(n^2m/p) + O(nm\log p)$
$\widehat{J}_{\bar{R},k}^{-1} r^k$	—	$O(n^2/p) + O(n\log p)$
Total	$O\left(\frac{m^2n}{p}\right) + O\left(m^3\right) + O\left(m\log p\right)$	$O(n^2m/p) + O(m^3) + O(nm\log p)$

the solver domains. To ensure this, we have to check that no interface computation becomes a severe bottleneck over a varying number of cores. Please note that, in our artificial test case, the number of cores only refers to the number of cores involved at the coupling surface such that the core count of the complete multi-physics corresponding to these interface cores is much higher. Scalability results for a complete fluid–structure–acoustic scenario realized with preCICE as a coupling tool are presented in the next book chapter.

3.1 Testcase Description

To isolate the coupling functionalities from solver effects, we developed a minimal coupling participant DummySolver. It does not solve any equations but only generates artificial values at the coupling surface that can then be used as input to preCICE. Thus, DummySolver produces only minimal computational load itself. It generates an equidistant Cartesian mesh of arbitrary size as coupling mesh. The mesh is linearized according to a linewise ordering and decomposed among the different processes by splitting the linear representation into partitions of similar size (cf. Fig. 10). In contrast to real-world solvers, we only generate nodes in a two-dimensional plane instead of a two-dimensional manifold in the full three-dimensional space.

For all experiments we use geometrically identical meshes emulating the surfaces of solver A and B. We have $n = \hat{n}^2$ unknowns at the coupling interface. The length and width of the coupling interface is fixed to $X = Y = 10$ resulting in a grid resolution of $10/\hat{n}$ in x- and y-direction. We show scaling results for different mesh

Fig. 10 DummySolver
decomposition of a 7×7
mesh into three partitions.
Arrows indicate the linewise
ordering of the mesh vertices

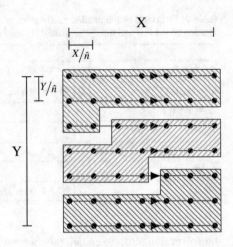

resolutions and different choices for all three building blocks in three test series:

(1) strong scaling for $n = 512^2 = 262{,}144$ unknowns on $p = 128, \cdots , 2048$ processors,[4]

(2) strong scaling for $n = 128^2 = 16{,}384$ unknowns on $p = 16, \cdots , 128$ processors,

(3) runtime complexity for $n = 16^2, \cdots , 128^2$ unknowns on $p = 32$ processors.

For all algorithmic building blocks, we measure both the *initialization* phase at the beginning of the simulation and the *work-per-time-step*. For the latter, we average over 10 time steps and 5 inner coupling iterations. Furthermore, we average the initialization of the communication over 3 runs as connection information is passed via files, resulting in some runtime uncertainty. Table 3 summarizes the ingredients for both phases and all three building blocks—communication, data mapping, and fixed-point equation solver. Both quasi-Newton schemes, ILS and IMVJ, do not reuse columns from old time steps. As radial basis functions, we use compact thin-plate splines with a support radius of twice the size of the grid resolution. We solve the RBF system up to a relative residual measure of 10^{-5}. All experiments were conducted on the thin nodes partition of SuperMUC, Leibniz Supercomputing Center in Garching (Sandy Bridge architecture with Xeon E5-2680 8C processors). Nodes comprise of 16 cores and are interconnected with an Infiniband FDR10.[5]

---

[4]We always refer to the number of processors per participant.

[5]For more details: https://www.lrz.de/services/compute/supermuc/systemdescription/.

Table 3 Summary and abbreviations for the algorithmic steps of all three building blocks—communication, data mapping, fixed-point equation solver—and separately the mesh re-partition, split up in *initialization* and in *work-per-time-step*

	Initialization	Work-per-timestep
Mesh re-partition (cf. Sect. 2.1.1)	- Gather coupling mesh (I)	
	- Communicate coupling mesh (II)	
	- Broadcast coupling mesh (III)	
	- Filter coupling mesh (IV)	
	- Feedback coupling mesh (V)	
Communication (cf. Sect. 2.1.2)	- Broadcast vertex distribution	- Point-to-point com. of data
	- Filter vertex distribution	
	- Establish all 1:N kernel com.	
Data mapping (cf. Sect. 2.1)	- proj.-map.: compute weights (NN)	- proj.-map.: copy data (NN)
	- RBF: assemble mapping	- RBF: solve mapping system
Fixed-point equation solver (cf. Sect. 2.3.2)	- ILS & IMVJ: set up data structures	- ILS & IMVJ: QR-decomposition
	- IMVJ: establish cyclic com.	- ILS & IMVJ: backsubstitution
		- IMVJ1: $\widetilde{J} = WZ$
		- IMVJ2: $\widetilde{W} = (W - J_{prev}V)$
		- IMVJ3: $\Delta x = J(-r)$

3.2   Strong Scaling for $n = 512^2 = 262,144$

For a total number of 262,144 unknowns at the coupling interface, we perform a strong scaling test series for the IQN-ILS method as well as for the communication, the mesh re-partition process and the nearest neighbor mapping. The IMVJ fixed-point acceleration method and the RBF-mapping are excluded due to large storage requirements.[6] Figure 11 shows the *initialization* timings. Despite the gather-scatter methodology, the mesh re-partition still remains in an acceptable range, in total around one second for 2048 processors. For such a number of processors the setup of the point-to-point communication, however, becomes increasingly costly. This is due to the file system access we use to exchange connection information. Substituting this by some proper publishing technique should simplify this step. The setup of the ILS fixed-point equation solver as well as the computation of the nearest neighbor mapping is almost negligible.

Figure 12 shows the *work-per-timestep* timings. The nearest neighbor mapping as well as the point-to-point communication are insignificant. The ILS fixed-point acceleration appears to be cheap as well, and might therefore remain a non-dominating and hence non-critical part in an overall multi-physics coupling. This confirms the findings of [12]. The time consumption of ILS stays rather constant over a varying number of processors. This is due to the low overall computational complexity of the method. The increase for $p = 2048$ processors is due to the small message size and the growing communication overhead. Further tests with an even larger N and, therefore, a better load to communication ratio show a nearly linear scaling. First tests with the IMVJ method pointed out various technical obstacles.

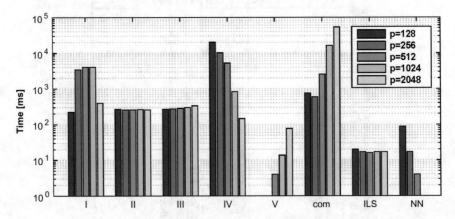

Fig. 11 Timings for the *initialization* of test series (1), $n = 512^2$, and all building blocks. Sizes under 1 ms are suppressed. Table 3 lists the corresponding abbreviations

---

[6]Optimizations of the memory requirements of IMVJ and RBF are possible and work in progress.

Fig. 12 Timings for the *work-per-timestep* of test series (1), $n = 512^2$, and all building blocks. The nearest-neighbor mapping showed timings below 0.01 ms for all runs

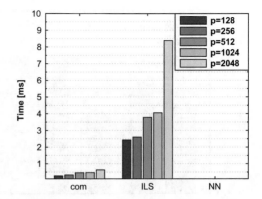

The cyclic communication runs into a well-known bug concerning the MPI Ports implementation on SuperMUC for more than 128 connections (Intel and IBM MPI). Building up the cyclic communication on TCP sockets allows to reach beyond 128 processors, but shows a rather poor performance. In the long run, we plan to directly reuse or duplicate the MPI communicator of the solver participant, though this leads to a software dependence of preCICE on MPI and, thus, to possible incompatibilities with commercial closed-source software. Furthermore, IMVJ is limited by its large storage consumption. We currently work on a subspace tracking method that exploits the low rank characteristics of the Jacobian.

3.3 Strong Scaling for $n = 128^2 = 16,384$

For a clean comparison of the two IQN methods ILS and IMVJ as well of the RBF mapping to the NN mapping, we study a further strong scaling series, but with an only moderate number of unknowns, $n = 16,384$. To rule out inter-dependencies, we compare both IQN approaches for a fixed setting, using only NN mappings, while we compare RBF to NN for a fixed setting with ILS only. Figure 13 shows the *initialization* timings for all configurations. The mesh re-partition remains at a low cost due to the small mesh sizes. Establishing the point-to-point communication as well as the cyclic communication for IMVJ comes at a certain cost, but appears tolerable as it is clearly below 1 s. The cyclic communication setup also explains the difference between ILS and IMVJ. For the sake of readability, we suppressed the RBF matrix assembly time in the graphs. It scales nearly quadratically with the inverse of the number of processors (due to the scaling of the number of matrix block entries), from $7.4 \cdot 10^5$ ms for 16 processors to $1.9 \cdot 10^4$ ms for 128 processors, but is, in general, too large. This is most probably due to the non-optimal usage of PETSc sparse matrices and is subject of upcoming optimization efforts.

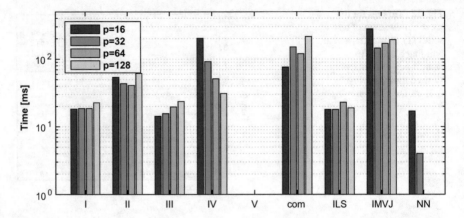

Fig. 13 Timings for the *initialization* of test series (2), $n = 128^2$, and all building blocks. Times below 1 ms are suppressed. RBF timings are not shown for sake of readability. They read: 741,350 ms ($p = 16$), 251,928 ms ($p = 32$), 74,690 ms ($p = 64$), and 19,061 ms ($p = 128$). Table 3 lists the corresponding abbreviations

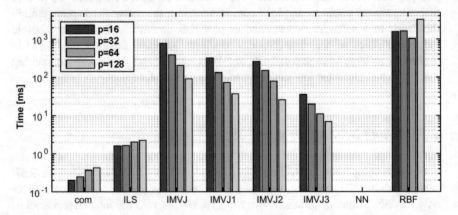

Fig. 14 Timings for the *work-per-timestep* of test series (2), $n = 128^2$, and all building blocks. The nearest-neighbor mapping showed timings below 0.1 ms for all runs

Figure 14 shows the *work-per-timestep*. As expected, the point-to-point commu-nication remains negligible. IMVJ comes with an approximately 100 times higher computational effort than ILS due to the higher complexity as well as the increased communication and memory requirements. Please note, that this difference is still subject to optimization as mentioned above and that it might be negligible for a costly solver. In such a situation, the numerics of the fixed-point equation solvers, meaning the number of necessary solver executions until termination, dominates and might render IMVJ favorable over ILS for certain cases. The RBF timings show a speed-up from 32 to 64 processors, but a worse performance for 128 processors, probably due to the small problem size. While the overall execution time is tolerable for some applications, it is surely not optimal and as well subject of current optimization efforts.

3.4 Varying Problem Size $n = 16^2, \cdots, 128^2$

For further insight, we finally study the runtime complexity depending on the number of unknowns at the interface for a fixed number of processors, $p = 32$. Figure 15 shows the *initialization* timings. The mesh re-partition rises, in general, with the number of unknowns, though the results appear to be dominated by memory latency due the rather small problem sizes. As expected, the setup of the point-to-point as well as the cyclic communication is almost independent from the problem size. The RBF matrix assembly explodes with the problem sizes, due the aforementioned reasons.

Figure 16, finally, lists the *work-per-timestep* for test series (3). The point-to-point communication, IQN and the nearest neighbor mapping show almost no dependence on the small problem size. Contrary, IMVJ grows significantly due the higher computational effort as well as the higher memory requirements. The RBF mapping also shows a reasonable complexity, despite the general high costs.

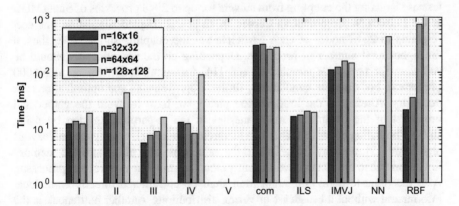

Fig. 15 Timings for the *initialization* of test series (3), $p = 32$, and all building blocks. Sizes under 1ms are suppressed. For sake of readability the logarithmic scale is not adjusted to the last RBF value (251,928 ms). Table 3 lists the corresponding abbreviations

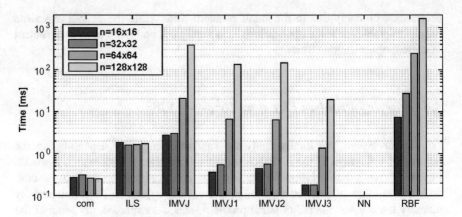

Fig. 16 Timings for the *work-per-timestep* of test series (3), $p = 32$, and all building blocks. The nearest-neighbor mapping showed timings below 0.1 ms for all runs

4   Conclusions

With this contribution, we could show that we can provide a scalable implementation of all basic coupling components required for a black-box partitioned simulation of surface-coupled multi-physics simulations. We scaled a pure interface testcase isolating the coupling from solvers for up to 2048 processes of SuperMUC at the Leibniz Supercomputing Center. In an actual coupled simulation this corresponds to 2048 nodes of each participant at the coupling surface and, thus, a far higher total number of processes. Depending on the scenario, this should be sufficient for peta-scale simulations (cf. [14] for simulations with up to 32,000 processes). For exascale simulations, further improvements and optimizations are necessary. Already at the initialization phase, we have to avoid the gather and broadcast of the complete surface meshes in the re-partitioning step by using suitable bounding boxes (possibly in a hierarchical manner), e.g., by the rendezvous algorithm [13], used by [15]. This at the same time also eliminates the memory limitation due to the fact that we currently store the complete mesh at one processor. Also the point-to-point communication shall be constructed directly from local information without a broadcast of vertex distributions. Another bottleneck is the publishing of connection information for the 1:N kernel communication that limits the number of cores in its file-based implementation. In terms of complexity of the actual numerical components, we have to reduce the storage and computational requirements for matrix assembly, matrix multiplication, and matrix storage in the RBF data mapping and the IMVJ interface quasi-Newton approach. Hereby, low-rank properties as well as matrix sparsity have to be fully exploited.

Summarizing the current state and ideas for further improvements, we can state that massively parallel and robust partitioned black-box simulations of surface coupled scenarios are possible. preCICE offers already a good basis for the complete set of required functionalities and will be further improved in current and future work.

Acknowledgements The financial support of the priority program 1648 Software for Exascale Computing (www.sppexa.de) of the German Research Foundation and of the Institute for Advanced Study (www.tum-ias.de) of the Technical University of Munich as well as provided computing time on the SuperMUC at the Leibniz Supercomputing Centre, are thankfully acknowledged.

References

1. Anderson, D.G.: Iterative procedures for nonlinear integral equations. J. ACM **12**(4), 547–560 (1965)
2. Balay, S., Abhyankar, S., Adams, M.F., Brown, J., Brune, P., Buschelman, K., Dalcin, L., Eijkhout, V., Gropp, W.D., Kaushik, D., Knepley, M.G., McInnes, L.C., Rupp, K., Smith, B.F., Zampini, S., Zhang, H.: PETSc users manual. Tech. Rep. ANL-95/11 - Revision 3.6, Argonne National Laboratory (2015). http://www.mcs.anl.gov/petsc
3. de Boer, A., van Zuijlen, A., Bijl, H.: Comparison of conservative and consistent approaches for the coupling of non-matching meshes. Comput. Method. Appl. Mech. Eng. **197**(49–50), 4284–4297 (2008).
4. Buhmann, M.: Radial basis functions. Acta Numer. **9**(January 2000), 1–38 (2000)
5. Bungartz, H.J., Lindner, F., Gatzhammer, B., Mehl, M., Scheufele, K., Shukaev, A., Uekermann, B.: preCICE – a fully parallel library for multi-physics surface coupling. Comput. Fliuds (2016)
6. Bungartz, H.J., Lindner, F., Mehl, M., Uekermann, B.: A plug-and-play coupling approach for parallel multi-field simulations. Comput. Mech. **55**(6), 1119–1129 (2015)
7. Deparis, S., Forti, D., Quarteroni, A.: A rescaled localized radial basis function interpolation on non-cartesian and nonconforming grids. SIAM J. Sci. Comput. **36**(6), A2745–A2762 (2014). http://dx.doi.org/10.1137/130947179
8. Fang, H.R., Saad, Y.: Two classes of multisecant methods for nonlinear acceleration. Numer. Linear Algebra **16**, 197–221 (2008)
9. Gatzhammer, B.: Efficient and flexible partitioned simulation of fluid-structure interactions. Phd thesis, Technische Universität München (2014)
10. Keyes, D., McInnes, L.C., Woodward, C.S., Gropp, W., Myra, E., Pernice, M., Bell, J., Brown, J., Clo, A., Connors, J., Constantinescu, E., Estep, D., Evans, K., Farhat, C., Hakim, A., Hammond, G., Hansen, G., Hill, J., Isaac, T., Jiao, X., Jordan, K., Kaushik, D., Kaxiras, E., Koniges, A., Lee, K., Lott, A., Lu, Q., Magerlein, J., Maxwell, R., McCourt, M., Mehl, M., Pawloski, R., Randles, A., Reynolds, D., Riviere, B., Rüde, U., Scheibe, T., Shadid, J., Sheehan, B., Shephard, M., Siegel, A., Smith, B., Tang, X., Wilson, C., Wohlmuth, B.: Multiphysics simulations: challenges and opportunities. Int. J. High Perform. Comput. Appl. **27**(1), 4–83 (2012)
11. Lindner, F., Mehl, M., Scheufele, K., Uekermann, B.: A comparison of various quasi-Newton schemes for partitioned fluid-structure interaction. In: Proceedings of 6th International Conference on Computational Methods for Coupled Problems in Science and Engineering, Venice, pp. 1–12 (2015)

12. Loffeld, J., Woodward, C.: Considerations and the implementation and use of Anderson acceleration on parallel computers. In: *Advances in the Mathematical Sciences: Research from the 2015 Association for Women in Mathematics Symposium.* AWM Springer Series (2016)
13. Plimpton, S.J., Hendrickson, B., Stewart, J.R.: A parallel rendezvous algorithm for interpolation between multiple grids. J. Parallel Distrib. Comput. **64**(2), 266–276 (2004)
14. Shukaev, A.K.: A fully parallel process-to-process intercommunication technique for preCICE. Master's thesis, Institut für Informatik, Technische Universität München (2015)
15. Slattery, S., Wilson, P., Pawlowski, R.: The data transfer kit: a geometric rendezvous-based tool for multiphysics data transfer. In: International Conference on Mathematics & Computational Methods Applied to Nuclear Science & Engineering (M&C 2013), pp. 5–9 (2013)
16. Smith, M.J., Cesnik, C.E.S., Hodges, D.H.: Evaluation of algorithms suitable for data transfer between noncontiguous meshes. J. Aerospace Eng. **13**(2), 52–58 (2000)
17. Uekermann, B., Bungartz, H.J., Gatzhammer, B., Mehl, M.: A parallel, black-box coupling for fluid-structure interaction. In: Idelsohn, S., Papadrakakis, M., Schrefler, B. (eds.) Computational Methods for Coupled Problems in Science and Engineering, COUPLED PROBLEMS 2013. Stanta Eulalia, Ibiza (2013). http://congress.cimne.com/coupled2013/proceedings/full/p559.pdf
18. Vierendeels, J., Degroote, J., Annerel, S., Haelterman, R.: Stability issues in partitioned FSI calculations. In: Bungartz, H.J., Mehl, M., Schäfer, M. (eds.) Fluid Structure Interaction II. Lecture Notes in Computational Science and Engineering, pp. 83–102. Springer, Berlin/Heidelberg (2010). http://link.springer.com/chapter/10.1007/978-3-642-14206-2_4
19. Walker, H.F., Ni, P.: Anderson acceleration for fixed-point iterations. SIAM J. Numer. Anal. **49**(4), 1715–1735 (Aug 2011). http://dx.doi.org/10.1137/10078356X
20. Yokota, R., Barba, L.A., Knepley, M.G.: PetRBF – a parallel O(N) algorithm for radial basis function interpolation with Gaussians. Comput. Method. Appl. Mech. Eng. **199**(25–28), 1793–1804 (2010).

Partitioned Fluid–Structure–Acoustics Interaction on Distributed Data: Numerical Results and Visualization

David Blom, Thomas Ertl, Oliver Fernandes, Steffen Frey, Harald Klimach, Verena Krupp, Miriam Mehl, Sabine Roller, Dörte C. Sternel, Benjamin Uekermann, Tilo Winter, and Alexander van Zuijlen

Abstract We present a coupled simulation approach for fluid–structure–acoustic interactions (FSAI) as an example for strongly surface coupled multi-physics problems. In addition to the multi-physics character, FSAI feature multi-scale properties as a further challenge. In our partitioned approach, the problem is split into spatially separated subdomains interacting via coupling surfaces. Within each subdomain, scalable, single-physics solvers are used to solve the respective equation systems. The surface coupling between them is realized with the scalable

---

D. Blom • A. van Zuijlen
Aerospace Engineering, Delft University of Technology, Delft, The Netherlands
e-mail: d.s.blom@tudelft.nl; A.H.vanZuijlen@tudelft.nl

B. Uekermann
Scientific Computing in Computer Science, Technical University of Munich, München, Germany
e-mail: uekerman@in.tum.de

H. Klimach • V. Krupp (✉) • S. Roller
Simulation Techniques and Scientific Computing, University of Siegen, Siegen, Germany
e-mail: harald.klimach@uni-siegen.de; verena.krupp@uni-siegen.de; sabine.roller@uni-siegen.de

M. Mehl
Institute for Parallel and Distributed Systems, University of Stuttgart, Stuttgart, Germany
e-mail: miriam.mehl@ipvs.uni-stuttgart.de

T. Ertl • O. Fernandes • S. Frey
VISUS, University of Stuttgart, Stuttgart, Germany
e-mail: thomas.ertl@visus.uni-stuttgart.de; oliver.fernandes@visus.uni-stuttgart.de; steffen.frey@visus.uni-stuttgart.de

D.C. Sternel
Institut for Scientific Computing, TU Darmstadt, Darmstadt, Germany
e-mail: doerte.sternel@hpc-Hessen.de

T. Winter
Institute of Numerical Methods in Mechanical Engineering, TU Darmstadt, Darmstadt, Germany,
e-mail: winter@fnb.tu-darmstadt.de

© Springer International Publishing Switzerland 2016   267
H.-J. Bungartz et al. (eds.), *Software for Exascale Computing – SPPEXA 2013-2015*, Lecture Notes in Computational Science and Engineering 113, DOI 10.1007/978-3-319-40528-5_12

open-source coupling tool preCICE described in the "Partitioned Fluid–Structure–Acoustics Interaction on Distributed Data: Coupling via preCICE". We show how this approach enables the use of existing solvers and present the overall scaling behavior for a three-dimensional test case with a bending tower generating acoustic waves. We run this simulation with different solvers demonstrating the performance of various solvers and the flexibility of the partitioned approach with the coupling tool preCICE. An efficient and scalable in-situ visualization reducing the amount of data in place at the simulation processors before sending them over the network or to a file system completes the simulation environment.

1   Introduction

The handling of fluid–structure–acoustics interactions (FSAI) in detailed, direct simulations is challenging and became feasible only recently on large scale computing systems. FSAI induces not only multiple physical domains but, at the same time, also different length and time scales. The possibility to consider the interaction of all those aspects in a single simulation enables us to gain a better understanding of the physical system and, thereby, to make more accurate predictions in engineering design processes. This can, for example, help to reduce the noise emission of technical devices such as aircraft, fans or wind turbines. Wind turbines are of increasing importance. While generating electricity, noise is emitted (due to complex rotor–wind–interaction). With their increasing size and growing number, limiting their noise emission becomes necessary. The multi-scale nature of the FSA interactions can be nicely seen in this example. Noise is generated in the boundary layers and resulting vortices at the moving geometry at a length scale in the order of centimeters. This whole turbine at the scale of meters while the noise emission is of relevance in a range of hundreds of meters up to a few kilometers. Simulating the entire domain while resolving the smallest turbulent scales and resolving the boundary layer adequately would require approximately 10^{18} degrees if freedom. This is out of reach for current systems.

Though tremendous computational power is available nowadays, such a simulation is still too demanding. In addition, acceptable software development times are possible only if we re-use existing scalable software based on decades of experience in each single-physics discipline. Therefore, we employ a partitioned coupling approach, i.e., we split the physical space into smaller domains, each covering a so-called single-physics subdomain. Within each of these subdomains, the specific physics are solved with a locally adapted resolution. The interaction between the subdomains is realized by exchanging data and applying suitable iterative solvers at the common surface. This coupled approach allows for adapted numerical methods and resolutions in each of the domains according to the physical requirements. By the adaptation of the numerical approximation in the individual domains, the computation of the complete interaction between fluid mechanics, structural mechanics, and acoustic wave propagation becomes feasible.

As the coupling mechanism needs to deal with various types of solvers and does not know or need to know anything about their internal numerical and technical details, this approach is also referred to as black-box coupling. A suitable tool to realize such a coupling in a parallel and scalable way is the open-source coupling library preCICE.[1] It handles without the data exchange, interpolations between non-matching meshes of adjacent domains, and iterative solvers for the coupling surface equations. As solvers, we use FEAP [17] or OpenFOAM[2] for structural mechanics, FASTEST [6] or OpenFOAM for the simulation of acoustic fluids, and Ateles[3] for the acoustic far field [1].

Naturally, this approach raises new numerical challenges such as stability issues due to the partitioned coupling and data interpolations between non-matching meshes. In addition, also the performance and scalability of the coupled black-box setup is a non-trivial task. The coupling library preCICE and the contributions to highly parallel partitioned coupling are described in "Partitioned Fluid–Structure–Acoustics Interaction on Distributed Data: Coupling via preCICE". In this chapter, we describe a real world application including not only the two-dimensional surface coupling, but the efficiency of the overall three-dimensional simulation. Section 2 presents the individual solvers and the solved equation systems. The two coupling strategies for fluid–structure interaction and fluid–acoustic interaction are described in Sect. 3. Section 4 deals with the visualization of the application with a focus on in-situ visualization and the developed data handling strategy. This is followed by a discussion of numerical results and scaling data for a three-dimensional testcase in Sect. 5.

2   Description of the Individual Solvers

In this section, we briefly present the physical background including the governing equations for each single-physics discipline. We then continue with a short description of the solvers for each part including the deployed numerical methods. The physical regimes we simulate are in the field of classical mechanics. We are interested in the motion of fluids and elastic structures. Acoustic waves are a special case of fluid motion, where small disturbances of the pressure in a compressible medium are considered.

The collection of solvers we use is a combination of inhouse solvers and open-source software. The focus is on testing the methodologies of the inhouse solvers in a complex coupled simulation environment on the one hand and to check the black-box parallel coupling concept of preCICE on the other hand. The solvers use sophisticated techniques that make them particularly suited for high performance simulations. For example, the flow and acoustic solver Ateles uses octree grids and high order Discontinuous Galerkin (DG) discretizations that are both known to perform well on massively parallel computers and model version of DG is

---

[1]http://www.precice.org
[2]http://www.openfoam.org/
[3]University of Siegen, STS

predestined for solving linear problem efficiently. FASTEST implements an internal volume coupling between an incompressible flow and an acoustic perturbation in order to be able to efficiently use space and time adaptivity corresponding to the different spatial and temporal scale of flow and acoustics. OpenFOAM and Feap are widely used open-source solvers, an important class of solvers that preCICE should be able to handle. In addition, they are used extensively in real-world applications in two of the authors' groups.

Whereas the splitting into an elastic structure subdomain and a fluid subdomain in an application scenario is obvious, the division between flow and acoustics is more involved: because acoustics is the phenomenon of travelling pressure waves in fluids, it is ultimately governed by the compressible fluid dynamics equations. However, only small perturbations in an otherwise constant fluid state are considered. Moreover, the resolution of all relevant scales for turbulent flows in a low Mach number regime with the speed of sound would be prohibitively expensive. Therefore, we suggest a domain splitting into acoustic near field and acoustic far field. The near field is the domain where acoustic pressure waves are generated by the fluid flow and, thus, coincides with the flow domain. The far field neglects the flow and takes into account only the acoustic pressure waves.

For the near field, we use two different approaches: the first approach uses the fully compressible fluid dynamics equations (using OpenFOAM), the second one is based on a splitting as, for example, proposed by Hardin and Pope [9]. They suggest (for comparably small flow velocities) to use an incompressible method for the calculation of the flow field, from which the acoustic sources can be deduced by the time derivative of the pressure. These sources can then be fed into an acoustic solver, which calculates the propagation of the acoustic waves. We have realized both approaches with the following setups:

(a) compressible flow (OpenFOAM) $\rightarrow$ acoustic far field (Ateles)
(b) incompressible flow (FASTEST) $\rightarrow$ acoustic near field (FASTEST) $\rightarrow$ acoustic far field (Ateles)

The arrows define the direction of interaction. In the following, we summarize first the governing fluid dynamics equations and the two solvers OpenFOAM (compressible) and FASTEST (incompressible), followed by a section on the acoustic equations both in the near field (used for approach (a)) and in the far field (used in approach (a) and (b)). Afterwards, the mechanics of elastic structures and the respective solvers OpenFOAM and FEAP are recapitulated.

2.1 Fluid Dynamics in the Acoustic Near Field

Fluid flow is governed by the compressible Navier-Stokes equations. These equations are the conservation of mass

$$\frac{\partial \rho^f}{\partial t} + \nabla \cdot \left(\rho^f \, \boldsymbol{v} \right) = 0 \,, \tag{1}$$

the balance of momentum

$$\rho^f \left(\frac{\partial v^f}{\partial t} + \left(\nabla \cdot v^f \right) v^f \right) = -\nabla p^f + \rho^f v \left(\Delta v^f + \frac{1}{3} \nabla (\nabla \cdot v^f) \right) + B \,, \qquad (2)$$

and the balance of energy

$$\frac{\partial \rho^f e}{\partial t} + \nabla \cdot \left(v^f \left(\rho^f e + p^f \right) \right) = 0 \,. \qquad (3)$$

In these equations, v^f denotes the velocity field, p^f the pressure field, ρ^f the density, v the kinematic viscosity, B an external force, and e the energy. The assumption of an ideal gas yields a relation between the pressure p^f and the energy e:

$$p^f = \rho^f R T = (\gamma - 1) \left(e - \frac{\rho^f v^f \cdot v^f}{2} \right) \,.$$

R is the ideal gas constant, T the temperature, and γ the isentropic coefficient. For an incompressible flow, the density is assumed to be constant and the conservation of mass in Eq. (1) is reduced to

$$\nabla \cdot v^f = 0 \,. \qquad (4)$$

Equation (3) become redundant in this case and (2) is reduced to

$$\frac{\partial v^f}{\partial t} + (\nabla \cdot v^f) v^f = -\nabla p^f + v \Delta v^f + B \,. \qquad (5)$$

2.1.1 OpenFOAM: Compressible Flow Solver

We use the foam-extend-3.1 package[4] for compressible flow simulations. It is a fork of the well known OpenFOAM package.[5]

foam-extend-3.1 uses second order finite volume discretization in space and a second order backward differencing time integration scheme. To solve the Navier-Stokes equations at all speeds, a coupled pressure-based algorithm [4, 5] is employed. Hence, the continuity and momentum equations are solved in a fully coupled implicit manner and, thereafter, the energy equation is solved in a segregated manner.

---

[4]http://www.extend-project.de/
[5]http://www.openfoam.org/

2.1.2 FASTEST: Incompressible Flow Solver

FASTEST [6] is a fully conservative finite volume code which is second order accurate in space and time. The block structured, boundary adjusted grid with free topology enables the use of an efficient algorithm and the application to moderately complex geometries.

2.2 Acoustic Wave Propagation

Acoustics can be considered as small perturbations of the flow field. We split the state variables of the flow into a background state denoted by the subscript 0 and the acoustic perturbations denoted by the superscript a, accordingly. For the density, we get $\rho^f = \rho^0 + \rho^a$, for the velocity $v^f = v^0 + v^a$ and for the pressure $p = p^0 + p^a$. Non-linear effects can be neglected for the acoustic perturbations ρ^a, v^a, and p^a. This leads to the linearized Euler equations given by the linearized equation for the conservation of mass

$$\frac{\partial \rho^a}{\partial t} + \nabla \cdot (v_0 \rho^a + \rho_0 v^a) = 0 , \tag{6}$$

the conservation of the velocity perturbation (neglecting friction)

$$\frac{\partial v^a}{\partial t} + \nabla \cdot \left(v_0 v^a + \frac{1}{\rho_0} p^a \right) = 0 \tag{7}$$

and the conservation of the pressure perturbation

$$\frac{\partial p^a}{\partial t} + \nabla \cdot (v_0 p^a + \gamma \, p_0 \, v^a) = q . \tag{8}$$

The background variables ρ_0, v_0, p_0 are constant variables which define the flow properties as the speed of sound $c = \sqrt{\frac{\gamma p_0}{\rho_0}}$, with which the acoustic perturbations are then transported through the far field. q denotes a source term, which generates the acoustic perturbations.

2.2.1 FASTEST: Acoustic Near Field

In FASTEST, the acoustic solver calculates the wave propagation in the acoustics near field in an integrated way (volume coupling) based on the incompressible background flow equations [11]. This integrated solver is based on the approach of Hardin and Pope [9] with the modifications of Shen and Sørensen [15, 16], which lead to a set of linearized Euler Eqs. (6), (7), and (8) with source term $q = -\frac{\partial p^0}{\partial t}$

(in this case p^0 is the incompressible pressure which is time variant) generating the acoustic perturbation in the near field. Considering this method, the state variables ρ^0, v^0, and p^0 are the incompressible flow variables.

To solve the linearized Euler equations, a finite volume scheme together with a dimensional splitting approach [18] is employed. For time stepping, an explicit Euler time discretization scheme is used. The flux computation is realized with a high resolution scheme that combines an exact Riemann solver and a Lax-Wendroff scheme with a suitable flux limiter. The details of the method are described in [11]. The internal coupling between the flow field and the acoustic perturbations is described in Sect. 3.2.

2.2.2 Ateles: Acoustic Far Field

The far field acoustic solver Ateles propagates acoustic waves solving (6), (7), and (8) based on boundary information provided at the coupling surface between acoustic near field and acoustic far field. For the far field, the source term $q = 0$, since acoustic waves are only propagation and not generated. The flow properties p_0, v_0, v_0 need to be given similar to the acoustic near field.

Ateles uses a modal high order Discontinuous Galerkin discretization as part of the APES framework [14]. The Discontinuous Galerkin (DG) method is based on a polynomial representation within each element and the flux calculation between elements [10]. Ateles allows for an arbitrary choice of the polynomial degree in the spatial representation and thus, an arbitrarily high order in space. This allows for a perfect tailoring to the physics and the computer architecture.

A high order approximation provides a low numerical dissipation and dispersion and a high accuracy with only few degrees of freedom. For nonlinear systems, high order implies increased computational costs, but for our linear system a modal scheme keeps the computational effort per degree of freedom constant over increased spatial orders and solves them efficiently. Such a discretization is, therefore, ideally suited for the simulation of the acoustic domain. The APES [14] framework, in which the solver is included, is designed such to take advantage of massively parallel systems.

2.3 Structural Dynamics

The configuration of the structure domain is described by the displacement u^s. An elastic and compressible structure is assumed. The governing equation in a Lagrangian description, i.e. with respect to the initial reference state Γ^s, is given by the balance of momentum

$$\rho^s \frac{\partial^2 u^s}{\partial t^2} = \nabla \cdot \left(J\sigma^s F^{-T}\right) + \rho^s g \quad \text{in } \Omega^s, \tag{9}$$

where the deformation gradient tensor F is defined as $F = I + \nabla u^s$, and the Jacobian J is the determinant of the deformation gradient tensor F. Applying the constitutive law for the St. Venant-Kirchhoff material, the Cauchy stress tensor σ^s becomes

$$\sigma^s = \frac{1}{J} F \left(\lambda^s (\operatorname{tr} E) I + 2\mu^s E \right) F^T \tag{10}$$

with the Young's modulus $E = \frac{1}{2} \left(F^T F - I \right)$, the relation of the shear modulus μ^s, and the Poisson's ratio ν given in [3].

2.3.1 OpenFOAM: Finite Volume Structure Solver

To facilitate implementation, the structure solver which is implemented within the foam-extend-3.1 framework, is used in combination with the compressible flow solver. Here, a finite volume discretization is used. The reader is referred to Cardiff et al. [3] for further details on the solid mechanics solver.

2.3.2 FEAP: Finite Element Structure Solver

In FSA simulations with FASTEST, the finite element code FEAP [17] is used. The spatial discretization is based on bilinear enhanced solid elements, and the time discretization scheme is the Newmark-Beta scheme [13].

3 Coupling

In this section, we describe the coupling between our three involved fields—fluid flow, elastic structure deformation, and acoustic wave propagation. As already mentioned above, this includes three coupling interfaces:

1. a surface coupling between structure and acoustic fluid (near field),
2. a surface coupling between acoustic fluid (near field) and acoustic wave propagation (far field),
3. a volume coupling between background flow and acoustic perturbations (near field).

The latter is realized in a monolithic way in our setup using the compressible flow solver in OpenFOAM and a software-internal volume coupling in the setup with FASTEST, respectively. For the two surface couplings, in the first step we assume that the interaction between near field and far field is unidirectional which is a feasible assumption also used in traditional approaches like [12] or [9]. An advantage in using a coupling library is that the interaction can be easily expanded to be bidirectional. Accordingly, a strong implicit coupling is used only at the

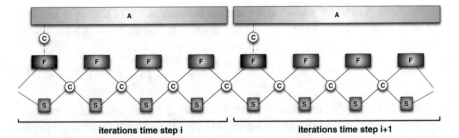

Fig. 1 Overview of the execution of the fluid–structure–acoustics simulations. Multiple calls to the fluid and solid solvers are performed, since an implicit coupling is applied for the fluid–structure interaction problem. A good load balancing can be achieved with the proper number of CPU cores for the acoustic domain. F: fluid, S: solid, A: acoustics, C: coupling

surface between the elastic structure and the near field whereas the acoustic far field simulation is executed only once per time step based on the current time step's data at the surface between near and far field. See Fig. 1 for an illustration. Thus, the fluid and structure solvers are executed multiple times per time step, while the acoustic far field solver is called only once per time step. An optimal load balancing can be obtained by choosing an appropriate number of CPU cores for each partition such that solving the acoustic domain takes approximately the same computational time as all fluid–structure iterations together. Each solver is considered to be black-box. Hence, only input and output information is accessible for the coupling. In the following, we describe the realization of the three coupling connections.

3.1 Coupling the Elastic Structure with the Acoustic Fluid

At the surface between the elastic structure and the acoustic fluid, structural displacements/velocities and forces exerted by the fluid on the structure are exchanged between the two involved solvers. The displacements/velocities are the input of the flow solver and the output of the structure solver, whereas the forces are the output of the flow solver and the input for the structure solver (Dirichlet-Neumann coupling). Therefore, we can shortly write the actions of the fluid solver F_f and the structure solver F_s at the coupling surface as at each time step the response of the fluid solver F_f is defined as

$$y = F_f(x) \quad \text{and} \quad x = F_s(y) , \tag{11}$$

where x denotes the displacement of the fluid–structure interface, and y denotes the force acting on the fluid–structure interface.

With this given, we can enforce the balance of stresses and the kinematic boundary conditions

$$\sigma^f n = \sigma^s n \quad \text{and} \quad v^f = v^s$$

for the stress tensors σ^f and σ^s at the surface between structure and fluid Γ^{fs} with the unit normal vector n by solving the staggered or the parallel fixed-point equation

$$x = F_s \circ F_f(x) \quad \text{or} \quad \begin{pmatrix} x \\ y \end{pmatrix} = \begin{pmatrix} 0 & F_s \\ F_f & 0 \end{pmatrix} \begin{pmatrix} x \\ y \end{pmatrix}. \tag{12}$$

The motivation for these fixed-point equations, a further alternative and more details are described in [19]. The coupling tool preCICE provides several robust and efficient acceleration methods for the respective fixed-point iterations. Details are given in "Partitioned Fluid–Structure–Acoustics Interaction on Distributed Data–Coupling via preCICE".

3.2 Coupling the Acoustic Near Field with the Far Field

As already introduced, the primitive variables (pressure, density, velocity) at the interface between near field and far field are transferred (via the coupling tool preCICE) from near to far field once per time step. This yields a unidirectional coupling as shown above in Fig. 1.

To avoid non-physical oscillations induced by the coupling, the data mapping at the coupling interface is crucial. OpenFOAM uses an unstructured mesh in the fluid domain allowing for a geometry-adapted mesh even for complex geometries. A mesh deformation technique is used to interpolate the displacement of the fluid–structure interface into the complete flow field. In the acoustic far field, a structured octree mesh is used with a high order DG discretization. Thus, the exchange points at the coupling interface are non-equidistant quadrature points. Hence, coupling the finite volume OpenFOAM solver with this discontinuous Galerkin solver leads to non-matching meshes at the interface and an interpolation method is required to transfer the density, velocities and pressure from one solver to the other. We use a radial basis function interpolation or nearest neighbor projection (described in the chapter "Partitioned Fluid–Structure–Acoustics Interaction on Distributed Data–Coupling via preCICE") in order to reduce the introduced numerical errors due to the partitioning as far as possible.

Coupling the implicit second order backward difference time integration scheme used by OpenFOAM for the compressible flow with an explicit second or fourth order Runge Kutta scheme used by Ateles for the acoustic far field reduces the overall accuracy to first order in time. Thus, better combinations of time stepping schemes and more sophisticated coupling patterns in time are work in progress. In addition, applying explicit coupling poses time step restrictions for both solvers in order to achieve a stable integration in the acoustic field.

3.3 Coupling the Incompressible Flow with Acoustic Perturbations

As the kinetic energy of the acoustic perturbations is much smaller than the kinetic energy of the flow, there is only a coupling from the flow field to the acoustics field. Thus not from the acoustic field to the flow field. The acoustic sources $s(p^a)$ are computed as the time derivative $\frac{\partial p_0}{\partial t}$ of the incompressible flow pressure and transferred to the acoustic mesh.

Crucial aspects for the splitting approach are the coupling in time and the choice of time step sizes and mesh resolutions. Acoustics with an underlying "slow" flow velocity in a low Mach range, is a multi-scale problem in time as the flow velocity of an incompressible flow–with a maximum of $Ma = 0.3$–is much smaller than the speed of sound c. Using an explicit time discretization scheme for the acoustic equation, its *CFL* condition limits the time step size Δt by

$$\Delta t < \frac{\Delta x}{c} \tag{13}$$

depending on the spatial mesh resolution Δx. The flow solver with its slower velocity allows for larger time steps. Because the computation of the flow field is by far more expensive than the calculation of the acoustic perturbations, the numbers of flow solver time steps should be as small as possible. Two methods decoupling the time steps for the incompressible flow and the acoustic perturbations are available for the integrated flow–acoustics solver:

1. Sub-cycling: $N = \Delta t^a / \Delta t^f$ acoustic time steps are carried out within one flow time step and the acoustic sources are only updated after every N^{th} acoustic time step.
2. Adapting spatial mesh resolutions: the infrastructure of the geometric multigrid can be easily used for the restriction to a hierarchical coarsened spatial mesh for the acoustic perturbations leading correspondingly to a larger time step according to the CFL condition.

A combination of both leads to a substantial saving in computational time [11].

4 Visualization

In this section, we briefly discuss the visualization developed for the large scale simulation of fluid–structure–acoustic interactions. To reduce the communication and IO bottlenecks typical for massively parallel simulations, we employ an algorithm running *in-situ*: during the runtime of the simulation, a solver code calls visualization routines to process the data currently available. It resumes simulation calculations when the visualization routine call returns presented in Sect. 4.1. This concept is realized by a customized in-situ visualization architecture (Sect. 4.2). It employs an intermediate volume representation that reduces the amount of data but still maintains possibilities for user interaction (Sect. 4.3). We finally discuss in Sect. 4.4 how this representation is generated, stored and utilized for interactive exploration in the context of our in-situ architecture.

4.1 In-Situ Visualization

In our approach, the pressure data of the flow field determined after a full time step are used to generate an intermediate, view-dependent representation of the scalar field. Re-arranging and appropriately quantizing the new representation with the goal to raise statistical redundancy in the data, allows for an additional efficient compression of the pressure volume data by a lossless encoding scheme. Since no internode communication is necessary for this (all visualization steps are performed utilizing data available locally to a process), the network load is significantly reduced by only sending the efficiently compressed representation to a front-end node (instead of the full volume data). This enables an interactive volume visualization. The reduced representation additionally facilitates storing the amounts of data produced by an exascale simulation. A secondary goal is minimization of the impact of in-situ visualization calculations, as not to interfere with the ongoing simulation. Note that the acoustic domain is not yet considered in the integrated fluid–structure visualization.

4.2 Simulation–Visualization Setup

Figure 2 gives an overview on the integration of the visualization calculations into the solver. While the *simulation* step continuously generates new pressure data, *visualization transform* converts these volume data into an intermediate visualization representation of reduced size without or with only minimally compromising output quality. The reduced representation generated on each node is then sent to a front-end node which merges all the received data, from which *visualization render* finally generates an image. In the following, we shortly describe the basic ingredients of all these steps.

Simulation The simulation setup used for testing purposes is the bending tower case as described in Sect. 5.1, running a fluid–structure simulation with OpenFOAM components for the fluid and the structure solver (omitting the far field acoustics). In each time step, the simulation yields a three-dimensional scalar field for the pressure, representing a time-dependent unstructured mesh. At this point, the visualization algorithm is executed on these data.

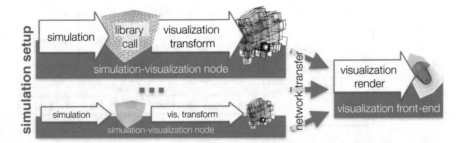

Fig. 2 The figure shows the embedding of the visualization into the simulation. The data gets transformed into an intermediate representation before being compressed by the TVDI algorithm on all simulation nodes. After being transferred and gathered on the front-end, the data from each node then gets reassembled and can be interactively explored

Visualization transform In *transform*, we generate an intermediate visualization representation with a direct interface to the solver. In particular, simulation results are not duplicated in memory, but the solver's own data object and access routines are employed for maximum performance and minimal resource strain. The core idea of visualization transform is to reduce the data set by partially already pre-processing the data as required by the chosen visualization method, yet not fully aggregating data to the final image in order to still maintain some flexibility. In the variant of the algorithm presented here, we employ a view-dependent representation that basically consists of volume rendered images in which the rays of a raycasting algorithm have not been fully composited to yield one resulting pixel color. Instead, while traversing the volume to gather samples, so-called segments are generated along the ray. For a detailed explanation on this intermediate representation, see Sect. 4.3.

Visualization render The visualization representation generated on each node is transferred to the front-end node where images are rendered, whereas certain parameters can be varied interactively. In the case of the algorithm presented here, the user can modify the selected time steps and arbitrary camera parameters. Only the transfer function used to determine a color from the scalar value is fixed.

4.3   Intermediate Representation: *Volumetric Depth Images*

Volumetric depth images (VDIs) are a condensed representation for classified volume data, providing high quality and reducing both render time and data size considerably. Instead of only saving one color value for each view ray as in standard images, VDIs store a set of so-called segments, each consisting of a geometry (depth range) with composited color and opacity. This compact representation is independent from the representation of the original data and can quickly be

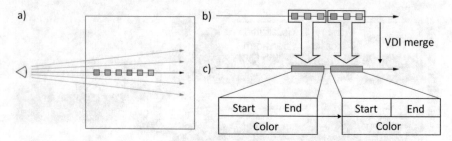

Fig. 3 VDI generation. (**a**) Raycasting is performed on pressure data. (**b**) Similar colors are merged to segments. (**c**) The list data structure is constructed

generated by a slight modification of existing raycaster codes. VDIs can be rendered efficiently at high quality with arbitrary camera configurations [8].

VDIs are generated during volumetric raycasting from a certain camera configuration by partitioning the samples along view rays according to similarity of the composited color, providing the additional possibility to skip 'empty' regions (see Fig. 3). These partitions are then stored as lists of so-called segments containing the bounding depth pair and respective accumulated color and opacity value. This reduces the amount of data in several ways: similar colors get merged and need to be stored only once per segment (as opposed to in every cell sampled). Domain knowledge can already be employed to limit data acquisition to an intuitively choosable camera frustum. Note that the raycasting is only performed on the data available to the process; no internode communication is performed.

4.4 Visualization Transform and Render

TVDI generation During the visualization transform stage, a VDI is generated by sampling the pressure data along viewing rays, which in turn are determined by a camera position, a view target and a size for the resulting "image" (i.e., the number of rays) chosen by the user. For each of these rays, the algorithm needs to traverse the time-dependent unstructured mesh of the simulation, generating the VDI as explained in Sect. 4.3.

Once the raycaster has completed its calculations, the newly obtained VDI data from the current time step are compared to the last step for all segments in a given ray. Considering that flow data typically vary rather smoothly over time, the segments stored in the current and the previous time steps will show a high degree of similarity, which can be exploited for compression. The so-called TVDI (time-clustered VDI) algorithm clusters the segments of a given ray over time by calculating the changes of VDI segments from the previous time step to the current. If similarity in color (a user-defined color space metric is employed) is determined, the segments get sorted into a so called region, in which all segments have the same color determined by the initial segment.

Exploiting the coherency between time steps mentioned before, the change in the segment geometry (starting/ending value) will be small between consecutive time steps and even constant over a few time steps, since flow pressure data typically vary approximately linear over a short time.

The changes in the geometry can be easily stored with minimal overhead (neighbor counts, for details see [7]) and are sufficient to reconstruct all segments at all time steps. Once a user-defined time step interval is reached, the current regions constructed in a given simulation node are sent over the network to the front-end node. An obvious gain is that the color for a time region only needs to be sent and stored once for all associated segments. Assuming that the geometry changes are constant (over a short time) and now have an optimal memory layout, we enable a lossless entropy encoder[6] to achieve a better compression, minimizing network load during data transfer.

Rendering (visualization render) On the front-end node, the regions constructed in the previous step are gathered, and the individual VDI representations for each time step are reconstructed. This is done by incrementally adding the changes stored for each segment to the previously constructed ones, with the first one having been stored as an absolute value. Each segment also gets assigned the regions color. The regions gathered from different processes are all incorporated into the same VDI data structure, giving a complete representation of each of the time steps within an output interval. Using the VDIs segment data and the origin and direction of the ray the segment was constructed from, a small frustum is constructed. Viewed from the origin of the constructing ray, the front and back square of this frustum exactly cover the pixel associated with the ray. The position of these frustum caps along the ray is just the front/end position stored in the segment as is the color. This is done for all segments in all rays, filling the space with these proxy geometry frustums. A second raycast, now taken from an arbitrary viewpoint, can then be run on this proxy geometry, exploiting the hardware acceleration associated with geometry than can be triangulated. Note that VDIs provide renderings identical to those of the original data when rendered from the view used for their generation, but provide high-quality approximations also for deviating views. The error in color stems from merging several similar segments across time, averaging their color. This error can be easily controlled by a user-defined threshold, trading quality versus compression rate.

5   The Three-Dimensional Bending Tower Testcase

In this section we present the physical and numerical results of our approach applied to a scenario involving all three discussed phenomena and their interaction. We have a look at the scalability of the coupled setup and the resulting solutions.

---

[6]http://bzip.org/

5.1 Testcase Description

To investigate the suitability of our approach with the coupling by preCICE and the different solvers for each domain, we use a bending tower as an example. The setup resembles for example the tower of a wind turbine, is fairly simple to understand, and includes three-dimensional effects. The tower is modeled with an elastic material and deforms according to the pressure forces by the surrounding fluid flow. Our computational setup is shown in Fig. 4. In the acoustic far field domain, only the propagation of the acoustic values is simulated, in the fluid domain the flow together with the acoustic perturbation as described in the previous sections. An overview to the physical parameters for each domain is given in Table 1 along with the initial conditions.

As already described shortly in Sect. 2, we apply two different sets of black-box solvers, the comparison of them is work in progress. In **set a**, OpenFOAM is used for the structure and for the acoustic fluid, which is considered compressible. For both regions, a finite volume discretization (2nd order) with a backward differencing time discretization (2nd order) is chosen. The structure is discretized with 4, 500 control volumes (CV), and the flow region with 3,095,500 CVs. For **set b**, the structural deflection is simulated with the finite element code FEAP, and the flow region with the finite volume flow solver FASTEST including the integrated finite volume approach for the acoustics perturbations (see Sect. 2). All discretization schemes are 2nd order accurate. For a first comparison, the numbers of elements and control

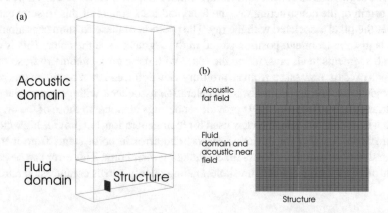

Fig. 4 Setup of our testcase scenario. (**a**) Domain description for a small near field acoustic fluid domain including the bending tower (*blue*) and a 4-fold larger acoustic far field domain on top. (**b**) Sketch of the mesh resolution in each domain, showing a matching (fluid–structure) and a non-matching (near field–far field) coupling interface. For production run, the structure is discretized with 3,600, the acoustic fluid domain in the near field with approximately 12.5 million control volumes and the acoustic domain in the far field with approximately 6 million elements of seventh order

Table 1 Overview of numerical set up for each domain in the three-dimensional bending tower testcase

Domain	Structure	Fluid	Acoustics
Material parameters	$E = 1.4 \cdot 10^6 \left[\frac{N}{m^2}\right]$	$\gamma = 1.4$	$\gamma = 1.4$
	$\nu = 0.4$	$\nu = 0.01 \left[\frac{m^2}{s}\right]$	$c = 11.8 \left[\frac{m}{s}\right]$
	$\rho = 1000 \left[\frac{kg}{m^3}\right]$	$R = 1 \left[\frac{m^2}{v^2 K}\right]$	$R = 1 \left[\frac{m^2}{v^2 K}\right]$
			$T = 100 \, [K]$
			$p_0 = 100 \left[\frac{N}{m^2}\right]$
			$\rho_0 = 1 \left[\frac{kg}{m^3}\right]$
			$v_0 = [2.3, 0, 0] \left[\frac{m}{s}\right]$
Initial condition	$u^s = [0, 0, 0] \, [m]$	$p^f = 100 \left[\frac{N}{m^2}\right]$	$p^a = 0 \left[\frac{N}{m}\right]$
		$\rho^f = 1 \left[\frac{kg}{m^3}\right]$	$\rho^a = 0 \left[\frac{kg}{m^3}\right]$
		$v^f = [2.3, 0, 0] \left[\frac{m}{s}\right]$	$v^a = [0, 0, 0] \left[\frac{m}{s}\right]$
		$B = 0 \, [N]$	
		$T = 100 \, [K]$	

volumes are chosen similar to those in **set a**. The pure acoustic domain in the far field is simulated with the acoustic solver Ateles in the APES framework.

As described in Sect. 3, a bi-directional implicit coupling is used for the fluid–structure interaction based on a Dirichlet-Neumann coupling. For the near field–far field interaction, we do one-way explicit coupling and the primitive flow variables density ρ, velocity v and pressure p are sent from OpenFOAM to Ateles via preCICE at the surface of the near field domain.

The derivative terms of the governing equations for the fluid domain are discretized using standard Gaussian finite volume integration, with linear interpolation from the cell centres to the cell faces. The same approach is used for the solid domain.

A relative tolerance of $1.0 \cdot 10^{-5}$ is set as a convergence criterion for the fluid–structure interface fixed-point equation including deviations both in displacement of the fluid–structure interface and the stresses acting on the interface. When convergence is reached for both variables, the data of the fluid–acoustics interface are communicated via preCICE to the acoustic domain and the next time step starts.

All simulations for the scaling results are done with **set a** on the IBM DataPlex machine SuperMUC at the Leibniz Computing Centre (LRZ) in Munich.

5.2 Numerical Results

A snapshot of a complete fluid–structure–acoustic interaction simulation, calculated
with **set a**, is given in Fig. 5. The displacement of the tower structure is monitored
at the center point at the top of the tower, as shown in Fig. 6. The displacement
in x-direction is shown with respect to the initial condition. A periodic motion
is observed for the simulated time span. Further damping of the motion of the
console is expected until a steady state situation is reached. The differences between
set a and **set b** are caused by the use of periodic boundary conditions in main
flow direction in **set a** and the use of inflow velocity and zero gradient boundary
conditions in **set b**. Figure 7 and 8 show the pressure, density, acoustics pressure,
and acoustics density at the monitoring points A ([3.2, 0.7, 0.0]), B ([3.2, 0.0, 0.8])
and C ([3.2, 0.0, 1.0]) for both sets. The results of **set a** and **set b** are quantitatively
in the same range. However, the pressure of the incompressible FASTEST solution
is shifted, as for an incompressible flow solver, the absolute pressure level is not
uniquely defined. Here, the acoustic pressure p^a has to be added to the pressure to
get the same values as for the compressible OpenFOAM solution.

5.3 Scaling Results

Figure 9 shows the results for a strong scaling study of the fluid–structure–acoustics
interaction testcase with **set a**. For this measurement, the fluid comprises 1.5 million
cells, the solid domain consists of 2304 cells, and the acoustics mesh has 6144

Fig. 5 Visualization of the pressure contours and Q-criterion of the velocity for the fluid flow of
the three-dimensional bending tower testcase

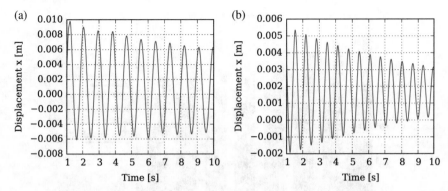

Fig. 6 Displacement of the center point of bending tower in x-direction—comparison of the results obtained for **set a** (**a**) and **set b** (**b**). (**a**) OpenFOAM setup. (**b**) FASTEST setup

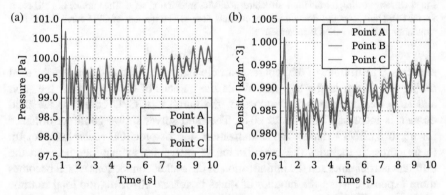

Fig. 7 Pressure and density at the monitoring points A, B and C for **set a** with OpenFOAM. (**a**) Pressure. (**b**) Density

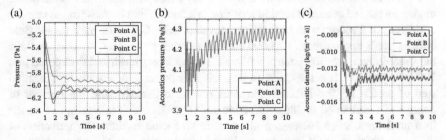

Fig. 8 Pressure, acoustic pressure and acoustic density at the monitoring points A, B and C for **set b** with FASTEST. (**a**) Pressure. (**b**) Acoustic pressure. (**c**) Acoustic density

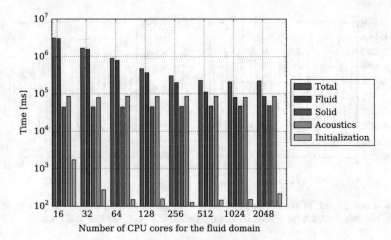

Fig. 9 Strong scaling for the fluid–structure–acoustics interaction **set a**. The number of CPU cores for the solid and acoustics domain is kept constant at two. The number of CPU cores for the flow simulation in the near field is increased

elements. The timings of the separate participants, of the initialization and the total timing are shown in the bar plot. While the number of CPU cores for the solid and acoustics domain is kept constant, the number of CPU cores for the fluid domain is increased from 16 to 2048. The fluid solver shows good performance for up to 512 MPI ranks. The initialization step of preCICE scales well for up to 64 cores. It includes the initialization of interpolation matrices between the different participants and the initialization of the socket connections which becomes more expensive when the number of ranks increases. However, the total runtime of the preCICE initialization is negligible compared to the total runtime of the simulation such that the bottleneck in this case is the scalability of the OpenFOAM compressible flow solver.

Figure 10 shows the results for a weak scaling study for **set a**. All domains are uniformly refined in each direction resulting in a factor eight increase in the number of degrees of freedom with each refinement step. Each CPU core used for the fluid domain holds approximately 25,000 control volumes and each CPU core used for the solid domain has 300 control volumes. The acoustics mesh is decomposed into 770 cells per core. It is important to note the poor performance of the initialization step of preCICE. The initialization does not scale linearly with the number of degrees of freedom on the fluid–structure and fluid–acoustics interface, and results in a bottleneck when increasing the domain size even further. The reduction of this complexity is work in progress. First ideas how to do this are presented in "Partitioned Fluid–Structure–Acoustics Interaction on Distributed Data: Coupling via preCICE".

Figures 9 and 10 show not optimal scaling behaviour. It is a three field coupling which includes different components. Each individual component has its advantages and disadvantages. Thus with OpenFOAM it is straightforward to implement a

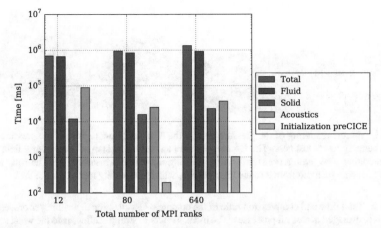

Fig. 10 Weak scaling for the fluid–structure–acoustics interaction **set a**. The number of degrees of freedom of each domain is increased proportional to the number of CPU cores

fluid–structure–acoustics interaction setup, but scalability is an issue. Other fluid solvers like Ateles, which is also capable of solving the governing flow equations, show better scaling results when coupled with preCICE [2], but does not yet include FSI. In such a complex setup, the disadvantages of one participant carry over to the complete fluid–structure–acoustics interaction simulation.

5.4 Visualization

To examine the performance of the visualization algorithm, several timing and data size measurements were taken. To judge the scalability, both a weak and a strong scaling test were performed for the bending tower testcase. For all tested configurations the compression ratio of the original data with lossless compressed TVDI was around 10:1. During the strong scaling test, high processor counts yielded a reduced compression of around 7:1. This is due to the extremely homogeneous data as can be seen in Fig. 11. The algorithm produces some minor overhead data for each process. If the data within a process are very similar (e.g., if there is only one large region), the overhead exceeds the gain by compression, and is noticeable in the reduced compression ratio.

The second measurement series took the total runtime of the visualization library call during the simulation. For all configurations, the relative runtime of the visualization relative to the simulation runtime was smaller than 1 %, and therefore achieved the goal of not impacting simulation calculations noticeably. The results presented in Table 2 show a strong scaling test (time$_s$) performed with a problem size of approximately 1.5 million cells. The weak scaling performance is shown in the last column denoted by time$_w$, using an appropriately resolved domain, ranging from 200,000 to 1.5 million cells.

Fig. 11 Volume rendering of the pressure field in the near field flow regime of the pressure field in the bending tower test case. The *left image* shows an early time step of the pressure field from the generating view, with a transfer function covering the whole pressure range. The *right image* shows the same scenario from a rotated perspective, which adequately represents the data

Table 2 Total time and compression ratio of the complete visualization algorithm per process and time step, averaged across all processes. The index s refers to strong scaling, w to the weak scaling test

# procs	time$_s$ (ms)	time$_w$ (ms)	Ratio$_s$
128	1291	271	10.2
256	691	280	10.4
512	402	312	10.5
1024	291	291	7.3

Judging from the measurements in Table 2, the calculation speed scales very well with the process count given a fixed problem size. However, for high process counts, a lot of unnecessary precomputations are done to traverse the data quickly during the raycast. If the raycasted domain contains too few cells, establishing the accelerating data structures costs more time than it saves and scaling diminishes.

As is to be expected, the timing stays nearly constant for the weak scaling test. This is not surprising considering the algorithm works exclusively on locally available data and does not employ any kind of internode communication. The variance still visible in the calculations time can be attributed to rather large fluctuations in storage access times. Note that averaging times across processes is admissible since the domain decomposition done by OpenFOAM yields evenly sized subdomains.

6 Conclusion and Outlook

We have shown the setup and numerical as well as scaling results for a complex fluid–structure–acoustics interaction testcase in a three-dimensional domain. Our partitioned simulation approach in combination with an efficient in-situ visualization has been proven to be highly flexible, efficient and scalable on parallel computer architectures.

Using a scalable coupling tool working on distributed data, presented in "Partitioned Fluid–Structure–Acoustics Interaction on Distributed Data: Coupling via preCICE", gives the benefits to efficiently reuse existing software which is adapted to the different physics (here OpenFOAM for structure and compressible flow, FEAP for structure, FASTEST for incompressible flow and acoustic near field, Ateles for acoustic far field). Using a monolithic approach is not feasible as resolving the small scales in a large domain would be highly demanding in terms of wall clock time, memory and efficiency. Exploiting appropriate methods for each physical discipline, e.g. higher order methods for the acoustic far field, facilitates such large scale simulations nowadays on supercomputers. Applying a static load-balancing based on heuristics, computational resources are used efficiently and the total time to solution is decreased. Certainly, in such a complex setup which consists of several participants, a bottleneck of one participant overshadows the whole simulation. Hence, e.g. point-to-point communication between parallel solver processes in the coupling tool described is indispensable. Further reductions of the complexity of numerical and initialization steps are work in progress such that also complex fixed-point acceleration methods and sophisticated data mapping become scalable on massively parallel architectures.

The comparison of results for simulations using different black-box solvers in terms of physical results, as well as in terms of compute time and scalability of the framework, has to be done next. For validation of the results, also experimental data shall be used.

The algorithm for the presented in-situ visualization features an easily adjustable and intuitive trade-off between image quality and runtime on a node level basis. The amount of rays used to perform the raycasting can be adjusted for each process individually. In addition, time intervals for off-loading a set of time regions can be of different length. This behavior can be exploited to further optimize the load balancing of the overall simulation: nodes known to have a high computational load from the simulation, as for example nodes at the coupling boundary, can run a low quality version of the algorithm, while internal nodes run the full quality variant. This way, idle processes can be avoided in favor of a higher accuracy and/or better compression. The current implementation would also allow for an adaptive quality adjustment, e.g., based on the idle time remaining during the last step.

Although we showed only the very simple example of a bending tower, the whole simulation framework comprising the solvers, the coupling tool and the visualization is prepared for more complex simulations such as wind turbines or fans. To actually realize such scenarios with realistic results, however, further numerical tests for easier cases have to be done. Applications with large geometry changes such as a rotating wind turbine in addition require either a different grid approach in the flow solver (Eulerian instead of ALE) or the introduction of a second mesh for the acoustic fluid domain that rotates with the turbine and is coupled to a surrounding fixed grid. A qualitative phenomenon that can be important in real-world scenarios where noise propagation in the acoustic far field includes the sound reflection at obstacles is the coupling of the acoustic far field back to the acoustic flow domain. We currently only consider a one-way coupling here. For the two-way coupling, preCICE provides a suite of coupling methods (as used for the fluid–structure interface). However, the correct modeling of coupling conditions for a two-way coupling still needs to be done. Summarizing, we can state that our simulation environment provides the technical components for even very complex simulations. The exact setup, however, has not yet been defined and modeled for all cases. The advantage over monolithic approaches that obviously offer more possibilities to taylor the methods for a specific application, is the high flexibility and generality of our approach.

Acknowledgements The financial support of the priority program 1648 Software for Exascale Computing (www.sppexa.de) of the German Research Foundation and of the Institute for Advanced Study (www.tum-ias.de) of the Technical University of Munich is thankfully acknowledged.

References

1. Blom, D.S., Krupp, V., van Zuijlen, A.H., Klimach, H., Roller, S., Bijl, H.: On parallel scalability aspects of strongly coupled partitioned fluid-structure-acoustics interaction. In: VI International Conference on Computational Methods for Coupled Problems in Science and Engineering – COUPLED PROBLEMS 2015 (2015)
2. Bungartz, H.J., Klimach, H., Krupp, V., Lindner, F., Mehl, M., Roller, S., Uekermann, B.: Fluid-acoustics interaction on massively parallel systems. In: Mehl, M., Bischoff, M., Schäfer, M. (eds.) International Workshop on Computational Engineering CE 2014. Lecture Notes in Computational Science and Engineering, pp. 151–165. Springer, Heidelberg/Berlin (2015)
3. Cardiff, P., Karač, A., Ivanković, A.: A large strain finite volume method for orthotropic bodies with general material orientations. Comput. Method. Appl. Mech. Eng. **268**, 318–335 (2014)
4. Darwish, M., Moukalled, F.: A fully coupled Navier-Stokes solver for fluid flow at all speeds. Numer. Heat Tr A.-Appl. **65**(5), 410–444 (2014)
5. Darwish, M., Sraj, I., Moukalled, F.: A coupled finite volume solver for the solution of incompressible flows on unstructured grids. J. Comput. Phys. **228**(1), 180–201 (2009)
6. FASTEST-Manual: Fachgebiet für Numerische Berechnungsverfahren im Maschinenbau, Technische Universität Darmstadt, 1st edn. (2005)
7. Fernandes, O., Frey, S., Sadlo, F., Ertl, T.: Space-time volumetric depth images for in-situ visualization. In: Proceedings of IEEE 4th Symposium on Large Data Analysis and Visualization (LDAV), pp. 59–65 (2014)
8. Frey, S., Sadlo, F., Ertl, T.: Explorable volumetric depth images from raycasting. In: Proceedings of the Conference on Graphics, Patterns and Images, pp. 123–130 (2013)
9. Hardin, J., Pope, D.: An acoustic/viscous splitting technique for computational aeroacoustics. Theor. Comput. Fluid Dyn. **6**, 323–340 (1994)
10. Hesthaven, J.S., Warburton, T.: Nodal Discontinuous Galerkin Methods: Algorithms, Analysis, and Applications, 1 edn. Springer, New York (2007)
11. Kornhaas, M., Schäfer, M., Sternel, D.: Efficient numerical simulation of aeroacoustics for low mach number flows interacting with structures. Comput. Mech. **55**(6), 1143–1154 (2015). http://dx.doi.org/10.1007/s00466-014-1114-1
12. Lighthill, M.: On sound generated aerodynamically. I. General theory. Proc. R. Soc. A **211**, 564–587 (1952)
13. Newmark, N.M.: A method of computation for structural dynamics. J. Eng. Mech. Div.-Asce. **85**(7), 67–94 (1959)

14. Roller, S., Bernsdorf, J., Klimach, H., Hasert, M., Harlacher, D., Cakircali, M., Zimny, S., Masilamani, K., Didinger, L., Zudrop, J.: An adaptable simulation framework based on a linearized octree. In: Resch, M., Wang, X., Bez, W., Focht, E., Kobayashi, H., Roller, S. (eds.) High Performance Computing on Vector Systems 2011, pp. 93–105. Springer, Berlin/Heidelberg (2012)
15. Shen, W.Z., Sørensen, J.N.: Aeroacoustic modelling of low-speed flows. Theor. Comput. Fluid Dyn. **13**, 271–289 (1999)
16. Shen, W., Sørensen, J.: Comment on the aeroacoustic formulation of Hardin and Pope. AIAA J. **37**(1), 141–143 (1999)
17. Taylor, R.L.: FEAP – A Finite Element Analysis Program – Version 7.5 User Manual. University of California (2003). citeseer.ist.psu.edu/taylor03feap.html
18. Toro, E.F.: Riemann Solvers and Numerical Methods for Fluid Dynamics, 2 edn. Springer, Berlin/Heidelberg (1999)
19. Uekermann, B., Bungartz, H.J., Gatzhammer, B., Mehl, M.: A parallel, black-box coupling for fluid-structure interaction. In: Idelsohn, S., Papadrakakis, M., Schrefler, B. (eds.) Computational Methods for Coupled Problems in Science and Engineering, COUPLED PROBLEMS 2013. Stanta Eulalia, Ibiza (2013). http://congress.cimne.com/coupled2013/proceedings/full/p559.pdf

Part VII
ESSEX: Equipping Sparse Solvers for Exascale

Towards an Exascale Enabled Sparse Solver Repository

Jonas Thies, Martin Galgon, Faisal Shahzad, Andreas Alvermann,
Moritz Kreutzer, Andreas Pieper, Melven Röhrig-Zöllner, Achim Basermann,
Holger Fehske, Georg Hager, Bruno Lang, and Gerhard Wellein

Abstract As we approach the exascale computing era, disruptive changes in the
software landscape are required to tackle the challenges posed by manycore CPUs
and accelerators. We discuss the development of a new 'exascale enabled' sparse
solver repository (the ESSR) that addresses these challenges—from fundamental
design considerations and development processes to actual implementations of some
prototypical iterative schemes for computing eigenvalues of sparse matrices. Key
features of the ESSR include holistic performance engineering, tight integration
between software layers and mechanisms to mitigate hardware failures.

1 Introduction

It is widely accepted that the step from peta- to exascale is qualitatively different
from previous advances in high performance computing and therefore poses urgent
questions. Considering applications that need these vast computing resources, which
algorithms expose such massive parallelism? What will the next generations of
supercomputers look like, and how can we write sustainable yet efficient software

J. Thies (✉) • M. Röhrig-Zöllner • A. Basermann
Simulation and Software Technology, German Aerospace Center (DLR), Köln, Germany
e-mail: Jonas.Thies@DLR.de

M. Galgon • B. Lang
School of Mathematics and Natural Sciences, University of Wuppertal, Wuppertal, Germany

A. Alvermann • A. Pieper • H. Fehske
Institute of Physics, University of Greifswald, Greifswald, Germany

M. Kreutzer • F. Shahzad • G. Hager • G. Wellein
Erlangen Regional Computing Center, Friedrich-Alexander-University Erlangen-Nuremberg,
Erlangen, Germany

© Springer International Publishing Switzerland 2016   295
H.-J. Bungartz et al. (eds.), *Software for Exascale Computing – SPPEXA
2013-2015*, Lecture Notes in Computational Science and Engineering 113,
DOI 10.1007/978-3-319-40528-5_13

for them? The ESSEX project[1] has developed the 'Exascale enabled Sparse Solver Repository' (ESSR) over the past three years, and in this paper we want to share our experiences and summarize our results in order to contribute to answering these questions. Besides reviewing the ESSEX project, the paper contributes a thorough presentation of a software architecture for iterative sparse solver libraries on heterogeneous supercomputers that overcomes some of the shortcomings of existing packages on the road to exascale.

The applications we study come from quantum physics and material science, and are directly or indirectly related to solving the Schrödinger equation. The Hamiltonian of the systems studied can be represented as a (very) large and sparse matrix, and the numerical task is to solve sparse eigenvalue problems in various flavors. The software we develop is intended as a blueprint for other applications of sparse linear algebra.

In the next few years, we expect no radical change in the architecture of supercomputers, so that a scaled up version of current petascale systems is used as target architecture for the ESSR. That is, a distributed memory cluster of (possibly heterogeneous) nodes. On the other hand, node-level programming will become much more challenging because of the strong increase in node level parallelism and complexity.[2] Due to the increasing node count, we do anticipate a much shorter mean time to failure (MTTF) on the full system scale, which has to be addressed for large simulations using substantial parts of an exascale system.

A key challenge in the efficient implementation of sparse matrix algorithms is the 'bandwidth bottleneck', the fact that in most modern architectures the amount of data that can be loaded per floating point operation is continually decreasing. To hide this gap, cache systems of increasing complexity and non-uniform cache/memory hierarchies are used. Another issue is the relative increase of the latency of global reduction/synchronization operations, which are central to many numerical schemes. In the ESSR we address these problems using block algorithms with tailored kernels (see also [27]) and communication hiding.

Three overarching principles guide the design of the ESSR: *disruptive changes of data structures* for node-level efficiency, *holistic performance engineering* to avoid accumulation of losses on various hardware or software levels, and *user-level fault tolerance* schemes to keep the overhead for guaranteeing stable runs as low as possible.

The various layers of the ESSR (application, algorithms and building blocks) were co-developed 'from scratch' within the past three years. This rapid process was only possible with a comprehensive software engineering approach, which we will describe in this paper. We use the term 'repository' rather than 'library' because of the young age of our effort. In the future, the ESSR components will be integrated

[1]Equipping Sparse Solvers for the Exascale, http://blogs.fau.de/essex, funded by the priority program "Software for Exascale Computing" (SPPEXA) of the German Research Foundation (DFG)

[2]see, e.g., https://www.olcf.ornl.gov/summit/

to form a complete software stack for extreme scale sparse eigenvalue computations and applications.

Related work A large number of decisions has to be made when designing basic linear algebra data structures such as classes for sparse matrices, (block) vectors or dense matrices. On the other hand, iterative algorithms may remain largely oblivious of these implementation details (e.g. the storage scheme for sparse matrices, the parallelization techniques used). In the past, iterative solver libraries were therefore often based on reverse communication interfaces (RCI, see, e.g., (P)ARPACK [30] or FEAST [32]), or simple callback functions that allowed the user only to provide the result of a matrix–vector product and possibly a preconditioning operation (as in PRIMME [42]). In such approaches, the user is bound to the parallelization technique prescribed by the solver library (i.e. pure MPI in the examples above), and the solver library can not exploit techniques like kernel fusion or overlapping of communication and computation across operations. Another library implementing sparse eigenvalue solvers is SLEPc [17]. Here the user has to adapt to the data structures of the larger software framework PETSc [3].

A more flexible approach is the concept of an interface layer in the Trilinos library Anasazi [2]. Solvers in this C++ library are templated on scalar data type and the 'multi-vector' and operator types. For each kernel library providing these objects, an 'adapter' has to be written. Apart from the operator application (which may wrap a sparse matrix–vector product), the kernel library implements a multi-vector class with certain functionality. For an overview of Trilinos, see [18, 19]. Our own approach is to use an interface layer which is slightly more extensive than the one in Anasazi, but puts less constraints on the underlying data structures (see Sect. 3.4).

The predicted range of MTTF for exascale machines (between hours and minutes [5]) necessitates the inclusion of fault tolerance capabilities in our applications, as they fall in the category of long running large jobs. The program can face various failures during its run, e.g. hardware faults, soft errors, Byzantine failures, software bugs, etc. [21]. According to [8], a large fraction of failures can be attributed to CPU and memory related issues which eventually lead to complete process failures. Such failures define the fault tolerance scope in this work.

Document structure We start out by describing the basic software architecture of the ESSR in Sect. 2, and a process that allows the concurrent development of sparse solvers and the building blocks they need to achieve optimal performance. Section 3 gives an overview of the software components available in the ESSR. In Sect. 4, three classes of algorithms studied in the ESSEX project are briefly discussed. The objective here is neither to present new algorithmic features or performance results, nor to study any particular application. Instead, we want to summarize the optimization techniques and implementation details we identified while developing these solvers. The fault tolerance capabilities explored in our applications are described in Sect. 5. Section 6 summarizes the paper and gives an outlook on future developments surrounding the ESSR.

2 ESSR Architecture and Development Process

It is a substantial effort to implement a scalable sparse solver library 'from scratch'. In this section we describe the architecture and development cycle of a set of tightly integrated software layers, that together form the 'Exascale enabled Sparse Solver Repository', ESSR. The actual implementation in terms of software packages is detailed further in Sect. 3.

2.1 Software Architecture

The ESSR consists of the following main parts, depicted in Fig. 1: an application layer, the computational core and a vertical integration pillar. An optional preconditioner can be used for better convergence. A final part is an extensive test suite, not shown here.

The computational core (or kernel library) has the task of providing highly optimized implementations of the kernels required by the algorithms and applications we study. It hides implementation details such as SIMD instructions, NUMA aware memory management and MPI communication from the other layers. It is a 'component' in the sense that it could be replaced by another implementation if the software is ported to radically different hardware, or if new applications with different requirements come up. The basic data structures it provides are classes for sparse matrices (sparseMats), tall and very skinny matrices (or 'multi-vectors', mVecs) and small and dense matrices (sdMats).

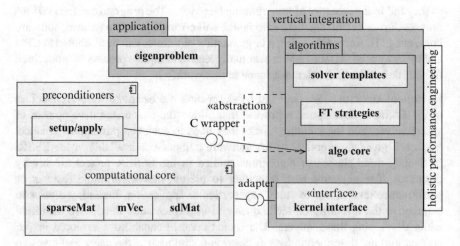

Fig. 1 ESSR software architecture

The vertical integration pillar is based on a clear interface to the computational core, subsequently referred to as 'kernel interface'. It defines the basic data structures and operations that the computational core has to provide. The 'algo core' layer implements common functionality useful for various high level algorithms. Examples include block orthogonalization, evaluating matrix polynomials and extracting Ritz values from a subspace. On top of the kernel interface and core functionality, iterative algorithms are implemented. Fault tolerance strategies are built into algorithms, and common concepts here are again implemented in the algorithmic core layer. The vertical integration pillar is designed to enable holistic performance engineering, as will be discussed below.

The application layer defines an eigenvalue problem and uses an algorithm to solve it. To set up the problem and pre-/postprocess the results, it may either use the simplified kernel interface or the full functionality of the computational core library. While the vertical pillar is connected to the computational core only via a clear interface, the degree to which an application can use another kernel library depends on its implementation and need for specific preconditioners and pre-/postprocessing. Simple applications that only need matrix/vector construction (or I/O) and standard operations can stay independent of the underlying implementation by using the kernel interface as the lowest level.

Preconditioners may be used to accelerate the solution of linear systems arising in an eigenvalue computation. These may either be algebraic schemes using the data structures of the kernel library, or 'physics-based' techniques that exploit specific knowledge of the problem at hand, like a mesh or spectral information. Third-party or own preconditioning software can easily be incorporated because the interface requires only two functions for setting up and applying the preconditioner.

Tightly connected to the vertical pillar is an extensive test framework (cf. Sect. 3.6), with a continuous integration process to ensure software quality. The largest number of tests targets the computational core, through the kernel interface. The algorithmic core is tested using synthetic cases (integration tests), and system tests (numerical test cases for the algorithm layer) are provided by matrix collections/generators and the application layer.

2.2 Concurrent Development of all Layers

The introduction of the kernel interface enables the use of established libraries while developing/implementing iterative methods. The core layers can thus be developed in parallel to the algorithms layer. The kernels required are defined dynamically during the development process and implemented in a test-driven process in the computational core, see Fig. 2. In a similar workflow, common functionality used in several solvers is identified and abstracted into the 'algo core' layer, where a numerically robust and fully optimized implementation is brought forth while algorithm development continues at a higher level. An example is the development

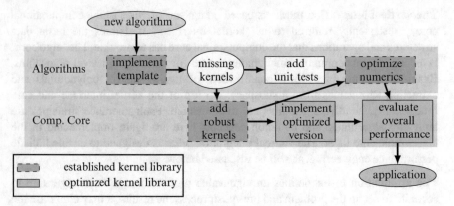

Fig. 2 Test-driven co-development of optimized algorithms in the ESSR

of a communication optimal and robust block orthogonalization scheme while implementing block Jacobi-Davidson (Sect. 4.3) based on a simple yet robust (iterated) modified Gram-Schmidt process.

2.3 Integration of Performance Engineering

While developing an iterative solver, all performance critical operations are identified and added to the kernel interface. As the number of relevant kernels is moderate, a combination of performance models and dedicated benchmarks can be used to ensure their near optimal performance. Many of these operations (such as the sparse matrix-vector multiplication, spMVM, or operations on mVecs), are bounded by the main memory bandwidth, such that the roofline model [49] gives a good indication of the quality of the implementation. To understand the performance of a complete algorithm, code instrumentation for performance analysis tools is used. This may reveal, e.g., overhead of thread synchronization or effects of non-uniform memory access (NUMA) which may not occur in isolated benchmarks. More details on how this concept is implemented can be found in Sect. 3.6.

Our primary focus here is node-level performance. The changes in CPU architecture are currently more dramatic than those concerning node interconnection, and any losses at the node level scale with the number of nodes in a supercomputer.

2.4 Fault Tolerance Strategy

The strategy followed in the ESSR to achieve fault tolerance w.r.t. hardware failures can be classified as an application-level checkpoint/restart (C/R) method. In this approach, algorithm-specific knowledge is exploited to store the minimum amount of data needed for restarting the computation. A highly optimized implementation of this approach (using e.g. asynchronous checkpointing and neighbor-level checkpoints) promises a low overhead for our long running iterative schemes on many nodes.

Due to the early development stage of fault tolerant communication libraries [29], our strategy is to evaluate various technical solutions in simple use cases before condensing them into a common feature of the ESSR solvers and applications in the 'algo core' layer. Section 5 gives an overview of our work in this area.

3 ESSR Software Landscape

The conceptual design discussed in the previous section is implemented in a collection of compatible software packages, which are publicly accessible under a BSD open source license.[3] Before discussing the software structure further, we will comment on the target computer architecture for the software.

3.1 Hardware and Execution Models Supported

Exascale computers are not available to date, and a competitive 'race of flops' is going on to develop this new generation of supercomputers. Based on the developments in the TOP500 list [45] over the past few years, we decided to develop software targeting machines that consist of many nodes with distributed memory. A node features several multi- or manycore CPUs with non uniform access to caches and main memory, and 'accelerator' hardware, e.g. multiple GPUs. At the lowest level, data parallelism is exploited by the hardware through SIMD/SIMT like techniques, compelling choices in data structures and low level implementation. Typical sparse matrix algorithms will continue to be memory-bound on these devices.

In this environment we employ the following execution model. Numerical algorithms are implemented as a sequence of function calls, executed transparently on a parallel heterogeneous machine (SPMD model). A distributed memory communication protocol (e.g. MPI) is used between processes running on complete nodes or parts of nodes of the cluster. Within a function we allow arbitrary multithreading techniques for flexible node utilization. The execution of functions may be interleaved using 'tasks' which use only a part of the resources available to the process. Data transfers between host CPU and accelerator devices must be handled explicitly by the computational algorithm between function calls where necessary (the underlying kernels do not 'know' if the CPU or device memory is up to date).

---

[3] see http://bitbucket.org/essex

3.2 ESSR Toolkits and Functionality

The ESSR is implemented in a number of co-developed software packages, also called *toolkits*. These toolkits do not necessarily implement one part of the architecture (Fig. 1) each. Rather, each partner in the ESSEX project has the responsibility for one of the toolkits, whereas the responsibility for the conceptual ESSR parts may be shared among several project partners. In the future, the repository will evolve into a set of libraries providing state-of-the-art, highly scalable and fault tolerant eigensolvers. This may lead to a redistribution of functionality according to the architecture depicted in Fig. 1.

The four toolkits are briefly characterized as follows:

- ESSEX-Physics, a quantum physics toolkit defining applications that we want to solve using the ESSR. It provides scalable sparse matrices from real-world applications and polynomial eigensolvers (see Sects. 3.3 and 4.1).
- GHOST (General, Hybrid and Optimized Sparse Toolkit) implements basic building blocks with a focus on optimal performance on heterogeneous supercomputers. This design goal is achieved by consequent application of performance engineering techniques. GHOST implements the 'computational core' of the ESSR in single or double precision, and in real or complex arithmetic [27, 28].
- PHIST (Pipelined Hybrid-parallel Iterative Solver Toolkit) implements the vertical integration pillar of Fig. 1, and adapters for several kernel libraries. It also hosts the test framework, and contributes Jacobi-Davidson type eigensolvers and Krylov methods for linear systems to the algorithms layer. To provide a more diverse spectrum of methods, we also included adapters for GHOST to the Trilinos libraries Anasazi and Belos.
- BEAST (Beyond fEAST) extends the algorithms layer of the ESSR by innovative projection-based eigensolvers which take up the idea of the contour integration-based FEAST method [32] (see Sect. 4.2).

We will now describe some of the features of the ESSR, with references to the toolkit where they can be found. The eigensolvers are described in more detail in Sect. 4.

3.3 Applications

Following the overall philosophy of the SPPEXA priority program,[4] our development of the ESSR components is closely guided by—but not restricted to—the intended application range in quantum physics and chemistry. Three different types of eigenvalue problems arise for the large sparse symmetric (or Hermitian) matrices

---

[4] see http://www.sppexa.de/

derived from the Schrödinger equation. The study of equilibrium properties, e.g., of the electronic states in a certain material, requires computation of either a few extremal eigenvalues (of the order 10–100) or many interior eigenvalues (100–1000) with the Jacobi-Davidson algorithm or BEAST, respectively. On the other hand, effectively all the eigenvalues contribute to the dynamic properties of highly excited or driven systems out of equilibrium, and expansion techniques such as the kernel polynomial method (KPM) and Chebyshev time propagation (ChebTP) come into play. These algorithms and their implementation are briefly discussed in Sect. 4. Thus, our target applications require solution of the entire range of large sparse symmetric eigenvalue problems.

Similarly, a variety of matrices occur in the applications: while stencil- and band-like matrices are characteristic for graphene and topological insulators, the tensor structure of quantum mechanical Hilbert space leads to intricate sparsity patterns with long thin subdiagonals or scattered small subblocks for correlated many-particle quantum systems. Also, spectral properties of the matrices differ widely, which allows for algorithmic developments and thorough testing without losing contact to the real application. For example, the appearance of a pseudo-gap in the density of states for topological insulators can be exploited for interior eigenvalue computations with polynomial filter functions [31]. Scalable matrix generation routines are included in the ESSEX-Physics library for correlated many-particle systems and new topological materials, all of which are research problems of current interest.

3.4 Kernel Interface

The algorithms summarized in Sect. 4 can be implemented with the three basic data structures introduced in Sect. 2, sparseMats, mVecs and sdMats. To maintain flexibility, we added a fourth, an abstract linear operator type (linearOp), which may be used to provide, e.g., preconditioning techniques or implement matrix-free methods. Inspired by the Petra object model employed by Trilinos [18], we also abstracted data distribution into a map object and inter-process communication into a comm object. Another Petra concept that is useful when implementing iterative solvers is a 'view' of (part of) an mVec or sdMat. A view is a light-weight object that only has meta data and provides (read and/or write) access to the elements of the 'viewed' object without copying them. Thus it is, e.g., possible to apply an operator or sparse matrix to selected columns of an mVec.

As mentioned in Sect. 1, the Anasazi interface layer resolves the problems of earlier techniques by allowing the sparse matrix and block vector implementations to be co-designed with matching parallelization techniques and data layouts. We adapted this idea to our needs, in PHIST, with the following main differences:

C interface Having to provide a C++ adapter may be a hassle for e.g. Fortran programmers. We restrict ourselves to four scalar data types (ST), single or double, real or complex, which can be implemented optionally. For each ST, a set of plain

C functions has to be provided, which accept objects as void pointers. Errors and flags are passed via the last (int*) argument, similar to the MPI interface. This minimalistic interface allows maximum flexibility for users of PHIST and providers of kernel libraries alike. The lack of type safety introduced by passing around objects as void* is alleviated by the test framework discussed in Sect. 3.6.

sdMat We require the kernel library to provide this object to increase flexibility. For instance, an sdMat may be replicated on host CPU and GPU, or it may be stored in higher precision to increase the numerical stability of reduction operations.

View concepts Allowing custom sdMats, we also require views of contiguous rows and columns in an sdMat. On the other hand, we only require views of contiguous and increasing columns of an mVec. This makes it easier to implement mVecs in row-major ordering for better performance [36]. Strided memory access leads to a significant performance penalty in that case, and restricting the interface therefore gives more uniform performance of the view objects supported.

Explicit data transfers for accelerators For compute platforms that have both a host processor and one or more accelerators, we support the data parallel execution model implemented in GHOST [28]. At least one MPI process is used for each component of a heterogeneous node, and a 'GPU process' has a management thread running on the host CPU. Special kernel interface functions exist to transfer the data of sdMats between host and device.

3.5 Computational Core

The mathematical simplicity of the objects and functions required by the kernel interface is misleading. Let us consider the operation $C = V^T W, C \in \mathbb{R}^{m \times k}, V \in \mathbb{R}^{n \times m}, W \in \mathbb{R}^{n \times k}$. If this operation is implemented using OpenMP inside each MPI process and Intel(R) AVX SIMD instructions, the data in the objects must be contiguous, correctly aligned and padded, which may not be the case if V, W and/or C are views of some parts of larger objects. The reduction operation must produce consistent results on all MPI processes, and if accelerators like GPUs are involved, data transfers must be managed explicitly. The constraints on data layout also hold for efficient GPU processing. All of these complexities are hidden in the ESSR library GHOST [28]. Automatically generated kernels are selected dynamically depending on data alignment, block size and CPU type. Shared memory parallelism on CPUs and the Intel(R) Xeon Phi is implemented using OpenMP, and Nvidia GPUs are supported by providing optimized CUDA kernels.

Another important component of GHOST is a lightweight, general purpose tasking mechanism that plays well within the standard data parallel execution model of 'MPI+X'. It is used in the ESSR for overlapping communication with computation, asynchronous checkpointing, etc. The PHIST library provides macros to simplify the use of this tool when implementing an algorithm.

Apart from GHOST, PHIST currently has adapters for the Trilinos libraries Epetra and Tpetra. Builtin Fortran/C99 kernels make PHIST self-contained in principle and are used for performance engineered prototypes of functionality not yet available in GHOST.

3.6 Verifying Software Correctness and Performance

Correctness tests The number of possible execution paths in GHOST is huge, because it uses automatically generated high-end kernels for fixed block sizes, allows mixing of row- and column-major dense matrices and real and complex arithmetic, etc. In order to keep the effort of testing the building blocks in ESSEX at a reasonable level, we therefore restrict ourselves to testing via the kernel interface.

The test framework in PHIST is based on Google Test,[5] with modifications to ensure correct behavior in a hybrid parallel setting with MPI+X. These modifications include broadcasting test errors to all MPI processes and assertions to verify that certain data is identical on all processes. The main point here is to decide what type of errors the tests should be able to detect, and under which conditions they should work correctly. For example, some communication errors with MPI cannot be detected by the test framework as it relies on MPI itself. Here one may run the tests in simplified settings (single/multiple thread(s), single/multiple MPI rank(s), GPU only etc.) to test each layer of parallelism separately. Various tools can support this kind of testing, e.g., the thread and address sanitizer included in recent versions of GCC,[6] the MPI checker MUST [7] or CUDA-MEMCHECK.[8]

Tests are automatically generated from single source files for different block sizes, vector lengths, and data types, and for views and standard objects where appropriate. They are executed in nightly builds for different configurations, which leads to a total of currently about 80,000 tests for each kernel library, compiler and MPI version tested. We use the continuous integration tool Jenkins[9] to obtain an overview of the results. Comparison with the comparatively stable Epetra and Tpetra implementations increases the confidence in the correctness of the tests themselves.

Performance testing Our adapters for the kernel interface and the functions of the core and algorithmic layers are instrumented to provide timing information and/or markers for the Likwid performance monitoring tool [46]. Another option that can be turned on at compile time is to include a simple performance model for memory bounded kernels. In this case, a small benchmark of the memory bandwidth is

---

[5]https://github.com/google/googletest

[6]https://github.org/google/sanitizers

[7]https://doc.itc.rwth-aachen.de/display/CCP/Project+MUST

[8]http://docs.nvidia.com/cuda/cuda-memcheck/

[9]https://jenkins-ci.org

run and the percentage of the roofline [49] performance achieved by each kernel function is printed at the end of a run.

There are two 'modes' of performance testing: one incorporates the actual data layout in memory and thus helps to verify that the underlying kernel library achieves the predicted performance for each operation, whether it involves views or not. The other mode only considers the amount of data. This reveals possible performance flaws in the design or implementation of algorithms. For example, if the main operations are performed with a single column view of a row major multi-vector of block size 2, less than 50 % of the roofline performance may be achieved on cache-based architectures.

4 Algorithms Implemented in the ESSR

In this section we want to give a broad overview of the algorithms studied in the ESSEX project, and summarize the lessons learned while developing their highly optimized implementations in the ESSR. For more details, numerical experiments and performance results on massively parallel systems, we refer to the publications cited below.

4.1 Algorithms Based on Chebyshev Polynomials

Algorithms based on the evaluation of polynomial matrix functions are a basic ESSR component. They are represented by the kernel polynomial method (KPM) [48] for spectral functions and eigenvalue densities, Chebyshev time propagation (ChebTP) [44, 47] for matrix exponentials $\exp[tA]$, and Chebyshev filter diagonalization (ChebFD) [31] for the computation of interior eigenvalues. The latter is available through the BEAST-P variant, see Sect. 4.2.

In contrast to, e.g., sparse factorizations or preconditioning that require explicit access to the matrix elements, polynomial algorithms address the matrix in question only through spMVM. Therefore, they are well-suited for situations where the former techniques do not work, or where the matrix is not stored explicitly but only constructed 'on-the-fly' in the spMVM routine. While from the mathematical point of view polynomial algorithms are inferior to algorithms based on rational matrix functions, they are often the only alternative for extremely large matrices.

The common idea behind KPM, ChebTP, and ChebFD is the expansion of a function $f(z) = \sum_{n=0}^{\infty} c_n p_n(z)$ into a series of polynomials $p_n(z)$, especially the Chebyshev polynomials $T_n(z)$ which are often the most favorable choice for numerical algorithms. The algorithms come in two variants: KPM computes the expansion coefficients c_n from scalar products $\langle y, p_n[A]x \rangle$ in order to (re-)construct the function $f(z)$, e.g., the eigenvalue density, while ChebTP and ChebFD use given coefficients c_n to accumulate a result vector $y = \sum_n c_n p_n[A]x$, either for

---

Algorithm 1 Polynomial matrix function evaluation

```
 1  for k = 1 to M do                                            ▷ First two recurrence steps
 2      u_k = α_1(A + β_1 𝟙)x_k                                              ▷ spmv ()
 3      w_k = α_2(A + β_2 𝟙)u_k + γ_2 x_k                                    ▷ spmv ()
 4      x_k = c_0 x_k + c_1 u_k + c_2 w_k                        ▷ axpy & scal (ChebTP, ChebFD)
 5      c_0^(k) = ⟨y, x_k⟩, c_1^(k) = ⟨y, u_k⟩, c_2^(k) = ⟨y, w_k⟩         ▷ dot or gemm (KPM)
 6  for n = 3 to N do                                            ▷ Remaining recurrence steps
 7      for k = 1 to M do
 8          swap(w_k, u_k)                                                  ▷ swap pointers
 9          w_k = α_n(A + β_n 𝟙)u_k + γ_n w_k                               ▷ spmv ()
10          x_k = x_k + c_n w_k                                  ▷ axpy (ChebTP, ChebFD)
11          c_n^(k) = ⟨y, w_k⟩                                              ▷ dot or gemm (KPM)
```

---

the matrix exponential $y = \exp[tA]x$ (ChebTP) or a subspace projection $y = Px$ (ChebFD). An important idea from approximation theory that features both in KPM and ChebFD is the use of so-called kernel polynomials to improve convergence of the expansion [22, 31, 48].

To achieve high execution speed with minimal memory requirements, the polynomials $p_n(z)$ are computed from a two term recurrence

$$x_{n+1} = \alpha_n(A + \beta_n \mathbb{1})x_n + \gamma_n x_{n-1} \tag{1}$$

for the vectors $x_n = p_n[A]x$, which gives the algorithmic core in Algorithm 1 of KPM, ChebTP, and ChebFD. Depending on which operations are used in lines 4/5 and 10/11, it serves two different purposes: replace x_k by $f[A]x_k$ (lines 4,10), or compute moments $\{c_n^{(k)}\}$ (lines 5,11). Algorithm 1 computes the polynomials $p_n[A]x_k$ for several vectors $x_1, \ldots, x_M$ simultaneously, as required in KPM and ChebFD. In addition to spMVM it uses only BLAS-1 vector operations within the two loops over k (vector index) and n (polynomial degree). Owing to this simplicity, the algorithmic core allows for effective performance engineering through straightforward optimizations such as loop-fusion. A particularly rewarding step is the combination of the individual spMVMs for $k = 1, \ldots, M$ into spMMVMs on block vectors, which improves cache utilization due to less erratic memory access patterns. Row-major storage of mVecs (as implemented in GHOST) is the key to reaping the full benefits of this optimization [25, 31]. With such node-level optimizations one can achieve decoupling of the algorithmic core performance from main memory bandwidth on modern CPU systems. Then, the overall performance depends only on the distributed sp(M)MVMs, i.e., is bounded by the inter-node communication bandwidth and latency.

Notice that Algorithm 1 has no internal synchronization points, because neither the dot products in lines 5/11 nor the vectors accumulated in lines 4/10 are used in the following iteration steps. Global synchronization can be delayed until after the execution of the entire algorithmic core, and thus does not affect scalability.

Apart from KPM, Algorithm 1 is normally executed repeatedly. In ChebTP intermediate computations between different executions usually consist of a few xDOT operations, and can be delegated to separate tasks. The results are not needed in the next iterations, and (global) synchronization still is not required. In ChebFD, however, vectors have to be orthogonalized between subsequent executions of the algorithmic core. We use communication-avoiding techniques such as TSQR [6] or SVQB [43] to mitigate the ensuing adverse effects on performance.

The potential of the ESSR implementations of KPM, ChebTP, and ChebFD was demonstrated in a series of papers [1, 25, 31]. With the fully heterogeneous CPU-GPU implementation of KPM [25] we computed the density of states of a matrix with dimension $D = 6.5 \times 10^9$ on 1024 hybrid nodes of the Piz Daint supercomputer.[10] Performance engineering resulted in a speedup of 3–5 at the single node level [1]. Recently, these computations were extended to 4096 nodes ($D = 10^{10}$) and achieved 0.5 Pflop/s sustained performance [26], which corresponds to 11 % of LINPACK efficiency. With the ChebFD implementation we could compute the 148 innermost eigenvalues of a matrix with dimension $D = 10^9$, using 512 nodes of SuperMUC[11] at 40 Tflop/s sustained performance [31]. With the full SuperMUC phase 2 we will be able to obtain inner eigenvalues for matrix dimensions 10^{10}, at an expected sustained performance level of 250 Tflop/s.

The only remaining bottleneck for our polynomial algorithms is the performance of the distributed sp(M)MVMs. In many quantum physics applications (see Sect. 3.3) the inter-node communication volume grows strongly with matrix dimension, and reduction of communication is the most crucial issue for scalability. For stencil type matrices, techniques such as octree ordering are used [36]. For more complex sparsity patterns, GHOST allows sparse matrix repartitioning by PT-Scotch [34]. Future versions of the ESSR will include scalable matrix reordering techniques tailored to the application matrices.

4.2 Beyond FEAST: Projection Based Methods

Consider the (generalized) eigenvalue problem $AX = \Lambda BX$. FEAST [32] is a subspace iteration method to compute all eigenvalues inside a user-defined interval I_λ, and their corresponding eigenvectors. In each step, a size-m search space Y is projected approximately onto the desired invariant subspace, and a Rayleigh-Ritz procedure is used to compute approximate eigenpairs. The computed eigenvectors serve as the new refined search space and the scheme is iterated until convergence. The projection is achieved by (numerical) integration of the resolvent $(zB - A)^{-1}B$ over a contour in the complex plane that encloses I_λ, but no other eigenvalues of (A, B); see [32] for more details and [33] for recent variants. The ESSEX project

---

[10]http://www.cscs.ch/computers/pizdaint/index.html

[11]https://www.lrz.de/services/compute/supermuc/

Algorithm 2 Basic BEAST projection-based eigensolver

Input: Interval I_λ, Matrix pair $A, B \in \mathbb{C}^{N \times N}$
Output: $\hat{m}$ eigenpairs (X, Λ) in I_λ
1   Estimate $\tilde{m} \approx \hat{m}$, choose random $Y \in \mathbb{C}^{N \times m}$ of rank $m > \tilde{m}$
2   **while** not $\tilde{m}$ pairs converged **do**
3   Compute $U = PY$ with suitable projector $P = P_{I_\lambda}(A, B)$
4   Compute Rayleigh quotients $A_U = U^* A U$ and $B_U = U^* B U$
5   Update estimate $\tilde{m}$ of $\hat{m}$ and adjust $m > \tilde{m}$
6   Solve EVP $A_U W = B_U W \Lambda$
7      $X \leftarrow UW$
8   Orthogonalize X against locked vectors and lock newly converged vectors
9      $Y \leftarrow BX$

---

has contributed to improving FEAST in two ways: by proposing techniques for solving or avoiding the linear systems that arise, and by improving robustness and performance of the algorithmic scheme.

Linear systems Our intended use of the FEAST adaptations in BEAST is computing up to 1 000 interior eigenpairs of very large and sparse Hermitian matrices. This use case is not well-supported by other FEAST implementations as they typically rely on direct sparse solvers for the linear systems that arise. We use two strategies to overcome this problem: (i) a robust and scalable iterative solver for the linear systems in contour integration-based BEAST (BEAST-C, [12]), and (ii) use of polynomial approximation as an alternative to contour integration (BEAST-P, [13]). A rough layout of algorithmic key steps in BEAST is presented in Algorithm 2; see [13] for a more detailed formulation.

The linear systems arising in BEAST-C have the form $(zB - A)X = F$, with a possibly large number of right-hand sides F. The complex shifts z get very close to the spectrum, making these systems very ill-conditioned. For interior eigenvalue computations, the system matrix also becomes completely indefinite. For these reasons, standard preconditioned iterative solvers typically fail in this context [12, 23]. In [12] we demonstrated that an accelerated parallel row-projection method called CARP-CG [15] is well-suited for highly indefinite systems arising in this context, and particularly apt at handling small diagonal elements, which are common in our applications. We also proposed a hybrid parallel implementation of the method, which is available as a prototype in the PHIST builtin kernels.

Matrix inversion can be avoided altogether if the projector can be acquired by means other than numerical integration or rational approximation. A common choice is spectral filtering using Chebyshev polynomials via the ChebFD scheme [31], see Sect. 4.1, in particular for the discussion of kernel functions for reducing Gibbs oscillations [23, 48]. This is implemented in the BEAST-P variant, available through PHIST and GHOST.

General improvements The size of the search space is crucial for the convergence of the method [23, 24, 31, 32]. In BEAST we compute a suitable initial guess of the number of eigenpairs in the target interval by integrating the density of states obtained by the KPM (cf. Sect. 4.1). The most recent version of the FEAST library uses a similar approach [7]. As iteration progresses, the search space size is controlled using singular value decomposition [11, 13, 23], that gives a more accurate estimation and consequentially a smaller search space. This lowers memory usage, which may be preferable for very large problems. A more generous search space size can be chosen to reduce the impact of the polynomial degree on convergence speed. The SVD is also used for other purposes like detecting empty intervals or undersized search spaces [10, 23].

Furthermore, a locking technique is implemented in BEAST. By excluding converged eigenpairs from the search space—at the cost of orthogonalizing the remaining vectors in each iteration—it is possible to reduce the cost of later iterations where only few eigenpairs have not yet converged [10, 13, 23].

The most influential parameters for the cost of an iteration in BEAST are the polynomial degree in BEAST-P and residual accuracy for the iterative linear solver in BEAST-C, respectively. These two parameters have different semantics for the progress of the method, though, and need separate consideration.

To minimize the overall work, BEAST-P finds a (problem-dependent) polynomial degree p that, in one BEAST iteration, achieves comparably large residual drop with respect to the number of spMVMs required to evaluate the polynomial [13]. It is then adjusted dynamically by inspecting the residual reduction versus p. This removes the necessity of an initial guess for a suitable degree and makes early iterations cheap since the optimal degree is approached from below. In BEAST-C, we reduce the target residual of the iterative linear solver [13] in early iterations. In later iterations, a higher accuracy is required to achieve a good overall approximation.

Future releases of BEAST will include extension of the method to multiple adjacent intervals (which requires careful orthogonalization and is currently in the testing stage), and the use of single-precision solves in early iterations. BEAST was successfully tested with matrices from graphene and topological insulator modeling of size up to 10^9, typically computing few hundred interior eigenpairs, using the BEAST-P variant with GHOST back end.

4.3 Block Jacobi-Davidson QR

The Jacobi-Davidson method [9] is a popular algorithm for computing a few eigenpairs of a large sparse matrix. It can be seen as a Rayleigh-Ritz procedure with subspace acceleration and deflation. Depending on some implementation details, such as the inner product used and the way eigenvalue approximations are extracted, it may be used for Hermitian and non-Hermitian, standard or generalized eigenproblems, and to find eigenpairs at the border or inside of the spectrum. The Jacobi-Davidson method has several attractive features: it exhibits locally cubic (quadratic) convergence for Hermitian (general) eigenvalue problems, and is very robust w.r.t. approximate solution of the linear systems that occur in each iteration. It also allows integrating preconditioning techniques, and the deflation of eigenvalues near the shift make the linear systems much more well-behaved than in the case of FEAST. For an overview of the Jacobi-Davidson method, see [20].

In [35, 36] we presented the implementation of a block Jacobi-Davidson QR (BJDQR) method which uses block operations to increase the arithmetic intensity and reduce the number of synchronization points (i.e. mitigate the latency of global reduction operations). Use cases for this ESSR solver include the computation of a moderate number of extremal eigenpairs of large, sparse, symmetric or nonsymmetric matrices. BJDQR is a subspace algorithm: in every iteration the search space V is extended by n_b new vectors, w_j, which are obtained by approximately solving a set of correction equations (2), and orthogonalized against all previous directions. The solution of the sparse linear systems (2) is done iteratively.

$$(I - \tilde{Q}\tilde{Q}^*)(A - \sigma_i I)(I - \tilde{Q}\tilde{Q}^*)\Delta q_i \approx -(A\tilde{q}_i - \tilde{Q}\tilde{r}_i), \quad i = 1 \ldots n_b . \tag{2}$$

The successful implementation of this method in PHIST goes hand-in-hand with the development of highly optimized building blocks in GHOST. The basic operations required are spMMVM ($Y_j \leftarrow AX_j$) and the dense matrix–matrix products $Y = X \cdot C$ and $C = X^H Y$, where X and Y denote mVecs and C an sdMat. For the full optimization, we added several custom kernels, including the 'in place' variant $X_{:,1:k} = X \cdot C, X \in \mathbb{C}^{n \times m}, C \in \mathbb{C}^{m \times k}$ and an spMMVM with varying shifts per column, $Y_j = AX_j + \sigma_j X_j$.

Two main observations guided the implementation of this algorithm:

1. row-major storage of mVecs leads to much better performance of both the spMMVM, see also [16], and the dense kernels;
2. accessing single columns in an mVec in row-major storage is disproportionally more expensive than in column-major storage because unnecessary data is loaded into the cache.

To avoid access to single vectors, 'blocked' implementations of the GMRES and MINRES solvers for the correction equation are used. These schemes solve k linear systems simultaneously with separate Krylov spaces, bundling inner products and spMVMs. The second important phase, orthogonalization of W against V, is performed using communication optimal algorithms like TSQR [6] or SVQB [43].

The final performance critical component for Jacobi-Davidson is a preconditioning step used to accelerate the inner solver. Preconditioning techniques typically depend strongly on details of the sparse matrix storage format. As we do not want to impose a particular format on the kernel library that provides the basic operations, PHIST views the preconditioner as an abstract operator (linearOp). This struct contains a pointer to a data object and an apply function, which the application can use to implement e.g. a sparse approximate inverse, an incomplete factorization or a multigrid cycle. The only preconditioned iteration implemented directly in PHIST is CARP-CG, used in the BEAST-C algorithm in ESSEX (Sect. 4.2). This method could also be used in the context of BJDQR, but this combination is not yet implemented.

It is well known that the block variant of JDQR increases the total number of operations (measured for instance in the number of spMVMs). The ESSEX results presented in [36] demonstrated for the first time that this increase is more than compensated by the performance gains in the basic operations, so that an overall speedup of about 20 % can be expected for a wide range of problems and problem sizes. The paper also shows that the only way to achieve this is by consequent performance engineering on all levels. On upcoming hardware, one can expect the benefits of the block variant over the single vector JDQR to grow because of the increasing gap between memory bandwidth and flop rate. Furthermore, the reduction in the number of synchronization points will increase this advantage on large scale systems. We will present results on the heterogeneous execution of this solver on large CPU/GPU clusters in the near future.

5   Fault Tolerance

This section describes our development and evaluation of strategies for efficient checkpointing and restarting of iterative eigenvalue solvers. The former can be done either by storing critical data on a parallel file system (PFS) or on a neighboring node. The latter depends highly on the availability of a fault tolerant communication library, and two options have been evaluated here.

Asynchronous checkpointing via dedicated threads We use the term 'asynchronous checkpointing' for application-level checkpointing where a dedicated thread is used to transfer the checkpoint data to the PFS while the application performs its computations. The benefits of this approach over synchronous PFS-level checkpointing have been demonstrated as proof of concept in [41]. In a first step, an asynchronous copy of the critical data is made in an application- (or algorithm-)specific checkpoint object. The task concept available in GHOST [28] is then used for asynchronously writing the backup file to a global file system. Critical data in the context of eigensolvers may, for instance, be a basis for the (nearly) converged eigenspace. We have implemented and tested this strategy for KPM, ChebTP, ChebFD, and Lanczos solvers. The detailed analysis of this approach for the Lanczos algorithm is presented in [39] where we used dedicated OpenMP-threads for asynchronous writing.

Node-level checkpointing using SCR A more scalable approach has been evaluated using the Scalable Checkpoint-Restart (SCR) library [37], which provides node-level checkpoint/restart mechanisms. Beside the local node-level checkpoints, SCR also provides the functionality to make partner-level and XOR-encoded checkpoints. In addition, occasional PFS-level checkpoints can be made to enable recovery from any catastrophic failures. This strategy introduces very little overhead to the application and is demonstrated in detail along with its comparison with asynchronous checkpointing in [39, 40]. Within the ESSR, we have equipped KPM, ChebTP, and Lanczos algorithms with this checkpointing strategy.

Automatic Fault Recovery The automatic fault recovery (AFR) concept is to enable the application to 'heal itself' after a failure. The basic building block of the concept is a fault-tolerant (FT) communication library. As an FT MPI implementation was not (yet) available, we used the GASPI communication layer [14] to evaluate the concept in a conjugate gradient (CG) solver [38].

As a next step, we evaluated a recent prototype of FT MPI—'User-Level Failure Mitigation' or ULFM [4]—in the context of the KPM with automatic fault recovery. In this implementation, we combined the AFR technique with node-level checkpointing using SCR. The failed processes are replaced by newly spawned ones which take over the identity (i.e., rank) of the failed processes in a rebuilt communicator. All processes then read a consistent copy of the checkpoint from the local or neighbor's memory and resume the computation. Experimental results on this approach are currently being prepared for publication.

6 Summary and Outlook

We have discussed the development of a new software repository for extreme scale sparse eigenvalue computations on heterogeneous hardware. One key challenge of the project was to co-design several interdependent software layers 'from scratch'. We described a simple layered software architecture and a flexible test-driven development process which enabled this. The scalability challenge is addressed by holistic performance engineering and redesigning algorithms for better data locality and communication avoidance. Techniques for mitigating hardware failure were investigated and implemented in prototypical iterative methods.

While this report focused on the software engineering process and algorithmic advancements, we have submitted a second report which demonstrates the parallelization strategy as well as hardware and energy efficiency of our basic building block library GHOST, see [27].

In order to achieve scalability beyond today's petascale computers, we are planning to investigate (among other) scalable communication reducing orderings for our application matrices, communication hiding using the tasking mechanism in our GHOST library, and scalable preconditioners in GHOST for accelerating BEAST-C and Jacobi-Davidson, for instance based on the prototype of CARP-CG in the PHIST builtin kernel library. Future applications will include non-Hermitian matrices and generalized eigenproblems, which requires extensions to some of the algorithms. We are also planning to further integrate our efforts and improve the software structure and documentation to bring forth an ESSL (Exascale Sparse Solver Library).

Acknowledgements This work was supported by the German Research Foundation (DFG) through the Priority Program 1648 "Software for Exascale Computing" under project ESSEX. We would like to thank Michael Meinel (DLR Simulation and Software Technology, software engineering group) for helpful comments on the manuscript.

References

1. Alvermann, A., Basermann, A., Fehske, H., Galgon, M., Hager, G., Kreutzer, M., Krämer, L., Lang, B., Pieper, A., Röhrig-Zöllner, M., Shahzad, F., Thies, J., Wellein, G.: ESSEX: equipping sparse solvers for exascale. In: Lopes, L., et al. (eds.) Euro-Par 2014: Parallel Processing Workshops. Lecture Notes in Computer Science, vol. 8806, pp. 577–588. Springer, Cham (2014). http://dx.doi.org/10.1007/978-3-319-14313-2_49
2. Baker, C.G., Hetmaniuk, U.L., Lehoucq, R.B., Thornquist, H.K.: Anasazi software for the numerical solution of large-scale eigenvalue problems. ACM Trans. Math. Softw. 36(3), 1–23 (2009). http://doi.acm.org/10.1145/1527286.1527287
3. Balay, S., Abhyankar, S., Adams, M.F., Brown, J., Brune, P., Buschelman, K., Dalcin, L., Eijkhout, V., Gropp, W.D., Kaushik, D., Knepley, M.G., McInnes, L.C., Rupp, K., Smith, B.F., Zampini, S., Zhang, H.: PETSc Web page (2015). http://www.mcs.anl.gov/petsc
4. Bland, W., Bouteiller, A., Herault, T., Hursey, J., Bosilca, G., Dongarra, J.: An evaluation of user-level failure mitigation support in MPI. In: Träff, J.L., Benkner, S., Dongarra, J. (eds.) Recent Advances in the Message Passing Interface. Lecture Notes in Computer Science, vol. 7490, pp. 193–203. Springer, Berlin/Heidelberg (2012)
5. Daly, J. et al.: Inter-Agency Workshop on HPC Resilience at Extreme Scale. Tech. rep. (Feb 2012)
6. Demmel, J., Grigori, L., Hoemmen, M., Langou, J.: Communication-optimal parallel and sequential QR and LU factorizations. SIAM J. Sci. Comput. 34(1), A206–A239 (2012)
7. Di Napoli, E., Polizzi, E., Saad, Y.: Efficient estimation of eigenvalue counts in an interval (2013). Preprint (arXiv:1308.4275), http://arxiv.org/abs/1308.4275
8. El-Sayed, N., Schroeder, B.: Reading between the lines of failure logs: understanding how HPC systems fail. In: Proceedings of the 2013 43rd Annual IEEE-IFIP International Conference on Dependable Systems and Networks (DSN '13), pp. 1–12. IEEE Computer Society, Washington, DC (2013)
9. Fokkema, D.R., Sleijpen, G.L.G., van der Vorst, H.A.: Jacobi–Davidson style QR and QZ algorithms for the reduction of matrix pencils. SIAM J. Sci. Comput. 20(1), 94–125 (1998)
10. Galgon, M., Krämer, L., Lang, B.: Counting eigenvalues and improving the integration in the FEAST algorithm (2012). Preprint BUW-IMACM 12/22, available from http://www.imacm.uni-wuppertal.de
11. Galgon, M., Krämer, L., Lang, B., Alvermann, A., Fehske, H., Pieper, A.: Improving robustness of the FEAST algorithm and solving eigenvalue problems from graphene nanoribbons. Proc. Appl. Math. Mech. 14(1), 821–822 (2014)
12. Galgon, M., Krämer, L., Thies, J., Basermann, A., Lang, B.: On the parallel iterative solution of linear systems arising in the FEAST algorithm for computing inner eigenvalues. J. Parallel Comput. 49, 153–163 (2015)
13. Galgon, M., Krämer, L., Lang, B.: Adaptive choice of projectors in projection based eigensolvers (2015), submitted. Available from http://www.imacm.uni-wuppertal.de/
14. GASPI project website: http://www.gaspi.de/en/project.html
15. Gordon, D., Gordon, R.: CARP-CG: A robust and efficient parallel solver for linear systems, applied to strongly convection dominated PDEs. J. Parallel Comput. 36(9), 495–515 (2010)
16. Gropp, W.D., Kaushik, D.K., Keyes, D.E., Smith, B.F.: Towards realistic performance bounds for implicit CFD codes. In: Ecer, A., et al. (eds.) Proceedings of Parallel CFD'99, pp. 233–240. Elesevier, New York (1999)
17. Hernandez, V., Roman, J.E., Vidal, V.: SLEPc: A scalable and flexible toolkit for the solution of eigenvalue problems. ACM Trans. Math. Softw. 31(3), 351–362 (2005)
18. Heroux, M.A., Bartlett, R.A., Howle, V.E., Hoekstra, R.J., Hu, J.J., Kolda, T.G., Lehoucq, R.B., Long, K.R., Pawlowski, R.P., Phipps, E.T., Salinger, A.G., Thornquist, H.K., Tuminaro, R.S., Willenbring, J.M., Williams, A., Stanley, K.S.: An overview of the Trilinos project. ACM Trans. Math. Softw. 31(3), 397–423 (2005), http://doi.acm.org/10.1145/1089014.1089021

19. Heroux, M.A., Willenbring, J.M.: A new overview of the Trilinos project. Sci. Program. **20**(2), 83–88 (2012)
20. Hochstenbach, M.E., Notay, Y.: The Jacobi-Davidson method. GAMM-Mitteilungen **29**(2), 368–382 (2006). http://mntek3.ulb.ac.be/pub/docs/reports/pdf/jdgamm.pdf
21. Hursey, J.: Coordinated checkpoint/restart process fault tolerance for MPI applications on HPC systems. Ph.D. thesis, Indiana University, Bloomington (2010)
22. Jackson, D.: On approximation by trigonometric sums and polynomials. Trans. Am. Math. Soc. **13**, 491–515 (1912)
23. Krämer, L.: Integration based solvers for standard and generalized Hermitian eigenvalue problems. Ph.D. thesis, Bergische Universität Wuppertal (2014). http://nbn-resolving.de/urn/resolver.pl?urn=urn:nbn:de:hbz:468-20140701-112141-6
24. Krämer, L., Di Napoli, E., Galgon, M., Lang, B., Bientinesi, P.: Dissecting the FEAST algorithm for generalized eigenproblems. J. Comput. Appl. Math. **244**, 1–9 (2013)
25. Kreutzer, M., Hager, G., Wellein, G., Pieper, A., Alvermann, A., Fehske, H.: Performance engineering of the kernel polynomial method on large-scale CPU-GPU systems. In: Parallel and Distributed Processing Symposium (IPDPS), 2015 IEEE International, pp. 417–426 (2015). http://arXiv.org/abs/1410.5242
26. Kreutzer, M., Pieper, A., Alvermann, A., Fehske, H., Hager, G., Wellein, G., Bishop, A.R.: Efficient large-scale sparse eigenvalue computations on heterogeneous hardware. In: Poster at the 2015 ACM/IEEE International Conference for High Performance Computing, Networking, Storage and Analysis (2015). http://sc15.supercomputing.org/sites/all/themes/SC15images/tech_poster/tech_poster_pages/post205.html.
27. Kreutzer, M., Thies, J., Pieper, A., Alvermann, A., Galgon, M., Röhrig-Zöllner, M., Shahzad, F., Basermann, A., Bishop, A., Fehske, H., Hager, G., Lang, B., Wellein, G.: Performance engineering and energy efficiency of building blocks for large, sparse eigenvalue computations on heterogeneous supercomputers. In: Bungartz, H.-J., et al. (eds.) Software for Exascale Computing – SPPEXA 2013–2015. Lecture Notes in Computational Science and Engineering, vol. 113. Springer (2016)
28. Kreutzer, M., Thies, J., Röhrig-Zöllner, M., Pieper, A., Shahzad, F., Galgon, M., Basermann, A., Fehske, H., Hager, G., Wellein, G.: GHOST: building blocks for high performance sparse linear algebra on heterogeneous systems (2015). Preprint (arXiv:1507.08101), http://arxiv.org/abs/1507.08101
29. Laguna, I., et al.: Evaluating user-level fault tolerance for MPI applications. In: Proceedings of the 21st European MPI Users' Group Meeting (EuroMPI/ASIA '14), pp. 57:57–57:62. ACM, New York (2014)
30. Lehoucq, R.B., Yang, C.C., Sorensen, D.C.: ARPACK users' guide: solution of large-scale eigenvalue problems with implicitly restarted Arnoldi methods. SIAM, Philadelphia (1998). http://opac.inria.fr/record=b1104502
31. Pieper, A., Kreutzer, M., Galgon, M., Alvermann, A., Fehske, H., Hager, G., Lang, B., Wellein, G.: High-performance implementation of Chebyshev filter diagonalization for interior eigenvalue computations (2015), submitted. Preprint (arXiv:1510.04895)
32. Polizzi, E.: A density matrix-based algorithm for solving eigenvalue problems. Phys. Rev. B **79**, 115112 (2009)
33. Polizzi, E., Kestyn, J.: High-performance numerical library for solving eigenvalue problems: FEAST eigenvalue solver v3.0 user guide (2015). http://arxiv.org/abs/1203.4031
34. (PT-)SCOTCH project website. http://www.labri.fr/perso/pelegrin/scotch/
35. Röhrig-Zöllner, M., Thies, J., Kreutzer, M., Alvermann, A., Pieper, A., Basermann, A., Hager, G., Wellein, G., Fehske, H.: Performance of block Jacobi-Davidson eigensolvers. In: Poster at 2014 ACM/IEEE International Conference on High Performance Computing Networking, Storage and Analysis (2014)
36. Röhrig-Zöllner, M., Thies, J., Kreutzer, M., Alvermann, A., Pieper, A., Basermann, A., Hager, G., Wellein, G., Fehske, H.: Increasing the performance of the Jacobi-Davidson method by blocking. SIAM J. Sci. Comput. **37**(6), C697–C722 (2015). http://elib.dlr.de/89980/

37. Sato, K. et al.: Design and modeling of a non-blocking checkpointing system. In: Proceedings of the Conference on High Performance Computing, Networking, Storage and Analysis, pp. 19:1–19:10. IEEE Computer Society Press, Los Alamitos (2012)

38. Shahzad, F., Kreutzer, M., Zeiser, T., Machado, R., Pieper, A., Hager, G., Wellein, G.: Building a fault tolerant application using the GASPI communication layer. In: Proceedings of the 1st International Workshop on Fault Tolerant Systems (FTS 2015), in conjunction with IEEE Cluster 2015, pp. 580–587 (2015)

39. Shahzad, F., Wittmann, M., Kreutzer, M., Zeiser, T., Hager, G., Wellein, G.: A survey of checkpoint/restart techniques on distributed memory systems. Parallel Process. Lett. **23**(04), 1340011-1–1340011-20 (2013). http://www.worldscientific.com/doi/abs/10.1142/S0129626413400112

40. Shahzad, F., Wittmann, M., Zeiser, T., Hager, G., Wellein, G.: An evaluation of different I/O techniques for checkpoint/restart. In: Proceedings of the 2013 IEEE International Parallel and Distributed Processing Symposium (IPDPS), pp. 1708–1716. IEEE Computer Society (2013). http://dx.doi.org/10.1109/IPDPSW.2013.145

41. Shahzad, F., Wittmann, M., Zeiser, T., Wellein, G.: Asynchronous checkpointing by dedicated checkpoint threads. In: Proceedings of the 19th European conference on Recent Advances in the Message Passing Interface (EuroMPI'12), pp. 289–290. Springer, Berlin/Heidelberg (2012)

42. Stathopoulos, A., McCombs, J.R.: PRIMME: preconditioned iterative multimethod eigensolver–methods and software description. ACM Trans. Math. Softw. **37**(2), 1–30 (2010)

43. Stathopoulos, A., Wu, K.: A block orthogonalization procedure with constant synchronization requirements. SIAM J. Sci. Comput. **23**(6), 2165–2182 (2002)

44. Tal-Ezer, H., Kosloff, R.: An accurate and efficient scheme for propagating the time dependent Schrödinger equation. J. Chem. Phys. **81**, 3967 (1984)

45. TOP500 Supercomputer Sites. http://www.top500.org, accessed: June 2015

46. Treibig, J., Hager, G., Wellein, G.: LIKWID: A lightweight performance-oriented tool suite for x86 multicore environments. In: Proceedings of the 2010 39th International Conference on Parallel Processing Workshops (ICPPW '10), pp. 207–216. IEEE Computer Society, Washington, DC (2010). http://dx.doi.org/10.1109/ICPPW.2010.38

47. Weiße, A., Fehske, H.: Chebyshev expansion techniques. In: Fehske, H., Schneider, R., Weiße, A. (eds.) Computational Many-Particle Physics. Lecture Notes Physics, vol. 739, pp. 545–577. Springer, Berlin/Heidelberg (2008)

48. Weiße, A., Wellein, G., Alvermann, A., Fehske, H.: The kernel polynomial method. Rev. Mod. Phys. **78**, 275–306 (2006). http://dx.doi.org/10.1103/RevModPhys.78.275

49. Williams, S., Waterman, A., Patterson, D.: Roofline: an insightful visual performance model for multicore architectures. Commun. ACM **52**(4), 65–76 (2009). http://doi.acm.org/10.1145/1498765.1498785

Performance Engineering and Energy Efficiency of Building Blocks for Large, Sparse Eigenvalue Computations on Heterogeneous Supercomputers

Moritz Kreutzer, Jonas Thies, Andreas Pieper, Andreas Alvermann,
Martin Galgon, Melven Röhrig-Zöllner, Faisal Shahzad, Achim Basermann,
Alan R. Bishop, Holger Fehske, Georg Hager, Bruno Lang,
and Gerhard Wellein

Abstract Numerous challenges have to be mastered as applications in scientific computing are being developed for post-petascale parallel systems. While ample parallelism is usually available in the numerical problems at hand, the efficient use of supercomputer resources requires not only good scalability but also a verifiably effective use of resources on the core, the processor, and the accelerator level. Furthermore, power dissipation and energy consumption are becoming further optimization targets besides time-to-solution. Performance Engineering (PE) is the pivotal strategy for developing effective parallel code on all levels of modern architectures. In this paper we report on the development and use of low-level

M. Kreutzer (✉) • F. Shahzad • G. Hager • G. Wellein
Erlangen Regional Computing Center, Friedrich-Alxander-University Erlangen-Nuremberg,
Erlangen, Germany
e-mail: moritz.kreutzer@fau.de; faisal.shahzad@fau.de; georg.hager@fau.de;
gerhard.wellein@fau.de

A. Alvermann • A. Pieper • H. Fehske
Institute of Physics, Ernst-Moritz-Arndt-Universität Greifswald, Greifswald, Germany
e-mail: alvermann@physik.uni-greifswald.de; pieper@physik.uni-greifswald.de;
fehske@physik.uni-greifswald.de

M. Galgon • B. Lang
Bergische Universität Wuppertal, Wuppertal, Germany
e-mail: galgon@math.uni-wuppertal.de; lang@math.uni-wuppertal.de

J. Thies • M. Röhrig-Zöllner • A. Basermann
German Aerospace Center (DLR), Simulation and Software Technology, Köln, Germany
e-mail: jonas.thies@dlr.de; melven.roehrig-zoellner@dlr.de; achim.basermann@dlr.de

A.R. Bishop
Theory, Simulation and Computation Directorate, Los Alamos National Laboratory, Los Alamos,
NM, USA
e-mail: arb@lanl.gov

© Springer International Publishing Switzerland 2016   317
H.-J. Bungartz et al. (eds.), *Software for Exascale Computing – SPPEXA*
2013-2015, Lecture Notes in Computational Science and Engineering 113,
DOI 10.1007/978-3-319-40528-5_14

parallel building blocks in the GHOST library ("General, Hybrid, and Optimized Sparse Toolkit"). We demonstrate the use of PE in optimizing a density of states computation using the Kernel Polynomial Method, and show that reduction of runtime and reduction of energy are literally the same goal in this case. We also give a brief overview of the capabilities of GHOST and the applications in which it is being used successfully.

1 Introduction

The supercomputer architecture landscape has encountered dramatic changes in the past decade. Heterogeneous architectures hosting different compute devices (CPU, GPGPU, and Intel Xeon Phi) and systems running 10^5 cores or more are dominating the Top500 top ten [33] since the year 2013. Since then, however, turnover in the top ten has slowed down considerably. A new impetus is expected by the "Collaboration of Oak Ridge, Argonne, and Livermore" (CORAL)[1] with multi-100 Pflop/s systems to be installed around 2018. These systems may feature high levels of thread parallelism and multiple compute devices at the node-level, and will exploit massive data parallelism through SIMD/SIMT features at the core level. The SUMMIT[2] and Aurora[3] architectures are instructive examples. State-of-the-art interconnect technologies will be used to build clusters comprising 10^3 to 10^5 compute nodes. While the former will be of heterogeneous nature with IBM Power9 CPUs and Nvidia Volta GPUs in each node, the latter is projected to be built of homogeneous Intel Xeon Phi manycore processors. Although two different approaches towards exascale computing are pursued here, commonalities like increasing SIMD parallelism and deep memory hierarchies can be determined and should be regarded when it comes to software development for the exascale era.

The hardware architecture of the CORAL systems, which are part of the DOE Exascale Computing Project, can be considered blueprints for the systems to be deployed on the way to exascale computing and thus define the landscape for the development of hardware-/energy-efficient, scalable, and sustainable software as well as numerical algorithms. Additional constraints are set by the continuously increasing power consumption and the expectation that mean-time-to-failure (MTTF) will steadily decrease. It is obvious that long-standing simulation software needs to be completely re-designed or new codes need to be written from scratch. The project "Equipping Sparse Solvers for Exascale" (ESSEX),[4] funded by the Priority Program "Software for Exascale Computing" (SPPEXA) of the German

---

[1] http://www.energy.gov/downloads/fact-sheet-collaboration-oak-ridge-argonne-and-livermore-coral

[2] https://www.olcf.ornl.gov/summit/

[3] https://www.alcf.anl.gov/articles/introducing-aurora

[4] http://blogs.fau.de/essex

Fig. 1 Basic ESSEX project organization: the classic boundaries of application, algorithms, and basic building blocks tightly interact via a holistic performance engineering process

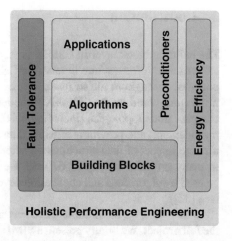

Research Foundation (DFG) is such an endeavor in the field of sparse eigenvalue solvers.

The ESSEX project addresses the above challenges in a joint software co-design effort involving all three fundamental layers of software development in computational science and engineering: basic building blocks, algorithms, and applications. Energy efficiency and fault tolerance (FT) form vertical pillars forcing a strong interaction between the horizontal activities (see Fig. 1 for overall project organization). The overarching goal of all activities is minimal time-to-solution. Thus, the project is embedded in a structured holistic Performance Engineering (PE) process that detects performance bottlenecks and guides optimization and parallelization strategies across all activities.

In the first funding period (2013–2015) the ESSEX project has developed the "Exascale enabled Sparse Solver Repository" (ESSR), which is accessible under a BSD open source license.[5]

The *application layer* has contributed various scalable matrix generation routines for relevant quantum physics problems and has used the ESSR components to advance research in the fields of graphene structures [8, 22, 23, 25] and topological materials [26].

In the *algorithms layer* various classic, application-specific and novel eigen-solvers have been implemented and reformulated in view of the holistic PE process. A comprehensive survey on the activities in the algorithms layer (including FT) is presented in [32]. There we also report on the software engineering process to allow for concurrent development of software in all three layers.

Work performed in the *basic building block layer*, which drives the holistic PE process, is presented in this report.

---

[5]https://bitbucket.org/essex

2 Contribution

The building block layer in ESSEX is responsible for providing an easy to use but still efficient infrastructure library (GHOST), which allows exploiting optimization potential throughout all software layers. GHOST is an elaborate parallelization framework based on the "MPI+X"[6] model, capable of mastering the challenges of complex node topologies (including ccNUMA awareness and node-level resource management) and providing efficient data structures and tailored kernels. In particular the impact of data structures on heterogeneous performance is still underrated in many projects. On top of GHOST we have defined an interface layer that can be used by algorithms and application developers for flexible software development (see [32]).

In this work we illustrate selected accomplishments, which are representative for the full project. We briefly present a SIMD/SIMT-friendly sparse matrix data layout, which has been proposed by ESSEX and gives high performance across all available HPC compute devices. As a sample application we choose the Kernel Polynomial Method (KPM), which will first be used to revisit our model-driven PE process. Then we demonstrate for the first time the impact of PE on improving the energy efficiency on the single socket level for the KPM. Using a coupled performance and energy model, we validate these findings qualitatively and can conclude that the achieved performance improvements for KPM directly correlate with energy savings.

Then we present a brief overview of the GHOST library and give an overview of selected solvers that use GHOST in ESSEX. We finally demonstrate that sustained petascale performance on a large CPU-GPGPU cluster is accessible for our very challenging problem class of sparse linear algebra.

3 Holistic Performance Engineering Driving Energy Efficiency on the Example of the Kernel Polynomial Method (KPM)

The KPM [36] is well established in quantum physics and chemistry. It is used for determining the eigenvalue density (KPM-DOS) and spectral properties of sparse matrices, exposing high optimization potential and the feasibility of petascale implementations. In the following study the KPM is applied to a relevant problem of quantum physics: the determination of electronic structure properties of a three-dimensional topological insulator.

---

[6]The term "MPI+X" denotes the combination of the Message Passing Interface (MPI) and a node-level programming model.

3.1 Performance Engineering for KPM

The naive version of the KPM as depicted in Algorithm 1 builds on several BLAS [17] level 1 routines and the Sparse BLAS [7] level 2 spmv (Sparse matrix–vector multiplication) kernel. The computational intensities of all involved kernels for the topological insulator application are summarized in Table 1. To classify the behavior of a kernel on a compute architecture it is useful to correlate the computational intensity with the machine balance which is the flops/byte ratio of a machine for data from main memory or, in other words, the ratio between peak performance and peak memory bandwidth. It turns out that for each kernel in Table 1 the computational intensity is smaller than the machine balance of any relevant HPC architecture. Even very bandwidth-oriented vector architectures like the NEC SX-ACE with a theoretical machine balance of 1 byte/flop, fail to deliver enough data per cycle from main memory to keep the floating point units busy. This discrepancy only gets more severe on standard multicore CPUs or GPGPUs.

The relative share of data volume assuming minimum data traffic for each kernel can also be seen in Table 1. As all kernels are strongly bound to main memory bandwidth, we can directly translate the relative data volume shares to relative runtime shares if we assume optimal implementations of all kernels and no excess data transfers. Hence, the spmv is the dominating operation in the naive KPM-DOS solver. This, together with the fact that BLAS level 1 routines offer only very limited performance optimization potential, necessitates a detailed examination of this kernel.

Algorithm 1 Naive version of the KPM-DOS algorithm with corresponding BLAS level 1 function calls

---

$\quad$ **for** $r = 0$ to $R - 1$ **do**

$\quad\quad |v\rangle \leftarrow |\text{rand}()\rangle$

$\quad\quad$ Initialization steps and computation of η_0, η_1

$\quad\quad$ **for** $m = 1$ to $M/2$ **do**

$\quad\quad\quad$ swap($|w\rangle, |v\rangle$) $\hfill \triangleright$ Not done explicitly

$\quad\quad\quad |u\rangle \quad \leftarrow H|v\rangle \hfill \triangleright$ spmv()

$\quad\quad\quad |u\rangle \quad \leftarrow |u\rangle - b|v\rangle \hfill \triangleright$ axpy()

$\quad\quad\quad |w\rangle \quad \leftarrow -|w\rangle \hfill \triangleright$ scal()

$\quad\quad\quad |w\rangle \quad \leftarrow |w\rangle + 2a|u\rangle \hfill \triangleright$ axpy()

$\quad\quad\quad \eta_{2m} \quad \leftarrow \langle v|v\rangle \hfill \triangleright$ nrm2()

$\quad\quad\quad \eta_{2m+1} \leftarrow \langle w|v\rangle \hfill \triangleright$ dot()

Table 1 Maximum computational intensities I_{max} in flops/byte and approximate minimum relative share of overall data volume in the solver for each kernel and the full naive KPM-DOS implementation (Algorithm 1) for the topological insulators application

Kernel	spmv	axpy	scal	nrm2	dot	KPM
I_{max}	0.317	0.167	0.188	0.250	0.250	0.295
$V_{min,rel}$ (%)	59.6	22.0	7.3	3.7	7.3	100

3.1.1   Sparse Matrix Data Format

Not only KPM-DOS but also many other sparse linear algebra algorithms are dominated by SpMV. This gave rise to intense research dealing with the performance of this operation. A common finding is that SpMV performance strongly depends on the sparse matrix data format. In the past there was an implicit agreement that an optimal choice of sparse matrix data format strongly depends on the compute architecture used. Obviously, this poses obstacles especially in the advent of heterogeneous machines we are facing today. This led to several efforts trying to either identify data formats that yield good performance on all relevant architectures or to alter the de facto standard format on CPUs (Compressed Sparse Row, or CSR) to enable high performance CSR SpMV kernels also on throughput-oriented architectures. The latter approach resulted in the development of ACSR [1], CSR-Adaptive [5, 9], and CSR5 [19]. The former approach was pursued by ESSEX, e.g., in [13] and led to the proposition of SELL-C-σ as a "catch-all" sparse matrix storage format for the heterogeneous computing era. Although re-balancing the sparse matrix between heterogeneous devices at runtime is not in the scope of this work, it probably is a wise decision in view of the future to choose an architecture-independent storage format if it does not diminish the performance of CPU-only runs. In ESSEX we decided for the SELL-C-σ storage format, which we will explain briefly in the following. Moreover, our preference of SELL-C-σ over CSR will be justified.

SELL-C-σ is a generalization of the Sliced ELLPACK [21] format. The sparse matrix is cut into chunks where each chunk contains C matrix rows, with C being a multiple of the architecture's SIMD width. Within a chunk, all rows are padded with zeros up to the length of the longest row. Matrix values and according column indices are stored along jagged diagonals and chunk after chunk. To avoid excessive zero padding, it may be helpful to sort σ successive matrix rows ($\sigma > C$) by their number of non-zeros before chunk assembly. In this case, also the column indices of matrix entries have to be permuted accordingly. Figure 2 demonstrates the assembly of a SELL-C-σ matrix from an example matrix. In contrast to CSR, SIMD processing is achieved along jagged diagonals of the matrix instead of rows. This enables effective vectorized processing for short rows (comparable to or shorter than the SIMD width), and it enhances the vectorization efficiency of longer rows compared to CSR due to the absence of a reduction operation.

Typically, even non-vectorized code yields optimal performance for bandwidth-bound kernels on a full multi-core CPU socket. However, a higher degree of vectorization usually comes with higher energy efficiency. Hence, we used SELL-C-σ for our experiments. Even if no performance gain over CSR can be expected on a full socket, we will demonstrate in Sect. 3.2 that SELL-C-σ turns out to be beneficial in terms of energy consumption. Due to the regular structure of the topological insulator system matrix, no row sorting has to be applied, i.e., $\sigma = 1$. The chunk height C was set to 32. While it is usually a good practice to choose C as small as possible (which would be C=4 in this case, cf. [13]) to avoid a loss of chunk occupancy in the SELL-C-σ matrix, we do not expect such problems for the present

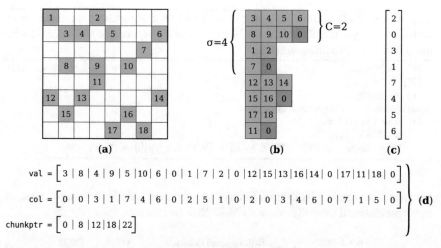

val = $\left[\, 3 \mid 8 \mid 4 \mid 9 \mid 5 \mid 10 \mid 6 \mid 0 \mid 1 \mid 7 \mid 2 \mid 0 \mid 12 \mid 15 \mid 13 \mid 16 \mid 14 \mid 0 \mid 17 \mid 11 \mid 18 \mid 0 \,\right]$

col = $\left[\, 0 \mid 0 \mid 3 \mid 1 \mid 7 \mid 4 \mid 6 \mid 0 \mid 2 \mid 5 \mid 1 \mid 0 \mid 2 \mid 0 \mid 3 \mid 4 \mid 6 \mid 0 \mid 7 \mid 1 \mid 5 \mid 0 \,\right]$ **(d)**

chunkptr = $\left[\, 0 \mid 8 \mid 12 \mid 18 \mid 22 \,\right]$

Fig. 2 SELL-C-σ matrix construction where the SELL-2-4 matrix (**b**) is created from the source matrix (**a**), which includes row permutation according to (**c**), and yields the final SELL-C-σ data structure for this matrix as shown in (**d**)

Algorithm 2 Enhanced version of the KPM-DOS algorithm using the augmented SpMV kernel, which covers all operations chained by '&'

---
for $r = 0$ to $R - 1$ **do**
  $|v\rangle \leftarrow |\mathrm{rand}()\rangle$
  Initialization and computation of η_0, η_1
  for $m = 1$ to $M/2$ **do**
    swap$(|w\rangle, |v\rangle)$
    $|w\rangle = 2a(H - b\mathbb{1})|v\rangle - |w\rangle$ & $\eta_{2m} = \langle v|v\rangle$ & $\eta_{2m+1} = \langle w|v\rangle$

---

test case due to the regularity of the system matrix. Hence, we opted for a larger C which turned out to be slightly more efficient due to a larger degree of loop unrolling.

3.1.2 Kernel Fusion and Blocking

The naive KPM-DOS implementation is strongly memory-bound as described in the introduction to Sect. 3.1. Thus, the most obvious way to achieve higher performance is to decrease the amount of data traffic.

As previously described in [15], a simple and valid way to do this is to fuse all involved kernels into a single tailored KPM-DOS kernel. Algorithm 2 shows the KPM-DOS algorithm with all operations fused into a single kernel. Taking the algorithmic optimization one step further, we can eliminate the outer loop by combining all random initial states into a block of vectors and operate on vector blocks in the fused kernel. The resulting fully optimized (i.e., fused and blocked)

Algorithm 3 Fully optimized version of the KPM-DOS algorithm combining kernel fusion (see Algorithm 2) and vector blocking; each η is a vector of R column-wise dot products of two block vectors

$$
\begin{aligned}
&|\mathbf{V}\rangle := |\mathbf{v}\rangle_{0..R-1} \qquad\qquad\qquad\qquad\qquad\qquad\qquad \triangleright \text{ Assemble vector blocks}\\
&|\mathbf{W}\rangle := |\mathbf{w}\rangle_{0..R-1}\\
&|\mathbf{V}\rangle \leftarrow |\text{rand}()\rangle\\
&\text{Initialization and computation of } \mu_0, \mu_1\\
&\textbf{for } m = 1 \text{ to } M/2 \textbf{ do}\\
&\quad \text{swap}(|\mathbf{W}\rangle, |\mathbf{V}\rangle)\\
&\quad |\mathbf{W}\rangle = 2a(H - b\mathbb{1})|\mathbf{V}\rangle - |\mathbf{W}\rangle \;\&\; \eta_{2m}[:] = \langle\mathbf{V}|\mathbf{V}\rangle \;\&\; \eta_{2m+1}[:] = \langle\mathbf{W}|\mathbf{V}\rangle
\end{aligned}
$$

kernel can be seen in Algorithm 3. Each of the proposed optimization steps increases the computational intensity of the KPM-DOS solver:

$$
I_{\max} = \frac{69}{234}\frac{\text{flops}}{\text{byte}} \approx 0.295 \frac{\text{flops}}{\text{byte}} \xrightarrow[\text{vector blocking}]{\text{kernel fusion \&}} \frac{69}{(130/R + 24)}\frac{\text{flops}}{\text{byte}} \tag{1}
$$

$$
\approx \begin{cases} 0.448 \frac{\text{flops}}{\text{byte}} & R = 1 \text{ (no blocking)}\\[4pt] 2.459 \frac{\text{flops}}{\text{byte}} & R = 32 \text{ (this work)}\\[4pt] 2.875 \frac{\text{flops}}{\text{byte}} & R \to \infty. \end{cases} \tag{2}
$$

Eventually, the fully optimized solver is decoupled from main memory bandwidth on the Intel Ivy Bridge architecture as we have demonstrated in [15].

3.2 Single-Socket Performance and Energy Analysis

3.2.1 Multi-Core Energy Modeling

The usefulness of analytic models that describe the runtime and power dissipation of programs and the systems they run on is obvious. Even if such models are often over-simplified, they can still predict and explain many important properties of hardware–software interaction. Bandwidth-based upper performance limits on the CPU level have been successfully used for decades [4, 12], but modeling power dissipation is more intricate. In [10] we have introduced a phenomenological power and energy consumption model from which useful guidelines for the energy-optimal operating point of a code (number of active cores, clock speed) could be derived. In the following we briefly review the model and its predictions as far as they are relevant for the application case of KPM.

The model takes a high-level view of energy consumption. It is assumed that the CPU chip dissipates a constant *baseline power* W_0, which is defined as the power at zero (extrapolated) clock speed. W_0 also contains contributions from cores in idle or deep sleep state, and it may also comprise other system components whose power dissipation is roughly constant. Every active core, i.e., when executing instructions, contributes additional *dynamic power*, which depends on the clock speed f. The power dissipation at n active cores is assumed as

$$W = W_0 + \left(W_1 f + W_2 f^2\right) n \,. \tag{3}$$

There is no cubic term in f since measurements on current multi-core CPUs show that the dynamic power is at most quadratic in f. The exact dependance on f is parameterized by W_1 and W_2. This is a consequence of the automatic adaptation of supply voltage to clock speed as imposed by the processor or the OS kernel [6]. Power- and energy-to-solution are connected by the program's runtime, which is work divided by performance. If F is the amount of work (e.g., in flops) we assume the following model for the runtime:

$$T(n,f) = \frac{F}{\min\left(nP_0(f), P_{\max}\right)} \,, \tag{4}$$

where P_0 is the single-core (i.e., sequential) performance and $P_{\max}$ is the maximum performance as given by a bandwidth-based limit (e.g., as given by the product of arithmetic intensity and memory bandwidth if the memory interface is a potential bottleneck). Assuming linear scalability up to a saturation point is justified on current multi-core designs if no other scaling impediments apply. In general P_0 will depend strongly on the clock speed since the serial execution time is dominated by intra-cache data transfers or in-core execution on modern CPUs with deep cache hierarchies. This is clearly described by our ECM performance model [31]. The energy-to-solution is thus

$$E(n,f) = F \cdot \frac{W_0 + \left(W_1 f + W_2 f^2\right) n}{\min\left(nP_0(f), P_{\max}\right)} \,. \tag{5}$$

There are several immediate conclusions that can be drawn from this model [10]. Here we restrict ourselves to the case of a fixed clock speed f. Then,

- if the performance saturates at some number of cores n_s, this is the number of active cores to use for minimal energy-to-solution.
- If the performance is linear in n one must use all cores for minimal energy-to-solution.
- Energy-to-solution is inversely proportional to performance, regardless of whether the latter is saturated or not.

We consider the last of these conclusions to be the most important one, since runtime (i.e., inverse performance) is the only factor in which energy is linear. This underlines that performance optimization is the pivotal strategy in energy reduction.

3.2.2 Measurements

In order to provide maximum insight into the connections between performance and energy in a multi-core chip we use what we call a *Z-plot*, combining performance in Gflop/s on the *x* axis with energy-to-solution in J on the *y* axis (see Fig. 3). One set of data points represents measurements for solving a fixed problem with a varying number of active cores on the chip. In a Z-plot, horizontal lines are "energy iso-lines," vertical lines are "performance iso-lines," and hyperbolas are "power iso-lines" (doubling performance, i.e., cutting the runtime in half, also halves energy). If a program shows saturating performance with respect to the number of cores, the curve bends upward at the saturation point, indicating that more resources (thus more power) are used without a performance gain, leading to growing energy-to-solution. For scalable programs the curve is expected to stay flat or keep falling if the power model described in Sect. 3.2.1 holds. The Z-plot has the further advantage that lines of constant energy-delay product (energy-to-solution multiplied by program runtime, EDP) are straight lines through the origin. This is convenient when EDP is used as an alternative target metric instead of plain energy.

All measurements shown in this section were performed on one node (actually a single socket with ten cores) of the "Emmy" cluster at RRZE, comprising Intel Ivy Bridge (Xeon E5-2660v2; "IVB") CPUs with 2.2 GHz base clock speed and 32 GB of RAM per socket. The clock frequency was set to 2.2 GHz, i.e., "Turbo Mode" was disabled. Energy measurements were done via the `likwid-perfctr` tool from the LIKWID tool suite [18, 34], leveraging Intel's on-chip RAPL infrastructure. No

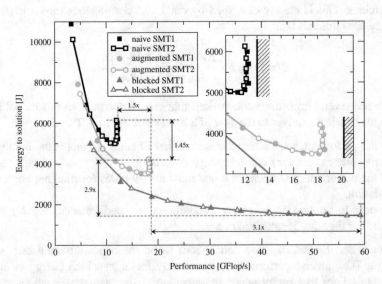

Fig. 3 Single-socket performance and energy Z-plot of naive (*squares*), augmented (*circles*), and blocked (*triangles*) versions on IVB, comparing one thread per core (*filled*) vs. two threads (*open*) using SELL-32-1. (*Inset*: enlarged region of saturation for naive and blocked versions with absolute upper performance limit)

significant variation in energy or performance was observed over multiple runs on the socket.

In Figure 3 we show package-level energy and performance data for the naive implementation of KPM (Algorithm 1) and the augmented and blocked versions (Algorithms 2 and 3) on one IVB socket at a fixed baseline frequency of 2.2 GHz. As expected from their low computational intensities (see Table 1 and Sect. 3.1.2), the naive and augmented variants show strong performance saturation at about 5 and 6 cores, respectively. The augmented kernel requires more cores for saturation since it performs more work per byte transferred from main memory. In the inset we show the bandwidth-based performance limits calculated by multiplying the maximum achievable memory bandwidth on the chip (45 GB/s) with the respective computational intensity. The measured saturated performance is only 6–7 % below this limit in both cases. Note that the maximum bandwidth was obtained using a read-only benchmark (likwid-bench load [35]) but the kernels do not exhibit pure load characteristics. Depending on the fraction of stored vs. loaded data, the maximum bandwidth delivered to the IVB chip can drop by more than 10 %. The blocked variant does not suffer from a memory bandwidth bottleneck on this processor and thus profits from all cores on the chip. As opposed to the naive and blocked versions, it also shows a significant speedup of 12 % when using both hardware threads per core (SMT2).

The energy-to-solution data in the figure was measured on the CPU package level, i.e., ignoring the rest of the system such as RAM, I/O, disks, etc. On the other hand, the particular IVB processor used for the benchmarks shows a low dynamic power compared to chips with higher clock speeds. As a consequence, performance improvements by algorithmic or implementation changes translate into almost proportional energy savings. This is demonstrated by the dashed lines in Fig. 3: Comparing full sockets, the naive version is 1.5× slower and takes 1.45× more energy than the augmented version. The blocked version is 3.1× faster and takes 2.9× less energy than the augmented version. This correspondence becomes only more accurate when adding the full baseline power contributions from all system components. Note that a further 20 % of package-level energy can be saved with the naive and blocked versions by choosing the minimum number of cores that ensures saturation.

The influence of SMT is minor in the saturating cases, which is expected since SMT cannot improve performance in the presence of a strong memory bottleneck. The 12 % performance boost for the blocked version comes with negligible energy savings. We must conclude that executing code on both hardware threads increases the power dissipation, which is also seen by the slight energy increase for SMT2 in the saturated case.

A performance-energy comparison of the SELL-1-1 (a.k.a. CSR) matrix storage format with SELL-32-1 is shown in Fig. 4 for all code versions. The energy advantage of SELL-32-1 in the saturating case is mainly due to the higher single-core performance and accordingly smaller number of required cores to reach the saturation point, leading to package-level energy savings of 8 % and 13 % for the naive and augmented kernels, respectively. We attribute the slight difference in

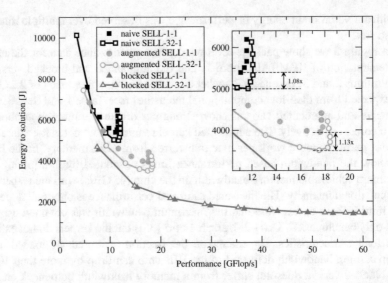

Fig. 4 Single-socket performance and energy Z-plot for the same kernel versions as in Fig. 3 but comparing the SELL-1-1 (CSR) matrix format (*filled symbols*) with SELL-32-1 (*open symbols*) at two threads per core

saturated performance to the different right-hand side data access patterns in the SpMV. The blocked variant shows no advantage (even a slight slowdown) for the SIMD-friendly data layout, which is expected since the access to the matrix data is negligible.

The conclusion from the socket-level performance and energy analysis is that optimization by performance engineering translates, to lowest order, into equivalent energy savings. Overall, the performance ratio between the fastest variant (blocked, with two threads per core) and the lowest (full-socket CSR-based naive implementation) is 5.1, at an energy reduction of 4.5×. At least on the Intel Ivy Bridge system studied here we expect similar findings for other algorithms investigated in the ESSEX project.

A comprehensive analysis of the power dissipation and energy behavior of the studied code variants and the changes for multi-socket and highly parallel runs is beyond the scope of this paper and will be published elsewhere.

4   An Overview of GHOST

The GHOST (General, Hybrid, and Optimized Sparse Toolkit) library summarizes the effort put into computational building blocks in the ESSEX project. A detailed description can be found in [16]. GHOST, a "physics" package containing several scalable sparse matrices, and a range of example applications are available for download.[7] GHOST features high performance building blocks for sparse linear algebra. It builds on the "MPI+X" programming paradigm where "X" can be one of either OpenMP+SIMD or CUDA. The development process of GHOST is closely accompanied by analytic performance modeling, which guarantees compute kernels with optimal performance where possible.

There are several software libraries available that offer some sort of heterogeneous execution capabilities. MAGMA [20], ViennaCL [30], PETSc [3], and Trilinos [11] are arguably the most prominent approaches, all of which have their strengths and weaknesses. PETSc and Trilinos are similar to GHOST as they also build on "MPI+X". MAGMA and ViennaCL, on the other hand, provide shared memory building blocks for different architectures but do not expose any distributed memory capabilities themselves. The most fundamental difference between GHOST and the aforementioned libraries is the possibility of data-parallel heterogeneous execution in GHOST (see below). GHOST has been designed from scratch with heterogeneous architecture in mind. This has to be viewed in contrast to the subsequent addition of heterogeneous computing features to originally homogeneous libraries such as, e.g., PETSc, for which a disclaimer says:[8] "WARNING: Using GPUs effectively is difficult! You must be dedicated and willing to get into the guts of GPU usage if you are serious about using GPUs."

GHOST is not intended to be a rival of the mentioned libraries, but rather a promising supplement and novel approach. Due to its young age, it certainly falls behind in terms of robustness and maturity. While other solutions focus on broad applicability, which often comes with sacrificing some performance, achieving optimal efficiency for selected applications without losing sight of possible broader applicability is clearly the main target of GHOST development. Within the ESSEX effort, we supply mechanisms to use GHOST in higher level software frameworks using the PHIST library [32]. To give an example, in [16] we have demonstrated the feasibility and performance gain of using PHIST to leverage GHOST for a Krylov-Schur algorithm as implemented in the Trilinos package Anasazi [2]. In the following we will briefly summarize the most important features of GHOST and how they influence the ESSEX effort.

A unique feature of GHOST is the capability of data-parallel execution across heterogeneous devices. MPI ranks can be assigned to arbitrary combinations of heterogeneous compute devices, as depicted in Fig. 5. A sparse system matrix is the central data structure in GHOST, and it is distributed row-wise among MPI ranks. In order to reflect heterogeneous systems in an efficient manner, the amount of matrix rows per rank can be arbitrarily set at runtime. Section 5.1 demonstrates possible performance gains due to this feature.

On top of "MPI+X", GHOST exposes the possibility for affinity-aware task-level parallelism. Users can create tasks, which are defined as arbitrary callback functions. OpenMP parallelism can be used inside those tasks and GHOST will take

---

[7]https://bitbucket.org/essex/

[8]http://www.mcs.anl.gov/petsc/features/gpus.html, accessed 02-16-2016

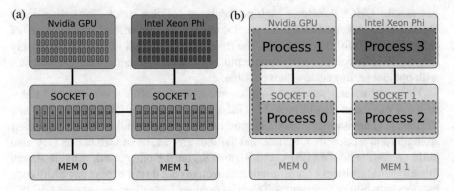

Fig. 5 Heterogeneous compute node and sensible process placement as suggested by GHOST (Figure taken from [16]). (**a**) Heterogeneous node. (**b**) Process placement

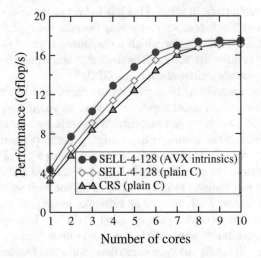

Fig. 6 Intra-socket performance on a single CPU showing the impact of vectorization on SpMV performance for different storage formats (Figure taken from [16])

care of thread affinity and resource management. This feature can be used, e.g., for communication hiding, asynchronous I/O, or checkpointing. In future work we plan to implement asynchronous preconditioning techniques based on this mechanism.

GHOST uses the SELL-C-σ sparse matrix storage format as previously described in Sect. 3.1.1. Note that this does not imply exclusion of CSR, since CSR is just a special case of SELL-C-σ with C=1 and σ=1. Selected kernels are implemented using compiler intrinsics to ensure efficient vectorization. This turned out to be a requirement for optimal performance of rather complex, compute-intensive kernels. However, vectorization may also pay off for kernels with lower computational intensity. Figure 6 backs up the findings of Sect. 3.2.2 in this regard. Not only the superior vectorization potential of SELL-C-σ over CSR, but also a manually

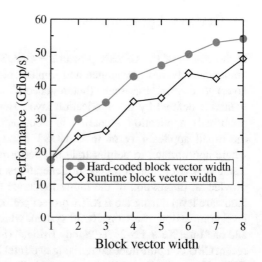

Fig. 7 The impact of hard-coded loop length on the SpMMV performance with increasing block vector width on a single CPU (Figure taken from [16])

vectorized implementation of the SELL-C-σ SpMV kernel yields a highly energy-efficient SpMV kernel.

Vector blocking, i.e., processing several dense vectors at once, is usually a highly appropriate optimization technique in sparse linear algebra due to the often bandwidth-limited nature of sparse matrix algorithms. GHOST addresses this by supporting efficient block vector operations for row- and column-major storage.

Block vector operations often lead to short loops due to a small number of vectors (i.e., in the order of tens) in a block. As short loops are often accompanied by performance penalties, it is possible to define a list of small dimensions at GHOST compile time. Block vector kernels will be automatically generated according to this list. This mechanism is used not only for block vectors, but also for the chunk height C in the SELL-C-σ sparse matrix format. Figure 7 illustrates the performance benefit observed due to generated block vector kernels for the sparse matrix multiple vector multiplication (SpMMV).

Another way to improve the computational intensity of sparse linear algebra algorithms is kernel fusion. In this regard, specialized kernels like the KPM-DOS operator are implemented in close collaboration with experts from the application domain. The specialization grade, i.e., the number and combination of fused operations, of those kernels can be gradually increased, which makes them potentially useful for applications beyond the ESSEX scope. In this regard it should be noted that kernel fusion, while certainly being a promising optimization approach, diminishes the potential for efficient task-parallel execution. This fact promotes the use of kernel fusion together with data parallelism as used in GHOST.

Among others, the described features enable very high performance on modern, heterogeneous supercomputers as demonstrated in our previous work [14, 15, 24, 29].

5 GHOST Applications

In the course of the ESSEX project the GHOST library has been used by several numerical schemes (developed and implemented in the computational algorithms layer) to enable large-scale (heterogeneous) computations for quantum physics scenarios defined by the application layer. Here we summarize selected (already published) application scenarios to demonstrate the capability, the state, and the broad applicability of the GHOST library. We have added measurements, where appropriate, to demonstrate the performance sustainability of the GHOST framework over several processor generations. Moreover these measurements also provide an impression of the rather moderate technological improvements on the hardware level during the ESSEX project period. In particular we focus on a node-level comparison of a Cray XC30 system, which hosts one Nvidia K20X GPGPU and one Intel Xeon E5-2670 "Sandy Bridge" (SNB) processor in each node, with a recent CPU compute node comprising two Intel Xeon E5-2695v3 "Haswell" (HSW) CPUs. While the Cray XC30 system (Piz Daint at CSCS Lugano) has entered the Top500 top ten list at the start of the ESSEX project and is still ranked as #7 (November 2015), Intel Haswell-based systems showed up first in the top ten in 2015.

5.1 Density of States Computations Using KPM-DOS

The basic algorithm (KPM-DOS) used in ESSEX to compute the density of states of large sparse matrices has been introduced in Sect. 3.1. In reference [15] we have presented the PE process and implementation details to enable fully heterogeneous (CPU+GPGPU) KPM-DOS computations and could achieve high node-level performance up to 1024 nodes in weak scaling scenarios. Since then we have extended our runs to up to 4096 nodes (which is approximately 80% of Piz Daint) to achieve 0.5 Pflop/s of sustained performance when computing the DOS of a topological insulator model Hamiltonian (see Fig. 8). The corresponding matrix has a dimension of 3×10^{10} and is extremely sparse with an average of 13 non-zero entries per row. On the node-level the optimizations described earlier have led to significant performance gains for both devices as shown in Fig. 9, and we expect similar energy efficiency improvements on the Cray XC30 system as demonstrated above. Note that during the optimization steps the performance bottleneck on the GPGPU changed from main memory saturation to the dot product. Extending the discussion to latest CPU hardware, we find the Haswell-based system being only 15% ahead of the Cray XC30 node.

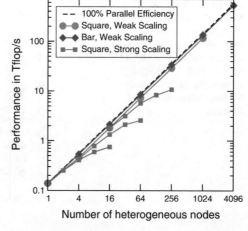

Fig. 8 Strong and weak scaling performance results for different geometries for the topological insulator test case on Piz Daint (measurements up to 1024 nodes have been presented in [15])

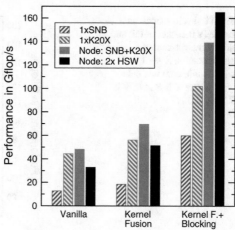

Fig. 9 Impact of optimization steps described in Sect. 3.1 on the node-level performance of Piz Daint (single device and heterogeneous) and a CPU-only node containing two HSW processors (Piz Daint numbers are taken from [15])

5.2 Inner Eigenvalue Computation with Chebyshev Filter Diagonalization (ChebFD)

Applying Chebyshev polynomials as a filter in an iterative subspace scheme allows for the computation of inner eigenpairs of large sparse matrices. The attractive feature of this well-known procedure is the close relation between the filter polynomial and the KPM-DOS scheme. Replacing the norm computation (`nrm2`) and the dot product in Algorithm 1 by a vector addition (`axpy`) yields the polynomial filter in our ChebFD scheme. For a more detailed description of ChebFD (which is also part of our BEAST-P solver), and the relation to KPM-DOS we refer to the report on the ESSEX solver repository [32] and to [24]. In ChebFD the polynomial filter is applied to a subspace of vectors and also optimization stage 2 (see Algorithm 3) can be applied. As compared to the KPM-DOS kernel, the lower computational

Fig. 10 Performance of
KPM-DOS kernel and
polynomial filter for the
topological insulator matrix
on the Nvidia K20m GPGPU,
a single SNB, and 2 HSW
sockets (the latter two only
for block vector width
$R = 32$)

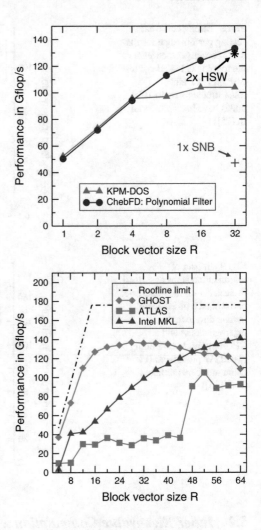

Fig. 11 Performance on a
single IVB socket of tall and
skinny matrix–matrix
multiplication $X \leftarrow V \times W$
with double complex data
type, where X is $R \times R$, V is
$R \times D$, W is $D \times R$ and
$D = 10^7$

intensity of the filter kernel reduces performance on the CPU architectures, while the
GPGPU benefits from the lack of reduction operations moving its bottleneck back to
data transfer (see Fig. 10). It is also evident that the Cray XC30 node (K20X+SNB)
outperforms the Intel Haswell node (2× HSW) on this kernel.

As a second part ChebFD requires a subspace orthogonalization step, which
basically leads to matrix–matrix multiplications involving "tall and skinny" matrices. The performance of widely used BLAS level 3 multi-threaded libraries such
as Intel MKL or ATLAS are often not competitive in the relevant parameter
space addressed by ESSEX applications as can be seen in Fig. 11. Up to a block
vector size of approximately 50 they may miss the upper performance bound
imposed by the memory bandwidth and the arithmetic peak performance by a large
margin. Hence, GHOST provides optimized kernels for these application scenarios
achieving typically 80 % of the maximum attainable performance (see Fig. 11). Note

that automatic kernel generation with compile time defined small dimensions as described in Sect. 4 also works for "tall and skinny" GEMM operations.

The corresponding cuBLAS calls show similar characteristics and thus ESSEX is currently preparing hand-optimized GPGPU kernels for "tall and skinny" dense matrix operations as well.

With the current ChebFD implementation we have computed 148 innermost eigenvalues of a topological insulator matrix (matrix dimension 10^9) on 512 Intel Xeon nodes on the second phase of SuperMUC[9] within 10 h (see [24] for details). Using all of the 3072 nodes we will be able to compute the relevant inner eigenvalues for a topological insulator matrix dimension of 10^{10} at a sustained performance of approximately 250 Tflop/s on that machine.

ChebFD is similar to the recent FEAST algorithm [27]. In FEAST the acceleration is not done with a matrix polynomial but by a contour integration of the resolvent, thus involving the solution of linear systems. FEAST can be faster if a very efficient (e.g., direct) solver is available for the ill-conditioned and highly indefinite linear systems and if high-degree polynomials must be employed in ChebFD. In other situations, ChebFD may be superior due to the high-performance kernels. ChebFD is limited to standard eigenvalue problems, whereas FEAST also can address generalized problems.

5.3 Block Jacobi-Davidson QR Method

The popular Jacobi-Davidson method has been chosen in ESSEX to compute a few low eigenpairs of large sparse matrices. A block variant (BJDQR) was implemented which operates on dense blocks of vectors and thus increases the computational intensity (similar to optimization stage 2 in Fig. 3) and decreases the amount of synchronization points (see [29] and our report on the ESSEX solver repository [32] for details).

The most time consuming operations in this algorithm are the SpMMV and various tall-skinny matrix–matrix products for a limited number of block sizes (e.g. 2,4 and 8). The implementation was tuned to make the best possible use of the highly optimized GHOST kernels (see Fig. 11), and in particular block vectors in row-major storage.

As soon as all optimized CUDA "tall and skinny" GEMM kernels are implemented in GHOST, BJDQR will also be available for fully heterogeneous computations. For a more detailed analysis of performance and numerical efficiency of our BJDQR solver we refer to [28, 29], where it was shown that GHOST delivers near optimal performance on an IVB system and is clearly superior to other implementations.

---

[9]https://www.lrz.de/services/compute/supermuc/systemdescription/

6 Summary and Outlook

We have given an overview of the building block layer in the ESSEX project, specifically the GHOST library. Using several examples of applications within the project (Kernel Polynomial Method [KPM], Chebyshev filter diagonalization [ChebFD], block Jacobi-Davidson QR [BJDQR]) we have shown that GHOST can address the challenges of heterogeneous, highly parallel architectures with its consistent "MPI+X" approach. GHOST implements the highly successful SELL-C-σ sparse matrix format, which contains several other popular formats such as CSR as special cases. We have demonstrated our model-driven Performance Engineering approach using the example of a KPM-DOS application, showing that improvements in the kernel implementation (including the choice of a SIMD-friendly data layout, loop fusion, and blocking) lead not only to the expected performance improvements but also to proportional savings in energy-to-solution on the CPU level, both validated using appropriate performance and power models. For KPM we have also shown the scalability on up to 4096 nodes on the Piz Daint supercomputer, delivering a sustained performance of 0.5 Pflop/s and 87% heterogeneous parallel efficiency on the node-level (CPU+GPGPU). The algorithmically more challenging ChebFD implementation benefited from the optimized tall skinny matrix multiplications in GHOST, which reach substantially higher (in fact, near-light speed, i.e., close to the roofline limit) socket-level performance than the vendor library (MKL) for small to medium block vector sizes. Finally, guided by the same PE approach as in the other cases we could improve the performance of our BJDQR implementation to yield a $3\times$ speedup compared with the Trilinos building block library Tpetra.

The first three years of research into sparse building blocks have already yielded effective ways of Performance Engineering, based on analytic models and insight into hardware-software interaction. Beyond the continued implementation and optimization of tailored kernels for the algorithmic and application-centric parts of the ESSEX project, we will in the future place more emphasis on optimized (problem-aware) matrix storage schemes, high-precision reduction operations with automatic error control, and on more advanced modeling and validation approaches. We have also just barely scratched the surface of the energy dissipation properties of our algorithms; more in-depth analysis is in order to develop a more detailed understanding of power dissipation on heterogeneous hardware.

Acknowledgements The research reported here was funded by Deutsche Forschungsgemeinschaft via the priority program 1648 "Software for Exascale Computing" (SPPEXA). The authors gratefully acknowledge support by the Gauss Centre for Supercomputing e.V. (GCS) for providing computing time on their SuperMUC system at Leibniz Supercomputing Centre through project pr84pi, and by the CSCS Lugano for providing access to their Piz Daint supercomputer. Work at Los Alamos is performed under the auspices of the USDOE.

References

1. Ashari, A., Sedaghati, N., Eisenlohr, J., Parthasarathy, S., Sadayappan, P.: Fast sparse matrix-vector multiplication on GPUs for graph applications. In: Proceedings of the International Conference for High Performance Computing, Networking, Storage and Analysis (SC '14), pp. 781–792. IEEE Press, Piscataway (2014)
2. Baker, C.G., Hetmaniuk, U.L., Lehoucq, R.B., Thornquist, H.K.: Anasazi software for the numerical solution of large-scale eigenvalue problems. ACM Trans. Math. Softw. 36(3), 13:1–13:23 (2009)
3. Balay, S., Abhyankar, S., Adams, M.F., Brown, J., Brune, P., Buschelman, K., Dalcin, L., Eijkhout, V., Gropp, W.D., Kaushik, D., Knepley, M.G., McInnes, L.C., Rupp, K., Smith, B.F., Zampini, S., Zhang, H.: PETSc Web page (2015). http://www.mcs.anl.gov/petsc
4. Callahan, D., Cocke, J., Kennedy, K.: Estimating interlock and improving balance for pipelined architectures. J. Parallel Distrib. Commun. 5(4), 334–358 (1988)
5. Daga, M., Greathouse, J.L.: Structural agnostic spmv: Adapting csr-adaptive for irregular matrices. In: 2015 IEEE 22nd International Conference on High Performance Computing (HiPC), pp. 64–74 (2015)
6. De Vogeleer, K., Memmi, G., Jouvelot, P., Coelho, F.: The energy/frequency convexity rule: modeling and experimental validation on mobile devices. In: Wyrzykowski, R., Dongarra, J., Karczewski, K., Waśniewski, J. (eds.) Parallel Processing and Applied Mathematics. Lecture Notes in Computer Science, vol. 8384, pp. 793–803. Springer, Berlin/Heidelberg (2014)
7. Duff, I.S., Heroux, M.A., Pozo, R.: An overview of the sparse basic linear algebra subprograms: the new standard from the BLAS technical forum. ACM Trans. Math. Softw. 28(2), 239–267 (2002)
8. Fehske, H., Hager, G., Pieper, A.: Electron confinement in graphene with gate-defined quantum dots. Phys. Status Solidi 252(8), 1868–1871 (2015)
9. Greathouse, J.L., Daga, M.: Efficient sparse matrix-vector multiplication on GPUs using the CSR storage format. In: Proceedings of the International Conference for High Performance Computing, Networking, Storage and Analysis, pp. 769–780 (SC '14). IEEE Press, Piscataway (2014)
10. Hager, G., Treibig, J., Habich, J., Wellein, G.: Exploring performance and power properties of modern multi-core chips via simple machine models. Concurr. Comput. 28(2), 189–210 (2014)
11. Heroux, M.A., Bartlett, R.A., Howle, V.E., Hoekstra, R.J., Hu, J.J., Kolda, T.G., Lehoucq, R.B., Long, K.R., Pawlowski, R.P., Phipps, E.T., Salinger, A.G., Thornquist, H.K., Tuminaro, R.S., Willenbring, J.M., Williams, A., Stanley, K.S.: An overview of the Trilinos project. ACM Trans. Math. Softw. 31(3), 397–423 (2005)
12. Hockney, R.W., Curington, I.J.: $f_{1/2}$: A parameter to characterize memory and communication bottlenecks. Parallel Comput. 10(3), 277–286 (1989)
13. Kreutzer, M., Hager, G., Wellein, G., Fehske, H., Bishop, A.R.: A unified sparse matrix data format for efficient general sparse matrix-vector multiplication on modern processors with wide SIMD units. SIAM J. Sci. Comput. 36(5), C401–C423 (2014)
14. Kreutzer, M., Pieper, A., Alvermann, A., Fehske, H., Hager, G., Wellein, G., Bishop, A.R.: Efficient large-scale sparse eigenvalue computations on heterogeneous hardware. In: Poster at 2015 ACM/IEEE International Conference on High Performance Computing Networking, Storage and Analysis (SC '15) (2015)
15. Kreutzer, M., Pieper, A., Hager, G., Alvermann, A., Wellein, G., Fehske, H.: Performance engineering of the kernel polynomial method on large-scale CPU-GPU systems. In: 29th IEEE International Parallel & Distributed Processing Symposium (IEEE IPDPS 2015), Hyderabad (2015)
16. Kreutzer, M., Thies, J., Röhrig-Zöllner, M., Pieper, A., Shahzad, F., Galgon, M., Basermann, A., Fehske, H., Hager, G., Wellein, G.: GHOST: building blocks for high performance sparse linear algebra on heterogeneous systems (2015), preprint. http://arxiv.org/abs/1507.08101

17. Lawson, C.L., Hanson, R.J., Kincaid, D.R., Krogh, F.T.: Basic linear algebra subprograms for Fortran usage. ACM Trans. Math. Softw. **5**(3), 308–323 (1979)
18. LIKWID: Performance monitoring and benchmarking suite. https://github.com/RRZE-HPC/likwid/. Accessed Feb 2016
19. Liu, W., Vinter, B.: CSR5: An efficient storage format for cross-platform sparse matrix-vector multiplication. In: Proceedings of the 29th ACM on International Conference on Supercomputing (ICS '15), pp. 339–350. ACM, New York (2015)
20. MAGMA: Matrix algebra on GPU and multicore architectures. http://icl.cs.utk.edu/magma/. Accessed Feb 2016
21. Monakov, A., Lokhmotov, A., Avetisyan, A.: Automatically tuning sparse matrix-vector multiplication for GPU architectures. In: Patt, Y., Foglia, P., Duesterwald, E., Faraboschi, P., Martorell, X. (eds.) High Performance Embedded Architectures and Compilers. Lecture Notes in Computer Science, vol. 5952, pp. 111–125. Springer, Berlin/Heidelberg (2010)
22. Pieper, A., Heinisch, R.L., Fehske, H.: Electron dynamics in graphene with gate-defined quantum dots. EPL **104**(4), 47010 (2013)
23. Pieper, A., Heinisch, R.L., Wellein, G., Fehske, H.: Dot-bound and dispersive states in graphene quantum dot superlattices. Phys. Rev. B **89**, 165121 (2014)
24. Pieper, A., Kreutzer, M., Galgon, M., Alvermann, A., Fehske, H., Hager, G., Lang, B., Wellein, G.: High-performance implementation of Chebyshev filter diagonalization for interior eigenvalue computations (2015), preprint. http://arxiv.org/abs/1510.04895
25. Pieper, A., Schubert, G., Wellein, G., Fehske, H.: Effects of disorder and contacts on transport through graphene nanoribbons. Phys. Rev. B **88**, 195409 (2013)
26. Pieper, A., Fehske, H.: Topological insulators in random potentials. Phys. Rev. B **93**, 035123 (2016)
27. Polizzi, E.: Density-matrix-based algorithm for solving eigenvalue problems. Phys. Rev. B **79**, 115112 (2009)
28. Röhrig-Zöllner, M., Thies, J., Kreutzer, M., Alvermann, A., Pieper, A., Basermann, A., Hager, G., Wellein, G., Fehske, H.: Performance of block Jacobi-Davidson eigensolvers. In: Poster at 2014 ACM/IEEE International Conference on High Performance Computing Networking, Storage and Analysis (2014)
29. Röhrig-Zöllner, M., Thies, J., Kreutzer, M., Alvermann, A., Pieper, A., Basermann, A., Hager, G., Wellein, G., Fehske, H.: Increasing the performance of the Jacobi–Davidson method by blocking. SIAM J. Sci. Comput. **37**(6), C697–C722 (2015)
30. Rupp, K., Rudolf, F., Weinbub, J.: ViennaCL – a high level linear algebra library for GPUs and multi-core CPUs. In: International Workshop on GPUs and Scientific Applications, pp. 51–56 (2010)
31. Stengel, H., Treibig, J., Hager, G., Wellein, G.: Quantifying performance bottlenecks of stencil computations using the execution-cache-memory model. In: Proceedings of the 29th ACM International Conference on Supercomputing (ICS '15), pp. 207–216. ACM, New York (2015)
32. Thies, J., Galgon, M., Shahzad, F., Alvermann, A., Kreutzer, M., Pieper, A., Röhrig-Zöllner, M., Basermann, A., Fehske, H., Hager, G., Lang, B., Wellein, G.: Towards an exascale enabled sparse solver repository. In: Proceedings of SPPEXA Symposium. Lecture Notes in Computational Science and Engineering. Springer (2016)
33. TOP500 Supercomputer Sites. http://www.top500.org. Accessed Feb 2016
34. Treibig, J., Hager, G., Wellein, G.: LIKWID: A lightweight performance-oriented tool suite for x86 multicore environments. In: Proceedings of the 2010 39th International Conference on Parallel Processing Workshops (ICPPW '10), pp. 207–216. IEEE Computer Society, Washington, DC (2010)
35. Treibig, J., Hager, G., Wellein, G.: likwid-bench: An extensible microbenchmarking platform for x86 multicore compute nodes. In: Brunst, H., Müller, M.S., Nagel, W.E., Resch, M.M. (eds.) Tools for High Performance Computing 2011, pp. 27–36. Springer, Berlin/Heidelberg (2012)
36. Weiße, A., Wellein, G., Alvermann, A., Fehske, H.: The kernel polynomial method. Rev. Mod. Phys. **78**, 275–306 (2006)

Part VIII
DASH: Hierarchical Arrays for Efficient and Productive Data-Intensive Exascale Computing

Expressing and Exploiting Multi-Dimensional Locality in DASH

Tobias Fuchs and Karl Fürlinger

Abstract DASH is a realization of the PGAS (partitioned global address space) programming model in the form of a C++ template library. It provides a multi-dimensional array abstraction which is typically used as an underlying container for stencil- and dense matrix operations. Efficiency of operations on a distributed multi-dimensional array highly depends on the distribution of its elements to processes and the communication strategy used to propagate values between them. Locality can only be improved by employing an optimal distribution that is specific to the implementation of the algorithm, run-time parameters such as node topology, and numerous additional aspects. Application developers do not know these implications which also might change in future releases of DASH. In the following, we identify fundamental properties of distribution patterns that are prevalent in existing HPC applications. We describe a classification scheme of multi-dimensional distributions based on these properties and demonstrate how distribution patterns can be optimized for locality and communication avoidance automatically and, to a great extent, at compile-time.

1 Introduction

For exascale systems the cost of accessing data is expected to be the dominant factor in terms of execution time as well as energy consumption [3]. To minimize data movement, applications have to consider initial placement and optimize both vertical data movement in the memory hierarchy and horizontal data movement between processing units. Programming systems for exascale must therefore shift from a compute-centric to a more data-centric focus and give application developers fine-grained control over data locality.

On an algorithmic level, many scientific applications are naturally expressed in multi-dimensional domains that arise from discretization of time and space.

T. Fuchs (✉) • K. Fürlinger
MNM-Team, Computer Science Department, Ludwig-Maximilians-Universität (LMU) München, München, Germany
e-mail: tobias.fuchs@nm.ifi.lmu.de; karl.fuerlinger@nm.ifi.lmu.de

© Springer International Publishing Switzerland 2016   341
H.-J. Bungartz et al. (eds.), *Software for Exascale Computing – SPPEXA*
2013-2015, Lecture Notes in Computational Science and Engineering 113,
DOI 10.1007/978-3-319-40528-5_15

However, few programming systems support developers in expressing and exploiting data locality in multiple dimensions beyond the most simple one-dimensional distributions. In this paper we present a framework that enables HPC application developers to express constraints on data distribution that are suitable to exploit locality in multi-dimensional arrays.

The DASH library [10] provides numerous variants of data distribution schemes. Their implementations are encapsulated in well-defined concept definitions and are therefore semantically interchangeable. However, no single distribution scheme is suited for every usage scenario. In operations on shared multi-dimensional containers, locality can only be maintained by choosing an optimal distribution. This choice depends on:

- the algorithm executed on the shared container, in particular its communication pattern and memory access scheme,
- run-time parameters such as the extents of the shared container, the number of processes and their network topology,
- numerous additional aspects such as CPU architecture and memory topology.

The responsibility to specify a data distribution that achieves high locality and communication avoidance lies with the application developers. These, however, are not aware of implementation-specific implications: a specific distribution might be balanced, but blocks might not fit into a cache line, inadvertently impairing hardware locality.

As a solution, we present a mechanism to find a concrete distribution variant among all available candidate implementations that satisfies a set of properties. In effect, programmers do not need to specify a distribution type and its configuration explicitly. They can rely on the decision of the DASH library and focus only on aspects of data distribution that are relevant in the scenario at hand.

For this, we first identify and categorize fundamental properties of distribution schemes that are prevalent in algorithms in related work and existing HPC applications. With DASH as a reference implementation, we demonstrate how optimized data distributions can then be determined automatically and, to a great extent, at compile-time.

From a software engineering perspective, we explain how our methodology follows best practices known from established C++ libraries and thus ensures that user applications are not only robust against, but even benefit from future changes in DASH.

The remainder of this paper is structured as follows: the following section introduces fundamental concepts of PGAS and locality in the context of DASH. A classification of data distribution properties is presented in Sect. 3. In Sect. 4, we show how this system of properties allows to exploit locality in DASH in different scenarios. Using the use case of SUMMA as an example, the presented methods are evaluated for performance as well as flexibility against the established implementations from Intel MKL and ScaLAPACK. Publications and tools related to this work are discussed in Sect. 6. Finally, Sect. 7 gives a conclusion and an outlook on future work where the DASH library's pattern traits framework is extended to sparse, irregular, and hierarchical distributions.

2 Background

This section gives a brief introduction to the Partitioned Global Address Space approach considering locality and data distribution. We then present concepts in the DASH library used to express process topology, data distribution and iteration spaces. The following sections build upon these concepts and present new mechanisms to exploit locality automatically using generic programming techniques.

2.1 PGAS and Multi-dimensional Locality

Conceptually, the Partitioned Global Address Space (PGAS) paradigm unifies memory of individual, networked nodes into a virtual global memory space. In effect, PGAS languages create a shared namespace for local and remote variables. This, however, does not affect physical ownership. A single variable is only located in a specific node's memory and local access is more efficient than remote access from other nodes. This is expected to matter more and more even within single (NUMA) nodes in the near future [3]. As locality directly affects performance and scalability, programmers need full control over data placement. Then, however, they are facing overwhelmingly complex implications of data distribution on locality.

This complexity increases exponentially with the number of data dimensions. Calculating a rectangular intersection might be manageable for up to three dimensions, but locality is hard to maintain in higher dimensions, especially for irregular distributions.

2.2 DASH Concepts

Expressing data locality in a Partitioned Global Address Space language builds upon fundamental concepts of process topology and data distribution. In the following, we describe these concepts as they are used in the context of DASH.

2.2.1 Topology: Teams and Units

In DASH terminology, a *unit* refers to any logical component in a distributed memory topology that supports processing and storage. Conventional PGAS approaches offer only the differentiation between local and global data and distinguish between private, shared-local, and shared-remote memory. DASH extends this model by a more fine-grained differentiation that corresponds to hierarchical machine models as units are organized in hierarchical *teams*. For example, a team at the top level could group processing nodes into individual teams, each again consisting of units referencing single CPU cores.

2.2.2 Data Distribution: Patterns

Data distributions in general implement a two-level mapping:

1. From index to process (*node-* or *process mapping*)
2. From process to local memory offset (*local order* or *layout*)

Index sets separate the logical index space as seen by the user from physical layout in memory space. This distinction and the mapping between index domains is usually transparent to the programmer.

Process mapping can also be understood as *distribution*, arrangement in local memory is also referred to as *layout* e.g. in Chapel [5].

In DASH, data decomposition is based on index mappings provided by different implementations of the DASH *Pattern* concept. Listing 1 shows the instantiation of a rectangular pattern, specifying the Cartesian index domain and partitioning scheme. Patterns partition a global index set into *blocks* that are then mapped to units. Consequently, indices are mapped to processes indirectly in two stages: from index to block (*partitioning*) and from block to unit (*mapping*). Figure 1 illustrates a pattern's index mapping as sequential steps in the distribution of a two-dimensional array. While the name and the illustrated example might suggest otherwise, blocks are not necessarily rectangular.

In summary, the DASH pattern concept defines semantics in the following categories:

Distribution	Well-defined distribution of indices to units, depending on properties in the subordinate categories:

	Partitioning	Grouping indices into blocks
	Mapping	Distributing blocks to units in a team
Layout	Arrangement of blocks and block elements in local memory	
Indexing	Operations related to index sets for iterating data elements in global and local scope	

Layout semantics specify the arrangement of values in local memory and, in effect, their order. Indexing semantics also include index set operations like slicing and intersection but do not affect physical data distribution.

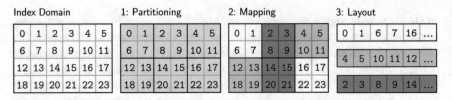

Fig. 1 Example of partitioning, mapping, and layout in the distribution of a dense, two-dimensional array

We define distribution semantics of a pattern type depending on the following set of operations:

$$\text{local}(i_G) \mapsto (u, i_L) \qquad\qquad \text{Index } i_G \text{ to unit } u \text{ and local offset } i_L$$

$$\text{global}(u, i_L) \mapsto i_G \qquad\qquad \text{Unit } u \text{ and local offset } i_L \text{ to global index } i_G$$

$$\text{unit}(i_G) \mapsto u \qquad\qquad \text{Index } i_G \text{ to unit } u$$

$$\text{local\_block}(i_G) \mapsto (u, i_{LB}) \qquad \text{Index } i_G \text{ to unit } u \text{ and local block index } i_{LB}$$

$$\text{global\_block}(i_G) \mapsto i_{GB} \qquad\qquad \text{Index } i_G \text{ to global block index } i_{GB}$$

with n-dimensional indices i_G, i_L as coordinates in the global/local Cartesian element space and i_{GB}, i_{LB} as coordinates in the global/local Cartesian block space. Instead of a Cartesian point, an index may also be declared as a point's offset in linearized memory order.

```
1  // Brief notation:
2  TilePattern<2> pattern(global_extent_x, global_extent_y,
3                  TILED(tilesize_x), TILED(tilesize_y));
4  // Equivalent full notation:
5  TilePattern<2, dash::default_index_t, ROW_MAJOR>
6    pattern(DistributionSpec<2>(
7            TILED(tilesize_x), TILED(tilesize_y),
8        SizeSpec<2, dash::default_size_t>(
9            global_extent_x, global_extent_y),
10       TeamSpec<1>(
11           Team::All())));
```

Listing 1 Explicit instantiation of DASH patterns

DASH containers use patterns to provide uniform notations based on view proxy types to express index domain mappings. User-defined data distribution schemes can be easily incorporated in DASH applications as containers and algorithms accept any type that implements the Pattern concept.

Listing 2 illustrates the intuitive usage of user-defined pattern types and the `local` and `block` view accessors that are part of the DASH container concept. View proxy objects use a referenced container's pattern to map between its index domains but do not contain any elements themselves. They can be arbitrarily chained to refine an index space in consecutive steps, as in the last line of Listing 2: the expression `array.local.block(1)` addresses the second block in the array's local memory space.

In effect, patterns specify local iteration order similar to the partitioning of iteration spaces e.g. in RAJA [11]. Proxy types implement all methods of their delegate container type and thus also provide `begin` and `end` iterators that specify the iteration space within the view's mapped domain. DASH iterators provide an intuitive notation of ranges in virtual global memory that are well-defined with respect to distance and iteration order, even in multi-dimensional and irregular index domains.

```
 1  CustomPattern pattern;
 2  dash::Array<double> a(size, pattern);
 3  double g_first = a[0]          // First value in global memory,
 4                                 // corresponds to a.begin()
 5  double l_first = a.local[0];   // First value in local memory,
 6                                 // corresponds to a.local.begin()
 7  dash::copy(a.block(0).begin(),          // Copy first block in
 8             a.block(0).end(),            // global memory to second
 9             a.local.block(1).begin());   // block in local memory
```

Listing 2 Views on DASH containers

3   Classification of Pattern Properties

While terms like *blocked*, *cyclic* and *block-cyclic* are commonly understood, the terminology of distribution types is inconsistent in related work, or varies in semantics. Typically, distributions are restricted to specific constraints that are not applicable in the general case for convenience.

Instead of a strict taxonomy enumerating the full spectrum of all imaginable distribution semantics, a systematic description of pattern properties is more practicable to abstract semantics from concrete implementations. The classification presented in this section allows to specify distribution patterns by categorized, unordered sets of properties. It is, of course, incomplete, but can be easily extended. We identify properties that can be fulfilled by data distributions and then group these properties into orthogonal *categories* which correspond to the separate functional aspects of the pattern concept described in Sect. 2.2.2: partitioning, unit mapping, and memory layout. This categorization also complies with the terminology and conceptual findings in related work [16].

DASH pattern semantics are specified by a configuration of properties in these dimensions:

$$\text{Global} \times \underbrace{\text{Partitioning} \times \text{Mapping}}_{\text{Distribution}} \times \text{Layout}$$

Details on a selection of single properties in all categories are discussed in the remainder of this section.

3.1   Partitioning Properties

Partitioning refers to the methodology used to subdivide a global index set into disjoint blocks in an arbitrary number of logical dimensions. If not specified otherwise by other constraints, indices are mapped into *rectangular* blocks. A partitioning is *regular* if it only creates blocks with identical extents and *balanced* if all block have identical size.

rectangular	Block extents are constant in every single dimension, e.g. every row has identical size.
minimal	Minimal number of blocks in every dimension, i.e. at most one block for every unit.
regular	All blocks have identical extents.
balanced	All blocks have identical size (number of elements).
multi-dimensional	Data is partitioned in at least two dimensions.
cache-aligned	Block sizes are a multiple of cache line size.

Note that with the classification, these properties are mostly independent: rectangular partitionings may produce blocks with varying extents, balanced partitionings are not necessarily rectangular, and so on. For example, partitioning a matrix into triangular blocks could satisfy the *regular* and *balanced* partitioning properties. The fine-grained nature of property definitions allows many possible combinations that form an expressive and concise vocabulary to express pattern semantics.

3.2 Mapping Properties

Well-defined mapping properties exist that have been formulated to define *multipartitionings*, a family of distribution schemes supporting parallelization of line sweep computations over multi-dimensional arrays.

The first and most restrictive multipartitioning has been defined based on the *diagonal* property [15]. In a multipartitioning, each process owns exactly one tile in each hyperplane of a partitioning so that all processors are active in every step of a line-sweep computation along any array dimension as illustrated in Fig. 2.

General multipartitionings are a more flexible variant that allows to assign more than one block to a process in a partitioned hyperplane. The generalized definition

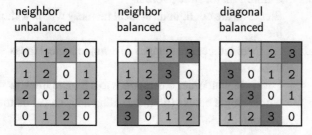

Fig. 2 Combinations of mapping properties. Numbers in blocks indicate the unit rank owning the block

subdivides the original diagonal property into the *balanced* and *neighbor* mapping properties [7] described below. This definition is more relaxed but still preserves the benefits for line-sweep parallelization.

balanced	The number of assigned blocks is identical for every unit.
neighbor	A block's adjacent blocks in any one direction along a dimension are all owned by some other processor.
shifted	Units are mapped to blocks in diagonal chains in at least one hyperplane.
diagonal	Units are mapped to blocks in diagonal chains in all hyperplanes.
cyclic	Blocks are assigned to processes like dealt from a deck of cards in every hyperplane, starting from first unit.
multiple	At least two blocks are mapped to every unit.

The constraints defined for multipartitionings are overly strict for some algorithms and can be further relaxed to a subset of its properties. For example, a pipelined optimization of the SUMMA algorithm requires a *diagonal shift* mapping [14, 18] that satisfies the diagonal property but is not required to be balanced. Therefore, the diagonal property in our classification does not imply a balanced mapping, deviating from its original definition.

3.3 Layout Properties

Layout properties describe how values are arranged in a unit's physical memory and, consequently, their order of traversal. Perhaps the most crucial property is storage order which is either row- or column-major. If not specified, DASH assumes row-major order as known from C. The list of properties can also be extended to give hints to allocation routines on the physical memory topology of units such as NUMA or CUDA.

row-major	Row major storage order, used by default.
col-major	Column-major storage order.
blocked	Elements are contiguous in local memory within a single block.
canonical	All local indices are mapped to a single logical index domain.
linear	Local element order corresponds to a logical linearization within single blocks (tiled) or within entire local memory (canonical).

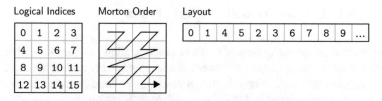

Fig. 3 Morton order memory layout of block elements

While patterns assign indices to units in logical blocks, they do not necessarily preserve the block structure in local index sets. After process mapping, a pattern's layout scheme may arrange indices mapped to a unit in an arbitrary way in physical memory. In *canonical* layouts, the local storage order corresponds to the logical global iteration order. *Blocked* layouts preserve the block structure locally such that values within a block are contiguous in memory, but in arbitrary order. The additional *linear* property also preserves the logical linearized order of elements within single blocks. For example, Morton order memory layout as shown in Fig. 3 satisfies the *blocked* property, as elements within a block are contiguous in memory, but does not grant the *linear* property.

3.4 Global Properties

The *Global* category is usually only needed to give hints on characteristics of the distributed value domain such as the *sparse* property to indicate the distribution of sparse data.

dense   Distributed data domain is dense.
sparse   Distributed data domain is sparse.
balanced The same number of values is mapped to every unit after
              partitioning and mapping.

It also contains properties that emerge from a *combination* of the independent partitioning and layout properties and cannot be derived from either category separately. The global *balanced* distribution property, for example, guarantees the same number of local elements at every unit. This is trivially fulfilled for balanced partitioning and balanced mapping where the same number of blocks b of identical size s is mapped to every unit. However, it could also be achieved in a combination of unbalanced partitioning and unbalanced mapping, e.g. when assigning b blocks of size s and $b/2$ blocks of size $2s$.

4 Exploiting Locality with Pattern Traits

The classification system presented in the last section allows to describe distribution pattern semantics using properties instead of a taxonomy of types that are associated with concrete implementations. In the following, we introduce *pattern traits*, a collection of mechanisms in DASH that utilize distribution properties to exploit data locality automatically.

As a technical prerequisite for these mechanisms, every pattern type is annotated with *tag* type definitions that declare which properties are satisfied by its implementation. This enables meta-programming based on the patterns' distribution properties as type definitions are evaluated at compile-time. Using tags to annotate type invariants is a common method in generic C++ programming and prevalent in the STL and the Boost library.[1]

```
1  template <dim_t NDim, ...>
2  class ThePattern {
3  public:
4    typedef mapping_properties<
5              mapping_tag::diagonal,
6              mapping_tag::cyclic >
7      mapping_tags;
8    ...
9  };
```

Listing 3 Property tags in a pattern type definition

4.1 Deducing Distribution Patterns from Constraints

In a common use case, programmers intend to allocate data in distributed global memory with the use for a specific algorithm in mind. They would then have to decide for a specific distribution type, carefully evaluating all available options for optimal data locality in the algorithm's memory access pattern.

To alleviate this process, DASH allows to automatically create a concrete pattern instance that satisfies a set of constraints. The function `make_pattern` returns a pattern instance from a given set of properties and run-time parameters. The actual type of the returned pattern instance is resolved at compile-time and never explicitly appears in client code by relying on automatic type deduction.

---

[1]http://www.boost.org/community/generic_programming.html

```
1   static const dash::dim_t NumDataDim = 2;
2   static const dash::dim_t NumTeamDim = 2;
3   // Topology of processes, here: 16x8 process grid
4   TeamSpec<NumTeamDim> teamspec(16, 8);
5   // Cartesian extents of container:
6   SizeSpec<NumDataDim> sizespec(extent_x, extent_y);
7   // Create instance of pattern type deduced from
8   // constraints at compile-time:
9   auto pattern =
10    dash::make_pattern<
11      partitioning_properties<
12        partitioning_tag::balanced >,
13      mapping_properties<
14        mapping_tag::balanced, mapping_tag::diagonal >,
15      layout_properties<
16        layout_tag::blocked >
17    >(sizespec, teamspec);
```

Listing 4 Deduction of an optimal distribution

To achieve compile-time deduction of its return type, make_pattern employs the *Generic Abstract Factory* design pattern [2]. Different from an *Abstract Factory* that returns a polymorphic *object* specializing a known base type, a Generic Abstract Factory returns an arbitrary type, giving more flexibility and no longer requiring inheritance at the same time.

Pattern constraints are passed as template parameters grouped by property categories as shown in Listing 4. Data extents and unit topology are passed as run-time arguments. Their respective dimensionality (*Num-DataDim, NumTeamDim*), however, can be deduced from the argument types at compile-time. Figure 4 illustrates the logical model of this process involving two stages: a type generator that resolves a pattern type from given constraints and argument types at compile-time and an object generator that instantiates the resolved type depending on constraints and run-time parameters.

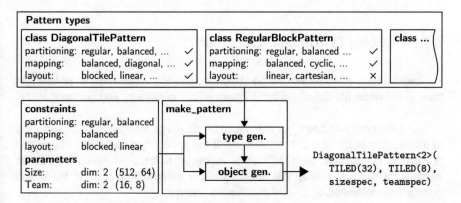

Fig. 4 Type deduction and pattern instantiation in dash::make_pattern

Every property that is not specified as a constraint is a degree of freedom in type selection. Evaluations of the GUPS benchmark show that arithmetic for dereferencing global indices is a significant performance bottleneck apart from locality effects. Therefore, when more than one pattern type satisfies the constraints, the implementation with the least complex index calculation is preferred.

The automatic deduction also is designed to prevent inefficient configurations. For example, pattern types that pre-generate block coordinates to simplify index calculation are inefficient and memory-intensive for a large number of blocks. They are therefore disregarded if the blocking factor in any dimension is small.

4.2   Deducing Distribution Patterns for a Specific Use Case

With the ability to create distribution patterns from constraints, developers still have to know which constraints to choose for a specific algorithm. Therefore, we offer shorthands for constraints of every algorithm provided in DASH that can be passed to make_pattern instead of single property constraints.

```
1  dash::TeamSpec<2> teamspec(16, 8);
2  dash::SizeSpec<2> sizespec(1024, 1024);
3  // Create pattern instance optimized for SUMMA:
4  auto pattern = dash::make_pattern<
5                      dash::summa_pattern_traits
6                  >(sizepec, teamspec);
7  // Create matrix instances using the pattern:
8  dash::Matrix<2, int> matrix_a(sizespec, pattern);
9  dash::Matrix<2, int> matrix_b(sizespec, pattern);
10 \ldots
11 auto matrix_c = dash::summa(matrix_a, matrix_b)
```

Listing 5 Deduction of a matching distribution pattern for a given use-case

4.3   Checking Distribution Constraints

An implementer of an algorithm on shared containers might want to ensure that their distribution fits the algorithm's communication strategy and memory access scheme.

The traits type pattern_constraints allows querying constraint attributes of a concrete pattern type at compile-time. If the pattern type satisfies all requested properties, the attribute satisfied is expanded to true. Listing 6 shows its usage in a static assertion that would yield a compilation error if the object pattern implements an invalid distribution scheme.

```
1  // Compile-time constraints check:
2  static_assert(
3    dash::pattern_constraints<
4        decltype(pattern),
5        partitioning_properties< ... >,
6        mapping_properties< ... >,
7        layout_properties< ... >
8    >::satisfied::value
9  );
10 // Run-time constraints check:
11 if (dash::check_pattern_constraints<
12        partitioning_properties< ... >,
13        mapping_properties< ... >,
14        indexing_properties< ... >
15     >(pattern)) {
16   // Object 'pattern' satisfies constraints
17 }
```

Listing 6 Checking distribution constraints at compile-time and run-time

Some constraints depend on parameters that are unknown at compile-time, such as data extents or unit topology in the current team.

The function check_pattern_constraints allows checking a given pattern object against a set of constraints at run-time. Apart from error handling, it can also be used to implement alternative paths for different distribution schemes.

4.4 Deducing Suitable Algorithm Variants

When combining different applications in a work flow or working with legacy code, container data might be preallocated. As any redistribution is usually expensive, the data distribution scheme is invariant and a matching algorithm variant is to be found.

We previously explained how to resolve a distribution scheme that is the best match for a known specific algorithm implementation. Pattern traits and generic programming techniques available in C++11 also allow to solve the inverse problem: finding an algorithm variant that is suited for a given distribution. For this, DASH provides adapter functions that switch between an algorithm's implementation variants depending on the distribution type of its arguments. In Listing 7, three matrices are declared using an instance of dash::TilePattern that corresponds to the known distribution of their preallocated data. In compilation, dash::multiply expands to an implementation of matrix–matrix multiplication that best matches the distribution properties of its arguments, like dash::summa in this case.

```
1  typedef dash::TilePattern<2, ROW_MAJOR>    TiledPattern;
2  typedef dash::Matrix<2, int, TiledPattern> TiledMatrix;
3  TiledPattern   pattern(global_extent_x,  global_extent_y,
4                         TILE(tilesize_x), TILE(tilesize_y));
5  TiledMatrix    At(pattern);
6  TiledMatrix    Bt(pattern);
7  TiledMatrix    Ct(pattern);
8  ...
9  // Use adapter to resolve algorithm suited for TiledPattern:
10 dash::multiply(At, Bt, Ct); // --> dash::summa(At, Bt, Ct);
```

Listing 7 Deduction of an algorithm variant for a given distribution

5 Performance Evaluation

We choose dense matrix–matrix multiplication (*DGEMM*) as a use case for evaluation as it represents a concise example that allows to demonstrate how slight changes in domain decomposition drastically affect performance even in highly optimized implementations.

In principle, the matrix–matrix multiplication implemented in DASH realizes a conventional blocked matrix multiplication similar to a variant of the SUMMA algorithm presented in [14]. For the calculation $C = A \times B$, matrices A, B and C are distributed using a blocked partitioning. Following the *owner computes* principle, every unit then computes the multiplication result

$$C_{ij} = A_{ik} \times B_{kj} = \sum_{k=0}^{K-1} A_{ik} B_{kj}$$

for all sub-matrices in C that are local to the unit.

Figure 5 illustrates the first two multiplication steps for a square matrix for simplicity, but the SUMMA algorithm also allows rectangular matrices and unbalanced partitioning.

We compare strong scaling capabilities on a single processing node against DGEMM provided by multi-threaded Intel MKL and PLASMA [1]. Performance of distributed matrix multiplication is evaluated against ScaLAPACK [8] for an increasing number of processing nodes.

Ideal tile sizes for PLASMA and ScaLAPACK had to be obtained in a large series of tests for every variation of number of cores and matrix size. As PLASMA does not optimize for NUMA systems, we also tried different configurations of numactl as suggested in the official documentation of PLASMA.

For the DASH implementation, data distribution is resolved automatically using the make_pattern mechanism as described in Sect. 4.2.

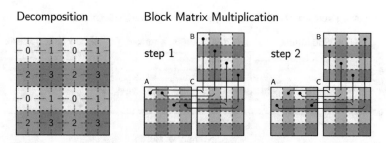

Fig. 5 Domain decomposition and first two block matrix multiplications in the SUMMA implementation. Numbers in blocks indicate the unit mapped to the block

5.1 Eperimental Setup

To substantiate the transferability of the presented results, we execute benchmarks on the supercomputing systems SuperMUC and Cori which differ in hardware specifications and application environments.

SuperMUC phase 2^2 incorporates an Infiniband fat tree topology interconnect with 28 cores per processing node. We evaluated performance for both Intel MPI and IBM MPI.

Cori phase 1^3 is a Cray system with 32 cores per node in an Aries dragonfly topology interconnect. As an installation of PLASMA is not available, we evaluate performance of DASH and Intel MKL.

5.2 Results

We only consider the best results from MKL, PLASMA and ScaLAPACK to provide a fair comparison to the best of our abilities.

In summary, the DASH implementation consistently outperformed the tested variants of DGEMM and PDGEMM on distributed and shared memory scenarios in all configurations (Fig. 6, 7, 8).

More important than performance in single scenarios, overall analysis of results in single-node scenarios confirms that DASH in general achieved predictable scalability using automatic data distributions. This is most evident when comparing results on Cori presented in Fig. 7: the DASH implementation maintained consistent scalability while performance of Intel MKL dropped when the number of processes was not a power of two, a good example of a system-dependent implication that is commonly unknown to programmers.

---

[2] https://www.lrz.de/services/compute/supermuc/systemdescription/

[3] http://www.nersc.gov/users/computational-systems/cori/cori-phase-i/

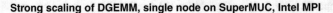

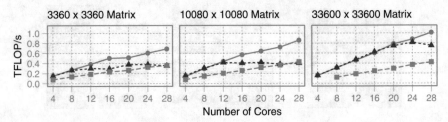

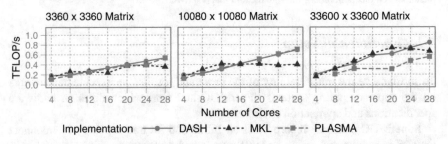

Fig. 6 Strong scaling of matrix multiplication on single node on SuperMUC phase 2, Intel MPI and IBM MPI, with 4 to 28 cores for increasing matrix size

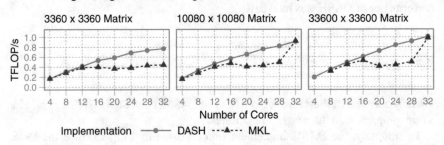

Fig. 7 Strong scaling of matrix multiplication on single node on Cori phase 1, Cray MPICH, with 4 to 32 cores for increasing matrix size

6   Related Work

Various aspects of data decomposition have been examined in related work that influenced the design of pattern traits in DASH.

The Kokkos framework [9] is specifically designed for portable multi-dimensional locality. It implements compile-time deduction of data layout depending on memory architecture and also specifies distribution traits roughly resembling some of the property categories introduced in this work. However, Kokkos targets intra-node locality focusing on CUDA- and OpenMP backends

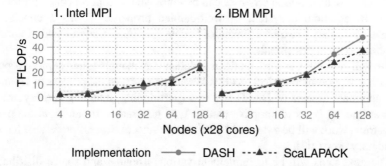

Fig. 8 Strong scaling of dash::summa and PDGEMM (ScaLAPACK) on SuperMUC phase 2 for IBM MPI and Intel MPI for matrix size 57344 × 57344

and does not define concepts for process mapping. It is therefore not applicable to the PGAS language model where explicit distinction between local and remote ownership is required.

UPC++ implements a PGAS language model and, similar to the array concept in DASH, offers local views for distributed arrays for rectangular index domains [12]. However, UPC++ does not provide a general view concept and no abstraction of distribution properties as described in this work.

Chapel's Domain Maps is an elegant framework that allows to specify and incorporate user-defined mappings [5] and also supports irregular domains. The fundamental concepts of domain decomposition in DASH are comparable to *DMaps* in Chapel with dense and strided regions like previously defined in ZPL [6]. Chapel does not provide automatic deduction of distribution schemes, however, and no classification of distribution properties is defined.

Finally, the benefits of hierarchical data decomposition are investigated in recent research such as TiDA, which employs hierarchical tiling as a general abstraction for data locality [17]. The Hitmap library achieves automatic deduction of data decomposition for hierarchical, regular tiles [4] at compile-time.

7 Conclusion and Future Work

We constructed a general categorization of distribution schemes based on well-defined properties. In a broad spectrum of different real-world scenarios, we then discussed how mechanisms in DASH utilize this property system to exploit data locality automatically.

In this, we demonstrated the expressiveness of generic programming techniques in modern C++ and their benefits for constrained optimization.

Automatic deduction greatly simplifies the incorporation of new pattern types such that new distribution schemes can be employed in experiments with minimal effort. In addition, a system of well-defined properties forms a concise and precise vocabulary to express semantics of data distribution, significantly improving testability of data placement.

We will continue our work on flexible data layout mappings and explore concepts to further support hierarchical locality. We are presently in the process of separating the functional aspects of DASH patterns (partitioning, mapping and layout) into separate policy types to simplify pattern type generators. In addition, the pattern traits framework will be extended by soft constraints to express preferable but non-mandatory properties.

The next steps will be to implement various irregular and sparse distributions that can be easily combined with view specifiers in DASH to support the existing unified sparse matrix storage format provided by SELL-C-σ [13]. We also intend to incorporate hierarchical tiling schemes as proposed in TiDA [17]. The next release of DASH including these features will be available in the second quarter of 2016.

Acknowledgements We gratefully acknowledge funding by the German Research Foundation (DFG) through the German Priority Program 1648 Software for Exascale Computing (SPPEXA).

References

1. Agullo, E., Demmel, J., Dongarra, J., Hadri, B., Kurzak, J., Langou, J., Ltaief, H., Luszczek, P., Tomov, S.: Numerical linear algebra on emerging architectures: The plasma and magma projects. J. Phys.: Conf. Ser. **180**, 012037 (2009). IOP Publishing
2. Alexandrescu, A.: Modern C++ Design: Generic Programming and Design Patterns Applied. Addison-Wesley, Boston (2001)
3. Ang, J.A., Barrett, R.F., Benner, R.E., Burke, D., Chan, C., Cook, J., Donofrio, D., Hammond, S.D., Hemmert, K.S., Kelly, S.M., Le, H., Leung, V.J., Resnick, D.R., Rodrigues, A.F., Shalf, J., Stark, D., Unat, D., Wright, N.J.: Abstract machine models and proxy architectures for exascale computing. In: Proceedings of the 1st International Workshop on Hardware-Software Co-design for High Performance Computing (Co-HPC '14), pp. 25–32. IEEE Press, Piscataway (2014)
4. de Blas Cartón, C., Gonzalez-Escribano, A., Llanos, D.R.: Effortless and efficient distributed data-partitioning in linear algebra. In: 2010 12th IEEE International Conference on High Performance Computing and Communications (HPCC), pp. 89–97. IEEE (2010)
5. Chamberlain, B.L., Choi, S.E., Deitz, S.J., Iten, D., Litvinov, V.: Authoring user-defined domain maps in Chapel. In: CUG 2011 (2011)
6. Chamberlain, B.L., Choi, S.E., Lewis, E.C., Lin, C., Snyder, L., Weathersby, W.D.: ZPL: A machine independent programming language for parallel computers. IEEE Trans. Softw. Eng. **26**(3), 197–211 (2000)
7. Chavarría-Miranda, D.G., Darte, A., Fowler, R., Mellor-Crummey, J.M.: Generalized multi-partitioning for multi-dimensional arrays. In: Proceedings of the 16th International Parallel and Distributed Processing Symposium. p. 164. IEEE Computer Society (2002)
8. Choi, J., Demmel, J., Dhillon, I., Dongarra, J., Ostrouchov, S., Petitet, A., Stanley, K., Walker, D., Whaley, R.C.: ScaLAPACK: A portable linear algebra library for distributed memory computers – Design issues and performance. In: Applied Parallel Computing Computations

in Physics, Chemistry and Engineering Science, pp. 95–106. Springer (1995)

9. Edwards, H.C., Sunderland, D., Porter, V., Amsler, C., Mish, S.: Manycore performance-portability: Kokkos multidimensional array library. Sci. Program. **20**(2), 89–114 (2012)

10. Fürlinger, K., Glass, C., Knüpfer, A., Tao, J., Hünich, D., Idrees, K., Maiterth, M., Mhedeb, Y., Zhou, H.: DASH: Data structures and algorithms with support for hierarchical locality. In: Euro-Par 2014 Workshops (Porto, Portugal). Lecture Notes in Computer Science, pp. 542–552. Springer (2014)

11. Hornung, R., Keasler, J.: The RAJA portability layer: overview and status. Tech. rep., Lawrence Livermore National Laboratory (LLNL), Livermore (2014)

12. Kamil, A., Zheng, Y., Yelick, K.: A local-view array library for partitioned global address space C++ programs. In: Proceedings of ACM SIGPLAN International Workshop on Libraries, Languages, and Compilers for Array Programming, p. 26. ACM (2014)

13. Kreutzer, M., Hager, G., Wellein, G., Fehske, H., Bishop, A.R.: A unified sparse matrix data format for efficient general sparse matrix-vector multiplication on modern processors with wide SIMD units. SIAM J. Sci. Comput. **36**(5), C401–C423 (2014)

14. Krishnan, M., Nieplocha, J.: SRUMMA: a matrix multiplication algorithm suitable for clusters and scalable shared memory systems. In: Proceedings of 18th International Parallel and Distributed Processing Symposium 2004, p. 70. IEEE (2004)

15. Naik, N.H., Naik, V.K., Nicoules, M.: Parallelization of a class of implicit finite difference schemes in computational fluid dynamics. Int. J. High Speed Comput. **5**(01), 1–50 (1993)

16. Tate, A., Kamil, A., Dubey, A., Größlinger, A., Chamberlain, B., Goglin, B., Edwards, C., Newburn, C.J., Padua, D., Unat, D., et al.: Programming abstractions for data locality. Research report, PADAL Workshop 2014, April 28–29, Swiss National Supercomputing Center (CSCS), Lugano (Nov 2014)

17. Unat, D., Chan, C., Zhang, W., Bell, J., Shalf, J.: Tiling as a durable abstraction for parallelism and data locality. In: Workshop on Domain-Specific Languages and High-Level Frameworks for High Performance Computing (2013)

18. Van De Geijn, R.A., Watts, J.: SUMMA: Scalable universal matrix multiplication algorithm. Concurr. Comput. **9**(4), 255–274 (1997)

Tool Support for Developing DASH Applications

Denis Hünich, Andreas Knüpfer, Sebastian Oeste, Karl Fürlinger,
and Tobias Fuchs

Abstract DASH is a new parallel programming model for HPC which is implemented as a C++ template library on top of a runtime library implementing various
PGAS (Partitioned Global Address Space) substrates. DASH's goal is to be an
easy to use and efficient way to parallel programming with C++. Supporting
software tools is an important part of the DASH project, especially debugging
and performance monitoring. Debugging is particularly necessary when adopting
a new parallelization model, while performance assessment is crucial in High
Performance Computing applications by nature. Tools are fundamental for a
programming ecosystem and we are convinced that providing tools early brings
multiple advantages, benefiting application developers using DASH as well as
developers of the DASH library itself. This work, first briefly introduces DASH and
the underlying runtime system, existing debugger and performance analysis tools.
We then demonstrate the specific debugging and performance monitoring extensions
for DASH in exemplary use cases and discuss an early assessment of the results.

1 Introduction

Developer tools are indispensable to develop complex and large applications. The
broad spectrum of existing tools ranges from simple autocompletion in IDEs[1] to
sophisticated suites for debugging or performance analysis. While no prior experience is needed to use autocompletion, debugging an application is significantly
more demanding and requires a certain degree of expertise.

---

[1]IDE—Integrated Development Environment.

D. Hünich (✉) • A. Knüpfer • S. Oeste
TU Dresden, Dresden, Germany
e-mail: denis.huenich@tu-dresden.de; andreas.knuepfer@tu-dresden.de;
sebastian.oeste@tu-dresden.de

K. Fürlinger • T. Fuchs
LMU München, München, Germany
e-mail: karl.fuerlinger@nm.ifi.lmu.de; tobias.fuchs@nm.ifi.lmu.de

© Springer International Publishing Switzerland 2016   361
H.-J. Bungartz et al. (eds.), *Software for Exascale Computing – SPPEXA*
2013-2015, Lecture Notes in Computational Science and Engineering 113,
DOI 10.1007/978-3-319-40528-5_16

HPC tools for debugging and performance analysis are of high relevance in particular. Unlike tools for conventional application development, HPC tools are required to maintain an arbitrary number of process states. A debugger for parallel applications has to manage all control streams (e.g. processes) and eventually their communication, for example.

Likewise, performance analysis tools have to collect metrics like the time spent in a function, the number of function invocations, or the communication time between processes. All these information have to be collected for every process, preferably without changing the program's behavior.

The acceptance for new libraries like DASH can be increased by providing tool support. The use and extension of existing tools has two significant benefits. First, users don't need to learn to use new tools. Second, the library developers make use of the experience of the tool developers and don't have to redevelop an already existing tool.

In DASH, we follow this approach and provide dedicated support to third-party solutions such as the GDB debugger [12], Score-P [22] and PAPI [5]. This paper describes the extensions and the challenges to incorporate these tools in the DASH template library and the DASH runtime (DART).

The remainder of this paper is structured as follows: publications and tools related to this work are presented in Sect. 2. Section 3 briefly introduces the DASH library and the underlying runtime. Challenges occurring when debugging parallel applications are discussed in Sect. 4. The following Sect. 5 explains extensions for Score-P and the DASH library to instrument and trace DART and DASH applications. A specific profiling application for the MPI backend is presented in Sect. 6. Section 7 describes the integration of the PAPI library in DASH. Finally, Sect. 8 concludes the work and looks out on the future work.

2 Related Work

2.1 DASH

The concept of the DASH library and the DART API is explained in detail in [13]. An MPI3 DART implementation can be found in [33].

Besides DASH, other PGAS approaches exist. UPC [30] and Co-array Fortran [25] extend C and Fortran with the PGAS concept. Other PGAS languages are Chapel [7], Fortress [3] and X10 [8]. All three projects were funded from the "DARPA High Productivity Computing Systems".

GASPI [17] and OpenSHMEM [27] are an effort to create a standardized API for PGAS programming. UPC++ [32] and Co-array C++ [20] extended C++ with PGAS. STAPL [6] implements a C++ template library that shares several concepts with DASH like local view on data and the representation of distributed containers but does not seem to target classic HPC applications.

2.2 Debugging

We decided to extend the debugger GDB [12] because it is a widely used open source debugger for sequential applications. LLDB [31] is also an alternative, but not as prevalent as GDB. Existing debuggers for parallel applications like Eclipse Parallel Tools [2], PGDB [10], Totalview [28] or DDT [4] are limited to MPI, OpenMP, CUDA, etc. and don't support the DASH library. Totalview and DDT support the most libraries but are proprietary and therefore, not the first choice for dedicated support in an open source project like DASH.

2.3 Performance Analysis

The Score-P^2 measurement system is a performance analysis software for parallel programs. Because performance analysis and optimization is important for a successful utilization of HPC resources, many tools for analyzing parallel programs are available. Some of them rely on sophisticated profiling techniques e.g. TAU [29] and HPC-Toolkit [1], others use event tracing.

Score-P is a joint effort of RWTH Aachen, ZIH TU Dresden, Jülich Supercomputing Centre, TU München, and GNS Braunschweig. Score-P can record a wide range of parallel subsystems such as MPI, OpenMP, SHMEM, CUDA and PThreads. Auxiliary instrumentation through compiler instrumentation and manual user-defined instrumentation are supported while sampling based recording is planned for future releases. Furthermore, Score-P can utilize CPU performance counters, provided by the PAPI, to identify bottlenecks and enables performance analysis based on event tracing (OTF2 [11] trace format) or profiling (CUBE format). Post mortem, tools like Vampir [21] or Scalasca [15] support the performance analysis by visualizing the traces and profiles. With Periscope [16] it is possible to make an online analysis.

3 Overview DASH

The main goal of the DASH project is the development of a C++ template library for distributed data structures that provides a convenient parallel programming model. DASH is divided into four layers (Fig. 1). The DASH runtime (DART) implements different communication substrates like MPI [23] and provides a uniform interface to the upper DASH C++ template library. This library provides containers for distributed data structures (e.g. arrays) which can be used similar

---

^{2}www.score-p.org

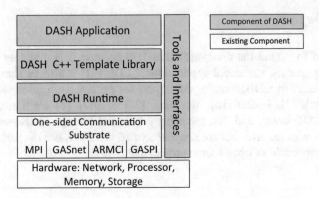

Fig. 1 The layered structure of the DASH project

to the containers of the C++ STL.[3] Furthermore, these containers can be used with algorithms provided by the C++ STL. The user application is located on top of the DASH template library, thus it is independent from any changes made in DART or the communication substrate. Additionally the application benefits from performance optimizations in DART. The tools and interface layer is connected to all aforementioned layers and supports the developers of each layer. Be it debugging (Sect. 4), performance analysis (Sect. 5), implementing/optimizing the MPI implementation in DART (Sect. 6), or getting direct support within the DASH template library for PAPI counters (Sect. 7).

In the following we briefly explain the DART and the DASH template library, more detailed information can be found in [13].

3.1 DART: The DASH Runtime

As mentioned before, DART abstracts the low-level communication libraries from the DASH template library. The DART API is written in plain C and sufficiently general enough to use any low-level communication substrate. Processes are called units in DART/DASH to be independent from other terms like threads, ranks or images. These units are all part of at least one team.

Units communicate directly with one-sided communication or collective communication operations. Blocking and non-blocking communication are provided as well as synchronization mechanisms.

Besides the communication, the DART API manages the distributed global memory which can be allocated symmetrically and team aligned or locally with global accessibility. The former allocates the same amount of storage for every

---

[3]STL—Standard Template Library.

unit that is part of the team. So, globally the memory looks like one big block, but locally each unit manages only a part of it. The latter allocates memory locally but is accessible from other units.

For now, DART supports MPI and GASPI [17] as communication substrate. Furthermore, a System V shared memory implementation for testing purposes and a prototype for GPGPUs exist.

3.2 DASH: Distributed C++ Template Library

The DASH template library is written in C++ which allows efficient PGAS [26] semantics. PGAS approaches often use local- and global-views on data. While in the global-view it is not directly clear whether the data is accessed remotely or locally. The local-view is limited to local data accesses only which need no communication, thus, they are much faster than remote accesses. The distributed data containers in the DASH template library use the local-/global-views and combine it with the C++ STL iterator concept. So, it is possible to iterate over local data only, all data, or only a part of it(mixture of global and local data accesses). Listing 1 demonstrates the construction of an 1D-array (line 2), a global iteration over all elements (lines 9–12) and local data accesses with the C++ STL function *fill* (line 20). This example shows a small set of possibilities to access the data elements of a container. To combine the distributed data structure concept with the C++ STL concepts, new functionality had to be added. In Sect. 5.2 we explain why C++ container, like our DASH container, are quite difficult to trace for performance analysis.

```
1   // allocates an array over all units
2   dash::array<double> my_array(100000, nodeteam);
3
4   // gets the unit's id
5   int myid = nodeteam.myID();
6
7   // The unit with id 0 iterates over all array elements and
8   // assigns them to the value 1
9   if( myid == 0){
10    for(auto it = my_array.begin(); it != my_array.end(); ++it)
11      *it = 1;
12  }
13
14  // blocks all units until every unit reached this routine
15  my_array.barrier();
16
17  // every unit fills its part of the array with its unit id
18  // lbegin() returns the beginning of the local memory while
19  // lend() returns the ending
20  std::fill(my_array.lbegin(), my_array.lend(), myid);
```

Listing 1 DASH array container used with global-view and local-view semantics

4   Debugging DASH Applications

To conveniently debug parallel DASH applications, it is necessary to interrupt the
parallel start up, provide a useful output for DASH data structures, and get access to
all data elements (local and remote) of a DASH container.

We decided to extend the widely used and extensible debugger GDB because no
debugger fulfilled all requirements. GDB doesn't support debugging multiprocess
applications but allows to attach to existing processes remotely. The initialization
process of DASH was modified to interrupt the start up and, in debugging mode,
traps the master process in an infinite loop while all other processes are waiting in a
barrier. Now, the user can attach GDB instances to all interesting DASH processes
and, for instance, set breakpoints. To continue debugging, the master process has to
be released from the infinite loop.

The concept of pretty-printers, provided by newer versions of the GDB, supports
the modification of the data structure output, printed in the debugging process.
We used this extension and defined pretty-printers for all DASH containers and
added them to the GDB environment. The following example shows the difference
between the default (Listing 2) and the pretty printed (Listing 3) output for a DASH
1-D array. The pretty printed version is obviously more helpful for most users,
because it provides more specific information.

It is necessary to get access to all data elements, local and global ones, to get
the shown pretty printed output. Therefore, we added, only in debugging mode, new
methods to the DASH containers which enabled GDB to access all data elements.

More detailed information, concerning the start up process and made modifications, can be found in [18].

```
$1 = (dash::Array<int, dash::PatternBlocked> &) @0x7fff6018f370:
    {m_team = @0x619ea0, m_size = 20, m_pattern = {<std::
    __shared_ptr<dash::Pattern, (__gnu_cxx::_Lock_policy)2>> = {
    _M_ptr = 0xac2de0, _M_refcount = {
        _M_pi = 0xac2dd0}}, <No data fields>}, m_ptr = {<std::
            __shared_ptr<dash::GlobPtr<int>, (__gnu_cxx::
            _Lock_policy)2>> = {_M_ptr = 0xab1310, _M_refcount =
            {_M_pi = 0xab1300}}, <No data fields>},
    m_acc = {<std::__shared_ptr<dash::MemAccess<int>, (__gnu_cxx::
    _Lock_policy)2>> = {_M_ptr = 0xac2e50, _M_refcount = {_M_pi
    = 0xac2e40}}, <No data fields>}}
```

Listing 2 Standard GDB output for a DASH 1-D array

```
$1 = Dash::Array<int, dash::PatternBlocked> with 20 elements -
    Team with dart id 0 and position 0 - dart global pointer: std
    ::shared_ptr (count 1, weak 0) 0xab1310 = {1, 1, 1, 1, 1, 2,
    2, 2, 2, 2, 3, 3, 3, 3, 3, 4, 4, 4, 4, 4}
```

Listing 3 Pretty printed GDB output for a DASH one-dimensional array

5 Using Score-P to Analyze DASH and DART

Performance analysis distinguishes between data acquisition and data presentation. Data acquisition describes techniques to get performance data during a program execution, such as Sampling, or Instrumentation. Data presentation deals with the visualization of the acquired data. This section explains two performance data presentation techniques (profiling, and tracing) and techniques used to support DASH/DART in Score-P.

Profiling A profile presents the performance data in summarized aggregations of performance metrics. These metrics can be the number of function invocations or the inclusive/exclusive time spent in a code region, e.g. function, or loop body. The simplest form of a profile provides information about all functions, the program spent the majority of time (*flat profile*), but no caller context. A profile can also offer information about program execution based on a call path which results in a *call path profile*. It is even possible to split the program execution into phases and create profile records of each phase (*phase profile*). Performance data for a profile can be gathered from instrumented regions or even from sample based data [19].

Tracing Tracing stores performance information in event traces during the program execution. All interesting events are hold in memory and flushed, process dependent, to disk. These flushes (*buffer-flush-event* [24]) only happen if not enough memory is available, or the program is finished. When the program finished, the separated event streams will be unified.

Performance data are stored in special trace formats, such as OTF2 [11] and are produced by instrumentation, or sample based data. They contain information about where, when, and what kind of event is recorded. Typical events can be simple enters or leaves of subroutines, or communication calls (e.g. MPI Point-to-Point communication). In contrast to profiling, tracing records time stamps for the events to later reconstruct the program behavior. Event tracing provides a great level of detail of a program's performance, but with the cost of a lot disk space and *buffer-flush-events* for large and long running applications.

5.1 DART

The DART function calls can be automatically instrumented by the Score-P adapter, using library wrapping. Library wrapping is a technique to link function calls to alternative functions instead of calling the original function. The alternative function can record the enter event; call the original library function and forward the parameters; record the leave event; and return the result of the original function to the caller. This approach needs the symbols of all interesting library functions the application is linked against. Library wrapping is a widely used and a well known concept for automatically instrumented libraries [9].

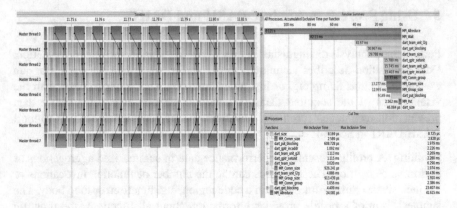

Fig. 2 Vampir screenshot. On the *left* a time frame of the recorded events for 8 processes is shown. The accumulated exclusive time of all measured functions is located *top right* and the function call graph at the *bottom right*

We extended Score-P with a specialized adapter, wrapping all defined functions of the DART API. Besides the adapter, the build system of Score-P and DART specific parameters were added. Now, Score-P records DART API functions which can be visualized in tools like Vampir.

DART API functions are independent from the implemented communication substrates.The recording of DART API only, supports especially the development process of the different DART implementations because it allows a comparison of the runtime/behavior of DART functions using different implementations, or even the native communication substrate.

The Vampir screenshot in Fig. 2 shows an instrumented application with 8 processes. Yellow events symbolize DART events and red ones MPI events. The call tree (on the bottom right) displays all recorded DART functions and the used MPI functions. The inclusive and exclusive time is given behind the function name. The graph on top shows the runtime of each recorded function. On the left, the timeline for all process in the chosen time frame is plotted and visualizes all events ordered by their time stamp.

5.2 DASH

DASH is a C++ template library, implemented as header only which means, all template declarations and definitions are located in the same header file. This is necessary because the compiler has to know both. The actual type of the template is first known when the user includes the header, containing the template definition, into the application source code. Further, symbol names of C++ functions and methods appear mangled in an object file. A mangled name encodes additional information of the symbol, for instance, type information of: the parameters, templates, namespaces, and classes. Symbols of the DASH library are not entirely determined before compilation of the actual template calls. In the library file, header only symbols do not exist. So, it is not possible to guess which symbols will actually appear in the application. For this kind of libraries the library wrapping approach, mentioned in Sect. 5.1, doesn't work.

Another approach to instrument the DASH library is the automatic compiler instrumentation. The compiler includes Score-P measurement function calls before and after each function call. But real C++ applications tend to have an enormous amount of small function calls (e.g. operators) which results in recording many events. Another drawback is the instrumentation of all functions or methods of other used libraries, such as internal functions of the STL. Internals of the STL with all the constructor calls and overloaded operators are probably not of interest for analysis of an application, but are recorded and increase: the number of events, the overhead in execution time, and the memory used to store the events. The Score-P measurement system provides function filter to reduce the overhead, but this costs the user maintenance and extra effort.

Fortunately, Score-P offers a user API to instrument the source code manually. This API generally provides a set of preprocessor directives (macros) which can be inserted at the beginning of a function or method. Expanded, a macro creates a new object on the stack to generate an enter event at construction time; when the function leaves, this object is destructed and triggers a leave event. Instrumenting every single function this way results in a similar behavior as the compiler instrumentation, in terms of runtime overhead and memory footprint. To avoid this, we used a level based approach for the instrumentation. According to the architecture of the DASH library the instrumentation is portioned into the following three levels:

- CONTAINER_LEVEL instrumentation
- PATTERN_LEVEL instrumentation
- ALL_LEVEL instrumentation

CONTAINER_LEVEL is the lightest level of instrumentation, only public methods of the DASH container classes are recorded. This level can show performance-critical methods of the DASH containers. PATTERN_LEVEL extends the container level with all methods of the DASH pattern classes and delivers additional information for the data distribution. Figure 3 shows Vampir's function summary plot for an instrumented DASH application with PATTERN_LEVEL instrumentation. All DASH functions and methods will be recorded with the level ALL_LEVEL.

The instrumentation and the concrete level can be enabled with the compiler flags -DDASH_SCOREP and -D<LEVELNAME>_LEVEL=true. If no level is set the default level is CONTAINER_LEVEL (only when instrumentation is enabled).

The used disk space of the produced traces for the different instrumentation levels are shown in Table 1. NO_DASH represents DART and MPI functions only (no DASH functions were recorded) while COMPILER recorded all functions (automatic compiler instrumentation). Table 1 shows, that the instrumentation of the DASH container functions already increased the trace sizes by about 10 GB.

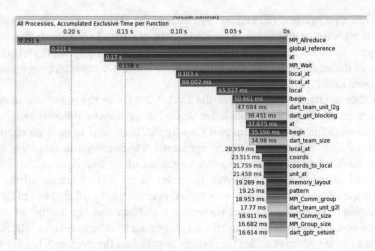

Fig. 3 Vampir screenshot. Accumulated exclusive time per function for all functions recorded with PATTERN_LEVEL

Table 1 Comparison of the used disk space for different instrumentation levels. Measured was a 2-D heat equation running on two nodes (each 4 cores) with a problem size of 64 × 64. The columns describe the different instrumentation levels. NO DASH means only MPI and DART calls are recorded. COMPILER used automatic compiler instrumentation

NO DASH	CONTAINER_LEVEL	PATTERN_LEVEL	ALL_LEVEL	COMPILER
6 GB	16 GB	22 GB	61 GB	304 GB

Table 2 Comparison of the runtime for a 2-D heat equation running on two nodes (each 4 cores) with a problem size of 64 × 64. The columns describe the different instrumentation levels. PLAIN represents the runtime for a no instrumentation, NO DASH means only DART and MPI calls were recorded, and COMPILER is the automatic compiler instrumentation

PLAIN	NO DASH	CONTAINER_LEVEL	PATTERN_LEVEL	ALL_LEVEL	COMPILER
9.5 s	30.6 s	59.2 s	79.6 s	188.6 s	868.9 s

Recording all DASH functions used 61 GB which is 10 times more than NO_DASH. However, compared to COMPILER, ALL_LEVEL uses significantly less disk space.

Additionally to the traces' sizes, we compared the runtimes of the instrumentation levels (Table 2). Therefore, we measured the whole application execution time which includes initialization and finalization of DASH.[4] The reason for the enormous difference of factor 3 between PLAIN and NO DASH is the tracing of DART and MPI events in NO DASH. This and the large number of iterations, cause the big runtime overhead. The difference between CONTAINER_LEVEL and PATTERN_LEVEL is quite acceptable, considering the higher degree of information. Table 2 also shows that the highest instrumentation level of DASH

---

[4]The unification process of Score-P was not included in the runtime measurements.

(ALL_LEVEL) produced significantly less overhead than the automatic compiler instrumentation (COMPILER).

Nevertheless, our approach has two drawbacks. First, using manual instrumentation macros in constexpr[5] is not possible. Second, the __func__ macro only provides the function/method name but no namespace and class information for manual instrumented functions. For future work some modern introspection techniques could be of interest, to improve the quality of the information of our manual instrumentation.

6 MPI Profiling

As an additional option for lightweight profiling on the MPI level we extended and adapted IPM [14]. IPM is a low-overhead performance profiling tool that provides a high-level overview of application activity and allows for an attribution of execution time to various event sources. IPM keeps its performance data in a hashtable and uses both the hash key and the hash value to store performance data. Upon encountering a relevant application event, a hash key is formed using a numeric event ID and context information. The hash values consist of first-order statistics of the events and their duration.

To work with DASH, IPM was extended to support MPI-3 one-sided operations. A total of 38 functions from the MPI-3 RMA (remote memory access) API were added to the event coverage of IPM. These functions deal with the setup of one-sided communication operations (window creation), the actual communication operations (put and get) as well as synchronization operations (flush, flush local). The event context information also captured by IPM consists of the size of data transfers (where appropriate, i.e., MPI_Put, MPI_Get, MPI_Accumulate, and similar) and communication and synchronization partner ranks (i.e., the origin or target ranks for the operation). For each monitored application event, a lookup is performed in the hashtable and the associated statistical data is updated. Per default IPM records the number of invocations for each (event, context) pair as well as the minimum, maximum and sum of all event durations encountered.

It is possible to extend the default context information for application events by including a callsite and a region identifier. The callsite ID is automatically generated by IPM by keeping track of the callstack for each invocation of an MPI operation and this allows the differentiation of MPI calls from different calling contexts (e.g., subroutines). Region identifiers can serve the same purpose, but are manually managed by the application developer, for example to differentiate phases of an application (e.g., initialization, main iteration loop, result analysis).

Data is kept in per-process local hashtables by IPM and on program termination all hashtables are merged at the master rank and application performance reports are generated. In its most basic form, this is an application performance report banner

---

[5]C++ specifier for a constant expression.

```
##IPMv2.0.3####################################################
#
# command    : ./gups.ipm
# start      : Thu Nov 26 11:42:23 2015    host      : vbox
# stop       : Thu Nov 26 11:42:25 2015    wallclock : 1.77
# mpi_tasks  : 4 on 1 nodes                %comm     : 84.34
# mem (GB)   : 0.18                        gflop/sec : 0.00
#
#          :        [total]        <avg>          min          max
# wallclock :           7.02         1.76         1.74         1.77
# MPI       :           5.92         1.48         1.45         1.49
# %wall     :
#   MPI     :                        84.35        81.99        85.73
# #calls    :
#   MPI     :         405674       101418        96298       116780
# mem (GB)  :           0.18         0.04         0.04         0.04
#
#                          [time]      [count]      <%wall>
# MPI_Win_flush             1.91        36864        27.20
# MPI_Wait                  1.70        36864        24.17
# MPI_Barrier               1.62           28        23.05
# MPI_Comm_group            0.15        73728         2.10
  ...
################################################################
```

Fig. 4 Example IPM banner output showing high level metrics such as the total time spent in MPI operations and individual contributing MPI calls

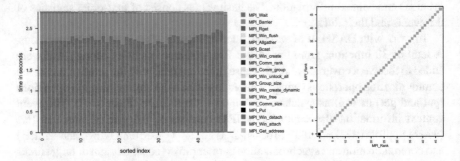

Fig. 5 Example performance profiling displays generated by IPM. Time spent in MPI calls, sorted by total MPI time across ranks (*left*) communication topology map (*right*)

written to stdout immediately after program termination. An example IPM report with full detail is shown in Fig. 4. Evidently this program (a simple communications benchmark) spent most of its time in communication and synchronization routines. More detailed data is recorded by IPM in an XML based performance data file suitable for archival. A parser reads the report file and generates a HTML profiling report that can be viewed using a web browser. The HTML profiling report includes displays such as a communication topology heatmap (the amount of data or time spent for communication between any pair of processes) and a breakdown of time spent in communication routines by transfer size. Examples for these displays are shown in Fig. 5.

The detail level of data collection and output can be specified in IPM by using environment variables. IPM_REPORT specifies the amount of information to be included in the banner output upon program termination (the available settings are none, terse, and full). The environment variable IPM_LOG specifies the amount of data to be included in the XML profiling log. The setting full means that all information, included in the hash table, are available in the XML file, whereas terse means that only summaries are available.

7 PAPI Support in DASH

Modern microprocessors provide registers that count the occurrence of events related to performance metrics. Hardware performance counters allow to query clock cycles and the occurrence of more specific events like cache misses. Their usage and semantics vary between processor architectures, however, as they usually can only be accessed using assembly instructions.

The PAPI (Performance API) project [5] specifies a standard application programming interface that allows portable access to hardware performance counters and has evolved to a de-facto standard in performance evaluation. The DASH library provides dedicated support of PAPI: its usage can be specified in build configuration and performance-related concepts have been designed with PAPI in mind.

Application developers using DASH can call PAPI functions directly. However, the DASH developer toolkit also provides its own interface for performance counters that switches to native fallback implementations for several architectures and platforms if PAPI is not available on a system. Typically, the DASH performance interface just acts as a light-weight wrapper of PAPI functions with a more convenient, intuitive interface.

As an example, we demonstrate the use of the DASH Timer class and discuss available fallback implementations.

7.1 The DASH Timer Class

Time measurements allow to derive the most conventional metrics for application performance tuning, such as latency and throughput. In this, time can be measured in clock-based and counter-based domains.

Clock-based time measurements refer to wall-clock time and are provided by the operating system. Functions like clock_gettime are reasonably convenient for end users and have standardized semantics. Their accuracy and precision depend on the operating system's implementation, however. To address a common misconception, note that a timestamp's resolution only represents its unit of measure but does not imply precision: a timestamp with 1 ns resolution might only grant 1 ms precision or less. Therefore, two consecutive timestamps might have identical values if both measurements fall into the same time sampling interval.

Counter-based time measurements require low-level interaction with hardware in client code. Dedicated counters in hardware like *RTDSC* achieve optimal accuracy and precision at instruction-level and return monotonic counter values. As an advantage, consecutive counter-based time measurements are guaranteed to yield different values. Machine instructions and semantics of counter registers are specific for each CPU architecture and tedious to use for end users, however. At the time of this writing, no standardized interface exists to access hardware counters; efforts like PerfCtr and Perfmon have not been accepted as parts of vanilla Linux. In addition, counter values can only be related to pseudo wall-clock time via frequency scaling, i.e. when dividing by a hypothetical constant CPU clock frequency.

The DASH Timer class represents a unified, concise interface for both clock- and counter-based measurements. In both cases, elapsed time between two measurements is automatically transformed to wall-clock time, applying frequency scaling when needed. Developers can still access raw timestamp values directly. Listing 4 shows how the scaling frequency for counter-based timers in DASH can be configured explicitly. Instances of the DASH Timer are then initialized for either clock- or counter-based measurement.

We apply the RAII scheme in the Timer class to provide convenient usage: in the constructor, a class member is initialized with the current timestamp. Elapsed time since creation of a Timer object can then be resolved without maintaining a start timestamp manually as shown in Listing 5.

```
1   dash::Timer<dash::Timer::Counter>::Calibrate(
2                                   2560.0 // bogo-MIPS
3                                   );
```

Listing 4 Configuring the DASH Timer

```
1   // Create timer based on time stamp counter:
2   dash::Timer<dash::Timer::Counter> timer_ct;
3   // Create timer based on wall-clock time:
4   dash::Timer<dash::Timer::Clock>   timer_ck;
5   // Operation to measure:
6   do_work();
7   // Get timestamp of construction of the timer:
8   auto   timestamp_start = timer_ct.start();
9   // Get current timestamp based on time stamp counter:
10  auto   timestamp_end   = timer_ct.now();
11  // Get the time in nanoseconds elapsed since construction
12  // of the timer instance. Counter value in timestamp is
13  // converted to real-time domain using frequency scaling:
14  double elapsed_ns = timer_ct.elapsed();
15  // same as:
16  double elapsed_ns = timer_ct.elapsed_since(
17                            timer_ct.start());
```

Listing 5 Usage of the DASH Timer class

7.2   Fallback Timer Implementations

Accuracy and precision of time measurements are tedious to verify and optimize for a specific architecture. PAPI implements stable time measurements for a wide range of platforms based on hardware counters like *RTDSC*.

The DASH developer toolkit includes own implementations to access time stamp counter registers and the operating system's time functions like `clock_gettime` on Unix. Therefore, DASH applications do not depend on the PAPI library for reliable time measurements, but it can still be easily integrated in an existing application as both variants are accessed using the same interface.

If DASH is configured without PAPI support, the use of fallback implementations is defined at compile time. The decision for a concrete fallback variant depends on system architecture, operating system, and available functions and flags in the linked C standard library. Clock-based timestamps are then typically obtained from the Linux function `clock_gettime` using the most precise mode available on the system. For counter-based timestamps, specific assembly instructions read values from counter registers like RDTSC on x86 architectures or PCCNT on ARM32.

8   Conclusion and Future Work

This paper described the current tool infrastructure of the DASH project. To understand what DASH is about, the project was briefly introduced and main components were explained. Then the debugging support for DASH, especially the challenges of controlling many processes and the use of new pretty-printers, were described. Following this, we explained the differences between profiling and tracing to demonstrate our developed extensions for the DASH/DART support in the Score-P measurement environment. The need to use different approaches to support DASH/-DART were discussed afterwards. Especially, the necessity for different instrumentation levels (recording only a few or all functions) in the DASH template library were demonstrated. At the end, we presented a profiling tool for MPI (supporting especially DART-MPI developers) and how PAPI counters are integrated in DASH.

The next step will be a performance monitoring environment for monitoring memory accesses in DASH containers which especially helps DASH/DART developers to verify distribution patterns and to evaluate the memory access behavior of implemented DART communication libraries. The monitoring environment can also be used to analyze and optimize existing patterns and algorithms using them.

Acknowledgements The DASH concept and its current implementation have been developed in the DFG project "Hierarchical Arrays for Efficient and Productive Data-Intensive Exascale Computing" funded under the German Priority Programme 1648 "Software for Exascale Computing" (SPPEXA).[6]

---

[6]See http://www.sppexa.de and http://www.dfg.de/foerderung/info_wissenschaft/info_wissens chaft_14_57/index.html

References

1. Adhianto, L., Banerjee, S., Fagan, M., Krentel, M., Marin, G., Mellor-Crummey, J., Tallent, N.R.: HPCToolkit: Tools for performance analysis of optimized parallel programs. Concurr. Comput.: Pract. Exper. **22**(6), 685–701 (2010)
2. Alameda, J., Spear, W., Overbey, J.L., Huck, K., Watson, G.R., Tibbitts, B.: The eclipse parallel tools platform: toward an integrated development environment for XSEDE resources. In: Proceedings of the 1st Conference of the Extreme Science and Engineering Discovery Environment: Bridging from the eXtreme to the Campus and Beyond (XSEDE '12), pp. 48:1–48:8. ACM, New York (2012). http://doi.acm.org/10.1145/2335755.2335845
3. Allen, E., Chase, D., Hallett, J., Luchangco, V., Maessen, J.-W., Ryu, S., Steele Jr, G.L., Tobin-Hochstadt, S., Dias, J., Eastlund, C., et al.: The fortress language specification. Sun Microsyst. 139, 140 (2005)
4. Allinea DDT: The global standard for high-impact debugging on clusters and supercomputers (2015). http://www.allinea.com/products/ddt Online; Accessed 12 Jan 2015
5. Browne, S., Dongarra, J., Garner, N., Ho, G., Mucci, P.: A portable programming interface for performance evaluation on modern processors. Int. J. High Perform. Comput. Appl. **14**(3), 189–204 (2000)
6. Buss, A., Papadopoulos, I., Pearce, O., Smith, T., Tanase, G., Thomas, N., Xu, X., Bianco, M., Amato, N.M., Rauchwerger, L., et al.: STAPL: standard template adaptive parallel library. In: Proceedings of the 3rd Annual Haifa Experimental Systems Conference, p. 14. ACM (2010)
7. Chamberlain, B.L., Callahan, D., Zima, H.P.: Parallel programmability and the Chapel language. Int. J. High Perform. Comput. Appl. **21**, 291–312 (2007)
8. Charles, P., Grothoff, C., Saraswat, V., Donawa, C., Kielstra, A., Ebcioglu, K., Von Praun, C., Sarkar, V.: X10: an object-oriented approach to non-uniform cluster computing. ACM Sigplan Notices **40**(10), 519–538 (2005)
9. Dietrich, R., Ilsche, T., Juckeland, G.: Non-intrusive performance analysis of parallel hardware accelerated applications on hybrid architectures. In: 2010 39th International Conference on Parallel Processing Workshops (ICPPW), pp. 135–143. IEEE (2010)
10. Dryden, N.: PGDB: A debugger for MPI applications. In: Proceedings of the 2014 Annual Conference on Extreme Science and Engineering Discovery Environment (XSEDE '14), pp. 44:1–44:7. ACM, New York (2014). http://doi.acm.org/10.1145/2616498.2616535
11. Eschweiler, D., Wagner, M., Geimer, M., Knüpfer, A., Nagel, W.E., Wolf, F.: Open Trace Format 2: The next generation of scalable trace formats and support libraries. In: Applications, Tools and Techniques on the Road to Exascale Computing. Advances in Parallel Computing, vol. 22, pp. 481–490. IOS Press (2012)
12. Free Software Foundation, Inc.: GDB: The GNU Project Debugger. http://www.gnu.org/software/gdb/ (2014). Online; Accessed 01 Nov 2015
13. Fürlinger, K., Glass, C., Gracia, J., Knüpfer, A., Tao, J., Hünich, D., Idrees, K., Maiterth, M., Mhedheb, Y., Zhou, H.: DASH: data structures and algorithms with support for hierarchical locality. In: Lopes, L., Žilinskas, J., Costan, A., Cascella, R., Kecskemeti, G., Jeannot, E., Cannataro, M., Ricci, L., Benkner, S., Petit, S., Scarano, V., Gracia, J., Hunold, S., Scott, S., Lankes, S., Lengauer, C., Carretero, J., Breitbart, J., Alexander, M. (eds.) Euro-Par 2014: Parallel Processing Workshops. Lecture Notes in Computer Science, vol. 8806, pp. 542–552. Springer, Cham (2014). http://dx.doi.org/10.1007/978-3-319-14313-2_46
14. Fürlinger, K., Wright, N.J., Skinner, D.: Performance analysis and workload characterization with IPM. In: Proceedings of the 3rd International Workshop on Parallel Tools for High Performance Computing, pp. 31–38. Springer, Dresden (2010)
15. Geimer, M., Wolf, F., Wylie, B.J., Ábrahám, E., Becker, D., Mohr, B.: The Scalasca performance toolset architecture. Concurr. Comput.: Pract. Exper. **22**(6), 702–719 (2010)
16. Gerndt, M., Ott, M.: Automatic performance analysis with periscope. Concurr. Comput.: Pract. Exper. **22**(6), 736–748 (2010)

17. Grünewald, D., Simmendinger, C.: The GASPI API specification and its implementation GPI 2.0. In: 7th International Conference on PGAS Programming Models, vol. 243 (2013)
18. Hünich, D., Knüpfer, A., Gracia, J.: Providing parallel debugging for DASH distributed data structures with GDB. Procedia Comput. Sci. **51**, 1383–1392 (2015). http://www.sciencedirect. com/science/article/pii/S1877050915011539. International Conference on Computational Science, ICCS 2015 Computational Science at the Gates of Nature
19. Ilsche, T., Schuchart, J., Schöne, R., Hackenberg, D.: Combining instrumentation and sampling for trace-based application performance analysis. In: Proceedings of the 8th International Parallel Tools Workshop, pp. 123–136. Springer (2014)
20. Johnson, T.A.: Coarray C++. In: 7th International Conference on PGAS Programming Models, Edinburgh (2013)
21. Knüpfer, A., Brunst, H., Doleschal, J., Jurenz, M., Lieber, M., Mickler, H., Müller, M.S., Nagel, W.E.: The Vampir performance analysis tool-set. In: Tools for High Performance Computing, pp. 139–155. Springer, Berlin/Heidelberg (2008)
22. an Mey, D., Biersdorf, S., Bischof, C., Diethelm, K., Eschweiler, D., Gerndt, M., Knüpfer, A., Lorenz, D., Malony, A., Nagel, W.E., et al.: Score-P: a unified performance measurement system for petascale applications. In: Competence in High Performance Computing 2010, pp. 85–97. Springer, Berlin/Heidelberg (2012)
23. MPI Forum: MPI: A Message-Passing Interface Standard. Version 3.0 (2012). Available at: http://www.mpi-forum.org (Sept 2012)
24. Müller, M.S., Knüpfer, A., Jurenz, M., Lieber, M., Brunst, H., Mix, H., Nagel, W.E.: Developing scalable applications with Vampir, VampirServer and VampirTrace. In: Parallel Computing: Architectures, Algorithms and Applications. Advances in Parallel Computing, vol. 18, pp. 637–644. John von Neumann Institute for Computing, Jülich (2007)
25. Numrich, R.W., Reid, J.: Co-array Fortran for parallel programming. SIGPLAN Fortran Forum **17**(2), 1–31 (1998)
26. Partitioned Global Address Space (2014). [Online] http://www.pgas.org
27. Poole, S.W., Hernandez, O., Kuehn, J.A., Shipman, G.M., Curtis, A., Feind, K.: OpenSHMEM - toward a unified RMA model. In: Padua, D. (ed.) Encyclopedia of Parallel Computing, pp. 1379–1391. Springer US (2011)
28. Rogue Wave Software, I.: TotalView debugger: faster fault isolation, improved memory optimization, and dynamic visualization for your high performance computing apps (2015). http://www.roguewave.com/products-services/totalview, [Online; Accessed 12 Jan 2015]
29. Shende, S.S., Malony, A.D.: The TAU parallel performance system. Int. J. High Perform Comput. Appl. **20**(2), 287–311 (2006)
30. UPC Consortium: UPC language specifications, v1.2. Tech Report LBNL-59208, Lawrence Berkeley National Lab (2005). http://www.gwu.edu/~upc/publications/LBNL-59208.pdf
31. at the University of Illinois at Urbana-Champaign, C.S.D.: The LLDB Debugger (2015). http://www.gnu.org/software/gdb/, [Online; Accessed 12 Jan 2015]
32. Zheng, Y., Kamil, A., Driscoll, M.B., Shan, H., Yelick, K.: UPC++: A PGAS extension for C++. In: 28th IEEE International Parallel & Distributed Processing Symposium, pp. 1105–1114. IEEE (2014)
33. Zhou, H., Mhedheb, Y., Idrees, K., Glass, C.W., Gracia, J., Fürlinger, K.: DART-MPI: An MPI-based Implementation of a PGAS Runtime System. In: Proceedings of the 8th International Conference on Partitioned Global Address Space Programming Models (PGAS '14), pp. 3:1–3:11. ACM, New York (2014). http://doi.acm.org/10.1145/2676870.2676875

Part IX
EXAMAG: Exascale Simulations of the Evolution of the Universe Including Magnetic Fields

Simulating Turbulence Using the Astrophysical Discontinuous Galerkin Code TENET

Andreas Bauer, Kevin Schaal, Volker Springel, Praveen Chandrashekar, Rüdiger Pakmor, and Christian Klingenberg

Abstract In astrophysics, the two main methods traditionally in use for solving the Euler equations of ideal fluid dynamics are smoothed particle hydrodynamics and finite volume discretization on a stationary mesh. However, the goal to efficiently make use of future exascale machines with their ever higher degree of parallel concurrency motivates the search for more efficient and more accurate techniques for computing hydrodynamics. Discontinuous Galerkin (DG) methods represent a promising class of methods in this regard, as they can be straightforwardly extended to arbitrarily high order while requiring only small stencils. Especially for applications involving comparatively smooth problems, higher-order approaches promise significant gains in computational speed for reaching a desired target accuracy. Here, we introduce our new astrophysical DG code TENET designed for applications in cosmology, and discuss our first results for 3D simulations of subsonic turbulence. We show that our new DG implementation provides accurate results for subsonic turbulence, at considerably reduced computational cost compared with traditional finite volume methods. In particular, we find that DG needs about 1.8 times fewer degrees of freedom to achieve the same accuracy and at the same time is more than 1.5 times faster, confirming its substantial promise for astrophysical applications.

A. Bauer • R. Pakmor
Heidelberg Institute for Theoretical Studies, Heidelberg, Germany
e-mail: andreas.bauer@h-its.org; ruediger.pakmor@h-its.org

K. Schaal • V. Springel (✉)
Heidelberg Institute for Theoretical Studies, Heidelberg, Germany

Astronomisches Recheninstitut, Zentrum für Astronomie der Universität Heidelberg, Heidelberg, Germany
e-mail: kevin.schaal@h-its.org; volker.springel@h-its.org

P. Chandrashekar
TIFR Centre for Applicable Mathematics, Bengaluru, India
e-mail: praveen@tifrbng.res.in

C. Klingenberg
Institut für Mathematik, Universität Würzburg, Würzburg, Germany
e-mail: klingenberg@mathematik.uni-wuerzburg.de

© Springer International Publishing Switzerland 2016
H.-J. Bungartz et al. (eds.), *Software for Exascale Computing – SPPEXA 2013-2015*, Lecture Notes in Computational Science and Engineering 113, DOI 10.1007/978-3-319-40528-5_17

1 Introduction

Turbulent flows are ubiquitous in astrophysical systems. For example, supersonic turbulence in the interstellar medium is thought to play a key role in regulating star formation [19, 20]. In cosmic structure formation, turbulence occurs in accretion flows onto halos and contributes to the pressure support in clusters of galaxies [28] and helps in distributing and mixing heavy elements into the primordial gas. Also, turbulence plays a crucial role in creating an effective viscosity and mediating angular momentum transport in gaseous accretion flows around supermassive black holes.

Numerical simulations of astrophysical turbulence require an accurate treatment of the Euler equations. Traditionally, finite volume schemes have been used in astrophysics for high accuracy simulations of hydrodynamics. They are mostly based on simple linear data reconstruction resulting in second-order accurate schemes. In principle, these finite volume schemes can also be extended to high order, with the next higher order method using parabolic data reconstruction, as implemented in piecewise parabolic schemes [9]. While a linear reconstruction needs only the direct neighbors of each cell, a further layer is required for the parabolic reconstruction. In general, with the increase of the order of the finite volume scheme, the required stencil grows as well. Especially in a parallelized code, this affects the scalability, as the ghost region around the local domain has to grow as well for a deeper stencil, resulting in larger data exchanges among different MPI processes and higher memory overhead.

An interesting and still comparatively new alternative are so-called discontinuous Galerkin (DG) methods. They rely on a representation of the solution through an expansion into basis functions at the subcell level, removing the reconstruction necessary in high-order finite volume schemes. Such DG methods were first introduced by [25], and later extended to non-linear problems [3, 5–8]. Successful applications have so far been mostly reported for engineering problems [4, 15], but they have very recently also been considered for astrophysical problems [21, 30]. DG methods only need information about their direct neighbors, independent of the order of the scheme. Furthermore, the computational workload is not only spent on computing fluxes between cells, but has a significant internal contribution from each cell as well. The latter part is much easier to parallelize in a hybrid parallelization code. Additionally, DG provides a systematic and transparent framework to derive discretized equations up to an arbitrarily high convergence order. These features make DG methods a compelling approach for future exascale machines. Building higher order methods with a classical finite volume approach is rather contrived in comparison, which is an important factor in explaining why mostly second and third order finite volume methods are used in practice.

As shown in [2], subsonic turbulence can pose a hard problem for some of the simulation methods used in computational astrophysics. Standard SPH in particular struggles to reproduce results as accurate as finite volume codes, and a far higher computational effort would be required to obtain an equally large inertial range as obtained with a finite volume method, a situation that has only been moderately improved by many enhancements proposed for SPH in recent years [1, 17, 18, 22, 24, 29]. In this work, we explore instead how well the DG methods implemented in our new astrophysical simulation code TENET [26] perform for simulations of subsonic turbulence. In this problem, the discontinuities between adjacent cells are expected to be small and the subcell representation within a cell can reach high accuracy. This makes subsonic turbulence a very interesting first application of our new DG implementation.

In the following, we outline the equations and main ideas behind DG and introduce our implementation. We will first describe how the solution is represented using a set of basis functions. Then, we explain how initial conditions can be derived and how they are evolved forward in time. Next, we examine how well our newly developed DG methods behave in simulating turbulent flows. In particular, we test whether an improvement in accuracy and computational efficiency compared with standard second-order finite volume methods is indeed realized.

2   Discontinuous Galerkin Methods

Galerkin methods form a large class of methods for converting continuous differential equations into sets of discrete differential equations [13]. Instead of describing the solution with averaged quantities $\mathbf{q}$ within each cell, in DG the solution is represented by an expansion into basis functions, which are often chosen as polynomials of degree k. This polynomial representation is continuous inside a cell, but discontinuous across cells, hence the name discontinuous Galerkin method. Inside a cell K, the state is described by a function $\mathbf{q}^K(\mathbf{x}, t)$. This function is only defined on the volume of cell K. In the following, we will use $\mathbf{q}^K$ to refer to the polynomial representation of the state inside cell K.

The polynomials of degree k form a vector space, and the state $\mathbf{q}^K$ within a cell can be represented using weights $\mathbf{w}_l^K$, where l denotes the component of the weight vector. Each $\mathbf{w}_l$ contains an entry for each of the five conserved hydrodynamic quantities. Using a set of suitable orthogonal basis functions $\phi_l^K(\mathbf{x})$, the state in a cell can be expressed as

$$\mathbf{q}^K(\mathbf{x}, t) = \sum_{l=1}^{N(k)} \mathbf{w}_l^K(t)\phi_l^K(\mathbf{x}) \ . \tag{1}$$

Note how the time and space dependence on the right hand side is split up into two functions. This will provide the key ingredient for discretizing the continuous partial differential equations into a set of coupled ordinary differential equations.

The vector space of all polynomials up to degree k has the dimension $N(k)$. The l-th component of the vector can be obtained through a projection of the state $\mathbf{q}$ onto the l-th basis function:

$$\mathbf{w}_l^K(t) = \frac{1}{|K|} \int_K \mathbf{q}(\mathbf{x}, t) \phi_l^K(\mathbf{x}) \, dV \,, \tag{2}$$

with $|K|$ being the volume of cell K and $\mathbf{w}_l^K = (w_{\rho,l}, \mathbf{w}_{p,l}, w_{e,l})$ being the l-th component of the weight vector of the density, momentum density and total energy density. The integrals can be either solved analytically or numerically using Gauss quadrature rules. By $w_{i,l}$ we refer to a single component of the l-th weight vector, i.e. $w_{0,0}$ and $w_{0,1}$ are the zeroth and first weights of the density field, which correspond to the mean density and a quantity proportional to the gradient inside a cell, respectively. If polynomial basis functions of degree k are used, a numerical scheme with spatial order $p = k + 1$ is achieved. However, near discontinuities such as shock waves, the convergence order breaks down to first order accuracy. A set of test problems demonstrating the claimed convergence properties of our implementation can be found in [26].

2.1 Basis Functions

We discretize the computational domain with a Cartesian grid and adopt a classical modal DG scheme, in which the solution is given as a linear combinations of orthonormal basis functions ϕ_l^K. For the latter we use tensor products of Legendre polynomials. The cell extensions are rescaled such that they span the interval from -1 to 1 in each dimension. The transformation is given by

$$\boldsymbol{\xi} = \frac{2}{\Delta x^K} \left(\mathbf{x} - \mathbf{x}^K \right) \,, \tag{3}$$

with $\mathbf{x}^K$ being the centre of cell K.

The full set of basis functions can be written as

$$\{\phi_l(\boldsymbol{\xi})\}_{l=1}^{N(k)} = \left\{ \tilde{P}_u(\xi_1) \tilde{P}_v(\xi_2) \tilde{P}_w(\xi_3) \,|\, u, v, w \in \mathbb{N}_0 \wedge u + v + w \leq k \right\} \,, \tag{4}$$

where $\tilde{P}_u$ are scaled Legendre polynomials of degree u. The sum of the degrees of the individual basis functions has to be equal or smaller than the degree k of the DG scheme. Thus, the vector space of all polynomials up to degree k has the dimensionality

$$N(k) = \sum_{u=0}^{k} \sum_{v=0}^{k-u} \sum_{w=0}^{k-u-v} 1 = \frac{1}{6}(k+1)(k+2)(k+3) \,. \tag{5}$$

2.2   Initial Conditions

To obtain the initial conditions, we have to find weight vectors $\mathbf{w}_l^K$ at $t = 0$ corresponding to the initial conditions $\mathbf{q}(\mathbf{x}, 0)$. The polynomial representation of a scalar quantity described by the weight vector is

$$q_i^K (\mathbf{x}, 0) = \sum_{l=1}^{N(k)} w_{i,l}^K(0)\phi_l^K (\mathbf{x}) . \tag{6}$$

The difference between the prescribed actual initial condition and the polynomial representation should be minimal, which can be achieved by varying the weight vectors $\mathbf{w}_l^K$ in each cell for each hydrodynamical component i individually:

$$\min_{\{w_{i,l}^K(0)\}_l} \int_K \left(q_i^K(\mathbf{x}, 0) - q_i(\mathbf{x}, 0)\right)^2 \, \mathrm{d}V . \tag{7}$$

Thus, the l-th component of the initial weights $\mathbf{w}_l^K$ is given by

$$\mathbf{w}_l^K(0) = \frac{1}{|K|} \int_K \mathbf{q}(\mathbf{x}, 0)\phi_l^K (\mathbf{x}) \, \mathrm{d}V . \tag{8}$$

Transformed into the ξ coordinate system, the equation becomes

$$\mathbf{w}_l^K(0) = \frac{1}{8} \int_{[-1,1]^3} \mathbf{q}(\boldsymbol{\xi}, 0)\phi_l(\boldsymbol{\xi}) \, \mathrm{d}\boldsymbol{\xi} . \tag{9}$$

In principle, the integral can be computed analytically for known analytical initial conditions. Alternatively, it can be computed numerically using a Gauss quadrature rule:

$$\mathbf{w}_l^K(0) \cong \frac{1}{8} \sum_{q=1}^{(k+1)^3} \mathbf{q}(\mathbf{x}_q, 0)\phi_l(\boldsymbol{\xi}_q)\omega_q , \tag{10}$$

using $(k + 1)^3$ sampling points $\mathbf{x}_q$ and corresponding quadrature weights ω_q. With $k + 1$ integration points polynomials of degree $\leq 2k + 1$ are integrated exactly by the Gauss quadrature rule. Therefore, the projection integral is exact for initial conditions in the form of polynomials of degree $\leq k$.

2.3   Time Evolution Equations

The solution is discretized using time-dependent weight vectors $\mathbf{w}_l^K(t)$. The time evolution equations for these weights can be derived from the Euler equation,

$$\frac{\partial \mathbf{q}}{\partial t} + \sum_{\alpha=1}^{3} \frac{\partial \mathbf{F}_\alpha(\mathbf{q})}{\partial x_\alpha} = 0 . \tag{11}$$

To obtain an evolution equation for the l-th weight, the Euler equation is multiplied with ϕ_l and integrated over the volume of cell K,

$$\frac{\mathrm{d}}{\mathrm{d}t} \int_K \mathbf{q}^K \phi_l^K \, \mathrm{d}V - \sum_{\alpha=1}^{3} \int_K \frac{\partial \mathbf{F}_\alpha(\mathbf{q})}{\partial x_\alpha} \phi_l^K \, \mathrm{d}V = 0 . \tag{12}$$

Integrating the second term by parts and applying Gauss's theorem leads to a volume integral over the interior of the cell and a surface integral with surface normal vector $\mathbf{n}$:

$$\frac{\mathrm{d}}{\mathrm{d}t} \int_K \mathbf{q}^K \phi_l^K \, \mathrm{d}V - \sum_{\alpha=1}^{3} \int_K \mathbf{F}_\alpha \frac{\partial \phi_l^K}{\partial x_\alpha} \, \mathrm{d}V + \sum_{\alpha=1}^{3} \int_{\partial K} \mathbf{F}_\alpha \phi_l^K n_\alpha \, \mathrm{d}A = 0 . \tag{13}$$

We will now discuss the three terms in turn, starting with the first one. Inserting the definition of $\mathbf{q}^K$ and using the orthogonality relation of the basis functions simplifies this term to the time derivative of the l-th weight:

$$\frac{\mathrm{d}}{\mathrm{d}t} \int_K \mathbf{q}^K \phi_l^K \, \mathrm{d}V = |K| \frac{\mathrm{d}\mathbf{w}_l^K}{\mathrm{d}t} . \tag{14}$$

We transform the next term into the ξ-coordinate system. The term involves a volume integral, which is solved using a Gauss quadrature rule:

$$\sum_{\alpha=1}^{3} \int_K \mathbf{F}_\alpha \left(\mathbf{q}^K(\mathbf{x}, t) \right) \frac{\partial \phi_l^K(\mathbf{x})}{\partial x_\alpha} \, \mathrm{d}V$$

$$= \frac{\left(\Delta x^K \right)^2}{4} \sum_{\alpha=1}^{3} \int_{[-1,1]^3} \mathbf{F}_\alpha \left(\mathbf{q}^K(\xi, t) \right) \frac{\partial \phi_l(\xi)}{\partial \xi_\alpha} \, \mathrm{d}\xi$$

$$\cong \frac{\left(\Delta x^K \right)^2}{4} \sum_{\alpha=1}^{3} \sum_{q=1}^{(k+1)^3} \mathbf{F}_\alpha \left(\mathbf{q}^K(\xi_q, t) \right) \left. \frac{\partial \phi_l}{\partial \xi_\alpha} \right|_{\xi_q} \omega_q . \tag{15}$$

The flux vector $\mathbf{F}_\alpha$ can be easily evaluated at the $(k+1)^3$ quadrature points $\boldsymbol{\xi}_q$ using the polynomial representation $\mathbf{q}^K(\boldsymbol{\xi}_q, t)$. An analytical expression can be obtained for the derivatives of the basis functions.

Finally, the last term is a surface integral over the cell boundary. Again, we transform the equation into the ξ-coordinate system and apply a Gauss quadrature rule to compute the integral:

$$
\sum_{\alpha=1}^{3} \int_{\partial K} \mathbf{F}_\alpha \phi_l^K(\mathbf{x}) n_\alpha \, \mathrm{d}A
$$

$$
= \frac{\left(\Delta x^K\right)^2}{4} \int_{\partial[-1,1]^3} \mathscr{F}\left(\mathbf{q}_L^K(\boldsymbol{\xi}, t), \mathbf{q}_R^K(\boldsymbol{\xi}, t)\right) \phi_l(\boldsymbol{\xi}) n_\alpha \, \mathrm{d}A'
$$

$$
\cong \frac{\left(\Delta x^K\right)^2}{4} \sum_{a \in \partial[-1,1]^3} \sum_{q=1}^{(k+1)^2} \mathscr{F}\left(\mathbf{q}_L^K(\boldsymbol{\xi}_{a,q}, t), \mathbf{q}_R^K(\boldsymbol{\xi}_{a,q}, t)\right) \phi_l(\boldsymbol{\xi}_q) \omega_{a,q} . \tag{16}
$$

Each of the interface elements a is sampled using $(k+1)^2$ quadrature points $\boldsymbol{\xi}_{a,q}$. The numerical flux $\mathscr{F}$ between the discontinuous states at both sides of the interface $\mathbf{q}_L^K$ and $\mathbf{q}_R^K$ is computed using an exact or approximative HLLC Riemann solver. Note that only this term couples the individual cells with each other.

Equations (15) and (16) can be combined into a function $\mathbf{R}_l^K\left(\mathbf{w}_1, \ldots, \mathbf{w}_{N(k)}\right)$. Combining this with Eq. (14) gives the following system of coupled ordinary differential equations for the weight vectors $\mathbf{w}_l^K$:

$$
\frac{\mathrm{d}\mathbf{w}_l^K}{\mathrm{d}t} + \mathbf{R}_l^K\left(\mathbf{w}_1, \ldots, \mathbf{w}_{N(k)}\right) = 0 . \tag{17}
$$

We integrate Eq. (17) with an explicit strong stability preserving (SSP) Runge-Kutta scheme [16]. We define $\mathbf{y} = \left(\mathbf{w}_1, \ldots, \mathbf{w}_{N(k)}\right)$ and thus we have to solve

$$
\frac{\mathrm{d}\mathbf{y}}{\mathrm{d}t} + R(\mathbf{y}) = 0 . \tag{18}
$$

A third order SSP Runge-Kutta scheme used in our implementation is given by

$$
\mathbf{y}^{(0)} = \mathbf{y}^n \tag{19}
$$

$$
\mathbf{y}^{(1)} = \mathbf{y}^{(0)} - \Delta t^n R(\mathbf{y}^{(0)}) \tag{20}
$$

$$
\mathbf{y}^{(2)} = \frac{3}{4}\mathbf{y}^{(0)} + \frac{1}{4}\left(\mathbf{y}^{(1)} - \Delta t^n R(\mathbf{y}^{(1)})\right) \tag{21}
$$

$$
\mathbf{y}^{(3)} = \frac{1}{3}\mathbf{y}^{(0)} + \frac{2}{3}\left(\mathbf{y}^{(2)} - \Delta t^n R(\mathbf{y}^{(2)})\right) \tag{22}
$$

$$
\mathbf{y}^{n+1} = \mathbf{y}^{(3)} \tag{23}
$$

with initial value $\mathbf{y}^n$, final value $\mathbf{y}^{n+1}$, intermediate states $\mathbf{y}^{(0)}, \mathbf{y}^{(1)}, \mathbf{y}^{(2)}$, and time step size Δt^n.

2.4 Time Step Calculation

The time step has to fulfill the following Courant criterium [6]:

$$\Delta t^K = \frac{C}{2k+1} \left(\frac{|v_1^K| + c^K}{\Delta x_1^K} + \frac{|v_2^K| + c^K}{\Delta x_2^K} + \frac{|v_3^K| + c^K}{\Delta x_3^K} \right)^{-1}, \tag{24}$$

with Courant factor C, components of the mean velocity v_i^K in cell K and sound speed c^K. The minimum over all cells is determined and taken as the global maximum allowed time step. Note the $(2k+1)^{-1}$ dependence of the time step, which leads to a reduction of the time step for high order schemes.

2.5 Positivity Limiter

Higher order methods usually require some form of limiting to remain stable. However there is no universal solution to this problem and the optimum choice of such a limiter is in general problem dependent. For our set of turbulence simulations we have decided to limit the solution as little as possible and adopt only a positivity limiter. This choice may lead to some oscillations in the solution, however, it achieves the most accurate result in terms of error measurements. We explicitly verified this for the case of shock tube test problems. At all times, the density ρ, pressure P and energy e should remain positive throughout the entire computational domain. However, the higher order polynomial approximation could violate this physical constraint in some parts of the solution. This in turn can produce a numerical stability problem for the DG solver if the positivity is violated at a quadrature point inside the cell or an interface. To avoid this problem, we use a so-called positivity limiter [14, 31]. By applying this limiter at the beginning of each Runge-Kutta stage, it is guaranteed that the density and pressure values entering the flux calculation are positive, as well as the mean cell values at the end of each RK stage. In addition, a strong stability preserving Runge-Kutta scheme and a positivity preserving Riemann solver is needed to guarantee positivity.

The set of points where positivity is enforced has to include the cell interfaces, because fluxes are computed there as well. A possible choice of integration points, which include the integration edges, are the Gauss-Lobatto-Legendre (GLL) points. In the following, we will be using tensorial products of GLL and Gauss points, where one coordinate is chosen from the set of GLL points and the remaining two are taken from the set of Gauss points:

$$S_x = \{(\hat{\xi}_r, \xi_s, \xi_t) : 1 \leq r \leq m, 1 \leq s \leq k+1, 1 \leq t \leq k+1\}, \tag{25}$$

$$S_y = \{(\xi_r, \hat{\xi}_s, \xi_t) : 1 \leq r \leq k+1, 1 \leq s \leq m, 1 \leq t \leq k+1\},$$

$$S_z = \{(\xi_r, \xi_s, \hat{\xi}_t) : 1 \leq r \leq k+1, 1 \leq s \leq k+1, 1 \leq t \leq m\}. \tag{26}$$

The full set of integration points is $S = S_x \cup S_y \cup S_z$, which includes all points where fluxes are evaluated in the integration step.

First, the minimum density at all points in the set S is computed:

$$\rho_{\min}^K = \min_{\xi \in S} \rho^K(\xi). \tag{27}$$

We define a reduction factor θ_1^K as

$$\theta_1^K = \min\left\{\left|\frac{\bar{\rho}^K - \epsilon}{\bar{\rho}^K - \rho_{\min}^K}\right|, 1\right\}, \tag{28}$$

with the mean density in the cell $\bar{\rho}^K$ (the 0-th density weight) and the minimum target density ϵ. All high order weights of the density are reduced by this factor

$$w_{j,1}^K \leftarrow \theta_1^K w_{j,1}^K, \quad j = 2, \ldots, N(k). \tag{29}$$

To guarantee a positive pressure P, a similar approach is taken:

$$\theta_2^K = \min_{\xi \in S} \tau^K(\xi), \tag{30}$$

with

$$\tau^K(\xi) = \begin{cases} 1 & \text{if } P^K(\xi) \geq \epsilon \\ \tau_* & \text{such that } P(\mathbf{q}^K(\xi) + \tau_*(\mathbf{q}^K(\xi) - \bar{\mathbf{q}}^K)) = \epsilon. \end{cases} \tag{31}$$

The equation for τ can not be solved analytically and has to be solved numerically. To this end we employ a Newton-Raphson method. Now, the higher order weights of all quantities are reduced by θ_2

$$w_{j,i}^K \leftarrow \theta_2^K w_{j,i}^K, \quad j = 2, \ldots, N(k), \quad i = 1, \ldots, 5. \tag{32}$$

Additionally the time step has to be modified slightly to

$$\Delta t^K = C \min\left(\frac{1}{2k+1}, \frac{\hat{w}_1}{2}\right)\left(\frac{|v_1^K| + c^K}{\Delta x_1^K} + \frac{|v_2^K| + c^K}{\Delta x_2^K} + \frac{|v_3^K| + c^K}{\Delta x_3^K}\right)^{-1}, \tag{33}$$

with the first GLL weight $\hat{w}_1$. For a second order DG scheme the first weight is $\hat{w}_1 = 1$, and $\hat{w}_1 = 1/3$ for a third and fourth order method.

3 Turbulence Simulations

We shall consider an effectively isothermal gas in which we drive subsonic turbulence through an external force field on large scales. The imposed isothermality prevents the buildup of internal energy and pressure through the turbulent cascade over time. Technically, we simulate an ideal gas but reset slight deviations from isothermality back to the imposed temperature level after every time step, allowing us to directly measure the dissipated internal energy.

We consider a 3D simulation domain of size $L = 1$. In the following, we will compare runs with a finite volume scheme and runs using our new DG hydro solver on a fixed Cartesian mesh. In the case of DG simulations we vary the resolution as well as the convergence order of the code. A summary of all of our runs is given in Table 1.

Note that we always state the convergence order, i.e. $\mathscr{O} = k + 1$ instead of k for our DG runs. At a fixed convergence order of 3, we vary the resolution from 32^3 up to 256^3, and at a fixed resolution of 128^3 we change the convergence order from 1 up to 4. This allows us to assess the impact of both parameters against each other. The number of basis functions is $N(0) = 1$ for a first order method, $N(1) = 4$ for a second order method, $N(2) = 10$ for a third order, and $N(3) = 20$ for a fourth order method. In Table 1 we also state the approximate number of degrees of freedom per dimension to better compare the impact of increasing the order versus increasing the resolution level. We compare against a second order MUSCL type finite volume method, using an exact Riemann solver.

Table 1 Summary of the turbulence simulations discussed in this article. The X in the name is a placeholder for the resolution level. As a reference solution we consider ordinary finite volume simulations with up to 512^3 resolution elements. In case of DG, we vary the resolution from 32^3 up to 256^3 for the third order code, as well as the convergence order from 1 up to 4 at a resolution of 128^3 cells. To better assess the impact of a higher order method, we state the number of degrees of freedom per cell per dimension. The number of degrees of freedom per cell are 1, 4, 10 and 20 (from 1 order up to 4 order) in the case of DG

Overview over our turbulence simulations				
Label	Numerical method	Conv. order $\mathscr{O}$	Resolution	(d.o.f./cell)$^{1/3}$
FV_X_1	Finite volume	1	$32^3 \ldots 512^3$	1
FV_X_2	Finite volume	2	$32^3 \ldots 512^3$	1
DG_X_1	Discontinuous Galerkin	1	128^3	1
DG_X_2	Discontinuous Galerkin	2	128^3	1.59
DG_X_3	Discontinuous Galerkin	3	$32^3 \ldots 256^3$	2.15
DG_X_4	Discontinuous Galerkin	4	128^3	2.71

3.1 Turbulence Driving

We use the same driving method as in [2], which is based on [10–12, 27] and [23]. We generate a turbulent acceleration field in Fourier space containing power in a small range of modes between $k_{min} = 6.27$ and $k_{max} = 12.57$. The amplitude of the modes is described by a paraboloid centered around $(k_{min} + k_{max})/2$. The phases are drawn from an Ornstein-Uhlenbeck (OU) process. This random process is given by

$$\boldsymbol{\theta}_t = f \, \boldsymbol{\theta}_{t-\Delta t} + \sigma \sqrt{(1-f^2)} \, \mathbf{z}_n \, , \tag{34}$$

with random variable $\mathbf{z}_n$ and decay factor f, given by $f = \exp(-\Delta t/t_s)$, with correlation length t_s. The phases are updated after a time interval of Δt. The variance of the process is set by σ. The expected mean value of the sequence is zero, $\langle \boldsymbol{\theta}_t \rangle = 0$, and the correlations between random numbers over time are $\langle \boldsymbol{\theta}_t \, \boldsymbol{\theta}_{t+\Delta t} \rangle = \sigma^2 f$. This guarantees a smooth, but frequent change of the turbulent driving field.

We want a purely solenoidal driving field, because we are interested in smooth subsonic turbulence in this study. A compressive part would only excite sound waves, which would eventually steepen to shocks if the driving is strong enough. These compressive modes are filtered out through a Helmholtz decomposition in Fourier space:

$$\hat{\mathbf{a}}(\mathbf{k})_i = \left(\delta_{ij} - \frac{k_i k_j}{|k|^2} \right) \hat{\mathbf{a}}_0(\mathbf{k})_j \, . \tag{35}$$

The acceleration field is incorporated as an external source term in the DG equations. The formalism is similar to adding an external gravitational field. We need to compute the following DG integrals for $\mathbf{a}_l^K$:

$$\begin{aligned}
\mathbf{a}_l^K(t) &= \int_K \mathbf{a}(\mathbf{x}, t) \phi_l^K(\mathbf{x}) \, dV \\
&= \frac{|K|}{8} \int_{[-1,1]^3} \mathbf{a}(\boldsymbol{\xi}, t) \phi_l(\boldsymbol{\xi}) \, d\boldsymbol{\xi} \\
&\cong \frac{|K|}{8} \sum_{q=1}^{(k+1)^3} \mathbf{a}(\boldsymbol{\xi}_q, t) \phi_l(\boldsymbol{\xi}_q) \omega_q \, ,
\end{aligned} \tag{36}$$

thus we have to evaluate the driving field for $(k + 1)^3$ inner quadrature points $\boldsymbol{\xi}$ for each Runge-Kutta stage. An additional evaluation at the cell center is required to compute the allowed time step size. A corresponding term is used to update the energy equation as well. The evaluation is done with a discrete Fourier sum over the few non-zero modes of the driving field. If the update frequency of the driving field is smaller than the typical time step size, storing the acceleration field for each inner quadrature point can speed up the computations. In case of the finite volume runs, we add the driving field through two half step kick operators at the beginning and end of a time step, like for ordinary gravity.

The overall amplitude of the acceleration field is rescaled such that a given Mach number is reached. Our target Mach number is $\mathcal{M} \sim 0.2$. The decay time scale is chosen as half the eddy turnover time scale, $t_s = \frac{1}{2}\frac{L}{\mathcal{M}c} = 2.5$ in our case. The acceleration field is updated 10 times per decay time scale, $\Delta t = 0.1t_s = 0.25$.

3.2 Dissipation Measurement

We use an adiabatic index of $\gamma = 1.01$ instead of the isothermal index $\gamma = 1$. The slight deviation from $\gamma = 1$ allows us to measure the dissipated energy while the dynamics of the fluid is essentially isothermal. After each time step, the expected specific internal energy is computed as

$$\epsilon = \frac{c^2}{\gamma - 1} \frac{\rho^{\gamma-1}}{\rho_0^{\gamma-1}}, \tag{37}$$

with sound speed c and reference density $\rho_0 = 1$. This specific internal energy is enforced at all quadrature points within a cell. Thus, the weights associated with the total energy density using the kinetic momentum and density field have to be adjusted:

$$
\begin{aligned}
w_{e,l}^K(t) &= \int_K \left(\frac{1}{2} \frac{\mathbf{p}(\mathbf{x},t)^2}{\rho(\mathbf{x},t)} + \rho(\mathbf{x},t)\epsilon(\mathbf{x},t) \right) \phi_l^K(\mathbf{x}) \, dV \\
&= \frac{|K|}{8} \int_{[-1,1]^3} \left(\frac{1}{2} \frac{\mathbf{p}(\boldsymbol{\xi},t)^2}{\rho(\boldsymbol{\xi},t)} + \rho(\boldsymbol{\xi},t)\epsilon(\boldsymbol{\xi},t) \right) \phi_l(\boldsymbol{\xi}) \, d\boldsymbol{\xi} \\
&\cong \frac{|K|}{8} \sum_{q=1}^{(k+1)^3} \left(\frac{1}{2} \frac{\mathbf{p}(\boldsymbol{\xi}_q,t)^2}{\rho(\boldsymbol{\xi}_q,t)} + \rho(\boldsymbol{\xi}_q,t)\epsilon(\boldsymbol{\xi}_q,t) \right) \phi_l(\boldsymbol{\xi}_q)\omega_q .
\end{aligned}
\tag{38}
$$

Afterwards, the average internal energy density in the cell can be recomputed as

$$\rho\epsilon = w_{e,0}^K - \frac{1}{2}\frac{{w_{p,0}^K}^2}{w_{\rho,0}^K} . \tag{39}$$

The dissipated energy is given by the difference between the average internal energy before and after adjusting the weights of the total energy density. Afterwards the positivity limiter is applied to guarantee non-negative values in our DG simulations.

3.3   Power Spectrum Measurement

The power spectrum of a scalar or vector field $w(\mathbf{x})$ is proportional to the Fourier transformed of the two point correlation function:

$$C_w(\mathbf{l}) = \langle w(\mathbf{x} + \mathbf{l})w(\mathbf{x})\rangle_{\mathbf{x}} . \tag{40}$$

Thus

$$E_w(\mathbf{k}) = (2\pi)^{3/2}\mathcal{F}(C_w(\mathbf{l})) = \int_V C_w(\mathbf{l}) \exp(-i\mathbf{k}\mathbf{l})\, \mathrm{d}^3\mathbf{l} \tag{41}$$

$$= |\hat{w}(\mathbf{k})|^2 , \tag{42}$$

where $\hat{w}$ is the Fourier transform of w.[1] Here, we are only interested in the 1D power spectrum, thus we average $E_w(\mathbf{k})$ over spherical shells:

$$E_w(k) = 4\pi k^2 \langle E_w(\mathbf{k})\rangle , \tag{43}$$

where $k = |\mathbf{k}|$. The overall normalization of the Fourier transformation is chosen such that the integral over the power spectrum is equivalent to the total energy:

$$\sigma^2 = \int w(x)\, \mathrm{d}\mathbf{x} = \int E_w(k)\, \mathrm{d}k = \frac{1}{(2\pi)^3 N^3} \sum_{i,j,k=0}^{N-1} |\hat{w}_{ijk}|^2 , \tag{44}$$

with $\hat{w}_{ijk}$ being the discrete Fourier transformation of the discretized continuous field w. Usually we show $kE(k)$ instead of $E(k)$ directly in log-log plots. This means a horizontal line in a log-log plot represents equal energy per decade and makes interpreting the area under a curve easier.

4   Results

In Fig. 1 we show a first visual overview of our simulation results at a resolution of 128^3 cells. The panels show the state at the final output time $t = 30$ for the magnitude of the velocity and the density in a thin slice through the middle of the box. Each cell is subsampled four times for this plot using the full subcell information present for each DG or finite volume cell. In the case of the finite volume scheme, we used the estimated gradients in subsampling the cells.

---

[1] We are using the convention of normalizing the Fourier transform symmetrically with $(2\pi)^{-3/2}$.

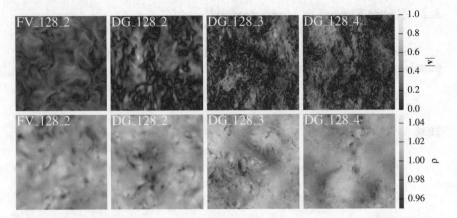

Fig. 1 Thin slices through the density and velocity field at $t = 30$. We compare the finite volume simulations against DG simulations of order 2 up to 4. Already 2nd order DG shows features which are finer than in the 2nd order finite volume run. The higher moments available in 3rd and 4th order DG allow for a representation of finer features without increasing the spatial resolution. The thin lines of zero velocity are much more pronounced in case of DG than in the finite volume case

The finite volume and DG results are similar at second order accuracy. However, already the second order DG run visually shows more small scale structure than the finite volume run. By increasing the order of accuracy and therefore allowing for more degrees of freedom within a cell, DG is able to represent considerably more structure at the same number of cells. Interestingly, the velocity field has regions of (almost) zero velocity. These thin stripes can be well represented in DG. The finite volume run shows the same features, but they are not as pronounced. Additionally, Fig. 2 shows a thin density slice for our highest resolution DG run DG_256_3. The high resolution and third order accuracy allows for more small scale details than in any other of our simulations.

4.1 Mach Number Evolution

All of our runs with a convergence order larger or equal to second order reach an average Mach number of $\mathcal{M} \sim 0.21$ after $t = 12$. The detailed history of the Mach number varies a bit from run to run. The differences between the different DG runs and finite volume runs are however insignificant. The same holds true for the other runs not shown in Fig. 3. Interestingly, both, the first order finite volume and DG runs fall substantially behind and can only reach a steady state Mach number of about $\mathcal{M} \sim 0.17$. The low numerical accuracy leads to a too high numerical dissipation rate in this case, preventing a fully established turbulent cascade. A similar problem was found in [2] for simulations with standard SPH even at comparatively high resolution, caused by a high numerical viscosity the noisy character of SPH.

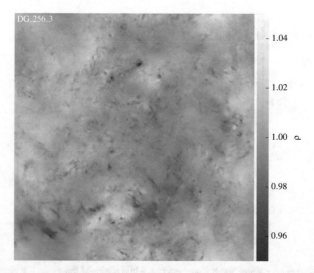

Fig. 2 A thin slice through the middle of our best resolved DG simulation at third order showing the density field. The field uses the subcell information given by the high order DG weights. Every cell is subsampled four times

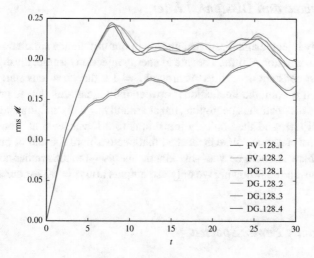

Fig. 3 Time evolution of the root mean square Mach number $\mathcal{M}$. The runs with a higher than first order convergence order agree well with each other and establish a Mach number of about $\mathcal{M} \sim 0.2$ at $t = 12$ in the quasi-stationary phase. However, the first order finite volume and DG runs do not manage to reach the same Mach number and fall substantially short of achieving a comparable kinetic energy throughout the entire run time

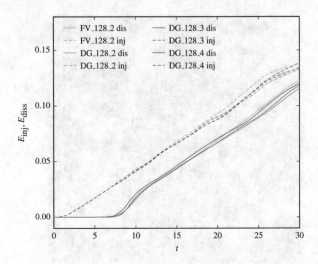

Fig. 4 The *dashed lines* show the injected energy, while the *solid lines* give the dissipated energy over time. Dissipation becomes only relevant after an initial start-up phase. Thereafter, a quasi-stationary state is established

4.2 Injected and Dissipated Energy

The globally injected and dissipated energy in our turbulence simulations is shown in Fig. 4 as a function of time. The rate of energy injection through the driving forces stays almost constant over time. At around $t = 12$, the variations start to increase slightly. At this point the fluctuations between individual runs start to grow as well. Initially, the dissipation is negligible, but at around $t = 8$ dissipation suddenly kicks in at a high rate, and then quickly transitions to a lower level at around $t = 12$, where a quasi stationary state is reached that persists until the end of our runs. The difference between both curves—the kinetic energy—remains rather constant after $t = 12$. Thus, in the following we only use outputs after $t = 12$ for our analysis.

4.3 Velocity Power Spectra

In Figs. 5 and 6, we show velocity power spectra of our runs. First, we focus on a resolution study of our third order DG and second order finite volume simulations in Fig. 5. In case of the finite volume runs, we show the power spectra up to the Nyquist frequency $k_n = 2\pi N/2L$, with N being the number of cells per dimension. For our DG runs we show the full power spectrum instead, obtained from the grid used in the Fourier transformation up to $k_g = 2\pi 4N/2L = 4k_n$. The finite volume runs have a second peak not shown here at modes above k_n, induced by noise resulting from

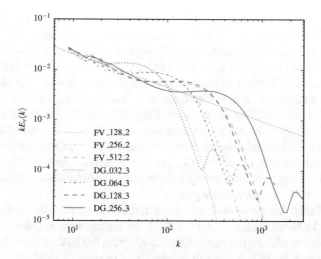

Fig. 5 Comparison of the velocity power spectra of our second order finite volume runs against our third order DG runs. Interestingly, the spectra of the DG runs match with the ones obtained from the finite volume runs at a quarter of the resolution. Thus, DG obtains similar results using only about half as many degrees of freedom per dimension as finite volume schemes. For comparison, the *grey line* shows the $k^{-5/3}$ Kolmogorov scaling

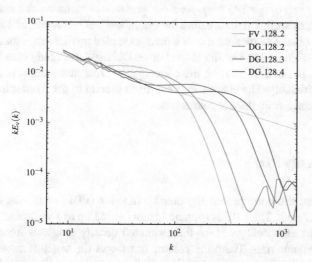

Fig. 6 Velocity power spectrum for our DG runs at different convergence order at a resolution of 128^3 cells. Already second order DG shows a large inertial range and a dissipation bottleneck at small scales. For comparison, the *grey line* shows the $k^{-5/3}$ Kolmogorov scaling

the discontinuities across cell boundaries. The third and higher order DG methods show a still declining power spectrum at k_n and only at even higher modes close to k_g start to show a noise induced rise. This is due to the available subcell information encoded in the DG weights.

All runs show an inertial range at scales smaller than the driving range on large scales. The inertial range is followed by a numerical dissipation bottleneck. This bottleneck is similar to the experimentally observed physical bottleneck effect, but appears to be somewhat stronger. The energy transferred to smaller scales can not be dissipated fast enough at the resolution scale and piles up there before it is eventually transformed to heat. The bottleneck feature moves to ever smaller scales as the numerical resolution is increased. Especially our highest resolution DG run DG_256_3 shows a quite large inertial range. However, the slope of the inertial range is measured slightly steeper than the expected $k^{-5/3}$ Kolmogorov scaling. We think a Mach number of $\mathcal{M} \sim 0.21$ and the associated density fluctuations are maybe already too high for a purely Kolmogorov-like turbulence cascade, which is only expected for incompressible gas.

Interestingly, the power spectra of our finite volume runs match those of our third order DG simulations, except that the finite volume scheme requires four times higher spatial resolution per dimension. Considering the 10 degrees of freedom per cell for third order DG, the effective number of degrees of freedom is still lower by a factor of 6.4 in the case of DG, which corresponds to a factor of 1.86 per dimension. This underlines the power of higher order numerical methods, especially if comparatively smooth problems such as subsonic turbulence are studied.

In Fig. 6 we compare the impact of the numerical convergence order on the power spectrum of our DG runs. As a comparison we include a second order finite volume run as well. All simulations have a numerical resolution of 128^3. Already the second order DG method shows a more extended inertial range than the second order finite volume run. But the second order DG method already uses four degrees of freedom per cell. Increasing the convergence order alone improves the inertial range considerably. The change in going from second to third order is a bit larger than the change from third to fourth order.

4.4 Density PDFs

In Fig. 7, we show the probability density function (PDF) of the density field for some of our runs. The PDF is averaged from $t = 12$ up to $t = 30$ and subsampled 4^3 times for each cell. We take the estimated density gradients into account for the finite volume runs. The finite volume run shows the smallest range of realized density values at the sampling points. Slightly more sampling points pile up at the extreme density values. This is due to the slope limited gradients used here, preventing more extreme density values. The DG runs show a more extended range of density values, with the range increasing with convergence order, because the higher order polynomial representations allow for a more detailed structure with more extrema within a cell. If only the mean values within the cells are considered, the PDFs are all rather similar to each other and not so different from the finite volume run shown.

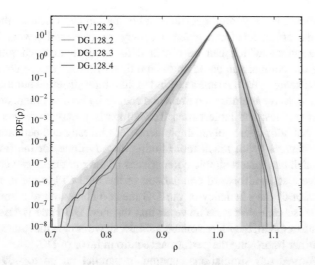

Fig. 7 The density PDF for our runs at a resolution of 128^3 cells. The PDF is obtained by subsampling each cell 4^3 times. In the finite volume case, we take the estimated density gradients into account. For DG, we use the full polynomial information present in each cell. The *shaded area* represents the standard deviation over time. Interestingly, finite volume schemes show a sharp drop off at the low and high density ends which is absent in this form in the DG calculations

5 Discussion

We presented the ideas and equations behind a new implementation of discontinues Galerkin hydrodynamics that we realized in the astrophysical simulation code TENET [26]. Unlike traditional finite volume schemes, DG uses subcell expansion into a set of basis functions, which leads to internal flux calculations in addition to surface integrals that are solved by a Riemann solver. Importantly, the reconstruction step needed in finite volume schemes is obsolete in DG. Instead, the coefficients of the expansion are evolved independently, and no information is 'thrown away' at the end of a time step, unlike done by the implicit averaging in finite volume schemes at the end of every step. This offers the prospect of a higher computational efficiency, especially at higher order where correspondingly more information is retained from step to step. Such higher order can be relatively easily achieved in DG approaches. Furthermore, stencils only involve direct neighbors in DG, even for higher order, thereby making parallelization on distributed memory systems comparatively easy and efficient.

These advantages clearly make DG methods an interesting approach for discretizing the Euler equations. However, at shocks, the standard method may fall back to first order accuracy unless sophisticated limiters are used. If the problem at hand is dominated by shocks and discontinuities, this might be a drawback. Ultimately, only detailed application tests, like we have carried out here, can decide which method proves better in practice.

In this regard, an important question for comparing numerical methods is their computational efficiency for a given accuracy, or conversely, what is the best numerical accuracy which can be obtained for a given invested total runtime. Obtaining a fair comparison based on the run time of a code can be complicated in general. For example, for the runs analyzed in this study, different numbers of CPUs had to be used, as the memory requirements change by several orders of magnitude between our smallest and largest runs. The comparison may be further influenced by the fact that both hydro solver implementations investigated here are optimized to different degrees (with much more tuning already done for the finite volume method), which can distort simple comparisons of the run times. Nevertheless, we opted to give a straightforward comparison of the total CPU time used as a first rough indicator of the efficiency of our DG method compared to a corresponding finite volume scheme. We note however that our new DG code is less optimized thus far compared with the finite volume module, so we expect that there is certainly room for further improving the performance ratio in favor of DG.

We performed our simulations running in parallel on up to 4096 cores on SUPERMUC. In part thanks to the homogeneous Cartesian mesh used in these calculations, our code showed excellent strong parallel scaling. We note that this is far harder to achieve when the adaptive mesh refinement (AMR) option present in TENET is activated.

We have generally found that the DG results for subsonic turbulence are as good as the finite volume ones, but only need slightly more than half as many degrees of freedom for comparable accuracy. Both the finite volume method at second order accuracy and the DG scheme at third order accuracy show very good weak scaling when increasing the resolution for the range of resolutions studied here. If we compare the run time for roughly equal turbulence power spectra, we find that the DG_032_3 run is about 1.14 times faster than the corresponding FV_128_2 run. This performance ratio increases if we improve the resolutions: The DG_064_3 is already 1.34 times faster than the FV_256_2 run. The DG_128_3 run is 1.53 times faster than the FV_512_2 run, which comes close to the factor 1.86 more degrees of freedom needed in the finite volume run to achieve the same accuracy. Thus, DG does not only need fewer degrees of freedom to obtain the same accuracy but also considerably less run time. This combination makes DG a very interesting method for solving the Euler equations.

Besides improving computational efficiency, DG has even more to offer. In particular, it can manifestly conserve angular momentum in smooth parts of the flow, unlike traditional finite volume methods. In addition, the $\text{div}\mathbf{B} = 0$ constraint of ideal magnetohydrodynamics (MHD) can be enforced at the level of the basis function expansion, opening up new possibilities to robustly implement MHD [21]. Combined with its computational speed, this reinforces the high potential of DG as an attractive approach for future exascale application codes in astrophysics, potentially replacing the traditional finite volume scheme that are still in widespread use today.

Acknowledgements We thank Gero Schnücke, Juan-Pablo Gallego, Johannes Löbbert, Federico Marinacci, Christoph Pfrommer, and Christian Arnold for very helpful discussions. The authors gratefully acknowledge the support of the Klaus Tschira Foundation. We acknowledge financial support through subproject EXAMAG of the Priority Program 1648 'SPPEXA' of the German Research Foundation, and through the European Research Council through ERC-StG grant EXAGAL-308037. KS and AB acknowledge support by the IMPRS for Astronomy and Cosmic Physics at the Heidelberg University. PC was supported by the AIRBUS Group Corporate Foundation Chair in Mathematics of Complex Systems established in TIFR/ICTS, Bangalore.

References

1. Abel, T.: rpSPH: a novel smoothed particle hydrodynamics algorithm. MNRAS **413**, 271–285 (2011)
2. Bauer, A., Springel, V.: Subsonic turbulence in smoothed particle hydrodynamics and moving-mesh simulations. MNRAS **423**, 2558–2578 (2012)
3. Cockburn, B., Hou, S., Shu, C.W.: The Runge-Kutta local projection discontinuous Galerkin finite element method for conservation laws. IV. The multidimensional case. Math. Comput. **54**, 545–581 (1990)
4. Cockburn, B., Karniadakis, G., Shu, C.: Discontinuous Galerkin Methods: Theory, Computation and Applications. Lecture Notes in Computational Science and Engineering. Springer, Berlin/Heidelberg (2011)
5. Cockburn, B., Lin, S.Y., Shu, C.W.: TVB Runge Kutta local projection discontinuous Galerkin finite element method for conservation laws III: one-dimensional systems. J. Comput. Phys. **84**, 90–113 (1989)
6. Cockburn, B., Shu, C.W.: TVB Runge-Kutta local projection discontinuous Galerkin finite element method for conservation laws. II. General framework. Math. Comput. **52**(186), 411–435 (1989)
7. Cockburn, B., Shu, C.W.: The Runge-Kutta local projection P^1-discontinuous-Galerkin finite element method for scalar conservation laws. RAIRO-Modélisation mathématique et analyse numérique **25**(3), 337–361 (1991)
8. Cockburn, B., Shu, C.W.: The Runge-Kutta discontinuous Galerkin method for conservation laws V. Multidimensional systems. J. Comput. Phys. **141**, 199–224 (1998)
9. Colella, P., Woodward, P.R.: The piecewise parabolic method (PPM) for gas-dynamical simulations. J. Comput. Phys. **54**, 174–201 (1984)
10. Federrath, C., Klessen, R.S., Schmidt, W.: The density probability distribution in compressible isothermal turbulence: solenoidal versus compressive forcing. ApJ **688**, L79–L82 (2008)
11. Federrath, C., Klessen, R.S., Schmidt, W.: The fractal density structure in supersonic isothermal turbulence: solenoidal versus compressive energy injection. ApJ **692**, 364–374 (2009)
12. Federrath, C., Roman-Duval, J., Klessen, R.S., Schmidt, W., Mac Low, M.M.: Comparing the statistics of interstellar turbulence in simulations and observations. A&A **512**, A81 (2010)
13. Galerkin, B.G.: On electrical circuits for the approximate solution of the laplace equation. Vestnik Inzh. **19**, 897–908 (1915)
14. Gallego-Valencia, P., Klingenberg, C., Chandrashekar, P.: On limiting for higher order discontinuous Galerkin methods for 2D Euler equations. Bull. Braz. Math. Soc. **47**(1), 335–345 (2016)
15. Gallego-Valencia, J.P., Löbbert, J., Müthing, S., Bastian, P., Klingenberg, C., Xia, Y.: Implementing a discontinuous Galerkin method for the compressible, inviscid Euler equations in the dune framework. PAMM **14**(1), 953–954 (2014)
16. Gottlieb, S., Shu, C.W., Tadmor, E.: Strong stability-preserving high-order time discretization methods. SIAM Rev. **43**(1), 89–112 (2001)

17. Heß, S., Springel, V.: Particle hydrodynamics with tessellation techniques. MNRAS **406**, 2289–2311 (2010)
18. Hopkins, P.F.: A general class of Lagrangian smoothed particle hydrodynamics methods and implications for fluid mixing problems. MNRAS **428**, 2840–2856 (2013)
19. Klessen, R.S., Heitsch, F., Mac Low, M.M.: Gravitational collapse in turbulent molecular clouds. I. Gasdynamical turbulence. ApJ **535**, 887–906 (2000)
20. Mac Low, M.M., Klessen, R.S.: Control of star formation by supersonic turbulence. Rev. Mod. Phys. **76**, 125–194 (2004)
21. Mocz, P., Vogelsberger, M., Sijacki, D., Pakmor, R., Hernquist, L.: A discontinuous Galerkin method for solving the fluid and magnetohydrodynamic equations in astrophysical simulations. MNRAS **437**, 397–414 (2014)
22. Price, D.J.: Modelling discontinuities and Kelvin Helmholtz instabilities in SPH. J. Comput. Phys. **227**, 10040–10057 (2008)
23. Price, D.J., Federrath, C.: A comparison between grid and particle methods on the statistics of driven, supersonic, isothermal turbulence. MNRAS **406**, 1659–1674 (2010)
24. Read, J.I., Hayfield, T., Agertz, O.: Resolving mixing in smoothed particle hydrodynamics. MNRAS **405**, 1513–1530 (2010)
25. Reed, W.H., Hill, T.R.: Triangularmesh methods for the neutron transport equation. Los Alamos Report LA-UR-73-479 (1973)
26. Schaal, K., Bauer, A., Chandrashekar, P., Pakmor, R., Klingenberg, C., Springel, V.: Astrophysical hydrodynamics with a high-order discontinuous Galerkin scheme and adaptive mesh refinement. MNRAS **453**, 4278–4300 (2015)
27. Schmidt, W., Hillebrandt, W., Niemeyer, J.C.: Numerical dissipation and the bottleneck effect in simulations of compressible isotropic turbulence. Comput. Fluids **35**(4), 353–371 (2006)
28. Schuecker, P., Finoguenov, A., Miniati, F., Böhringer, H., Briel, U.G.: Probing turbulence in the Coma galaxy cluster. A&A **426**, 387–397 (2004)
29. Wadsley, J.W., Veeravalli, G., Couchman, H.M.P.: On the treatment of entropy mixing in numerical cosmology. MNRAS **387**, 427–438 (2008)
30. Zanotti, O., Fambri, F., Dumbser, M.: Solving the relativistic magnetohydrodynamics equations with ADER discontinuous Galerkin methods, a posteriori subcell limiting and adaptive mesh refinement. MNRAS **452**, 3010–3029 (2015)
31. Zhang, X., Shu, C.W.: On positivity-preserving high order discontinuous Galerkin schemes for compressible Euler equations on rectangular meshes. J. Comput. Phys. **229**, 8918–8934 (2010)

Part X
FFMK: A Fast and Fault-Tolerant Microkernel-Based System for Exascale Computing

FFMK: A Fast and Fault-Tolerant Microkernel-Based System for Exascale Computing

Carsten Weinhold, Adam Lackorzynski, Jan Bierbaum, Martin Küttler, Maksym Planeta, Hermann Härtig, Amnon Shiloh, Ely Levy, Tal Ben-Nun, Amnon Barak, Thomas Steinke, Thorsten Schütt, Jan Fajerski, Alexander Reinefeld, Matthias Lieber, and Wolfgang E. Nagel

Abstract In this paper we describe the hardware and application-inherent challenges that future exascale systems pose to high-performance computing (HPC) and propose a system architecture that addresses them. This architecture is based on proven building blocks and few principles: (1) a fast light-weight kernel that is supported by a virtualized Linux for tasks that are not performance critical, (2) decentralized load and health management using fault-tolerant gossip-based information dissemination, (3) a maximally-parallel checkpoint store for cheap checkpoint/restart in the presence of frequent component failures, and (4) a runtime that enables applications to interact with the underlying system platform through new interfaces. The paper discusses the vision behind FFMK and the current state of a prototype implementation of the system, which is based on a microkernel and an adapted MPI runtime.

C. Weinhold (✉) • A. Lackorzynski • J. Bierbaum • M. Küttler • M. Planeta • H. Härtig
Department of Computer Science, TU Dresden, Dresden, Germany
e-mail: carsten.weinhold@tu-dresden.de

A. Shiloh • E. Levy • T. Ben-Nun • A. Barak
Department of Computer Science, The Hebrew University of Jerusalem, Jerusalem, Israel
e-mail: amnon@cs.huji.ac.il

T. Steinke • T. Schütt • J. Fajerski • A. Reinefeld
Zuse Institute Berlin, Berlin, Germany
e-mail: ar@zib.de

M. Lieber • W.E. Nagel
Center for Information Services and HPC, TU Dresden, Dresden, Germany
e-mail: wolfgang.nagel@tu-dresden.de

© Springer International Publishing Switzerland 2016   405
H.-J. Bungartz et al. (eds.), *Software for Exascale Computing – SPPEXA 2013-2015*, Lecture Notes in Computational Science and Engineering 113,
DOI 10.1007/978-3-319-40528-5_18

1   Exascale Challenges

Many reports and research papers, e.g. [12, 14, 19, 25], highlight the role of systems software in future exascale computing systems. It will gain importance in managing dynamic applications on heterogeneous, massively parallel, and unreliable platforms—a burden that cannot be the responsibility of application developers alone anymore, but has to shift to the operating system and runtime (OS/R). The starting point for the design of FFMK is the expectation that these major challenges have to be addressed by systems software for exascale systems:

Dynamic Applications Current high-end HPC systems are tailored towards extremely well-tuned applications. Tuning of these applications often includes significant load balancing efforts [11, 23, 38]. We believe a major part of this effort will have to shift from programmers to OS/Rs because of the complexity and dynamics of future applications. Additionally, exascale applications will need to expose more fine-grained parallelism, leading to new challenges in thread management. A number of runtime systems already addresses these challenges, notably Charm++ [1] and X10 [26]. We further believe that an exascale operating system must accommodate *elastic application partitions* that extend and shrink during their runtime. Still, the commonly used batch schedulers assume fixed size partitioning of hardware resources and networks. FFMK plans to provide interfaces for the cooperation between applications and their runtime to coordinate application-level balancing with overall system management.

Increasing Heterogeneity of Hardware Many current high-end HPC systems consist of compute nodes with at most two types of computing elements, a general purpose CPU (like x86) and an accelerator (like GPGPUs). These elements are assumed and selected to perform very regularly. We assume future hardware will have less regular performance due to fabrication tolerances and thermal concerns. This will add to the unbalanced execution of applications. We also assume that not all compute elements can be active at all time (dark silicon). In addition we assume that other types of computing elements can be expected, for example FPGAs. We believe that systems software can be adapted to such hardware more easily, if the lowest level of software is a small light-weight kernel (LWK) instead of a large and complex system like the Linux kernel.

Higher Fault Rates The sheer size of exascale computers with an unprecedented number of components will have significant impact on the failure-in-time rate for applications. Some OS/Rs already address this concern by enabling incremental and application-specific checkpoint/recovery and by using on-node memory to store checkpoint data. We believe a systems software design for exascale machines must contain a coordinated approach across system layers. For example, runtime checkpointing routines should be able to make use of memory management mechanisms at the OS level to support asynchronous checkpoints.

Deeper Memory Hierarchies We expect more types of memory that differ in aspects like persistence, energy requirements, fault tolerance, and speed. Important examples are on-node non-volatile memory (phase-change memory, flash, etc.) and stacked DRAM. A highly-efficient checkpoint store requires an integrated architecture that makes optimal use of these different types of memory.

Energy Constraints We understand, that provision and running cost of energy will become a—if not the—dominating cost and feasibility factor. To address this problem, we postulate that systems software should be based on an energy model of the complete system. The model should enable a design where each resource management decision can be controlled based on energy/utility functions for resources. For example, an on-node scheduler may choose between running one core at higher speed than others to balance execution times of compute processes. The scheduler's decision should be based on knowledge about which option provides the required cycles at the lowest energy and automatically-inferred predictions of how much time and memory certain computations (e.g., time steps) require.

2   FFMK Architecture Overview

We believe that a systems software design for exascale machines that addresses the challenges described above must be based on a coordinated approach across all layers, including applications. The platform architecture as shown in Fig. 1 uses an L4 microkernel [24] as the light-weight kernel (LWK) that runs on each node.

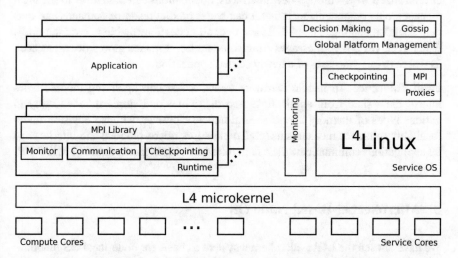

Fig. 1 FFMK software architecture. Compute processes with performance-critical parts of (MPI) runtime and communication driver execute directly on L4 microkernel; non-critical functionality split out into proxy processes on Linux, which also hosts global platform management

All cores are controlled by this minimal common foundation; the microkernel itself is supported by few extra services that provide higher-level OS functionality such as memory management (not shown in the figure). Additionally, an instance of a *service OS* is running on top of it, but only on a few dedicated cores we refer to as "service cores". In our case the system is a full-featured virtualized Linux.

Applications Applications on the system are started by the service OS and can use any functionality offered by it, including device drivers, such as for InfiniBand and network, as well as libraries and programming environments such as MPI. To exercise execution control over the HPC applications, the applications are *decoupled* from the service OS and run independently on the LWK. Any requests of the application to the service OS, such as system calls, are forwarded and handled.

Dynamic Platform Management In the presence of frequent component failures, hardware heterogeneity, and dynamic demands, applications can no longer assume that compute resources are assigned statically. Instead, load and health monitoring is part of the node OS and the platform as a whole is managed by a load distribution service. The necessary monitoring and decision making is done at three levels: (1) on each multi-core node, (2) per application/partition among nodes, and (3) based on a global view of a master management node.

Node-local thread schedulers take care of (1); scalable gossip algorithms disseminate information required to handle (2) and (3). Using gossip, the nodes build up a distributed, inherently fault tolerant, and scalable bulletin board that provides information on the status of the system. Nodes have partial knowledge of the whole system: they know about only a subset of the other nodes, but enough of them in order to make decisions on how to balance load and how to react to failures in a decentralized way. Through new interfaces, applications can pass hints to the local management component, such that it can better predict resource demands and thus help decision making. The global view over all nodes is available to a master node, which receives gossip messages from some nodes. It makes global decisions such as where to put processes of a newly started application.

Fault Tolerance To handle hardware faults, a fast checkpointing module takes intermediate state from applications and distributes and stores it redundantly in various types of memory across several nodes. However, we also envision node-local fault tolerance mechanisms (e.g., replication, micro-reboots) and interfaces to let applications communicate their fault tolerance requirements to the FFMK OS/R.

3   Microkernel-Based Node OS

We have chosen the L4Re microkernel system as basis for node-local OS functionality. For a detailed description of L4, we refer to [24]. In this document, we restrict ourselves to a short intro.

L4 Microkernel L4 had been designed for extensibility rather than as a minimized Unix. As such, it provides few basic abstractions: address spaces, threads, and inter-process communication (IPC). Key ingredient to enabling extensibility is a design that enables both IPC and unblocking of threads to be fast. The IPC mechanism is not only used to transmit ordinary data but also grant access rights to resources, such as capabilities and memory, to other address spaces. On L4, policies are implemented in user-level components. One example is memory management where so-called "pagers" manage the virtual address space of applications and implement any required policy. The microkernel itself only provides the mechanism to grant memory pages.

The fast and simple IPC mechanism enables us to build a componentized FFMK-OS that can achieve high performance. An important feature in this context is that the L4 kernel maps hardware interrupts to IPC messages. As a result, IPC messages can directly wake currently blocked application processes with low latency not only when required input is computed by another process on the same node, but also by processes running on other nodes when messages arrive over the HPC system's interconnect.

Virtualized Linux Our system also runs Linux as a service OS on each node to provide and reuse functionality that is not performance critical such as system initialization. We chose L⁴Linux, a modified Linux kernel that runs in a virtual machine on the microkernel; it is binary compatible to standard Linux and therefore capable of running unmodified Linux applications.

On the FFMK platform, HPC applications are ordinary Linux programs, too. They are loaded by the service OS and they can use all functionality offered by it, including device drivers and Linux-based runtime environments such as MPI. However, the underlying L4 microkernel is better suited, when applications perform their most "critical" work, which in the context of HPC and exascale systems means "critical to performance". For example, the microkernel can switch context faster than Linux and it provides much better control over what activities run on which core. The latter property is essential to let applications execute undisturbed from the various management and housekeeping tasks that a commodity OS performs in the background.

Decoupled Thread Execution To isolate HPC applications from such "noise", the FFMK OS allows their threads to be *decoupled* from the service OS and run undisturbed on dedicated compute cores. This novel mechanism leverages the tight integration of the paravirtualized commodity kernel and the L4 microkernel. L⁴Linux uses different L4 address spaces for the Linux kernel and each application process running on top of it. To virtualize CPU cores, it uses a vCPU execution model [20]. Such a vCPU is a special variant of an L4 thread. The Linux scheduler maps all Linux threads to one or more vCPUs, which then migrate between address spaces as they execute either kernel code during Linux system calls or user code of any of the Linux processes. However, since each process on top of L⁴Linux is backed by its own L4 address space, the code and data contained in it are accessible from all cores in the system, not just those assigned to the service OS.

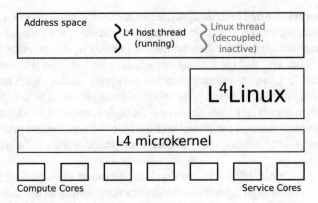

Fig. 2 Split execution model: the paravirtualized L⁴Linux kernel supports handing off thread execution of Linux programs to the underlying L4 microkernel, such that they can perform computations free of "OS noise" on cores controlled by the L4 microkernel. Decoupled threads are moved back temporarily to a service core assigned to Linux, whenever the program performs a Linux system call

To decouple a thread of a user process from unpredictable Linux behavior, L⁴Linux creates an additional L4 host thread to execute the application's code. Whenever the application is executing on the host thread, the corresponding Linux thread is detached from the scheduler of the service OS. Since this host thread is put on a separate compute core, which is controlled by L4 directly, it can thus execute in parallel to vCPUs of the service Linux (see Fig. 2). Thus, a noise-sensitive HPC application can run undisturbed and will not be subject to scheduling decisions of Linux, nor will it be interrupted by incoming interrupts.

Decoupled Linux programs can still perform Linux system calls. When doing so, the host thread causes an exception that is forwarded to L⁴Linux, which then reactivates the decoupled Linux thread and performs the requested operation in its context. Returning from the system call causes the thread to be decoupled again.

Device Access A key advantage of the decoupling mechanism apart from noise reduction is that it fits naturally into high-performance I/O stacks. For example, the InfiniBand driver stack consists of a Linux kernel driver and several user-space libraries (`libibverbs` and `libmlx5` in the case of recent Mellanox InfiniBand cards). These libraries contain the functionality that is on the performance-critical paths, which is why the user-space driver in `libmlx5` has direct access to I/O memory of the host-channel adapter (HCA) without having to call the kernel. Most of the management tasks (e.g., creating queue pairs, registering memory regions) are implemented in the kernel module; the user-space libraries communicate with the in-kernel driver, which is accessible through the system call forwarding as described in the preceding paragraph.

FFMK Node OS The previously described components and mechanisms form the basis of the FFMK node OS. It also hosts a decentralized platform management service which will be described in the next sections.

4   Dynamic Platform Management

FFMK addresses applications with varying resource demands and hardware platforms with variable resource availability (e.g. due to thermal limits or hardware faults). Although the FFMK OS/R is currently limited to node-local scheduling, we envision the full-featured version to dynamically optimize the usage of the application's resources by rebalancing its workload, optimizing network usage, and reacting to changing demands when its *elastic partition* shrinks or expands. Elastic partitions enable the FFMK platform to allocate resources to an application dynamically during the lifetime of the application (see Fig. 3b, c). The main task of the dynamic platform management is to continuously optimize the utilization of the system by means of an economic model. This economic model will include various

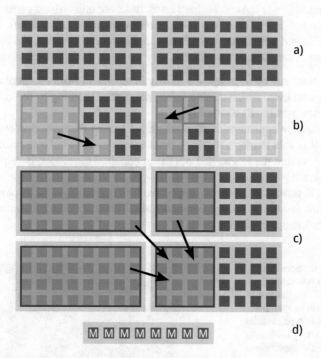

Fig. 3 Dynamic platform management. (**a**) Multicore nodes are organized in colonies. (**b**) Elastic applications partitions can expand and shrink. (**c**) Partitions can span mutliple colonies and expand to new colonies. (**d**) A redundant set of master nodes monitors and controls the system

aspects such as throughput and energy efficiency, fairness among applications, resiliency, and quality of service. However, its details are still subject to research.

The dynamic platform management consists of two basic components: monitoring and decision making. To achieve the scalability and resilience required for exascale systems, we decided to use gossip algorithms for all cross-node information dissemination of the monitoring component (see Sect. 4.2) and make decisions decentralized where possible (see Sect. 4.3).

4.1 Application Model

To support dynamic management of applications on our platform, we require an application model that is more flexible than the coarse-grained and static division of work that common MPI implementations impose. In our model, the decomposition of an application's workload is decoupled from the number execution units. The units of decomposition are migratable *tasks* that communicate with each other (see Fig. 4). For example, a core may run multiple tasks (one after each other) by preempting at blocking communication calls—a principle called overdecomposition [1]. At an abstract level, tasks are units which generate load for different hardware resources (e.g. cores, caches, memory, and network bandwidth) and the OS/R can map them to the hardware in order to optimize the application's performance. There are several reasons why we think this approach makes sense:

- Applications can be decomposed mostly independent from the number of nodes the program uses, which allows sizing the tasks according to the cache size or application-specific data structures.
- If the resource consumption of tasks varies among the tasks and over runtime, the OS/R is able to map and remap tasks intelligently to balance resource usage. This means that the OS/R, and not the application developer, is responsible for load balancing.
- The OS/R shrinks and expands applications to optimize global throughput.
- The OS/R is able to reduce communication costs by doing a communication-aware (re)mapping of tasks to nodes.

Fig. 4 Applications are decomposed into tasks. Multiple tasks are mapped to a node and can be migrated by the OS/R to expand/shrink the application's partition, to load balance the application, and to optimize communication

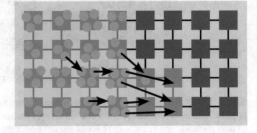

- Tasks waiting for a message are not scheduled to a core (i.e. busy waiting is avoided). This allows other tasks to run and to overlap communication with computation. Additionally, the OS/R is able to prioritize tasks that other tasks wait for.
- The OS/R may place tasks of different applications on the same node. Co-locating applications with different resource demands may increase the system utilization and throughput [42].

If, for example, bandwidth is the limiting resource on a node, the OS/R may increase the bandwidth available to the tasks by running fewer of them concurrently and migrating some of the tasks to another node. Additionally, the OS/R may either turn off unneeded cores (to reduce energy consumption) or co-locate bandwidth-insensitive tasks, possibly belonging to another application.

4.2 Monitoring and Gossip-Based Information Dissemination

To be capable of dynamic platform management, the system needs to collect status information about available resources of the nodes and their usage. The status information should contain:

- Current load on the node (cores, caches, memory, memory and network bandwidth)
- Maximum load the node can carry (i.e. available resources, may vary due to faults and thermal limits)
- Communication partners of the tasks running on that node.

The OS/R will use online monitoring (e.g. based on hardware counters) to gather the information on each node. We currently disseminate across node boundaries only information describing the overall resource state of a node. If that turns out to be too coarse-grained, we consider adding information about resource demands of individual tasks. Additionally, applications may pass hints to the runtime that enable a better prediction of future application behavior. The collected and disseminated information is the basis for making decisions as mentioned in the previous section.

Randomized Gossip As briefly introduced in Sect. 2, we will use randomized gossip algorithms to disseminate the resource information and build up the distributed bulletin board. In randomized gossip algorithms each node periodically sends messages with the latest information about other nodes to randomly selected nodes. Received information is merged with the local bulletin board by selecting the newest entry for each node. Thus, each node accumulates local information about the other nodes over time.

We have shown that these algorithms are resilient and they scale to exascale-size systems [5]. Scalability is achieved by dividing the system into *colonies*, each containing in the order of 1000 nodes. The colonies should consist of topologically nearby nodes, see Fig. 3. For the time being we assume that colonies are fixed and independent of the elastic application partitions. We run the gossip algorithm within each colony independently such that each node knows the status of all other nodes in the same colony; the colonies form the lower level of a gossip hierarchy.

Hierarchical Gossip One level above the colonies, a set of redundant master nodes maintains the global view on all nodes. The masters receive gossip messages from random nodes of each colony to obtain a complete picture of the resource usage and availability of the system. For decentralized decisions concerning multiple colonies (e.g. load balancing of a multi-colony application), the masters additionally send gossip messages with summary information about all colonies back to some colony nodes, which then disseminate it within the colony.

Quality of Information and Overhead Recent results of our research have shown the scalability and resiliency of the randomized gossip algorithms [5]. They work well even when some nodes fail, without the need for any recovery protocol, which is an advantage over tree-based approaches [2]. We developed formal expressions for approximating the average age (i.e., quality of information) of the local information at each node and the information collected by the master. These results closely match the results of simulations and measurements on up to 8192 nodes of a Blue Gene/Q system, as shown in Fig. 5.

We also investigated the overhead of gossip algorithms on the performance of HPC applications sharing network and compute resources [22]. The measurement results for two applications running concurrently to gossip with large information records per node (1024 bytes) are shown in Fig. 6. Sending gossip messages at an interval of 256 ms and above does not cause noticeable overhead, except for extremely communication-intensive codes like MPI-FFT (fast fourier transform).

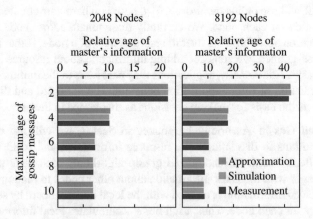

Fig. 5 Average age of the master's information using different age thresholds for gossip message entries (sending only newest information). The age is given relative to the interval of gossip messages. Approximations, simulations, and measurements on Blue Gene/Q match very well

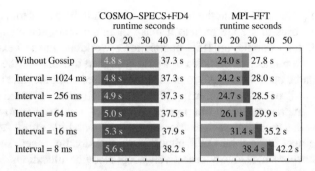

Fig. 6 Runtime overhead of gossip on two benchmark applications on 8192 Blue Gene/Q nodes when varying the interval of gossip messages. The inner red part indicates the MPI portion

4.3 Decision Making

Deciding on how to optimize system utilization is performed at three levels: within each node, decentralized between nodes for each application partition, and centralized at the master nodes. Each level is responsible for a part of the dynamic management of applications as outlined in Sect. 4.1. In the following, we explain the three levels top–down.

- **Whole system:** the master nodes optimize elastic partitions (i.e., shrinking and expanding them), multi-application resource assignment, placement of new partitions, and handling of failures. The master assigns nodes to partitions, but does not care about the mapping of individual tasks to nodes.
- **Per application partition:** gossiping nodes perform decentralized load balancing and communication optimization by migrating tasks within the partition. We will focus on scalable, distributed algorithms that act on small node neighborships or pairs of nodes. Depending on the application behavior, different algorithms will be considered (e.g., Diffusion-based [13], MOSIX [3]). Additionally, partition optimization decisions from the master are realized on the task level, e.g. decide which tasks to migrate to new nodes of an expanded partition.
- **Within each node:** the scheduler of the node OS assigns tasks to cores, taking into account data dependencies and arrival of messages from the network. It also performs dynamic frequency scaling and decides on which execution units to power up (dark silicon).

The FFMK OS makes load management decisions using local knowledge that each node acquired through monitoring and gossip-based information dissemination as described in Sect. 4.2. This information is always about the past, which is not always a good forecast of future behavior of highly dynamic workloads. Therefore, we plan to use techniques to predict resource consumption, like those employed by ATLAS [33]. ATLAS is an auto-training look-ahead scheduler that correlates observed behavior (e.g., execution times, cache misses) and application-provided information ("metrics") about the next chunk of work to be executed. Applications pass these metrics to the OS to help it make more accurate predictions of future behavior. If, for example, an HPC application's workload in the next time step depends on the number of particles in a grid cell, then this metric (the number of particles) can be used by ATLAS to predict the required compute time to complete the time step; it does so by inferring this information from observed correlation of previous (metric, execution time) pairs. We expect—and hear from application developers—that providing such metrics can be done with little effort. Additionally, applications may inform the OS/R about future workload changes, such that the platform management is able to proactively adapt resource allocations.

5 MPI Runtime

The FFMK architecture is designed such that it can support different runtimes on the LWK at the same time, such as MPI, X10 or Charm++. Due to limited resources and because MPI is the foundation of the vast majority of applications, we focus on dynamizing this traditional HPC runtime such that the FFMK OS can perform load balancing at the OS/R level.

5.1 MPI and Load Balancing

Load balancing applications for exascale HPC systems is a major challenge [14, 25]. For example, in the case of MPI-based applications, each of the participating MPI processes is usually mapped to its own core. If a few MPI processes reach a synchronization point later than the others, the majority of cores become effectively idle, thereby wasting resources. Unfortunately, load imbalances are typical for many important classes of applications, including atmospheric simulations [41], particle simulations [38], and fluid dynamics [17].

Load Balancing by Overdecomposition As explained in the previous sections, the common approach for tackling these load balancing issues is to (1) *overdecompose* by splitting the problem into more parts (i.e., tasks) than cores available, (2) assign the parts to cores, and (3) adapt this mapping dynamically during runtime so as to minimize both imbalance and communication costs. Typically, this method of dynamic load balancing is implemented at the application and library level [23, 38], because MPI implementations do not provide any built-in load management mechanism. This means that the mapping of MPI processes to cores remains static and the application itself is responsible for redistributing workload among ranks to maintain the balance. Even though this approach proved very effective in reducing imbalances and thereby improving performance, it is most often tailored to a specific application or problem domain and cannot be applied to arbitrary workloads easily. Thus, developers are forced to "reinvent the wheel" over and over again.

Adaptive MPI (AMPI) To save developer effort, one could overdecompose at the level of MPI ranks by just creating more ranks than cores available. AMPI [1] is an example of an MPI implementation that does exactly this. It is based on Charm++ [18] and maps each MPI rank to a "chare", which is the Charm++ equivalent of a task. This approach enables the underlying Charm++ runtime system to perform load balancing and migration of MPI ranks transparently. However, chares are not OS-level processes, but C++ objects encapsulating all code and data. Thus, MPI ranks in AMPI share the same address space of a single Charm++ runtime process on each node. Therefore, most MPI applications have to be modified to work on top of AMPI, because global variables are disallowed. Also, multithreaded MPI ranks cannot be supported, because chares are single threaded entities.

5.2 OS/R Support for Oversubscription

Adaptive MPI's compatibility limitations can be overcome by actually creating more MPI compute processes—and thereby more threads—which are subject to a system-level load balancer.

Requirements Analysis The advantage of MPI overdecomposition is that it enables automatic load balancing for MPI applications without having to modify their code. However, it comes at the cost of additional management and communication overhead due to the increased number of ranks. Furthermore, current MPI implementations cause any process that waits for a message transfer to complete to occupy a core, because polling is used. Such busy waiting causes unacceptable overhead in combination with oversubscription, because it effectively prevents overlapping computation and communication. In order for process-level oversubscription to work, waiting must be performed in a blocking fashion instead and the additional overhead must be kept at a minimum to allow for real performance gains. Thus, the OS/R has to provide light-weight message and thread management that allows for fast unblocking of a rank once a message for this rank arrives. Ideally, the system also takes communication dependencies into account when making scheduling decisions: it should prioritize those communication partners that other processes are waiting for so as to keep message latency low.

Preliminary Study To assess the potential of this approach, we conducted a preliminary study where we used MVAPICH2 [29] for oversubscribed runs of the weather simulation code COSMO-SPECS+FD4 [23] and the atomistic simulation CP2K [30]. Both are prone to load imbalances.[1] We used a small FDR InfiniBand test cluster with four nodes that ran a standard GNU/Linux system, since Linux

---

[1]COSMO-SPECS+FD4 has an internal load balancer, which we disabled in the experiments described here.

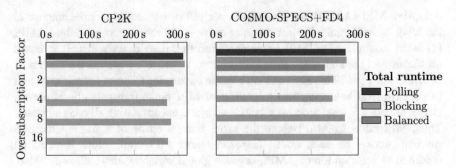

Fig. 7 Preliminary oversubscription study with the applications CP2K and COSMO-SPECS+FD4 using MVAPICH2 on a 16-core/4-node InfiniBand test cluster

kernels preinstalled on HPC systems are typically tuned to not migrate processes between cores. MVAPICH2 does not only support native InfiniBand as a communication back-end but also allows for blocking communication, where the library will block in the kernel until a new message arrives instead of actively polling for messages.

We found that blocking causes only a small overhead compared to busy waiting, as shown in Fig. 7 for the two applications: the purple bars show the runtime when using polling (traditional MPI behavior), the orange bars below show the same benchmark with blocking enabled. However, the results also indicate that overdecomposition and oversubscription of MPI processes can indeed improve performance. Compared to the configurations at the top of the diagrams, which show the total runtime with one MPI process per core (i.e., oversubscription factor of 1), we can see significant improvements in those runs where we oversubscribed the cores by a factor of up to 16 times. The workload remained the same in all cases; we just increased the number of MPI ranks working on the problem.

The MPI library was configured to block in the kernel when waiting for messages; no busy waiting was performed in MPI routines. This allows the scheduler of the Linux OS to migrate threads among cores in order to utilize all cores equally, thereby overlapping wait times with computations in other MPI processes.

For comparison, we also give the runtime of COSMO-SPECS+FD4 with its internal load balancer enabled (green bar labeled "balanced"). We can see that OS-level oversubscription still does not achieve the same performance, but it gets within 7 % at 4× oversubscription. The improvement in the oversubscribed configuration is achieved with no effort from the developer's side; in contrast, several person years went into COSMO-SPECS+FD4's load balancer.

More results of oversubscription experiments, also showing the benefit of multiple applications sharing the same nodes, are described in a tech report [37].

6 Migration

The FFMK prototype does not support inter-node process migration yet. It can only balance load within each node, where the OS scheduler migrates threads among cores. Nevertheless, we regard migration as the "swiss army knife" of an exascale OS/R: this mechanism can be used to (1) further improve load balancing, for (2) proactive fault tolerance as described in Sect. 7, and (3) as a tool for achieving better energy efficiency.

The Case for Migration Migration of MPI processes within a single node is taken care of by the local scheduler of the node OS. However, this approach to load balancing is no longer optimal, if the total amount of work per node varies within the application partition (i.e., the processes on some nodes take longer than on others). An example of this situation is shown for CP2K in Fig. 8. It visualizes how much time each of the 1024 MPI ranks spent doing useful computation in each time step. Green indicates a high computation/communication ratio, whereas yellow and red areas of the heatmap show that most of the time is spent in MPI waiting for communication to finish.

To reduce the load imbalance, nodes hosting "green" processes need to migrate some MPI ranks to nodes that are mostly red and yellow. Fortunately, our analysis of CP2K and other applications such as COSMO-SPECS+FD4 revealed that the

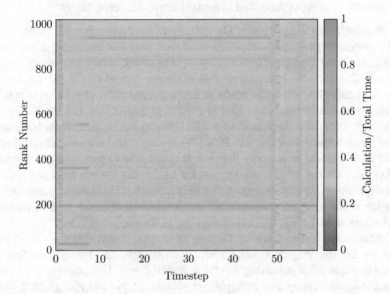

Fig. 8 Load imbalances in CP2K. Colors show computation vs communication ratio of each MPI process *(Y axis)* per time step *(X axis)*. *Yellow* and *red* indicate short computation time vs long waiting for other MPI ranks; a small number of overloaded processes delay all others, because they need significantly longer to compute their chunk of work in a time step *(green areas)*

load caused by each process changes rather slowly, if at all. This observation is encouraging, because inter-node migration takes much more time than migrating a thread within the same node, but can be performed less frequently.

Migration Obstacles Inter-node migration is complicated due to the static nature of communication back-ends such as InfiniBand and MPI itself. For the benefit of performance, implementations are designed such that after an initial setup phase, modifications to the partners involved in a communication are not easily possible. The RDMA-based job migration scheme [31] by Ouyang et. al. addresses this problem by tearing down all communication endpoints prior to migration and re-establishes them when the application resumes. The approach [4] taken by Barak et al. only works with TCP/IP-based communication. Despite these research efforts and others in the area [36], migration has never been integrated into production MPI libraries, even though the MPI standard [28] does not prohibit this feature.

Checkpoint-Migrate-Restart Given that transparent inter-node migration is hard with state-of-the-art communication stacks, and since it is needed only infrequently, we consider a simpler solution to the load-balancing problem that is based on coordinated checkpoint/restart (C/R): to migrate individual MPI processes, we (1) checkpoint at a convenient time (e.g., after completing a time step) the current state of the whole application, (2) terminate all processes, and then (3) restart them, but with certain processes assigned to previously underloaded nodes. The new placement of "migrated" processes is determined based on system monitoring and decision making as described in Sects. 4.2 and 4.3, respectively.

Checkpoint/Restart Approach The efficacy of the approach relies on the ability of the system to perform checkpoint/restart with very low overhead. A key metric to optimize is the amount of data that needs to be checkpointed and/or sent over the network. Compared to system-level C/R solutions such as BLCR [8], application-assisted checkpointing usually produces much smaller state. The reason is that they serialize just the internal state that is needed to restart, but not the contents of entire address spaces. Application-specific C/R support is common in HPC codes. There are also frameworks such as SCR [27] that support multi-level checkpointing, where data is stored in memory before it is transferred to persistent storage in the background. On the other hand, support for BLCR-like system-level solutions has been deprecated recently, or removed entirely from major MPI implementations. We therefore focus on application-assisted checkpoint/restart as the process-migration mechanism in the FFMK OS, but system-level C/R would work, too.

Furthermore, earlier work by Ouyang et al. [31] found that the restart phase takes by far the longest time in this migration scheme. We can confirm that re-initialization after restarting is still a major factor, but also one that leaves room for optimizations. For example, we found that, in MVAPICH2, `MPI_Init` spends several hundred milliseconds to obtain topology information about the local node using the hwloc library. Older versions of the MPI library also called initialization routines of the InfiniBand driver stack multiple time. This overhead can be eliminated by caching results or removing any redundant function calls; we submitted patches that fix the latter performance issue to the MVAPICH2 authors.

Finally, to achieve the level of performance for C/R to be usable as a migration mechanism, we employ in-memory checkpointing to make serialized application state accessible from any node where processes are migrated to. The next section on fault tolerance techniques covers requirements for a suitable checkpoint store.

7 Fault Tolerance

The HPC research community expects that the total number of individual components in exascale machines will grow dramatically. It is already becoming increasingly common to add more levels of node-local memory (e.g., SSDs), and heterogeneous architectures using accelerators are state of the art. This increased complexity and the expectation of higher failure rates for individual components and the whole system require a much more sophisticated approach to fault tolerance. In the following paragraphs, we give an overview of the key techniques and how they fit into the FFMK architecture.

Protecting Applications: Checkpoint/Restart The state-of-the-art mechanism to protect applications from crashes and other fail-stop errors is to make their execution state recoverable using checkpoint/restart (C/R) [10, 34]. FFMK aims at integrating a high-performance C/R system that utilizes the distributed storage built into all nodes of an exascale system, instead of relying on a traditional parallel file system that is connected to the supercomputer via a small number of I/O nodes. The general approach has been shown to scale extremely well with the number of nodes, for example in work by Rajachandrasekar et al. [32].

The FFMK project implements scalable C/R based on XtreemFS [40]. Due to space constraints, we do not discuss this distributed file system in detail, but give only a brief summary: XtreemFS supports storing erasure-coded file contents (e.g., checkpointed application state) in local memory of (potentially all) nodes of an HPC system. Erasure coding ensures that data is still accessible even if multiple nodes fail; at the same time, it minimizes both the network bandwidth required to transmit checkpoint data over the network and the amount of on-node storage that is required.

Proactive Fault Handling The FFMK OS' automatic load management and migration support (see Sect. 6) can also be used for proactive fault tolerance similar to [36]. By migrating all processes away from a node that is about to fail, the system can keep applications running without having to restart them from a checkpoint. To this end, FFMK leverages the hardware monitoring and information dissemination support described in Sect. 4.2: if a node observes critical CPU temperatures or correctable bit flips in a failing memory bank, it can initiate migration of all local processes to another node. We also consider partial node failures, where, for example, a single core becomes unreliable, but all other cores continue working properly. In both cases, the system may temporarily oversubscribe healthy resources (other nodes or unaffected local cores) by migrating processes. We consider any slowdowns caused by such "emergency evacuations" a special case of load imbalance, which can be resolved either by the FFMK load balancing system or by assigning replacement nodes to the application.

Resilient Gossip Algorithms At the system level, however, the FFMK OS relies on fault-tolerant algorithms. The most important ones are the randomized gossip algorithms, which are used to propagate information about the health of each node. Furthermore, they indirectly allow the system to identify nodes that stopped responding (e.g., due to an unexpected crash or network failures). The algorithms themselves are inherently fault-tolerant and they provide good quality of information even when some of the participating nodes failed; details of the theoretical foundations and simulation are discussed in [5].

The overview on fault tolerance concludes the presentation of the FFMK architecture. In the next section, we discuss related work.

8 Related Work

There exist several other projects that build operating systems for future HPC systems. In the following, we will characterize the projects from our point of view and emphasize the differences.

Argo and Hobbes The first two OSes, Hobbes [9] and Argo [6], are based on a general architecture similar to ours. They include a node OS as basis, global platform management, and an intermediate runtime providing a light weight thread abstraction. To our knowledge, the global management in both cases is based on MRNet [2], a fault-tolerant tree management structure, whereas FFMK uses gossip algorithms [5] for their inherent fault tolerance properties. The Argo consortium includes the research group behind Charm++ [18] to provide a versatile load balancing and resource management together with a light weight thread-like abstraction. The philosophy behind Charm++ is similar to our task-based application model. Argo uses Linux as the basis of their node OS. Hobbes is based on a newly built microkernel named Kitten [21]. In contrast to L4, Kitten's interface resembles the Unix interface, but is cut down and tailored towards enabling Linux applications to run directly on the microkernel. As does FFMK, Hobbes also relies on virtualization technology to support Linux applications that require features not provided by the microkernel; system calls not supported by Kitten are forwarded to Linux.

mOS The mOS project [39] at Intel is also based on a light-weight kernel (LWK) that runs colocated with a fully-fledged Linux. System calls that are not supported by the LWK are forwarded to the Linux kernel. However, in contrast to the FFMK platform, the mOS LWK controls only compute cores, whereas the L4 microkernel of our OS platform is in control of all cores.

Manycore OS Riken's OS [35] developed under Yukata Ishikawa also is a hybrid system. To the best of our knowledge, the main difference compared to FFMK is the fact that the microkernel can run on accelerators such as Xeon Phi, but remains under control of a Linux. The system pioneered splitting the InfiniBand driver stack, such that processes running on the accelerators can reuse the functionality hosted on Linux by way of communication between Linux and the microkernel.

9 Summary and Future Work

State of the Union In this paper, we described the challenges that future HPC systems pose to system and application developers. Based on these challenges, we motivated an architecture for an exascale operating system and runtime (OS/R): the microkernel-based FFMK OS. We described the current state of our prototype implementation, which, at the time of this writing, is capable of running unmodified MPI applications. The implementation of the node OS consists of an L4 microkernel, which is supported by a virtualized Linux kernel that we use as a service OS. While our gossip algorithms are well-studied and found to be suitable, the decision making algorithms that build on top are not yet implemented; gobal platform management is therefore not part of the prototype. However, the node OS has been successfully tested on a 112-node InfiniBand cluster across 1,344 Intel Xeon cores.

Future Work Our short-term agenda focuses on evaluating process-level overdecomposition and oversubscription of MPI applications (see Sect. 5). Furthermore, our work on the "decoupled thread" execution model presented in Sect. 3 is currently under peer review. The FFMK project is funded for three more years, during which we plan to finalize and integrate those building blocks of the architecture that are not yet complete. This includes especially the checkpoint/restart layer and cross-node migration support.

A key area of future work in the long term is research into novel interfaces between applications and the OS/R. We already have experience with schedulers [33] that can make better decisions based on application-provided hints about future behavior. We also investigated "programming hints" for optimizing memory accesses in GPU-based applications [7]. Application-level hints seem also promising for fault tolerance: HPC application developers [15] are already researching fault-tolerant versions of the core algorithms used in their HPC codes. Such codes may be able to handle node failures without restarting from a checkpoint, provided that the application can inform the OS/R about its fault tolerance requirements through a suitable interface.

Acknowledgements This research and the work presented in this paper is supported by the German priority program 1648 "Software for Exascale Computing" via the research project FFMK [16]. We also thank the cluster of excellence "Center for Advancing Electronics Dresden" (*cfaed*). The authors acknowledge the Jülich Supercomputing Centre, the Gauss Centre for Supercomputing, and the John von Neumann Institute for Computing for providing compute time on the JUQUEEN supercomputer.

References

1. Acun, B., Gupta, A., Jain, N., Langer, A., Menon, H., Mikida, E., Ni, X., Robson, M., Sun, Y., Totoni, E., Wesolowski, L., Kale, L.: Parallel programming with migratable objects: Charm++ in practice. In: Proceedings of the Supercomputing 2014, Leipzig, pp. 647–658. IEEE (2014)
2. Arnold, D.C., Miller, B.P.: Scalable failure recovery for high-performance data aggregation. In: Proceedings of the IPDPS 2010, Atlanta, pp. 1–11. IEEE (2010)
3. Barak, A., Guday, S., Wheeler, R.: The MOSIX Distributed Operating System: Load Balancing for UNIX. Lecture Notes in Computer Science, vol. 672. Springer, Berlin/New York (1993)
4. Barak, A., Margolin, A., Shiloh, A.: Automatic resource-centric process migration for MPI. In: Proceedings of the EuroMPI 2012. Lecture Notes in Computer Science, vol. 7490, pp. 163–172. Springer, Berlin/New York (2012)
5. Barak, A., Drezner, Z., Levy, E., Lieber, M., Shiloh, A.: Resilient gossip algorithms for collecting online management information in exascale clusters. Concurr. Comput. Pract. Exper. **27**(17), 4797–4818 (2015)
6. Beckman, P., et al.: Argo: an exascale operating system. http://www.argo-osr.org/. Accessed 20 Nov 2015
7. Ben-Nun, T., Levy, E., Barak, A., Rubin, E.: Memory access patterns: the missing piece of the multi-GPU puzzle. In: Proceedings of the Supercomputing 2015, Newport Beach, pp. 19:1–19:12. ACM (2015)
8. Berkeley Lab Checkpoint/Restart. http://ftg.lbl.gov/checkpoint. Accessed 20 Nov 2015
9. Brightwell, R., Oldfield, R., Maccabe, A.B., Bernholdt, D.E.: Hobbes: composition and virtualization as the foundations of an extreme-scale OS/R. In: Proceedings of the ROSS'13, pp. 2:1–2:8. ACM (2013)
10. Bronevetsky, G., Marques, D., Pingali, K., Stodghill, P.: Automated application-level check-pointing of MPI programs. ACM Sigplan Not. **38**(10), 84–94 (2003)
11. Burstedde, C., Ghattas, O., Gurnis, M., Isaac, T., Stadler, G., Warburton, T., Wilcox, L.: Extreme-scale AMR. In: Proceedings of the Supercomputing 2010, Tsukuba, pp. 1–12. ACM (2010)
12. Cappello, F., Geist, A., Gropp, W., Kale, S., Kramer, B., Snir, M.: Toward exascale resilience: 2014 update. Supercomput. Front. Innov. **1**(1), 5–28 (2014)
13. Corradi, A., Leonardi, L., Zambonelli, F.: Diffusive load-balancing policies for dynamic applications. IEEE Concurr. **7**(1), 22–31 (1999)
14. Dongarra, J., et al.: The international exascale software project roadmap. Int. J. High Speed Comput. **25**(1), 3–60 (2011)
15. EXAHD – An Exa-Scalable Two-Level Sparse Grid Approach for Higher-Dimensional Problems in Plasma Physics and Beyond. http://ipvs.informatik.uni-stuttgart.de/SGS/EXAHD/index.php. Accessed 29 Nov 2015
16. FFMK Website. http://ffmk.tudos.org. Accessed 20 Nov 2015
17. Harlacher, D.F., Klimach, H., Roller, S., Siebert, C., Wolf, F.: Dynamic load balancing for unstructured meshes on space-filling curves. In: Proceedings of the IPDPSW 2012, pp. 1661–1669. IEEE (2012)

18. Kale, L.V., Zheng, G.: Charm++ and AMPI: adaptive runtime strategies via migratable objects. In: Parashar, M., Li, X. (eds.) Advanced Computational Infrastructures for Parallel and Distributed Adaptive Applications, chap. 13, pp. 265–282. Wiley, Hoboken (2009)
19. Kogge, P., Shalf, J.: Exascale computing trends: adjusting to the "New Normal" for computer architecture. Comput. Sci. Eng. 15(6), 16–26 (2013)
20. Lackorzynski, A., Warg, A., Peter, M.: Generic virtualization with virtual processors. In: Proceedings of the 12th Real-Time Linux Workshop, Nairobi (2010)
21. Lange, J., Pedretti, K., Hudson, T., Dinda, P., Cui, Z., Xia, L., Bridges, P., Gocke, A., Jaconette, S., Levenhagen, M., Brightwell, R.: Palacios and Kitten: new high performance operating systems for scalable virtualized and native supercomputing. In: Proceedings of the IPDPS 2010, Atlanta, pp. 1–12. IEEE (2010)
22. Levy, E., Barak, A., Shiloh, A., Lieber, M., Weinhold, C., Härtig, H.: Overhead of a decentralized gossip algorithm on the performance of HPC applications. In: Proceedings of the ROSS'14, Munich, pp. 10:1–10:7. ACM (2014)
23. Lieber, M., Grützun, V., Wolke, R., Müller, M.S., Nagel, W.E.: Highly scalable dynamic load balancing in the atmospheric modeling system COSMO-SPECS+FD4. In: Proceedings of the PARA 2010. Lecture Notes in Computer Science, vol. 7133, pp. 131–141. Springer, Berlin/New York (2012)
24. Liedtke, J.: On micro-kernel construction. In: Proceedings of the 15th ACM Symposium on Operating Systems Principles (SOSP'95), Copper Mountain Resort, pp. 237–250. ACM (1995)
25. Lucas, R., et al.: Top ten exascale research challenges. DOE ASCAC subcommittee report. http://science.energy.gov/~/media/ascr/ascac/pdf/meetings/20140210/Top10reportFEB14.pdf (2014). Accessed 20 Nov 2015
26. Milthorpe, J., Ganesh, V., Rendell, A.P., Grove, D.: X10 as a parallel language for scientific computation: practice and experience. In: Proceedings of the IPDPS 2011, Anchorage, pp. 1080–1088. IEEE (2011)
27. Moody, A., Bronevetsky, G., Mohror, K., de Supinski, B.: Detailed modeling, design, and evaluation of a scalable multi-level checkpointing system. Technical report LLNL-TR-440491, Lawrence Livermore National Laboratory (LLNL) (2010)
28. MPI: A message-passing interface standard, version 3.1. http://www.mpi-forum.org/docs (2015). Accessed 20 Nov 2015
29. Mvapich: Mpi over infiniband. http://mvapich.cse.ohio-state.edu/. Accessed 20 Nov 2015
30. Open Source Molecular Dynamics. http://www.cp2k.org/. Accessed 20 Nov 2015
31. Ouyang, X., Marcarelli, S., Rajachandrasekar, R., Panda, D.K.: RDMA-based job migration framework for MPI over Infiniband. In: Proceedings of the IEEE CLUSTER 2010, Heraklion, pp. 116–125. IEEE (2010)
32. Rajachandrasekar, R., Moody, A., Mohror, K., Panda, D.K.: A 1 PB/s file system to checkpoint three million MPI tasks. In: Proceedings of the HPDC'13, New York, pp. 143–154. ACM (2013)
33. Roitzsch, M., Wachtler, S., Härtig, H.: Atlas: look-ahead scheduling using workload metrics. In: Proceedings of the RTAS 2013, Philadelphia, pp. 1–10. IEEE (2013)
34. Sato, K., Maruyama, N., Mohror, K., Moody, A., Gamblin, T., de Supinski, B.R., Matsuoka, S.: Design and modeling of a non-blocking checkpointing system. In: Proceedings of the Supercomputing 2012, Venice, pp. 19:1–19:10. IEEE (2012)
35. Sato, M., Fukazawa, G., Yoshinaga, K., Tsujita, Y., Hori, A., Namiki, M.: A hybrid operating system for a computing node with multi-core and many-core processors. Int. J. Adv. Comput. Sci. 3, 368–377 (2013)
36. Wang, C., Mueller, F., Engelmann, C., Scott, S.L.: Proactive process-level live migration and back migration in HPC environments. J. Par. Distrib. Comput. 72(2), 254–267 (2012)
37. Wende, F., Steinke, T., Reinefeld, A.: The impact of process placement and oversubscription on application performance: a case study for exascale computing. Technical report 15–05, ZIB (2015)

38. Winkel, M., Speck, R., Hübner, H., Arnold, L., Krause, R., Gibbon, P.: A massively parallel, multi-disciplinary Barnes-Hut tree code for extreme-scale N-body simulations. Comput. Phys. Commun. **183**(4), 880–889 (2012)
39. Wisniewski, R.W., Inglett, T., Keppel, P., Murty, R., Riesen, R.: mOS: an architecture for extreme-scale operating systems. In: Proceedings of the ROSS'14, Munich, pp. 2:1–2:8. ACM (2014)
40. XtreemFS – a cloud file system. http://www.xtreemfs.org. Accessed 20 Nov 2015
41. Xue, M., Droegemeier, K.K., Weber, D.: Numerical prediction of high-impact local weather: a driver for petascale computing. In: Bader, D.A. (ed.) Petascale Computing: Algorithms and Applications, pp. 103–124. Chapman & Hall/CRC, Boca Raton (2008)
42. Zheng, F., Yu, H., Hantas, C., Wolf, M., Eisenhauer, G., Schwan, K., Abbasi, H., Klasky, S.: Goldrush: resource efficient in situ scientific data analytics using fine-grained interference aware execution. In: Proceedings of the Supercomputing 2013, Eugene, pp. 78:1–78:12. ACM (2013)

Fast In-Memory Checkpointing with POSIX API for Legacy Exascale-Applications

Jan Fajerski, Matthias Noack, Alexander Reinefeld, Florian Schintke, Torsten Schütt, and Thomas Steinke

Abstract Exascale systems will be much more vulnerable to failures than today's high-performance computers. We present a scheme that writes erasure-encoded checkpoints to other nodes' memory. The rationale is twofold: first, writing to memory over the interconnect is several orders of magnitude faster than traditional disk-based checkpointing and second, erasure encoded data is able to survive component failures. We use a distributed file system with a tmpfs back end and intercept file accesses with LD_PRELOAD. Using a POSIX file system API, legacy applications which are prepared for application-level checkpoint/restart, can quickly materialize their checkpoints via the supercomputer's interconnect without the need to change the source code.

Experimental results show that the LD_PRELOAD client yields 69 % better sequential bandwidth (with striping) than FUSE while still being transparent to the application. With erasure encoding the performance is 17 % to 49 % worse than striping because of the additional data handling and encoding effort. Even so, our results indicate that erasure-encoded memory checkpoint/restart is an effective means to improve resilience for exascale computing.

1 Introduction

The path towards exascale computing with 10^{18} operations per second is paved with many obstacles. Three challenges are being tackled in the DFG project '*A Fast and Fault-Tolerant Microkernel-Based System for Exascale Computing*' (FFMK) [21]: (1) the vulnerability to system failures due to transient and permanent errors, (2) the performance losses due to workload imbalances in applications running on hundreds of thousands of cores, and (3) the performance degradation caused by interactions and noise of the operating system.

J. Fajerski (✉) • M. Noack • A. Reinefeld • F. Schintke • T. Schütt • T. Steinke
Zuse Institute Berlin (ZIB), Berlin, Germany
e-mail: fajerski@zib.de; noack@zib.de; reinefeld@zib.de; schintke@zib.de; schuett@zib.de; steinke@zib.de

© Springer International Publishing Switzerland 2016
H.-J. Bungartz et al. (eds.), *Software for Exascale Computing – SPPEXA 2013-2015*, Lecture Notes in Computational Science and Engineering 113, DOI 10.1007/978-3-319-40528-5_19

This paper addresses the first challenge, that is, to improve the fault tolerance of exascale systems. Such systems consist of hundreds of thousands of components, each designed to be reliable by itself. But running them all together will render node failures a common event applications have to cope with [4, 7]. Several mechanisms for improving fault tolerance in HPC have been suggested, like fault-tolerant communication layers and checkpoint/restart (C/R) techniques.

C/R typically uses fast parallel file systems like Lustre,[1] GPFS [17], or Pana-sas [10] to materialize the checkpoints on disks. Unfortunately, C/R will reach its limits as the applications' memory footprint grows faster than the parallel I/O bandwidth. Writing a checkpoint to disk will make the system processors idle for a growing fraction of time, which becomes increasingly uneconomic. On a typical HPC system like the Cray XC40 at ZIB [2] it takes more than half an hour to write the main memory's capacity to the parallel Lustre file system.[2] Thus, reducing the time of checkpointing is of vital importance. It does not only improve the efficiency, but it will become a necessity when the mean time between failure (MTBF) becomes shorter than the time needed to persist a checkpoint to disk.

2 Related Work

In-memory checkpointing [12] has been known for a long time. Several schemes like Charm++ [23, 24] and FTI [3, 8] have been successfully deployed. SCR[3] implements multi-level checkpointing, where checkpoints are written to different media like RAM, flash or rotating disks. It uses a simple RAID5 encoding to be able to cope with additional component failures.

Unfortunately, the mentioned approaches are difficult to apply to legacy appli-cations. They are either limited to the use of specific object-oriented programming languages like Charm++ or they require source code modifications to use specific APIs like the one used by SCR for reading and writing checkpoints. The BLCR [5] checkpoint framework is able to checkpoint unmodified applications but requires support from the MPI library, because it can only create a consistent checkpoint of MPI applications when no messages are in flight.

Our approach, in contrast, is based on POSIX which makes it suitable for legacy applications, since many applications are prepared to write and read their checkpoints using POSIX file system operations.

---

[1]http://wiki.lustre.org/

[2]The Cray XC40 'Konrad' is operated at ZIB as part of the North German Supercomputer Alliance. It comprises 1872 nodes (44.928 cores), Cray Aries network, 120 TB main memory, and a parallel Lustre file system of 4.5 PB capacity and 52 GB/s bandwidth.

[3]https://computation.llnl.gov/project/scr/

3   In-Memory Checkpointing with POSIX API

3.1   Implementation with XtreemFS

We use XtreemFS [19], a scalable distributed file system developed at ZIB, as a basis for our in-memory checkpointing mechanism. XtreemFS supports POSIX semantics for file operations while—transparently for the application—providing fault tolerance via file replication on distributed servers. We modified XtreemFS to perform I/O operations in-memory rather than on disk. This was possible because XtreemFS is a user-space file system and is therefore not as tightly integrated into the operating system kernel as other parallel file systems. Hence, it is well-suited for providing in-memory checkpointing on top of the L4-microkernel used in the FFMK project [21].

An instance of XtreemFS comprises three services: the *Directory Service (DIR)* is a central registry for all XtreemFS services and is used for service discovery. The *Metadata and Replica Catalog (MRC)* stores the directory tree and file metadata, and it manages user authentication and file access authorization. The *Object Storage Device (OSD)* stores the actual file data as objects. Figure 1 illustrates the architecture of XtreemFS. Clients and servers (MRC, OSD, DIR) are connected via some network with no specific requirements in terms of security, fault tolerance and

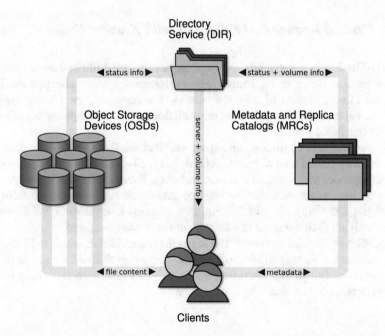

Fig. 1 XtreemFS architecture illustrating the three services (OSD, MRC and DIR) and the communication patterns between them and a client

performance. The separation of the metadata management in the MRCs from the I/O-intensive management of file content in the OSDs is a design principle found in many object-based file systems [10, 20]. To maximize scalability, metadata and storage servers are loosely coupled. They have independent life cycles and do not directly communicate with each other.

OSDs store data in their local directory tree. The underlying file system is only required to offer a POSIX compliant interface. We configured the OSDs to use the tmp directories of their respective nodes for data storage. The tmp directory is a *tmpfs* file system that exports Linux' disk caching subsystem as a RAM-based file system with an overflow option into swap space when the main memory capacity is exhausted. Thereby, all data sent to an OSD is stored solely in the node's main memory.

To create an XtreemFS instance, at least one OSD, MRC and DIR are needed. Though to make full use of XtreemFS' fault tolerance and scalability features, several OSDs should be started on different servers. Once all desired services are started and a volume is created, it can be mounted on any number of clients. All clients will see the same file system with the same directory tree. An XtreemFS volume is mounted through the *FUSE* kernel module[4] to provide a virtual file system. All operations in this file system are passed to the XtreemFS client library, which distributes the data to the OSD devices.

3.2 Fault-Tolerance and Efficiency with Erasure Codes

Data replication is a frequently used option in distributed file systems to provide fault tolerance. However, replication implies storage and communication overhead compared to the number of tolerated failures. The commonly used 3-way replication [18, 19] causes a $2/3$ overhead of the available raw storage capacity but can only tolerate one failed replica.

Erasure codes (EC) offer a more space-efficient solution for fault tolerance and have recently gained a lot of attention [1, 6, 9, 11, 14–16]. EC are a family of error correction codes that stripe data across k chunks. Every k data words are encoded to $k + m = n$ words such that the original data words can be recovered from any subset $\{s \mid s \in \mathscr{P}(n), |s| \geq k\}$ (cf. Fig. 2). Compared to replication, EC offers the same or higher fault tolerance at a fraction of the storage overhead.

An EC is considered *systematic* if it keeps the original k data words in its original representation. This property is desirable for storage applications since a fault-free read operation can simply consider the data striped across k objects. They can be read in parallel and no decoding is necessary.

---

[4]FUSE—Filesystem in Userspace allows the creation of a file system without changing Linux kernel code.

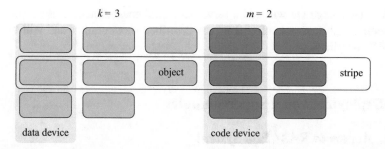

$k = 3$ $m = 2$

object   stripe

data device   code device

Fig. 2 Scheme of a systematic erasure code where data is striped across three devices and two code words per stripe are stored on two coding devices. Note that the last stripe only contains two data words

In order to exploit these advantages we implemented generic $k+m$ erasure coding in XtreemFS using the *Jerasure* library [13]. The encoding and decoding operation is implemented in the client library of XtreemFS.

File system operations are passed to the client library which then translates the request into a number of object requests, that are sent to the OSDs. A read request is performed optimistically by the client as it tries to read only the necessary data objects. If all OSDs with the corresponding data objects answer, the replies are concatenated and returned. If one or more OSDs fail to reply, the client will send out the necessary requests to OSDs storing coding objects. As soon as at least k out of n objects per stripe have been received, the data can be decoded and returned.

The implementation of a write operation is more complex. Since the encoding takes place in the client (i.e. on the client system) it needs a sufficient amount of data to calculate the coding objects. This establishes two requirements for write operations: (1) the size of a write operation should be $k * object\_size$ or a multiple thereof and (2) a write operation should align with stripes, i.e. the operation's offset should be a multiple of $k * object\_size$. These two requirements would diminish the POSIX compliance as POSIX does not put any restrictions on a write operation's size or offset.

This problem can be solved by implementing a *read-modify-write (rmw)* cycle in the client. When the client library receives a write request that violates the requirements above, it simply reads the necessary data to fulfill both requirements. The write request is then padded with the additional data, encoded and sent to the OSDs for storage. A write to the end of a file can simply be padded with zeros, since they act as neutral elements in the encoding operation.

However, we decided not to implement a rmw cycle, because we observed that writing checkpoint data is usually an append operation to the end of a file rather than updating random file locations. This means that both requirements of our client side implementation can be satisfied by simply caching write operations until a full stripe can be encoded and written to the OSDs or the file is closed. In the latter case the left over data is padded with zeros, encoded and then written to the OSDs.

At a later stage we will add a server side implementation of erasure coding to XtreemFS that does not suffer from the described shortcomings and will be fully POSIX compliant.

4   Deployment on a Supercomputer

4.1   Access to RAM File System

For Linux and Unix systems, there are two client solutions: the FUSE-based client that allows to mount an XtreemFS volume like any other file system, and the *libxtreemfs* library for C++ and Java, which allows application developers to directly integrate XtreemFS support into applications.

Since many HPC systems use a Linux-based operating system, the FUSE-client of XtreemFS would be a natural choice to use for our C/R system. But for performance reasons Linux configurations on HPC systems are often optimized and kernel modules like FUSE are typically disabled. We therefore developed a third client that intercepts and substitutes calls to the file system with the LD_PRELOAD mechanism. LD_PRELOAD allows to load libraries that are used to resolve dynamically linked symbols before other libraries (e.g. *libc*) are considered.

We implemented a *libxtreemfs_preload* that can be specified via the environment variable LD_PRELOAD. It intercepts and substitutes file system calls of an application. If an intercepted call relates to an XtreemFS volume or file, it is translated into its corresponding *libxtreemfs* call, which is similar to what the FUSE adapter does. Otherwise calls are passed through to the original *glibc* function, which would have handled it without the pre-load mechanism in place. Whether or not XtreemFS should be used is determined via a configurable path prefix, that can be thought of as a virtual mount point. For example, copying a file to an XtreemFS volume via FUSE using cp as application would be performed as follows:

```
$> mount.xtreemfs my.dir.host/myVolume /xtreemfs
$> cp myFile /xtreemfs
$> umount.xtreemfs /xtreemfs
```

The same operation with the LD_PRELOAD and *libxtreemfs_preload* instead of FUSE could be achieved with the following command:

```
$> XTREEMFS_PRELOAD_OPTIONS="my.dir.host/myVolume /xtreemfs" \
   LD_PRELOAD="libxtreemfs_preload.so" \
   cp myFile /xtreemfs
```

This example can be easily generalized into a shell script that wraps cp or other applications, such that the environment setup is hidden.

Figure 3 shows all three client solutions in comparison. The LD_PRELOAD client combines the transparency of the FUSE client with the potential performance benefits of directly using the *libxtreemfs*. Section 5 provides benchmark results on the different XtreemFS client solutions.

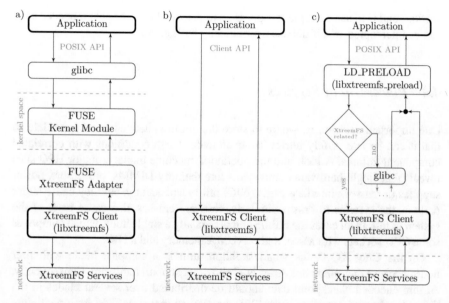

Fig. 3 Three different client solutions for XtreemFS. (**a**) The application interacts transparently with XtreemFS via FUSE. (**b**) The direct use of the client library avoids overhead but is intrusive. (**c**) The interception with LD_PRELOAD is non-intrusive and avoids FUSE as a bottleneck

Note that a similar approach is available with the *liblustre*[5] library in the Lustre parallel file system. Here, data of a Lustre volume can be accessed via LD_PRE-LOAD without the need of mounting. However, *liblustre* targets more at portability than performance.

4.1.1 Issues with LD_PRELOAD

The LD_PRELOAD mechanism is only able to intercept calls to dynamically linked functions. In most cases this works fine for the low-level file system calls of interest. However, there are situations where some of the calls are inaccessible. If we wanted, for instance, to intercept all close() calls made by an application there are two possible situations: the application either directly calls close() or it uses a higher-level operation like fclose() which then calls close() indirectly. The second case is problematic, since both calls are inside *glibc* and the inner close() call could have been inlined or statically linked, depending on the *glibc*-build. If so, it can not be intercepted by the LD_PRELOAD mechanism. One possible workaround is to also intercept the higher-level calls, but this would mean re-implementing and maintaining large parts of the *glibc*, which is not a good choice. A more practical

---

[5]http://wiki.lustre.org/index.php/LibLustre_How-To_Guide

workaround is to use a simple test-program to detect whether or not all needed calls can be intercepted, and if not use a specifically built *glibc*.

4.2 Placement of Services

One important decision is, where to store the erasure encoded checkpoint data so that it can be later safely retrieved—even under harsh conditions with correlated component failures. A look into the operators' machine books at major HPC sites reveals that single hardware components like memory DIMMs, processors, power supplies or network interface cards (NIC) fail independently, but they cause larger parts of the systems to crash. Similarly, software crashes also cause parts of the system to fail. In all cases the failure unit is typically a single node, which comprises some CPU sockets with associated ccNUMA memory and a NIC.

On the Cray XC40, the smallest failure unit is a compute blade with four nodes because they share one Aries NIC and other support hardware. Consequently, erasure encoded checkpoint data should be distributed over several blades in the Cray. As shown in another work [22], the latency increases only by a negligible amount when writing data from one blade to another in the same electrical unit via the Aries network (Fig. 4). This is also true when using the longer-distance fiber optics cables between different electrical units (pairs of cabinet). Moreover, the bandwidths are almost homogeneous across the entire system.

When an application crashed due to a node failure, additional resources are need to resume the application from a checkpoint. If all resources in the system are fully utilized, it may be difficult to provide additional nodes for the crashed application. On exascale systems, however, we expect the cost of reserving a few spare nodes per job to be negligible. Alternatively, the system provider could provide spare nodes from job fragmentation. The job will then be restarted with the same number of nodes but a slightly different job placement. As shown in [22] the cost for different placements varies only by a few percent.

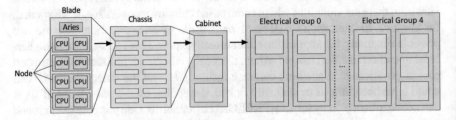

Fig. 4 Cray XC40 component hierarchy: node, blade, chassis, cabinet, electrical group

4.3 Deployment on a Cray XC40

Jobs on a Cray XC40 may be either run in extreme scalability mode (ESM) or in cluster compatibility mode (CCM). ESM is designed for scalable high performance applications. Only a minimal amount of system services are running on the compute nodes to minimize interference. Applications need to use Cray MPI and to be started with the command-line tool aprun. After the application finishes, the node-checker checks the node for errors and cleans up all remaining traces of the previous job. The ESM mode is not suited for our approach, because we need to run both, XtreemFS and the application on the same node. Additionally, we need to restart the application on these nodes and read the checkpoint from XtreemFS resp. the RAM-disk which is impossible with ESM.

In CCM, the reserved compute nodes can be treated like a traditional cluster. Standard system services, like an ssh daemon, are available. However, Cray MPI is not available in CCM. Instead, Cray provides an InfiniBand (IB) verbs emulation on top of the Aries network. For our experiments, we ran OpenMPI over the IB verbs emulation.

For a proof of concept we used the parallel quantum chromodynamics code BQCD.[6] It has a built-in checkpoint/restart mechanism. At regular intervals, it writes a checkpoint of its state to the local disk. In case of a crash, it can be restarted from these checkpoints.

We used ssh to start the XtreemFS services on the nodes in CCM mode. The services were distributed as follows: DIR, MRC and one OSD on the first node, and one OSD on all other nodes. We used the LD_PRELOAD client and OpenMPI to start BQCD. In this setup BQCD's snapshots are written to XtreemFS. We manually killed the BQCD job and successfully restarted it from the memory checkpoint.

5 Experimental Results

First, we evaluate the three XtreemFS client solutions described in Sect. 4.1. In order to compare the cost of the different data paths depicted in Fig. 3, we performed micro-benchmarks of the read and write operations to an XtreemFS volume with each solution. The different XtreemFS services ran on a single node, so there is no actual network traffic that might pose a bottleneck. A node has two Intel Xeon E5-2630v3 with 64 GB main memory and runs a Ubuntu 14.04 with a 3.13 kernel. Caching mechanisms of the kernel and FUSE were disabled by using the direct_io option of FUSE. This ensures that all requests reach the XtreemFS client and the OSDs, and are not just cached locally, which is especially important

---

[6]https://www.rrz.uni-hamburg.de/services/hpc/bqcd.html

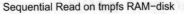

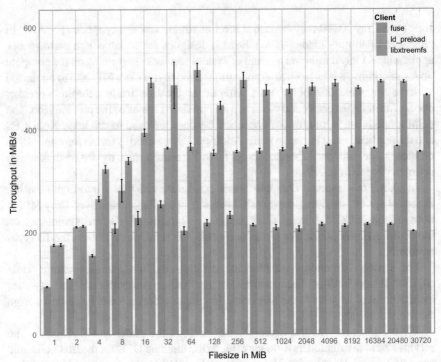

Fig. 5 Sequential read performance using the three different client approaches (RAM-disk)

for checkpoint data. All result values are averages over multiple runs, the error bars visualize the standard error.

Figures 5 and 6 show the results for sequential reading and writing, respectively. In both cases, the results match our expectations: *libxtreemfs* is faster than LD_PRELOAD which is faster than FUSE. For reading, LD_PRELOAD is between 35 % and 91 % faster than FUSE with an average of 69 %. Compared to *libxtreemfs*, it is around 21 % slower on average (between 0.5 % and 29 %). Writing performance is similar. LD_PRELOAD compared to FUSE is around 74 % faster (between 44 % and 91 %), and 23 % (between 6 % and 33 %) slower when compared to *libxtreemfs*. The results show that the newly developed LD_PRELOAD client approach yields a better sequential bandwidth than FUSE while still being transparent to the application. Regardless of performance, LD_PRELOAD is the only solution for applications that run in an environment where FUSE is not available and where modifying the application code is not possible.

The absolute throughput values of these micro-benchmarks are limited by the synchronous access pattern and the use of only a single data stream and a single client. In a real world scenario, there would be at least one client or data stream (i.e.

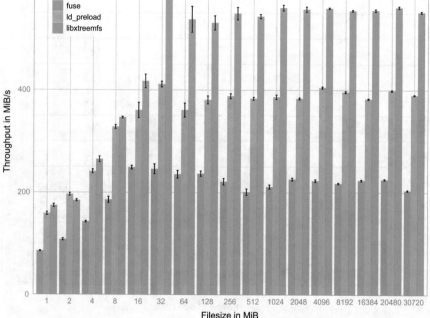

Fig. 6 Sequential write performance using the three different client approaches (RAM-disk)

file) per rank or thread, whose throughput would add up to an overall throughput that would be limited by some physical bound of the underlying hardware (i.e. memory, disk, or network bandwidth).

In a second experiment we compared the sequential throughput and scaling characteristics of the existing striping implementation and the client side erasure-coding solution. We used a distributed XtreemFS setup with 2–13 data OSDs. On each data OSD runs one IOR[7] process that reads/writes 1 GiB of data via the FUSE interface. MRC and DIR run on a separate machine. All machines have two Intel Xeon E5-2630v3 and 64 GB main memory, and all machines are interconnected with 10 Gbit/s. In the erasure-coding experiment, the XtreemFS instance has two additional OSDs for coding data and thus provides a RAID6-like configuration. All result values are averages over 10 runs with error bars that visualize the standard error.

---

[7]IOR is a I/O micro benchmark software by NERSC. https://www.nersc.gov/users/computational-systems/cori/nersc-8-procurement/trinity-nersc-8-rfp/nersc-8-trinity-benchmarks/ior/

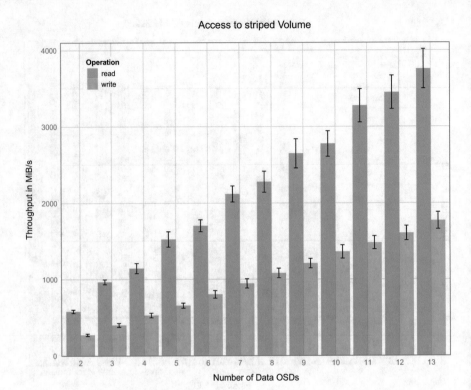

Fig. 7 Sequential read/write performance on a striped XtreemFS volume with an increasing number of OSDs and clients

Figures 7 and 8 show the results for reading and writing to variably sized XtreemFS instances. Both the striping and erasure-coding configuration exhibit good scaling characteristics. Compared to the striping configuration, writes to the erasure coded volume are 17–49 % slower, which reveals the overhead caused by the additional coding data. This corresponds roughly to the 15 % to 50 % data overhead the coding induces. For reference, a replicated setup that provides the same level of fault tolerance would induce a 200 % data overhead. The read operation exhibits a slowdown between 5 % and 14 % in the erasure-coding configuration.

The results show a performance penalty for using erasure codes in both reading and writing. For writes this slowdown was to be expected since each write operation creates a coding data overhead. When the achieved fault tolerance is taken into consideration the overhead appears insignificant.

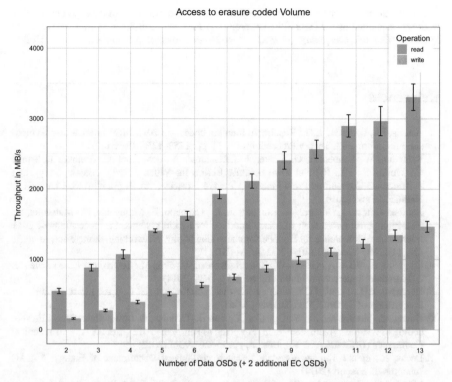

Fig. 8 Sequential read/write performance on an erasure coded XtreemFS volume with an increasing number of OSDs and clients

6 Summary

Checkpoint/Restart is a viable means to increase failure tolerance on supercomputers. We presented results on the implementation of a POSIX based checkpoint/restart mechanism. Checkpoints are stored in a RAM based distributed file system using XtreemFS. For fault tolerance checkpoints are encoded using erasure codes.

We evaluated our solution on a Cray XC40 with the quantum chromodynamics code *BQCD* which is already prepared for application-level checkpointing. XtreemFS provides three different clients: a FUSE based client, LD_PRELOAD and *libxtreemfs*. The first requires the FUSE kernel module to be loaded, which is typically not available on supercomputer environment. The last client, *libxtreemfs*, requires the application code to be modified and is therefore also not a good choice. The LD_PRELOAD client results in performance improvements for sequential access and extends the number of supported platforms and applications. The new client implementation transparently bypasses the operating system overhead by intercepting POSIX file system calls and redirecting them to *libxtreemfs*.

Acknowledgements We thank Johannes Dillmann who performed some of the experiments. This work was supported by the DFG SPPEXA project '*A Fast and Fault-Tolerant Microkernel-Based System for Exascale Computing*' (FFMK) and the North German Supercomputer Alliance HLRN.

References

1. Asteris, M., Dimakis, A.G.: Repairable fountain codes. In: 2012 IEEE International Symposium on Information Theory Proceedings (ISIT), pp. 1752–1756. IEEE (2012)
2. Baumann, W., Laubender, G., Läuter, M., Reinefeld, A., Schimmel, C., Steinke, T., Tuma, C., Wollny S.: HLRN-III at Zuse Institute Berlin. In: Vetter, J. (ed.) Contemporary High Performance Computing: From Petascale Toward Exascale, vol. 2, pp. 85–118. Chapman & Hall/CRC Press (2014)
3. Bautista-Gomez, L., Tsuboi, S., Komatitsch, D., Cappello, F., Maruyama, N., Matsuoka, S.: FTI: high performance fault tolerance interface for hybrid systems. In: Proceedings of 2011 International Conference for High Performance Computing, Networking, Storage and Analysis (SC'11), New York, pp. 32:1–32:32. ACM (2011)
4. Cappello, F., Geist, A., Gropp, W., Kale, S., Kramer, B., Snir, M.: Toward exascale resilience: 2014 update. Supercomput. Front. Innov. **1**(1), 1–28 (2014)
5. Hargrove, P.H., Duell, J.C.: Berkeley lab checkpoint/restart (BLCR) for Linux clusters. In: Proceedings of SciDAC 2006, Denver (2006)
6. Huang, C., Simitci, H., Xu, Y., Ogus, A., Calder, B., Gopalan, P., Li, J., Yekhanin, S.: Erasure coding in Windows Azure storage. In: Presented as Part of the 2012 USENIX Annual Technical Conference (USENIX ATC 12), Boston, pp. 15–26. ACM (2012)
7. Lucas, R., et al.: Top ten exascale research challenges. Department of Energy ASCAC subcommittee report (2014)
8. Moody, A., Bronevetsky, G., Mohror, K.K., de Supinski, B.R.: Design, modeling, and evaluation of a scalable multi-level checkpointing system. In: Proceedings of ACM/IEEE International Conference for High Performance Computing, Networking, Storage and Analysis (SC'10), New York. ACM (2010)
9. Mu, S., Chen, K., Wu, Y., Zheng, W.: When Paxos meets erasure code: reduce network and storage cost in state machine replication. In: Proceedings of the 23rd International Symposium on High-Performance Parallel and Distributed Computing (HPDC'14), New York, pp. 61–72. ACM (2014)
10. Nagle, D., Serenyi, D., Matthews, A.: The Panasas activescale storage cluster: delivering scalable high bandwidth storage. In: Proceedings of the SC'04, Pittsburgh, p. 53. ACM (2004). http://dl.acm.org/citation.cfm?id=1049998
11. Peter, K., Reinefeld, A.: Consistency and fault tolerance for erasure-coded distributed storage systems. In: Proceedings of the Fifth International Workshop on Data-Intensive Distributed Computing Date (DIDC'12), New York, pp. 23–32. ACM (2012)
12. Plank, J., Li, K.: Diskless checkpointing. IEEE Trans. Parallel Distrib. Syst. **9**(10), 972–986 (1998)
13. Plank, J.S., Simmerman, S., Schuman, C.D: Jerasure: a library in C facilitating erasure coding for storage applications. Technical report CS-07-603, University of Tennessee Department of Electrical Engineering and Computer Science (2007)
14. Rashmi, K.V., Shah, N.B., Gu, D., Kuang, H., Borthakur, D., Ramchandran, K.: A "Hitch-hiker's" guide to fast and efficient data reconstruction in erasure-coded data centers. SIG-COMM Comput. Commun. Rev. **44**(4), 331–342 (2014)
15. Rashmi, K.V., Nakkiran, P., Wang, J., Shah, N.B., Ramchandran, K.: Having your cake and eating it too: jointly optimal erasure codes for I/O, storage, and network-bandwidth. In: 13th USENIX Conference on File and Storage Technologies (FAST 15), Santa Clara, pp. 81–94. USENIX Association (2015)

16. Sathiamoorthy, M., Asteris, M., Papailiopoulos, D., Dimakis, A.G., Vadali, R., Chen, S., Borthakur, D.: XORing elephants: novel erasure codes for big data. Proc. VLDB Endow. **6**(5), 325–336 (2013)
17. Schmuck, F., Haskin, R.: GPFS: a shared-disk file system for large computing clusters. In: Proceedings of the USENIX FAST'02, Monterey. USENIX Association (2002)
18. Shvachko, K., Kuang, H., Radia, S., Chansler, R.: The Hadoop distributed file system. In: Proceedings of the 2010 IEEE 26th Symposium on Mass Storage Systems and Technologies (MSST) (MSST'10), Washington, DC, pp. 1–10. IEEE Computer Society (2010)
19. Stender, J., Berlin, M., Reinefeld, A.: XtreemFS – a file system for the cloud. In: Kyriazis, D., Voulodimos, A., Gogouvitis, S., Varvarigou, T. (eds.) Data Intensive Storage Services for Cloud Environments. IGI Global (2013)
20. Weil, S.A., Brandt, S.A., Miller, E.L., Long, D.D.E., Maltzahn, C.: Ceph: a scalable, high-performance distributed file system. In: 7th Symposium on Operating Systems Design and Implementation (OSDI'06), Seattle, pp. 307–320. ACM (2006)
21. Weinhold, C., Lackorzynski, A., Bierbaum, J., Küttler, M., Planeta, M., Härtig, H., Shiloh, A., Levy, E., Ben-Nun, T., Barak, A., Steinke, T., Schütt, T., Fajerski, J., Reinefeld, A., Lieber, M., Nagel, W.E.: FFMK: a fast and fault-tolerant microkernel-based system for exascale computing. In: Proceedings of SPPEXA Symposium, Garching. Springer (2016)
22. Wende, F., Steinke, T., Reinefeld, A.: The impact of process placement and oversubscription on application performance: a case study for exascale computing. In: Exascale Applications and Software Conference (ESAX-2015), Edinburgh (2015)
23. Zheng, G., Shi, L., Kalé, L.V.: FTC-Charm++: an in-memory checkpoint-based fault tolerant runtime for Charm++ and MPI. In: 2004 IEEE International Conference on Cluster Computing, San Diego, pp. 93–103. IEEE (2004)
24. Zheng, G., Ni, X., Kalé, L.V.: A scalable double in-memory checkpoint and restart scheme towards exascale. In: Proceedings of the 2nd Workshop on Fault-Tolerance for HPC at Extreme Scale (FTXS), Boston, pp. 1–6. IEEE (2012)

Part XI
CATWALK: A Quick Development Path for Performance Models

Automatic Performance Modeling of HPC Applications

Felix Wolf, Christian Bischof, Alexandru Calotoiu, Torsten Hoefler, Christian Iwainsky, Grzegorz Kwasniewski, Bernd Mohr, Sergei Shudler, Alexandre Strube, Andreas Vogel, and Gabriel Wittum

Abstract Many existing applications suffer from inherent scalability limitations that will prevent them from running at exascale. Current tuning practices, which rely on diagnostic experiments, have drawbacks because (i) they detect scalability problems relatively late in the development process when major effort has already been invested into an inadequate solution and (ii) they incur the extra cost of potentially numerous full-scale experiments. Analytical performance models, in contrast, allow application developers to address performance issues already during the design or prototyping phase. Unfortunately, the difficulties of creating such models combined with the lack of appropriate tool support still render performance modeling an esoteric discipline mastered only by a relatively small community of experts. This article summarizes the results of the Catwalk project, which aimed to create tools that automate key activities of the performance modeling process, making this powerful methodology accessible to a wider audience of HPC application developers.

F. Wolf • C. Bischof • A. Calotoiu (✉) • C. Iwainsky • S. Shudler
Technische Universität Darmstadt, Darmstadt, Germany
e-mail: wolf@cs.tu-darmstadt.de; christian.bischof@cs.tu-darmstadt.de;
calotoiu@cs.tu-darmstadt.de; christian.iwainsky@sc.tu-darmstadt.de;
shudler@cs.tu-darmstadt.de

T. Hoefler • G. Kwasniewski
ETH Zurich, Zurich, Switzerland
e-mail: htor@inf.ethz.ch; gkwasnie@inf.ethz.ch

B. Mohr • A. Strube
Jülich Supercomputing Center, Juelich, Germany
e-mail: b.mohr@fz-juelich.de; a.strube@fz-juelich.de

A. Vogel • G. Wittum
Goethe Universität Frankfurt, Frankfurt, Germany
e-mail: andreas.vogel@gcsc.uni-frankfurt.de; wittum@gcsc.uni-frankfurt.de

© Springer International Publishing Switzerland 2016   445
H.-J. Bungartz et al. (eds.), *Software for Exascale Computing – SPPEXA*
2013-2015, Lecture Notes in Computational Science and Engineering 113,
DOI 10.1007/978-3-319-40528-5_20

1 Motivation

When scaling their codes to larger numbers of processors, many HPC application developers face the situation that all of a sudden a part of the program starts consuming an excessive amount of time. Unfortunately, discovering latent scalability bottlenecks through experience is painful and expensive. Removing them requires not only potentially numerous large-scale experiments to track them down, prolonged by the scalability issue at hand, but often also major code surgery in the aftermath. All too often, this happens at a moment when the manpower is needed elsewhere. This is especially true for applications on the path to exascale, which have to address numerous technical challenges simultaneously, ranging from heterogeneous computing to resilience. Since such problems usually emerge at a later stage of the development process, dependencies between their source and the rest of the code that have grown over time can make remediation even harder. One way of finding scalability bottlenecks earlier is through analytical performance modeling. An analytical scalability model expresses the execution time or other resources needed to complete the program as a function of the number of processors. Unfortunately, the laws according to which the resources needed by the code change as the number of processors increases are often laborious to infer and may also vary significantly across individual parts of complex modular programs. This is why analytical performance modeling—in spite of its potential—is rarely used to predict the scaling behavior before problems manifest themselves. As a consequence, this technique is still confined to a small community of experts.

If today developers decide to model the scalability of their code, and many shy away from the effort [19], they first apply both intuition and tests at smaller scales to identify so-called *kernels*, which are those parts of the program that are expected to dominate its performance at larger scales. This step is essential because modeling a full application with hundreds of modules manually is not feasible. Then they apply reasoning in a time-consuming process to create analytical models that describe the scaling behavior of their kernels more precisely. In a way, they have to solve a chicken-and-egg problem: to find the right kernels, they require a pre-existing notion of which parts of the program will dominate its behavior at scale—basically a model of their performance. However, they do not have enough time to develop models for more than a few pre-selected candidate kernels, inevitably exposing themselves to the danger of overlooking unscalable code.

This article summarizes the results of the Catwalk project, which set out to improve this situation. The main goal of Catwalk was to make performance modeling more attractive for a broader audience of HPC application developers by providing tools that support key activities of the modeling process in a simpler and more intuitive way than existing tools.

2 Overview of Contributions

From a functional perspective, the most important goal was to give a good estimation of relative performance between different parts of a program when scaled to larger processor configurations. In this way, scalability bottlenecks of applications can be anticipated at a very early stage long before they become manifest in actual measurements on the same or on future platforms. From a non-functional perspective, we strove to maximize flexibility and ease of use. The latter is especially important—given that a larger fraction of codes is still exclusively developed by domain scientists as opposed to multidisciplinary teams with dedicated performance engineers. Our main accomplishments, which are illustrated in Fig. 1, can be summarized as follows:

- A method to automatically generate scaling models of applications from a limited set of profile measurements, which allows the quick identification of scalability bugs even in very complex codes with thousands of functions [10]. The method was integrated into the production-level performance-analysis tool Scalasca (Fig. 3).
- Numerous application case studies, in which we either confirm earlier studies with hand-crafted performance models or discover the existence of previously unknown scalability bottlenecks. This series of studies includes one with UG4 [40], an unstructured-grid package developed by one of our partners.

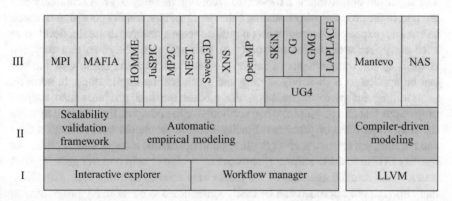

Fig. 1 Main accomplishments of Catwalk are separated into three tiers: **(I)** Infrastructure; **(II)** Core methods; **(III)** Application case studies. The core methods we developed include both an automatic empirical method and compiler-driven methods. The empirical method rests on infrastructure components developed by project partners, the compiler-driven methods leverage LLVM, an external open-source compiler infrastructure. The empirical method was later extended to also allow scalability validation of codes with known theoretical expectations. Both sets of methods have been successfully applied in a number of application studies. The unstructured-grid package UG4 *(highlighted)* is developed by one of the project partners, all other test cases were external to the project team

- A scalability analysis of several state-of-the-art OpenMP implementations, in which we highlight scalability limitations in some of them [22].
- A scalability test framework that combines the above method with performance expectations to systematically validate the scalability of libraries [32]. Using this framework, we conducted an analysis of several state-of-the-art MPI libraries, in which we identify scalability issues in some of them.
- A compilation and modeling framework that automatically instruments applications to generate performance models during program execution [7]. These automatically generated performance models can be used to quickly assess the scaling behavior and potential bottlenecks with regard to any input parameter and the number of processes of a parallel application.
- A fully static analysis technique to derive numbers of loop iterations from loops with affine loop guards and transfer functions [21]. This method can be used to limit the performance model search space and increase the accuracy of performance modeling in general.

In the remainder of the article, we describe the above contributions in more detail, followed by a review of related work.

3   Automatic Empirical Performance Modeling

The key result of Catwalk is a method to identify *scalability bugs*. A scalability bug is a part of the program whose scaling behavior is unintentionally poor, that is, much worse than expected. As computing hardware moves towards exascale, developers need early feedback on the scalability of their software design so that they can adapt it to the requirements of larger problem and machine sizes. Our method can be applied to both strong scaling and weak scaling applications. In addition to searching for performance bugs, the models our tool produces also support projections that can be helpful when applying for the compute time needed to solve the next larger class of problems. Finally, because we model both execution time and requirement metrics such as floating-point operations alongside each other, our results can also assist in software-hardware co-design or help uncover growing wait states. For a detailed description, the reader may refer to Calotoiu et al. [10]. Note that although our approach can be easily generalized to cover many programming models, we initially focused on MPI programs, later also on OpenMP programs.

The input of our tool is a set of performance measurements on different processor counts $\{p_1, \ldots, p_{max}\}$ in the form of parallel profiles. As a rule of thumb, we use five or six different configurations. The execution of these experiments is supported by a workflow manager. The output of our tool is a list of program regions, ranked by their predicted execution time at a chosen target scale or by their asymptotic execution time. We call these regions *kernels* because they define the code granularity at which we generate our models.

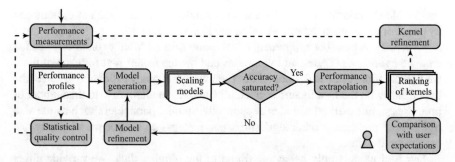

Fig. 2 Workflow of scalability-bug detection. *Solid boxes* represent actions or transformations, and *banners* their inputs and outputs. *Dashed arrows* indicate optional paths taken after user decisions

Figure 2 gives an overview of the different steps necessary to find scalability bugs. To ensure a statistically relevant set of performance data, profile measurements may have to be repeated several times—at least on systems subject to jitter (usually between three and five times). This is done in the optional statistical quality control step. Once this is accomplished, we apply regression to obtain a coarse performance model for every possible program region. These models then undergo an iterative refinement process until the model quality has reached a saturation point. To arrange the program regions in a ranked list, we extrapolate the performance either to a specific target scale p_t or to infinity, which means we use the asymptotic behavior as the basis of our comparison. A scalability bug can be any region with a model worse than a given threshold, such as anything scaling worse than linearly. Alternatively, a user can compare the model of a kernel with his own expectations to determine if the performance is worse than expected. Finally, if the granularity of our program regions is not sufficient to arrive at an actionable recommendation, performance measurements, and thus the kernels under investigation, can be further refined via more detailed instrumentation.

Model Generation When generating performance models, we exploit the observation that they are usually composed of a finite number n of predefined terms, involving powers and logarithms of p:

$$f(p) = \sum_{k=1}^{n} c_k \cdot p^{i_k} \cdot log_2^{j_k}(p) \,. \tag{1}$$

This representation is, of course, not exhaustive, but works in most practical scenarios since it is a consequence of how most computer algorithms are designed. We call it the *performance model normal form* (PMNF). Moreover, our experience suggests that neither the sets $I, J \subset \mathbb{Q}$ from which the exponents i_k and j_k are chosen nor the number of terms n have to be arbitrarily large or random to achieve a good fit. Thus, instead of deriving the models through reasoning, we only need to make reasonable choices for n, I, and J and then simply try all assignment options one by one. For example, a default we often use is $n = 3, I = \{\frac{0}{2}, \frac{1}{2}, \frac{2}{2}, \frac{3}{2}, \frac{4}{2}, \frac{5}{2}, \frac{6}{2}\}$, and $J = \{0, 1, 2\}$. A possible assignment of all i_k and j_k in a PMNF expression is called a *model hypothesis*. Trying all hypotheses one by one means that for each of them we find coefficients c_k with optimal fit. Then we apply cross-validation to select the hypothesis with the best fit across all candidates. As an alternative to the number of processes p, our method can also support other model parameters such as the size of the input problem or other algorithmic parameters—as long as we vary only one parameter at a time.

Our tool models only behaviors found in the training data. We provide direct feedback information regarding the number of runs required to ensure statistical significance of the modeling process itself, but there is no automatic way of determining at what scale particular behaviors start manifesting themselves. We expect that our method will be most effective for regular problems with repetitive behavior, whereas irregular problems with strong and potentially non-deterministic dynamic effects will require enhancements.

Integration into Scalasca The model generator has been integrated under the name *Extra-P* into Scalasca [14], a widely used performance analysis tool set. Figure 3 shows how the results of the model generator can be interactively explored. Instead of measured values such as the time spent in a particular call path, the GUI now annotates each call path with a performance model. The formula represents a previously selected metric as a function of the number of processes. The user can select one or more call paths and plot their models on the right. In this way, the user can visually compare the scalability of different application kernels.

The profiles needed as input for the model generator are created in a series of performance experiments. To relieve the user from the burden of manually submit-

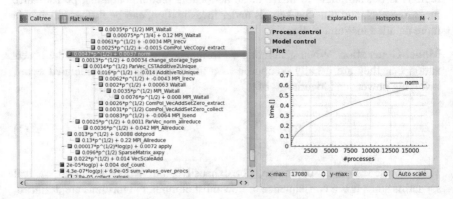

Fig. 3 Interactive exploration of performance models in Scalasca with Extra-P. The screen shot shows performance models generated for call paths in UG4, a multigrid framework developed by one of the partners. The call tree on the *left* allows the selection of models to be plotted on the *right*. The *color of the squares* in front of each call path highlights the complexity class. The call path ending in kernel norm has a measured complexity of $O(\sqrt{p})$

ting large numbers of jobs and collating their results, we use the Jülich Benchmark Environment (JUBE) [24], a workflow manager developed at Forschungszentrum Jülich.

The Extra-P performance-modeling tool has been made available online under an open-source license.[1] Users have access not only to the software but also to documentation material describing both our method and its implementation.[2] We are already in contact with researchers at several organizations, either planing to use Extra-P (High Performance Computing Center Stuttgart and TU Darmstadt) or actively using it (Lawrence Livermore National Laboratory and University of Washington). Moreover, the tool was presented at two conference tutorials, one at EuroMPI 2015 in Bordeaux, France and one at SC15 in Austin, Texas. Following a 90 min theoretical explanation of the method and the tool, users were able to model the performance of two example applications, SWEEP3D and BLAST, in a 90 min practical session. Using previously prepared measurement data, they were able to generate models for the entire codes, evaluate the results, and understand the scaling behavior of the two applications. With this knowledge, attendees are able to apply Extra-P to their own applications, once the required performance measurements have been gathered. Because Extra-P is compatible with Score-P, an established infrastructure for performance profiling, even collecting these measurements is straightforward.

Application Case Studies We analyzed numerous real-world applications from various fields, including HOMME [13] (climate), JuSPIC [25] (particle in cell), MILC [3] (quantum chromodynamics), MP2C [35] (soft matter physics and hydrodynamics), NEST [15] (brain), Sweep3D [45] (neutron transport), UG4 [39] (unstructured grids), and XNS [4] (fluid dynamics).

The worst-scaling kernels of our test cases are shown in Table 1. We were able to identify a scalability issue in codes that are known to have such issues (Sweep3D, XNS) but did not identify any scalability issue in codes that are known to have none (MILC, juSPIC, MP2C, NEST). In the cases of Sweep3D and Milc, we were able to confirm hand-crafted models reported in the literature [20, 43]. Moreover, we were able to identify two scalability issues in codes that were thought to have only one (HOMME, UG4). Cases which present potential scalability bottlenecks are marked with a danger sign in Table 1. In the other cases, the worst scaling kernels are not deemed likely to become scalability bottlenecks even at extreme scales, due to either slow growth rates, small coefficients, or a combination of both. Below, we discuss two applications in more detail and present another case study involving several state-of-the-art OpenMP runtime systems.

Case Study: UG4 UG4 is a package for unstructured grids developed by Goethe University Frankfurt, one of the project partners. The first bottleneck we found was previously unknown and occurs when UG4 defines subcommunicators depending

---

[1] http://www.scalasca.org/software/extra-p/download.html

[2] http://www.scalasca.org/software/extra-p/documentation.html

Table 1 Performance modeling case studies summary presenting the worst-scaling kernels for the analyzed metrics. In a number of cases a class of worst-scaling kernels is found rather than one specific kernel, indicated in the Kernel(s) column by using the label 'Multiple'. p denotes the number of processes and n the problem size. The *danger sign* indicates behavior likely to become a bottleneck at larger scales

Application	Metric	Kernel(s)	Model	Scalability bottleneck
Process scaling—weak (avg. across all processes)				
HOMME	Execution time	box_rearrange $\rightarrow$ MPI_Reduce	$10^{-12} \cdot p^3$	⚠
		Multiple	$10^{-7} \cdot p^2$	⚠
Sweep3D	Execution time	sweep $\rightarrow$ MPI_Recv	$\sqrt{p}$	⚠
UG4	bytes sent	InitLevels $\rightarrow$ MPI_AllReduce	$80 \cdot p$	⚠
	Execution time	CG::norm	$10^{-3} \cdot \sqrt{p}$	⚠
	Execution time	GMG::smooth	$10^{-2} \cdot \log p$	
	FLOPs	Multiple	$\log p$	
MILC	Execution time	g_vecdoublesum $\rightarrow$ MPI_AllReduce	$10^{-6} \cdot \log^2 p$	
	FLOPs	Multiple	$10^3 \cdot \log p$	
NEST	FLOPs	Multiple	$10^{-3} \cdot p^2$	
JuSPIC	FLOPs	Multiple	$10^5 \cdot p$	
MP2C	FLOPs	Multiple	$10 \cdot \log p$	
Process scaling—strong (sum across all processes)				
XNS	Execution time	ewdgennprm $\rightarrow$ MPI_Recv	$10^{-2} \cdot p^2$	⚠
Problem scaling—#processes constant (avg. across all processes)				
MILC	FLOPs	Multiple	$10^3 \cdot n \log n$	
MP2C	FLOPs	Multiple	$10^2 \cdot n$	
JuSPIC	FLOPs	Multiple	$10^9 \cdot n$	
NEST	FLOPs	Multiple	$10^2 \cdot n$	

on whether a process holds a part of the multigrid hierarchy or not. The second bottleneck is a known weakness of the CG method, one of the solver options UG4 provides, whose iteration count increases by a factor of two with each grid refinement. A detailed discussion of this case study is presented in Vogel et al. [40].

Case Study: HOMME To showcase how our tool helps to find hidden scalability bugs in a production code for which no performance model was available, we applied it to HOMME [13], the dynamical core of the Community Atmospheric Model (CAM) being developed at the National Center for Atmospheric Research. HOMME, which was designed with scalability in mind, employs spectral element and discontinuous Galerkin methods on a cubed sphere tiled with quadrilateral elements.

Table 1 lists two issues discovered in the code. Multiple kernels show a dependence of p^2 (with a small factor). After looking at the number of times any of the quadratic kernels was invoked at runtime, a metric we also measure and model, the quadratic growth was found to be the consequence of an increasing number of iterations inside a particular subroutine. The developers were aware of this issue and had already developed a temporary solution, involving manual adjustments of their production code configurations.

In contrast to the previous problem, the cubic growth of the time spent in the reduce function was previously unknown. The reduction is needed to funnel data to dedicated I/O processes. The reason why this phenomenon remained unnoticed until today is that it belongs to the initialization phase of the code that was not assumed to be performance relevant in larger production runs. While still not yet crippling in terms of the overall runtime, which is in the order of days for production runs, the issue already cost more than 1 h in the large-scale experiments we conducted. The example demonstrates the advantage of modeling the entire application instead of only selected candidate kernels expected to be time intensive. p^3 was the limit of our search space but we argue that the growth rate presented by the cubic power already constitutes a significant scalability issue that needs immediate addressing, and as such trying to categorize severity beyond p^3 would offer diminishing returns.

Figure 4 summarizes our two findings and compares our predictions with actual measurements. While the quadratically growing iteration count seems to be more urgent now, the reduce might become the more serious issue in the future.

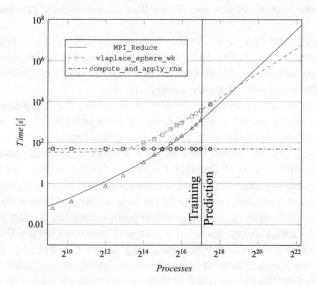

Fig. 4 Runtime of selected kernels in HOMME as a function of the number of processes. The graph compares predictions (*dashed or contiguous lines*) to measurements (*small triangles, squares, and circles*). The vlaplace_sphere_wk kernel is part of the group which scales quadratically, MPI_Reduce is the initialization issue that grows cubically, while compute_and_apply_rhs is a representative of the majority of constant computation kernels

Case Study: OpenMP Exascale systems will exhibit much higher degrees of parallelism not only in terms of the number of nodes but also in terms of the number of cores per node. OpenMP is a widely used standard for exploiting parallelism on the level of individual nodes. Here, not only the software must scale to the upcoming thread counts, but also the supporting OpenMP runtimes must scale. Hence, we investigated widely-used OpenMP implementations on several hardware-platforms to determine potential scalability limitations of various OpenMP constructs. Here we report our findings for an Intel Xeon multi-board system with 128 hardware threads, called *BCS*, an Intel Xeon Phi system with 240 threads and an IBM BlueGene/Q node.

Given the jitter present in our measurements, we only allowed one active term plus a constant and refrained from modeling behaviors past the leading term.

We ran our tests with the EPCC OpenMP micro-benchmark suite [9], a comprehensive collection of benchmarks that covers almost all OpenMP constructs. We modified the ECPP measurement system to directly interface with our model generator. Since the EPCC benchmarks do not measure the costs of individual OpenMP constructs directly, the resulting timings are much more prone to noise. To measure the costs of OpenMP constructs in isolation, we therefore had to develop additional benchmarks that allow the time spent in an OpenMP construct to be measured on a per-thread basis.

In our initial measurements, we observed strongly fluctuating runtimes when increasing the number of threads. These fluctuations initially prevented a good model fit.

We then realized that the observable runtimes showed a likely superposition of multiple behaviors, depending on how many threads were used. Dividing the set of thread counts into subsets finally enabled us to separate these behaviors and create stable runtime models for OpenMP constructs.

Table 2 shows the resulting performance models. Note that many models exhibit quadratic or higher growth rates, which can at some point become a considerable cost factor, especially if the trend towards fine-grained parallelism continues. For the sake of brevity, we restrict the detailed discussion to the BCS node. In many cases, the execution time of OpenMP constructs showed unfavorable growth and numerous scalability issues with current implementations became apparent. Neither of the evaluated compilers proved to be the best implementation in all situations. The Intel compiler showed the best absolute performance and scaling behavior for most of the metrics in our tests, but it was still surpassed by the PGI compiler on two occasions. Considering the increasing degree of intra-node parallelism, OpenMP compilers will have to tackle theses scalability issues in the future. Our benchmarking method is designed to support this process, as it can be used to continuously evaluate implementations as their scalability is improved.

Table 2 Runtime scaling models for entering a parallel region with thread creation, barrier, parallel for with different loop schedules, and firstprivate initialization. Models are presented separately for different subsets of thread counts (PO2, ODD, LINEAR, 8X) all in spread configuration. Measurements with a † were generated using EPCC, measurements with ⋆ were generated using our supplemental benchmarks. Each row showing models is followed by a row with the corresponding adjusted coefficient of determination ($\hat{R}^2$) as a quality indicator. Since we are only interested in the general scaling trend, we show only the leading terms and their coefficient rounded to powers of ten. The models describe the growth of the runtime in seconds

	Parallel (open)⋆	Barrier⋆	Dynamic 16†	Static 16†	Guided 16†	Firstprivate†
BCS – GNU						
PO2	$10^{-7} \cdot x^{1.25}$	$10^{-7} \cdot x^{1.33} \log x$	$10^{-7} \cdot x^{1.25} \log x$	$10^{-8} \cdot x^{1.33} \log x$	$10^{-6} \cdot x^{0.75} \log x$	$10^{-6} \cdot x$
$\hat{R}^2$	0.99	0.99	0.99	0.98	0.99	0.99
ODD	$10^{-6} \cdot x^{0.67}$	$10^{-5} \cdot x^{0.5}$	$10^{-5} \cdot x^{0.67} \log x$	$10^{-8} \cdot x^{1.25} \log x$	$10^{-5} \cdot x^{0.5} \log x$	$10^{-5} \cdot x^{0.5} \log x$
$\hat{R}^2$	0.95	0.93	0.96	0.94	0.93	0.98
BCS – Intel						
PO2	$10^{-6} \cdot \log x$	$10^{-5} \cdot x^{0.25}$	$10^{-6} \cdot x$	$10^{-6} \cdot \log x$	$10^{-5} \cdot x$	$10^{-6} \cdot \log x$
$\hat{R}^2$	0.78	0.98	0.99	0.84	0.99	0.94
BCS – PGI						
PO2	$10^{-6} \cdot x^{0.67} \log x$	$10^{-6} \cdot \log^2 x$	$10^{-6} \cdot x^{1.25} \log x$	$10^{-6} \cdot \log x$	$10^{-6} \cdot x^{1.67}$	$10^{-5} \cdot x^{0.67}$
$\hat{R}^2$	0.99	0.95	0.99	0.62	0.99	0.99
ODD	$10^{-7} \cdot x \log x$	$10^{-6} \cdot x^{0.5} \log x$	$10^{-7} \cdot x^{1.5} \log x$	$10^{-11} \cdot x^{2.5}$	$10^{-6} \cdot x^{1.67}$	$10^{-6} \cdot x^{0.5} \log x$
$\hat{R}^2$	0.97	0.90	0.99	0.50	0.99	0.89
XeonPhi – Intel						
PO2	$10^{-7} \cdot x^{0.67}$	$10^{-6} \cdot x^{0.5}$	$10^{-5} \cdot x^{0.25}$	$10^{-8} \cdot x^{1.5}$	$10^{-6} \cdot x$	$10^{-6} \cdot x^{0.67}$
$\hat{R}^2$	0.97	0.99	0.65	0.75	0.99	0.98
LINEAR	$10^{-7} \cdot x^{0.67}$	$10^{-6} \cdot x^{0.5}$	$10^{-6} \cdot \log x$	$10^{-9} \cdot x^{2.33}$	$10^{-7} \cdot x$	$10^{-7} \cdot x^{0.67}$
$\hat{R}^2$	0.95	0.94	0.55	0.30	0.99	0.96
8X	$10^{-7} \cdot \log^2 x$	$10^{-6} \cdot \log x$	$10^{-8} \cdot x^{1.25} \log x$	$10^{-7} \cdot x^{0.75} \log x$	$10^{-7} \cdot x$	$10^{-7} \cdot \log^2 x$
$\hat{R}^2$	0.95	0.86	0.92	0.70	0.99	0.94
Blue Gene/Q – IBM XL						
PO2	$10^{-7} \cdot x^{1.25}$	$10^{-9} \cdot x^{1.33} \log x$	$10^{-8} \cdot x^{2.33}$	$10^{-8} \cdot x^{2.33}$	$10^{-8} \cdot x^2$	$10^{-7} \cdot x^{1.25}$
$\hat{R}^2$	0.99	0.99	0.99	0.99	0.99	0.99

4 Scalability Validation Framework

Library developers are confronted with the challenge of comparing the actual scalability of their code base with their expectations. In cases where the library encapsulates complex algorithms that are the product of years of research, such expectations often exist in the form of analytical performance models [12, 37, 41]. However, translating such abstract models into expressions that can be verified in performance experiments is hard because it requires knowing all constants and restricts function domains to performance metrics that are effectively measurable on the target system. If only the asymptotic complexity is known, a very common case, this is even impossible. And if such a verifiable expression exists, it must be adapted every time the test platform is replaced and performance metrics and constants change.

To mitigate this situation, we combined our modeling approach with performance expectations in a novel scalability test framework [32]. Similar to performance assertions [38], our framework supports the user in the specification and validation of performance expectations. However, rather than formulating precise analytical expressions involving measurable metrics, the user would just provide the asymptotic growth rate of the function/metric pair in question, making this a simple but effective solution for future exascale library development. The goal of our scalability framework is to provide insights into the scaling behavior of a library with as little effort as possible. The framework, which is illustrated in Fig. 5, allows the user

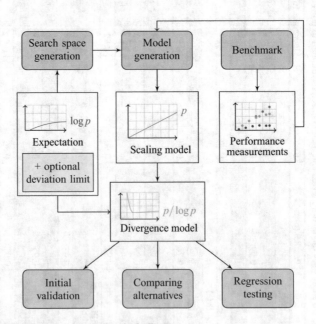

Fig. 5 Scalability framework overview including use cases

to evaluate whether the observed behavior of libraries corresponds to the expected behavior. Although the approach relies on automated performance modeling, it is different from our earlier work in that it assumes that the user provides the expected model and then constructs the model search space around the provided expectation. This enables the scalability framework to support models beyond the PMNF introduced in Eq. (1) in Sect. 3. Use cases include initial validation, regression testing, and benchmarking to compare implementation and platform alternatives.

The scalability framework workflow consists of four phases: (i) *define expectations*, (ii) *design benchmark*; (iii) *generate scaling models*; (iv) *validate expectations*. In the last phase, the framework calculates *divergence models* by dividing the expected models by the generated ones. A divergence model characterizes how severe the divergence of the expectation from the observed behavior is. In this phase, the framework also ranks the generated model as a full match, approximate match, or no match by checking whether the model falls within previously specified deviation limits. One of the strongest advantages of the scalability framework is that we provide a tool-chain that automates large parts of our four-phase workflow and is ready for immediate use by performance engineers. Because MPI is a fundamental building block in most HPC applications, we conducted a detailed case study involving several MPI implementations, which is summarized below.

Case Study: MPI MPI is probably one of the most widely used HPC libraries. It has clear performance expectations, as well as many commercially mature and well-tested implementations. Based on these important factors we chose it as our main case study for the scalability evaluation framework. Following the four phases of the framework workflow, we started with the definition of the expectations. Our focus is the runtime of collective operations and memory consumption since they are two of the major potential scalability obstacles in MPI. We specifically looked at the runtime of MPI_Barrier, MPI_Bcast, MPI_Reduce, MPI_Allreduce, MPI_Gather, MPI_Allgather, and MPI_Alltoall. We also measured the memory overhead of MPI_Comm_create, MPI_Comm_dup, MPI_Win_create, and MPI_Cart_create, as well as the estimated MPI memory consumption. To complete the second phase, we designed a benchmark to measure the runtimes and memory overheads for increasing numbers of MPI processes. The last two phases were automatic and produced the performance models along with the divergence models.

We ran the experiments and tested the scalability framework on three machines. The first one was Juqueen, a Blue Gene/Q machine built by IBM. It is the capability supercomputer at Forschungszentrum Jülich (FZJ). The second one was Juropa, which was at the time of the study the capacity machine at FZJ and which is based on Intel architecture, and the third machine was Piz Daint, an x86-based Cray-XC30 machine at the Swiss National Supercomputing Centre (CSCS). Tables 3 and 4 present the results of the experiments. Each table has the same structure in which the rows are divided into three compartments corresponding to Juqueen, Juropa, and Piz Daint results, respectively. Each compartment shows the generated model,

Table 3 Generated (empirical) runtime models of MPI collective operations on Juqueen, Juropa, and Piz Daint alongside their expected models

	Barrier	Bcast	Reduce	Allreduce	Gather	Allgather	Alltoall
Juqueen							
Expectation	$\mathcal{O}(\log p)$	$\mathcal{O}(\log p)$	$\mathcal{O}(\log p)$	$\mathcal{O}(\log p)$	$\mathcal{O}(p)$	$\mathcal{O}(p)$	$\mathcal{O}(p \log p)$
Model	$\mathcal{O}(\log p)$	$\mathcal{O}(\log p)$	$\mathcal{O}(\log p)$	$\mathcal{O}(\log p)$	$\mathcal{O}(p)$	$\mathcal{O}(p)$	$\mathcal{O}(p)$
$\bar{R}^2$	0.99	0.86	0.93	0.87	0.99	0.99	0.99
$\delta(p)$	$\mathcal{O}(1)$	$\mathcal{O}(1)$	$\mathcal{O}(1)$	$\mathcal{O}(1)$	$\mathcal{O}(1)$	$\mathcal{O}(1)$	$\mathcal{O}(\frac{1}{\log p})$
Match	✓	✓	✓	✓	✓	✓	≈
Juropa							
Expectation	$\mathcal{O}(\log p)$	$\mathcal{O}(\log p)$	$\mathcal{O}(\log p)$	$\mathcal{O}(\log p)$	$\mathcal{O}(p)$	$\mathcal{O}(p)$	$\mathcal{O}(p \log p)$
Model	$\mathcal{O}(p^{0.67} \log p)$	$\mathcal{O}(\sqrt{p})$	$\mathcal{O}(\sqrt{p} \log p)$	$\mathcal{O}(\sqrt{p})$	$\mathcal{O}(p)$	$\mathcal{O}(p)$	$\mathcal{O}(p^{1.25})$
$\bar{R}^2$	0.99	0.98	0.99	0.99	0.99	0.98	0.99
$\delta(p)$	$\mathcal{O}(p^{0.67})$	$\mathcal{O}(\frac{\sqrt{p}}{\log p})$	$\mathcal{O}(\sqrt{p})$	$\mathcal{O}(\frac{\sqrt{p}}{\log p})$	$\mathcal{O}(1)$	$\mathcal{O}(1)$	$\mathcal{O}(\frac{p^{0.25}}{\log p})$
Match	✗	≈	≈	≈	✓	✓	≈
Piz Daint							
Expectation	$\mathcal{O}(\log p)$	$\mathcal{O}(\log p)$	$\mathcal{O}(\log p)$	$\mathcal{O}(\log p)$	$\mathcal{O}(p)$	$\mathcal{O}(p)$	$\mathcal{O}(p \log p)$
Model	$\mathcal{O}(p^{0.33})$	$\mathcal{O}(\sqrt{p})$	$\mathcal{O}(\sqrt{p} \log p)$	$\mathcal{O}(p^{0.67} \log p)$	$\mathcal{O}(p)$	$\mathcal{O}(p^{1.25})$	$\mathcal{O}(p^{1.33})$
$\bar{R}^2$	0.99	0.94	0.94	0.99	0.99	0.99	0.99
$\delta(p)$	$\mathcal{O}(\frac{p^{0.33}}{\log p})$	$\mathcal{O}(\frac{\sqrt{p}}{\log p})$	$\mathcal{O}(\sqrt{p})$	$\mathcal{O}(p^{0.67})$	$\mathcal{O}(1)$	$\mathcal{O}(p^{0.25})$	$\mathcal{O}(\frac{p^{0.33}}{\log p})$
Match	≈	≈	≈	✗ ⚠	✓	≈	≈

Table 4 Generated (empirical) runtime models of memory overheads on Juqueen, Juropa, and Piz Daint alongside their expected models. The memory sizes for all the cases, except MPI memory, are specified in bytes

	MPI memory [MB]	Comm_create	Comm_dup	Win_create	Cart_create
Juqueen					
Expectation	$\mathcal{O}(1)$	$\mathcal{O}(p)$	$\mathcal{O}(1)$	$\mathcal{O}(p)$	$\mathcal{O}(p)$
Model	$10.7 \cdot 10^{-3} \cdot \log p$	$2.2 \cdot 10^5 + 24 \cdot p$	$2.2 \cdot 10^5$	$96 \cdot p$	$2.2 \cdot 10^5 + 52 \cdot p$
$\bar{R}^2$	0.72	1	–	1	0.99
$\delta(p)$	$\mathcal{O}(\log p)$	$\mathcal{O}(1)$	$\mathcal{O}(1)$	$\mathcal{O}(1)$	$\mathcal{O}(1)$
Match	≈	✓	✓	✓	✓
Juropa					
Expectation	$\mathcal{O}(1)$	$\mathcal{O}(p)$	$\mathcal{O}(1)$	$\mathcal{O}(p)$	$\mathcal{O}(p)$
Model	$16 + 55 \cdot p$	$264 + 28 \cdot p$	256	$256 + 60 \cdot p$	$356 + 24 \cdot p$
$\bar{R}^2$	1	1	–	1	1
$\delta(p)$	$\mathcal{O}(p)$	$\mathcal{O}(1)$	$\mathcal{O}(1)$	$\mathcal{O}(1)$	$\mathcal{O}(1)$
Match	✗	✓	✓	✓	✓
Piz Daint					
Expectation	$\mathcal{O}(1)$	$\mathcal{O}(p)$	$\mathcal{O}(1)$	$\mathcal{O}(p)$	$\mathcal{O}(p)$
Model	$46 + 1.35 \cdot \log p$	$3770 + 46 \cdot p$	$3770 + 18 \cdot p$	$3287 + 118 \cdot p$	$2545 + 63 \cdot p$
$\bar{R}^2$	0.23	0.99	0.99	0.99	0.99
$\delta(p)$	$\mathcal{O}(\log p)$	$\mathcal{O}(1)$	$\mathcal{O}(p)$	$\mathcal{O}(1)$	$\mathcal{O}(1)$
Match	≈	✓	✗	✓	✓

the adjusted coefficient of determination $\bar{R}^2$ (a standard statistical fit factor), the divergence model $\delta(p)$, and the match classification. A checkmark $\checkmark$ indicates total match, a $\approx$ shows that the generated model falls within the deviation limits, and a solid **x** represents unquestionable mismatch. On Juqueen, the runtime of collective operations was generally better than on the other machines, and we found that almost all of our expectations were met. The same is true for communicator memory overheads. On Juropa and Piz Daint, on the other hand, there were far greater discrepancies between expected and observed behavior, which suggest potential scalability issue. We can highlight three particular issues:

1. MPI memory consumption on Juropa is linear in the number of processes, and thus prohibitive.
2. Communicator duplication (MPI_Comm_dup) on Piz Daint is linear in the number of processes, whereas it is constant in other machines.
3. The runtime of MPI_Allreduce on Piz Daint is predicted to be slower than the combination of MPI_Reduce and MPI_Bcast (marked by a warning sign next to the match classification).

In conclusion, the MPI case study shows the effectiveness of the scalability framework, which allows us to identify and pinpoint scalability issues in thought-to-be-scalable libraries. The use case demonstrates how MPI developers can use our framework to spot scalability bugs early on, before commencing full-scale tests on a target supercomputer.

In addition to MPI, we also tested the scalability framework on the MAFIA (Merging of Adaptive Finite IntervAls) code, a sequential data-mining program utilizing a collection of key routines. The model parameter in this case was the cluster dimensionality, thereby providing an example of algorithmic modeling and with exponential models outside the scope of the PMNF. For further details, please refer to Shudler et al. [32].

5   Compiler-Driven Performance Modeling

Another possibility is to include the source code into the performance-model generation. We developed a static technique that can handle affine loop nests and then added a statically improved dynamic method to model the remainder of the program. The source-code analysis provides crucial insights—it can find relations between a program's kernels, provide analytical execution models, and determine the influence of input parameters. This information is computed statically and passed on to the dynamic analyzer, which otherwise either would not be able to determine such information, or the results would be imprecise or computationally expensive. We showed that a static analysis is a powerful tool alone [18] and combined with a dynamic approach increases both precision and performance [6].

Here we present a short summary of our work. The figures and examples are the same as in the aforementioned papers.

Static Analysis Our static analysis focuses on loop modeling. In scientific applications, most of the time is spent executing large loop nests [5]. Here, we consider *affine loops*. This means that all functions that determine the loop iteration count are affine with respect to iteration variables and program parameters. Such loops can be arbitrarily nested in a tree-like structure. This approach provides a precise model even for non-constant increments (e.g. i=i*2), although Blanc et al. [8] explicitly exclude such statements.

Some loops cannot be modeled in compile time due to their dynamic nature. The loop bounds or starting conditions can depend nondeterministically on the input or on random elements. Those loops are marked as *undefined*. This information is propagated in the nesting tree, as it may influence other loops.

We analyze the code using an LLVM pass [27]. We recursively trace the loop parameters to find their relation with the program input. As an example, the NAS CG benchmark code contains the following loop: do j=1,lastrow-firstrow+1. Our tool determines that the expression lastrow − firstrow +1 is equal to the program parameter row_size = $\frac{na}{nprows}$.

Hybrid Approach In our tool, we utilize all the information gathered by the static analysis to aid the dynamic profiler. We create a program model that can be expressed using an *extended performance model normal form* (EPMNF):

$$p = \{\iota_i^k \log^l \iota_i^k, \ k, l \in \mathbb{R}, \ \iota_i \in I\} \tag{2}$$

where I represents the set of program input parameters. The functions p are called *predictors*. Our performance model generation technique works by selecting, from an *initial* search space of predictors, the ones that most accurately model the performance of program kernels. This is done in two steps:

1. Construct the *initial* search space of predictors using the result from the static analysis.
2. Using runtime profiling information, run an online variant of LASSO regression [6] to select the most significant predictors and generate the performance model.

Profiling data is reduced further by a batched model-update technique and by adaptively changing the measurement frequency. These operations increase the information gain from a single update. By analyzing the convergence of the model, we regulate the size and frequency of the collected information.

Evaluation We used our tool to model the NAS [2] and MILC [30] benchmark suites. Our hybrid technique can improve the accuracy of the performance models on average by 4.3 %(maximum 10 %) and can reduce the overhead by 25 % (maximum 65 %) compared to a purely dynamic approach. Chosen insights are shown in Fig. 6. There, we present the goodness of fit of the generated models for

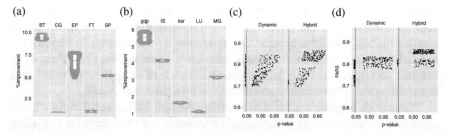

Fig. 6 (**a**) and (**b**) Predicted Adjusted R-Square (PARS) improvement of the new hybrid method over the dynamic approach. (**c**) and (**d**) P-value from F-test vs. PARS plot for the previous dynamic and hybrid approach for benchmarks. gqp and ksr stand for gp_quark_prop and kid_su3_rmd respectively

selected NAS and MILC kernels. To measure the fit, we use the modified Adjusted R-squared coefficient (Predicted Adjusted R-squared—PARS [6]). In the plots (a) and (b), we show the improvement of the PARS coefficient. In (c) and (d) we plot the p-value from the F-test vs. PARS value. One can see that the number of models that suffer from the significant lack-of-fit (p-value below 0.05) is greatly reduced, while the quality of the fit (high PARS value) is increased.

6   Related Work

Performance analysis and prediction of real-world application workloads is most important in high-performance computing. Performance tools such as HPCToolkit [1] allow the programmer to observe the performance of real-world applications at impressive scales but are often limited to observations of the current configuration and do not provide insight into their behavior when being scaled further.

Such insights can be obtained with the help of analytical performance models, which have a long history. Early manual models showed to be very effective in describing application performance characteristics [26] and understanding complex behaviors [31]. Hoefler et al. established a simple six-step process to guide the (manual) creation of analytical performance models [20]. The resulting models lead to interesting insights into application behavior at scale and on unknown systems [3]. The six-step process formed the blueprint of our own approach.

Various automated performance modeling methods exist. Tools such as PALM [36] use extensive and detailed per-function measurements to build structural performance models of applications. The creation of structural models is also supported by dedicated languages such as Aspen [34]. These methods are powerful but require the prior manual annotation of the source code.

Hammer et al. combine static source-code analysis with cache-access simulations to create ECM and roofline models of steady-state loop kernels [17]. While their approach uses hardware information gathered on the target machine, it does not actually run the code but relies on static information instead. Lo et al. create roofline models for entire applications automatically and attempt to identify the optimal configuration to run an application on a given system [28]. Extra-P, in contrast, identifies scalability bugs in individual parts of an application rather than determining the optimal runtime configuration on a particular system.

Vuduc et al. propose a method of selecting the best implementation for a given algorithm by automatically generating a large number of candidates for a selected kernel and then choosing the one offering the best performance according to the results of an empirical search [42]. Our approach generates performance models for *all* kernels in a given application to channel the optimization efforts to where they will be most effective. Zaparanakus et al. analyze and group loops and repetitions in applications towards automatically creating performance profiles for sequential algorithms [46]. Goldsmith et al. use clustering and linear regression analysis to derive performance model coefficients from empirical measurements [16]. This approach requires the user to define either a linear or power law expectation for the performance model unlike the greater freedom offered by the performance model normal form defined in our approach. Jayakumar et al. predict runtimes of entire applications automatically using machine-learning approaches [23].

Zhai, Chen, and Zheng extrapolate single-node performance of applications with a known regular structure to complex parallel machines via simulation [47], but require the entire memory that would be needed at the target scale to correctly extrapolate performance. Wu and Müller [44] showed how to predict the communication behavior of stencil codes at larger scales by extrapolating their traces. While still requiring an SPMD-style parallel execution paradigm, Extra-P has proven to work with general OpenMP or MPI codes beyond pure stencil codes.

Carrington et al. introduced a model-based performance prediction framework for applications on different computers [11]. Marin and Mellor-Crummey utilize semi-automatically derived performance models to predict performance on different architectures [29]. Siegmund et al. analyze the interaction of different configuration options and model how this affects performance of an application as a whole rather than looking at its individual components [33].

7 Conclusion

Our results confirm that automated performance modeling is feasible and that automatically generated models are accurate enough to identify scalability bugs. In fact, in those cases where hand-crafted models existed in the literature we found our models to be competitive. Obviously, the principles of mass production can also be exploited for performance models. On the one hand, approximate models are acceptable as long as the effort to create them is low and they do not mislead the user. On the other hand, being able to produce many of them helps drastically improve code coverage, which is as important as model accuracy. Having approximate models for all parts of the code can be more useful than having a model with 100 % accuracy for just a tiny portion of the code or no model at all. Finally, after the public release of the Extra-P software and two conference tutorials where Extra-P was introduced, we have not only developed this powerful technology on a conceptual level, but also put it into the hands of HPC application developers for immediate use.

References

1. Adhianto, L., Banerjee, S., Fagan, M.W., Krentel, M.W., Marin, G., Mellor-Crummey, J., Tallent, N.R.: HPCToolkit: tools for performance analysis of optimized parallel programs. Concurr. Comput. Pract. Exper. **22**(6), 685–701 (2010)
2. Bailey, D.H., Barszcz, E., Barton, J.T., Browning, D.S., Carter, R.L., Dagum, L., Fatoohi, R.A., Frederickson, P.O., Lasinski, T.A., Schreiber, R.S., Simon, H.D., Venkatakrishnan, V., Weeratunga, S.K.: The NAS parallel benchmarks–summary and preliminary results. In: Proceedings of the 1991 ACM/IEEE Conference on Supercomputing (SC), Albuquerque, pp. 158–165. ACM (1991)
3. Bauer, G., Gottlieb, S., Hoefler, T.: Performance modeling and comparative analysis of the MILC lattice QCD application su3_rmd. In: Proceedings of the CCGrid, Ottawa, pp. 652–659. IEEE (2012)
4. Behr, M., Nicolai, M., Probst, M.: Efficient parallel simulations in support of medical device design. NIC Ser. **38**, 19–26 (2008)
5. Benabderrahmane, M.W., Pouchet, L.N., Cohen, A., Bastoul, C.: The polyhedral model is more widely applicable than you think. In: Gupta, R. (ed.) Compiler Construction. LNCS, vol. 6011, pp. 283–303. Springer (2010). http://dx.doi.org/10.1007/978-3-642-11970-5_16
6. Bhattacharyya, A., Kwasniewski, G., Hoefler, T.: Using compiler techniques to improve automatic performance modeling. In: Accepted at the 24th International Conference on Parallel Architectures and Compilation (PACT'15), San Francisco. ACM (2015)
7. Bhattacharyya, A., Hoefler, T.: PEMOGEN: automatic adaptive performance modeling during program runtime. In: Proceedings of the 23rd International Conference on Parallel Architectures and Compilation Techniques (PACT'14). ACM, Edmonton (2014)
8. Blanc, R., Henzinger, T.A., Hottelier, T., Kovacs, L.: ABC: algebraic bound computation for loops. In: Clarke, E., Voronkov, A. (eds.) Logic for Programming, Artificial Intelligence, and Reasoning. LNCS, vol. 6355, pp. 103–118 (2010). http://dx.doi.org/10.1007/978-3-642-17511-4_7
9. Bull, J.M., O'Neill, D.: A microbenchmark suite for OpenMP 2.0. ACM Comput. Architech. News **29**(5), 41–48 (2001)
10. Calotoiu, A., Hoefler, T., Poke, M., Wolf, F.: Using automated performance modeling to find scalability bugs in complex codes. In: Proceedings of the ACM/IEEE Conference on Supercomputing (SC13), Denver, pp. 1–12. ACM (2013)
11. Carrington, L., Snavely, A., Wolter, N.: A performance prediction framework for scientific applications. Future Gener. Comput. Syst. **22**(3), 336–346 (2006). http://dx.doi.org/10.1016/j.future.2004.11.019
12. Chan, E., Heimlich, M., Purkayastha, A., van de Geijn, R.: Collective communication: theory, practice, and experience. Concurr. Comput. Pract. Exp. **19**(13), 1749–1783 (2007)
13. Dennis, J.M., Edwards, J., Evans, K.J., Guba, O., Lauritzen, P.H., Mirin, A.A., St-Cyr, A., Taylor, M.A., Worley, P.H.: CAM-SE: a scalable spectral element dynamical core for the community atmosphere model. Int. J. High Perform. Comput. **26**(1), 74–89 (2012). http://hpc.sagepub.com/content/26/1/74.abstract
14. Geimer, M., Wolf, F., Wylie, B.J.N., Ábrahám, E., Becker, D., Mohr, B.: The Scalasca performance toolset architecture. Concurr. Comput. Pract. Exp. **22**(6), 702–719 (2010)

15. Gewaltig, M.O., Diesmann, M.: Nest (neural simulation tool). Scholarpedia J. **2**(4), 1430 (2007)

16. Goldsmith, S.F., Aiken, A.S., Wilkerson, D.S.: Measuring empirical computational complexity. In: Proceedings of the 6th Joint Meeting of the European Software Engineering Conference and the ACM SIGSOFT Symposium on the Foundations of Software Engineering (ESEC-FSE '07), New York, pp. 395–404. ACM (2007). http://doi.acm.org/10.1145/1287624.1287681

17. Hammer, J., Hager, G., Eitzinger, J., Wellein, G.: Automatic loop kernel analysis and performance modeling with kerncraft. In: Proceedings of the 6th International Workshop on Performance Modeling, Benchmarking, and Simulation of High Performance Computing Systems (PMBS '15), New York, pp. 4:1–4:11. ACM (2015). http://doi.acm.org/10.1145/2832087.2832092

18. Hoefler, T., Kwasniewski, G.: Automatic complexity analysis of explicitly parallel programs. In: Proceedings of the 26th ACM Symposium on Parallelism in Algorithms and Architectures (SPAA'14), Prague. ACM (2014)

19. Hoefler, T., Snir, M.: Performance engineering: a must for petaflops and beyond. In: Proceedings of the Workshop on Large-Scale System and Application Performance (LSAP), in Conjunction with HPDC, San Jose. ACM (2011)

20. Hoefler, T., Gropp, W., Kramer, W., Snir, M.: Performance modeling for systematic performance tuning. In: State of the Practice Reports (SC '11), pp. 6:1–6:12. ACM (2011). http://doi.acm.org/10.1145/2063348.2063356

21. Hoefler, T., Kwasniewski, G.: Automatic complexity analysis of explicitly parallel programs. In: Proceedings of the 26th ACM Symposium on Parallelism in Algorithms and Architectures (SPAA '14), New York, pp. 226–235. ACM (2014). http://doi.acm.org/10.1145/2612669.2612685

22. Iwainsky, C., Shudler, S., Calotoiu, A., Strube, A., Knobloch, M., Bischof, C., Wolf, F.: How many threads will be too many? On the scalability of OpenMP implementations. In: Proceedings of the 21st Euro-Par Conference, Vienna. LNCS, vol. 9233, pp. 451–463. Springer (2015)

23. Jayakumar, A., Murali, P., Vadhiyar, S.: Matching application signatures for performance predictions using a single execution. In: 2015 IEEE International Parallel and Distributed Processing Symposium (IPDPS), Hyderabad, pp. 1161–1170. IEEE (2015)

24. JuBE – Jülich Benchmarking Environment (2016). http://www.fz-juelich.de/jsc/jube

25. JuSPIC – Jülich Scalable Particle-in-Cell Code (2016). http://www.fz-juelich.de/ias/jsc/EN/Expertise/High-Q-Club/JuSPIC/_node.html

26. Kerbyson, D.J., Alme, H.J., Hoisie, A., Petrini, F., Wasserman, H.J., Gittings, M.: Predictive performance and scalability modeling of a large-scale application. In: Proceedings of the ACM/IEEE Conference on Supercomputing (SC'01), Denver, p. 37. ACM (2001)

27. LLVM home page (2016). http://llvm.org/

28. Lo, Y.J., Williams, S., Van Straalen, B., Ligocki, T.J., Cordery, M.J., Wright, N.J., Hall, M.W., Oliker, L.: Roofline model toolkit: a practical tool for architectural and program analysis. In: High Performance Computing Systems. Performance Modeling, Benchmarking, and Simulation, New Orleans, pp. 129–148. Springer (2014)

29. Marin, G., Mellor-Crummey, J.: Cross-architecture performance predictions for scientific applications using parameterized models. SIGMETRICS Perform. Eval. Rev. **32**(1), 2–13 (2004). http://doi.acm.org/10.1145/1012888.1005691

30. MILC Code Version 7 (2016). http://www.physics.utah.edu/~detar/milc/milc_qcd.html

31. Pllana, S., Brandic, I., Benkner, S.: Performance modeling and prediction of parallel and distributed computing systems: a survey of the state of the art. In: Proceedings of the 1st International Conference on Complex, Intelligent and Software Intensive Systems (CISIS), Vienna, pp. 279–284. IEEE (2007)

32. Shudler, S., Calotoiu, A., Hoefler, T., Strube, A., Wolf, F.: Exascaling your library: will your implementation meet your expectations? In: Proceedings of the 29th ACM on International Conference on Supercomputing (ICS '15), New York, pp. 165–175. ACM (2015). http://doi.acm.org/10.1145/2751205.2751216

33. Siegmund, N., Grebhahn, A., Apel, S., Kästner, C.: Performance-influence models for highly configurable systems. In: Proceedings of the 2015-10th Joint Meeting on Foundations of Software Engineering (ESEC/FSE 2015), New York, pp. 284–294. ACM (2015). http://doi. acm.org/10.1145/2786805.2786845
34. Spafford, K.L., Vetter, J.S.: Aspen: a domain specific language for performance modeling. In: Proceedings of the International Conference on High Performance Computing, Networking, Storage and Analysis (SC '12), Los Alamitos, pp. 84:1–84:11. IEEE Computer Society Press (2012). http://dl.acm.org/citation.cfm?id=2388996.2389110
35. Sutmann, G., Westphal, L., Bolten, M.: Particle based simulations of complex systems with mp2c: hydrodynamics and electrostatics. In: International Conference of Numerical Analysis and Applied Mathematics 2010 (ICNAAM 2010), Rhodes, vol. 1281, pp. 1768–1772. AIP Publishing (2010)
36. Tallent, N.R., Hoisie, A.: Palm: easing the burden of analytical performance modeling. In: Proceedings of the 28th ACM International Conference on Supercomputing (ICS '14), NewYork, pp. 221–230. ACM (2014). http://doi.acm.org/10.1145/2597652.2597683
37. Thakur, R., Rabenseifner, R., Gropp, W.: Optimization of collective communication operations in mpich. Int. J. High Perform. Comput. 19(1), 49–66 (2005)
38. Vetter, J., Worley, P.: Asserting performance expectations. In: Proceedings of the ACM/IEEE Conference on Supercomputing, Baltimore, pp. 1–13. ACM (2002)
39. Vogel, A., Reiter, S., Rupp, M., Nägel, A., Wittum, G.: UG 4: a novel flexible software system for simulating PDE based models on high performance computers. Comput. Vis. Sci. 16(4), 165–179 (2013)
40. Vogel, A., Calotoiu, A., Strube, A., Reiter, S., Nägel, A., Wolf, F., Wittum, G.: 10,000 performance models per minute – scalability of the ug4 simulation framework. In: Proceedings of the 21st Euro-Par Conference, Vienna. LNCS, vol. 9233, pp. 519–531. Springer (2015)
41. Vömel, C.: ScaLAPACK's MRRR algorithm. ACM T. Math. Softw. 37(1), 1:1–1:35 (2010)
42. Vuduc, R., Demmel, J.W., Bilmes, J.A.: Statistical models for empirical search-based performance tuning. Int. J. High Perform. Comput. 18(1), 65–94 (2004). http://dx.doi.org/10.1177/1094342004041293
43. Wasserman, H., Hoisie, A., Lubeck, O., Lubeck, O.: Performance and scalability analysis of teraflop-scale parallel architectures using multidimensional wavefront applications. Int. J. High Perform. Comput. 14, 330–346 (2000)
44. Wu, X., Müller, F.: Scalaextrap: trace-based communication extrapolation for SPMD programs. ACM T. Lang. Sys. 34(1), 113–122 (2012)
45. Wylie, B.J.N., Geimer, M., Mohr, B., Böhme, D., Szebenyi, Z., Wolf, F.: Large-scale performance analysis of Sweep3D with the Scalasca toolset. Parallel Process. Lett. 20(4), 397–414 (2010)
46. Zaparanuks, D., Hauswirth, M.: Algorithmic profiling. Sigplan Not. 47(6), 67–76 (2012). http://doi.acm.org/10.1145/2345156.2254074
47. Zhai, J., Chen, W., Zheng, W.: Phantom: predicting performance of parallel applications on large-scale parallel machines using a single node. Sigplan Not. 45(5), 305–314 (2010). http://doi.acm.org/10.1145/1837853.1693493

Automated Performance Modeling of the UG4 Simulation Framework

Andreas Vogel, Alexandru Calotoiu, Arne Nägel, Sebastian Reiter,
Alexandre Strube, Gabriel Wittum, and Felix Wolf

Abstract Many scientific research questions such as the drug diffusion through
the upper part of the human skin are formulated in terms of partial differential
equations and their solution is numerically addressed using grid based finite element
methods. For detailed and more realistic physical models this computational
task becomes challenging and thus complex numerical codes with good scaling
properties up to millions of computing cores are required. Employing empirical
tests we presented very good scaling properties for the geometric multigrid solver in
Reiter et al. (Comput Vis Sci 16(4):151–164, 2013) using the UG4 framework that is
used to address such problems. In order to further validate the scalability of the code
we applied automated performance modeling to UG4 simulations and presented
how performance bottlenecks can be detected and resolved in Vogel et al. (10,000
performance models per minute—scalability of the UG4 simulation framework. In:
Träff JL, Hunold S, Versaci F (eds) Euro-Par 2015: Parallel processing, theoretical
computer science and general issues, vol 9233. Springer, Springer, Heidelberg,
pp 519–531, 2015). In this paper we provide an overview on the obtained results,
present a more detailed analysis via performance models for the components of the
geometric multigrid solver and comment on how the performance models coincide
with our expectations.

A. Vogel (✉) • A. Nägel • S. Reiter • G. Wittum
Goethe Universität Frankfurt, Frankfurt, Germany
e-mail: andreas.vogel@gcsc.uni-frankfurt.de; arne.naegel@gcsc.uni-frankfurt.de;
sebastian.reiter@gcsc.uni-frankfurt.de; wittum@gcsc.uni-frankfurt.de

A. Calotoiu • F. Wolf
Technische Universität Darmstadt, Darmstadt, Germany
e-mail: calotoiu@cs.tu-darmstadt.de; wolf@cs.tu-darmstadt.de

A. Strube
Jülich Supercomputing Center, Germany
e-mail: a.strube@fz-juelich.de

© Springer International Publishing Switzerland 2016   467
H.-J. Bungartz et al. (eds.), *Software for Exascale Computing – SPPEXA
2013-2015*, Lecture Notes in Computational Science and Engineering 113,
DOI 10.1007/978-3-319-40528-5_21

1 Introduction

The mathematical description for many important scientific and industrial questions is given by a formulation in terms of partial differential equations. Numerical simulations of the modeled systems via finite element and finite volume discretizations (e.g., [6, 10, 16]) can help to better understand the physical behavior by comparing with measured data and ideally provide the possibility to predict physical scenarios. Using detailed computational grids the discretization thereby leads to large sparse systems of equations and these matrix equations can be resolved using advanced methods of optimal order—such as the multigrid method (e.g., [6, 15]).

Looking at the variety of applications and the constantly growing computing resources on modern supercomputers the efficient solution of partial differential equations is an important challenge and it is advantageous to address the numerous problems with a common framework. Ideally, the framework should provide scalable and reusable components that can be applied in all of the fields of interest and serve as a common base for the construction of applications for concrete problems. To this end the UG software framework has been developed and a renewed implementation has been given in the current version 4.0 [35, 37] that pays special attention to parallel scalability.

In order to validate the scaling properties of the software framework on such architectures we carried out several scalability studies. Starting with a hand-crafted analysis we presented close to optimal weak scaling properties of the geometric multigrid solver in [27]. However, the study focused only on a few coarse-grained aspects leaving room for potential performance bottlenecks, that are not visible at current scales due to a small execution constant, but may become dominant at largest scales due to bad asymptotic behavior. Therefore, in a subsequent study we analyzed entire *UG* 4 runs in [38] applying an automated performance modeling approach by Calotoiu et al. [7] to *UG* 4 simulations. The modeling approach creates performance models at a function level granularity and uses few measurement runs at smaller core counts in order to predict the asymptotic behavior of each code kernel at largest scales. By detecting bad asymptotic behavior for code kernels in the grid setup phase we were able to detect and remove a performance bottleneck.

In this paper we focus on more detailed models for the geometric multigrid solver and explain how the observed performance models meet our expectations. Since the geometric multigrid solver is one of the crucial aspects for simulation runs in terms of scalability, we have evaluated in more depth the models for fine-grained kernels of the employed geometric multigrid solver and compare the observed behavior to the intended implementation.

The main aspects of this report are:

- Summarize the automated modeling approach and obtained results for its application to the simulation framework *UG* 4.
- Provide a detailed analysis for components of the geometric multigrid solver.
- Validation of the scaling behavior for the multigrid solver components.

The remainder of this paper is organized as follows. In Sect. 2, the *UG* 4 simulation environment is presented with focus on the parallelization aspects of the parallel geometric multigrid (Sect. 2.2) and the skin permeation problem (Sect. 2.3) used in the subsequent studies. Section 3 outlines the performance modeling approach. In Sect. 4 we briefly summarize previously obtained analysis results for entire simulation runs and then present a detailed performance modeling for the geometric multigrid solver used in a weak scaling study for the skin problem. Sections 5 and 6 are dedicated to related work and concluding remarks.

2 The UG4 Simulation Framework

As a real world target application code for the performance modeling approach we will focus on the simulation toolbox *UG* 4 [37]. The software framework is written in C++ and uses grid-based methods to numerically address the solution of partial differential equations via finite element or finite volume methods. With the main goal to address questions from biology, technology, geology and finance with one common effort, several components are reused in all types of applications and thus the performance modeling for those program parts provides insight into the performance of all these applications. In the following we give a brief overview on the used numerical methods and especially comment on the parallelization aspects.

2.1 Concepts and Numerical Methods

In order to construct the required geometries the meshing software ProMesh [25, 26] is used that shares code parts with the *UG* 4 library. Meshes can be composed of different element types (e.g., tetrahedron, pyramid, prism and hexahedron in 3d) and subset assignment is used to distinguish parts of the domain with different physics or where boundary conditions are to be set. Once loaded in *UG* 4 meshes are further processed to create distributed, unstructured, adaptive multigrid hierarchies with or without hanging nodes. Implemented load-balancing strategies [25] range from simple but fast bisection algorithms to more advanced strategies including usage of external algorithms such as ParMetis [22]. In this study, however, we restrict ourselves to a 3d hexahedral grid hierarchy generated through globally applied anisotropic refinement (cf. [38]). A study for adaptive hierarchies with hanging nodes is work in progress and will be considered in a subsequent study.

A flexible and combinable discretization module allows to combine different kinds of physical problems discretized by finite element and finite volume methods (e.g., [6, 10, 16]) and boundary conditions in a modular way to build a new physical problem selecting from basic building blocks [36, 37]. As algebraic structures for the discretized solutions and associated matrices, block vectors and a *CSR* (compressed sparse row) matrix implementation are provided. For the parallel solution of such matrix equations several solvers are implemented, including Krylov methods such as CG and BiCGStab and preconditioners such as Jacobi, Gauss-Seidel, incomplete LU factorization, ILUT and block versions of these types (e.g., [28]). In addition, a strong focus is on multigrid methods (e.g. [15]) and geometric and algebraic multigrid approaches [17, 27].

The parallelization for the usage on massively parallel computing clusters with hundred thousands of cores is achieved using MPI. The separate library called PCL (parallel communication layer, [27]) builds on top of MPI and is used to ease the graph-based parallelization. Both, the parallelization of the computing grid—assigning a part of the multigrid hierarchy to each process—and of the algebraic structures are programmed based on the PCL. By storing parallel copies on each process in a well-defined order in interface containers identification is performed in an efficient way [25, 27, 37], and global IDs are dispensable.

In order to hide parallelization aspects and ease the usage for beginners the scripting language Lua [18] is used as end-user interface. A flexible plugin system allows to add additional functionality if required.

2.2 Parallel Hierarchical Geometric Multigrid

The multigrid method [16] is used to solve large sparse systems of equations that arise typically by the discretization of some partial differential equation. We briefly recap the idea of the algorithm and our modifications and implementation [27] for the parallel version. Given the linear equation system $A_L x_L = b_L$ on the finest grid level L, the desired solution x_L is computed iteratively: starting with some arbitrary initial guess x_L, in every iteration the defect $d_L = b_L - A_L x_L$ is used to compute a multigrid correction $c_L = M_L(d_L)$, where M_L is the multigrid operator, that is added to the approximate solution $x_L := x_L + c_L$. In order to compute the correction c_L not only the fine grid matrix A_L is used but several auxiliary coarse grid matrices $A_l, L_B \leq l \leq L$, are employed, where L_B denotes the base level. The multigrid cycle is then defined in a recursive manner: given a defect d_l on a certain level l the correction is first partly computed via a smoothing operator (e.g. Jacobi iteration). Then the defect is transferred to the next coarser level, where the algorithm is applied to the restricted defect d_{l-1}. The thereby computed coarse grid correction c_{l-1} is then prolongated to the finer level and added to the correction on level l, followed by some postsmoothing. Once the algorithm reaches the base level L_B, the correction is computed exactly as $c_l = A_l^{-1} d_l$ by, e.g., using LU factorization. Algorithm 1 summarizes this procedure.

The matrix equations for complex problems can easily grow beyond the size of billions of unknowns. In order to solve such problems, massively parallel linear solvers with optimal complexity have to be used. The multigrid algorithm only depends linearly on the number of unknowns and therefore good weak scaling properties are to be expected. As demonstrated in [27] geometric multigrid solvers

---

Algorithm 1 $\mathbf{c}_l = \mathbf{M}_l(\mathbf{d}_l)$ [16, 27]

Requirement: $\mathbf{d}_l = \mathbf{b}_l - \mathbf{A}_l\mathbf{x}_l$
if $l = L_B$ **then**
    Base solver: $\mathbf{c}_l = \mathbf{A}_1^{-1}\mathbf{d}_l$
    return $\mathbf{c}_l$
else
    Initialization: $\mathbf{d}_l^0 := \mathbf{d}_l, \mathbf{c}_l^0 := 0$
    (Pre-)Smoothing for $k = 1, \ldots, \nu_1$:
        $\mathbf{c} = S_l(\mathbf{d}_l^{k-1})$,
        $\mathbf{d}_l^k = \mathbf{d}_l^{k-1} - \mathbf{A}_l\mathbf{c}, \quad \mathbf{c}_l^k = \mathbf{c}_l^{k-1} + \mathbf{c}$
    Restriction: $\mathbf{d}_{l-1} = \mathbf{P}_l^T\mathbf{d}_l^{\nu_1}$
    Coarse grid correction: $\mathbf{c}_{l-1} = \mathbf{M}_{l-1}(\mathbf{d}_{l-1})$
    Prolongation:
        $\mathbf{c}_l^{\nu_1+1} = \mathbf{c}_l^{\nu_1} + \mathbf{P}_l\mathbf{c}_{l-1}$,
        $\mathbf{d}_l^{\nu_1+1} = \mathbf{d}_l^{\nu_1} - \mathbf{A}_l\mathbf{P}_l\mathbf{c}_{l-1}$
    (Post-)Smoothing for $k = 1, \ldots, \nu_2$:
        $\mathbf{c} = S_l(\mathbf{d}_l^{\nu_1+k})$,
        $\mathbf{d}_l^{\nu_1+1+k} = \mathbf{d}_l^{\nu_1+k} - \mathbf{A}_l\mathbf{c}, \quad \mathbf{c}_l^{\nu_1+1+k} = \mathbf{c}_l^{\nu_1+k} + \mathbf{c}$
    return $\mathbf{c}_l^{\nu_1+1+\nu_2}$

---

can exhibit nearly perfect weak scalability when employed in massively parallel environments with hundred thousands of computing cores.

To this end, the components of the algorithm must be parallelized. The basic idea is to construct a distributed multigrid hierarchy as follows:

1. Start with a coarse grid on a small number of processes.
2. Refine the grid several times to create additional hierarchy levels.
3. Redistribute the finest level of the hierarchy to a larger set of processes.
4. Repeat at (2) until the desired grid resolution is obtained. At this point all active processes should contain a part of the finest level of the multigrid hierarchy.

Refining the grid, new levels of the multigrid hierarchy are created and after some refinements the finest grid level is distributed to a larger set of processes and communication structures (called *vertical interfaces*) are established. This process can be iterated, successively creating a tree structure of processes holding parts of the hierarchical grid. Figure 1 shows a process hierarchy for a distributed multigrid hierarchy on four processes (cf. [25, 27]). The communication structures in vertical direction are used to parallelize the transfer between the grid levels, i.e. to implement the transfer of data between grid levels at restriction and prolongation phases within a multigrid cycle. However, if no vertical interface is present the transfer operators act completely process-locally. For the communication within multigrid smoothers on each grid level additional *horizontal interfaces* are required. These interfaces will be used to compute the level-wise correction in a consistent way. An illustration for the resulting hierarchy distribution and interfaces is given in Fig. 2 (cf. [25, 27, 36, 38]). In order to compute the required coarse grid matrices, each process calculates the contribution of the grid part assigned to the process itself.

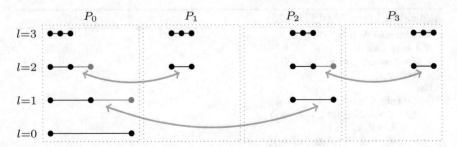

Fig. 1 Illustration for a 1d process hierarchy on 4 processes. Ghost elements (*red*) are sent during redistribution. Data is communicated between ghosts and actual elements through vertical interfaces (*orange*) (cf. [25, 27])

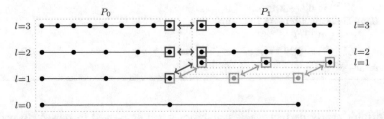

Fig. 2 Illustration for a 1d parallel multigrid hierarchy distributed onto two processes. Parallel copies are identified via horizontal (*blue*) and vertical interfaces (*orange*) (From [38], cf. also [27])

Thus, the matrices are stored in parallel in an additive fashion and no communication is required for this setup.

A Jacobi smoother has very good properties regarding scalability, however it may not be suitable for more complicated problems (e.g. with anisotropic coefficients or anisotropic grids). To handle this issue for anisotropic problems, we use anisotropic refinement in order to construct grid hierarchies with isotropic elements from anisotropic coarse grids: refining only those edges in the computing grid that are longer than a given threshold, and halving this threshold in each step, the approach yields a grid hierarchy which contains anisotropic elements on lower levels and more and more isotropic elements on higher levels. An illustration for a resulting hierarchy is shown in Fig. 3. The used refinement strategy produces non-adaptive grids, i.e. meshes that fully cover the physical domain. This eases the load-balancing compared to adaptive meshes where huge differences in the spatial resolution and thereby element distribution may occur during refinement and redistribution is necessary. In this work we focused on the non-adaptive strategy only, however, plan to report on the adaptive case in future works.

Reconsidering the hierarchical distribution approach described above, lower levels of the multigrid hierarchy are only contained on a smaller number of processes. This is well suited for fast parallel smoothing, prolongation and restriction operations thanks to maintaining a good ratio between computation and communication costs on all levels. A smoothing operation on coarser levels with

Fig. 3 Grid hierarchy created
by anisotropic refinement for
the 3d brick-and-mortar
model (in exploded view).
The aspect ratio of the grid
elements improve with every
refinement step

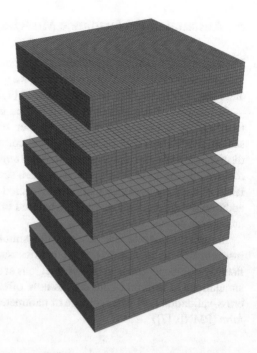

only few inner unknowns would be dominated by the communication and thus the
work is agglomerated to fewer processes resulting in idle processes on coarse levels.
However, the work on finer grid levels is dominating the overall runtime.

2.3 Application: Human Skin Permeation

As an exemplary application from the field of computational pharmacy we focus on
the permeation of substances through the human skin. These simulations consider
the outermost part of the epidermis, called stratum corneum, and are used to estimate
the throughput of chemical exposures. An overview on different descriptions of the
biological and geometric approaches to simulate such a setting can be found in [20]
and the references therein. For this study, we use the same setup as used in [38]: the
transport in two subdomains $s \in \{cor, lip\}$ (corneocyte, lipid) is described by the
diffusion equation

$$\partial_t c_s(t, \mathbf{x}) = \nabla \cdot (\mathbf{D}_s \nabla c_s(t, \mathbf{x})) ,$$

using a subdomain-wise constant diffusion coefficient $\mathbf{D}_s$. We use a 3d brick-and-
mortar model consisting of highly anisotropic hexahedral elements with aspect
ratios as bad as $1/300$ in the coarse grid. Employing anisotropic refinements we
construct a grid hierarchy with better and better aspect ratios on finer levels. The
resulting grid hierarchy is displayed in Fig. 3. For a more detailed presentation we
refer to [38].

3   Automated Performance Modeling

The automated modeling approach used to analyze the *UG* 4 framework has been
presented by Calotoiu et al. in [7, 8]. Here, we give a brief overview on the procedure
and ideas. For further details please refer to [7, 30, 38, 40].

Automated performance modeling is used to empirically determine the asymptotic scaling behavior for a large number of fine-grained code kernels. These
scaling models can then be inspected and compared to the expected complexity: a
discrepancy indicates a potential *scalability bug* that can be addressed and hopefully
removed by the code developers. If no such scalability issues are found this can be
taken as a strong evidence that no unexpected scalability problems are present. In
addition the created models can also be used to predict the resource consumption at
larger core counts if required.

In order to create the models, the simulation framework *UG* 4 has been
instrumented to measure relevant metrics such as time, bytes sent/received or
floating-point operations in program regions at a function level granularity. Running
simulations at different core counts now offers the opportunity to determine via
cross-validation [24] which choice of parameters in the *performance model normal
form* (PMNF, [7])

$$ f(p) = \sum_{k=1}^{n} c_k \cdot p^{i_k} \cdot log_2^{j_k}(p), $$

with $i_k, j_k \in I, J \subset \mathbb{Q}$, best fits the measurements. The approach is applicable to
strong and weak scaling. In this study, however, we have focused on weak scaling
only. In order to account for jitter, several runs for every core count have to be
executed. The required effort for this approach therefore is to run the application a
few times at a few core counts.

For the correct analysis of the multigrid algorithm in weak scaling studies,
a more careful approach than just analyzing the code kernels directly has to be
taken [38]. This is due to the following observation: within a weak scaling study
the problem size has to be increased and for multigrid approaches this leads to
an increase in the number of grid levels. The multigrid algorithm—traversing the
multigrid hierarchy top-down and then again bottom-up, applying smoothers and
level transfers for every level—will thus create a different call tree at different
core counts for multigrid functions due to the varying numbers of grid levels.
However, the performance modeling approach usually assumes the same call tree
for all core counts. Therefore, we preprocess the call tree: only kernels present in
all measurements remain in the modified call tree and measurement data of code
kernels not present in modified call trees is added to the parent kernel. This approach
is not only limited to multigrid settings but may be useful for all codes that use
recursive calls whose invocation count increases within a scaling study.

4   Results

The automated performance modeling (Sect. 3) has been applied to entire simulation runs of the *UG* 4 simulation framework (Sect. 2) and several aspects of the analysis have been reported in [38]. In order to analyze and validate the code behavior the proceeding and reasoning is as follows: we run several simulations at different core counts p and measure detailed metric information (times, bytes sent) at a function level granularity. By this, we receive fine grained information for small code kernels. For all of these kernels and all available metrics we create performance models and then rank these by their asymptotic behavior with respect to the core count. All code kernels with constant or only logarithmical dependency are considered optimal. However, if some code kernel, e.g., in the multigrid method would show a linear or quadratic dependency, this would not match our performance expectations and we consider it a scalability bug that has to be addressed and removed. Inspecting all measured code kernels thus provides us with a fine grained information for different parts of the simulation code. Given that all code kernels show an optimal dependency we finally obtain a validation of the expected scaling properties.

Here, we first briefly give an overview on the results presented in [38] and then show more detailed results focusing on the multigrid kernels and their scaling properties.

4.1   Analysis for Grid Hierarchy Setup and Solver Comparison

In a first test, we analyzed entire runs in a weak scaling study for the human skin permeation simulating the steady-state concentration distribution on a three-dimensional brick-and-mortar skin geometry. A scalability issue has been detected by the performance modeling that can be explained and resolved [38]: at the initialization of the multigrid hierarchy an MPI_Allreduce operation for an array of length p was used to inform each process about its intended communicator group membership. The resulting $p \cdot \mathcal{O}(\text{MPI\_Allreduce})$ dependency has been addressed by using MPI_Comm_split instead that can be implemented with a $\mathcal{O}(\log^2 p)$ [31] behavior. By this, we were able to remove this potential bottleneck at this stage of code development.

In a second study, we provided a comparison for two different types of solvers: the geometric multigrid solver has been compared in a weak scaling study to the unpreconditioned conjugate gradient (CG) method. The unpreconditioned conjugate gradient method is known to have unpleasant weak-scaling properties due to the increase by a factor of two for the iteration count resulting in a $\mathcal{O}(\sqrt{p})$ dependency (see [38] for a detailed theoretical analysis). Due to the created models we confirmed that the theoretical expectations are met by our implementation of the parallel solvers.

4.2 Scalability of Code Kernels in the Geometric Multigrid

In this section we give a more detailed view on the code kernels for the geometric multigrid. For the analysis of the multigrid solver we consider the human skin permeation model: we compute the steady-state of the concentration distribution for the brick-and-mortar model described in Sect. 2.3 and choose the diffusion parameter to $\mathbf{D}_{cor} = 10^{-3}$ and $\mathbf{D}_{lip} = 1$. For the solution of the arising linear system of equations, the geometric multigrid solver is used. As acceleration an outer conjugate gradient method is applied. For the smoothing a damped Jacobi is employed with three smoothing steps. As cycle type the V-cycle is used and as base solver we use a LU factorization. The stopping criterion for the solver is the reduction of the absolute residuum to 10^{-10}. The anisotropic refinement as laid out in Sect. 2.2 is used to enhance the aspect ratios of the hierarchy from level to level. Once satisfactory ratios are reached, this level is used as base level for the multigrid algorithm.

In Fig. 4 we present the accumulated wallclock times for exemplary coarse-grain kernels of the multigrid method and provide information on the number of used cores and the size of the solved matrix system (degrees of freedom). Please note that the iteration counts are bounded as expected for a multigrid method. Since the assembling for the matrix is an inherent parallel process without any communication it can be performed—given an optimal load-balancing—with constant wall-clock time in the weak scaling. This is confirmed by the generated performance model. All other shown aspects of the multigrid method show a logarithmical dependency. This is due to the fact that the number of involved coarse grid levels $L = \mathscr{O}(\log p)$ depends on the number of processes in a weak scaling. We consider this logarithmical dependency still as optimal since even allreduce operations implemented in a tree-like fashion will show the same behavior and are used to check for convergence.

The performance models for several code kernels are shown in Table 1. Please note that all code kernels in our measurements have shown constant or logarithmical dependency with respect to the number of processes. Here, we show some selected kernels in order to give more details on the parallelization aspects of the multigrid method. For a more detailed description on the mathematical algorithm and parallelization aspects we refer to Sect. 2 and [27].

The presmoothing is performed in a two step fashion: first, the Jacobi iteration is applied on process-wise data structures resulting in no data transfer (CG → GMG → PreSmooth → Jacobi → apply → step). In a second phase update information is exchanged between nearest neighbors in order to gain a consistent update resulting in data transfer (PreSmooth → Jacobi → apply → AdditiveToConsistent → MPI_Isend). All behaviors are found to depend logarithmically due to the increase in grid levels that are using this method.

The grid transfer is performed process-wise as well (Restrict → apply). No communication is needed unless vertical interfaces are present. The setup phase (Restrict → init) simply assembles the transfer operators into a matrix structure on each process and a constant time within a weak scaling is thus observed.

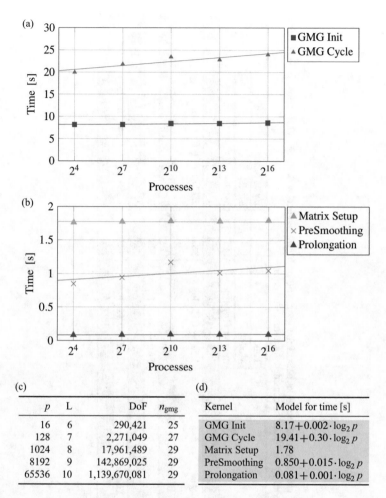

Fig. 4 Measured accumulated wallclock times (*marks*) and models (*lines*) for the skin 3d problem (self time and subroutines). (**a**) (cf. [38]) Initialization of the multigrid solver and time spent in the multigrid cycle. (**b**) Times for coarse matrix assembling, smoothing and prolongation. (**c**) (From [38]) Number of grid refinements (*L*), degrees of freedom (*DoF*) and number of iterations of the solver (n_{gmg}). (**d**) Performance models for the kernels shown in the graphs

Finally, we show some kernels for the outer CG iteration. In order to check for convergence, the norm of a defect vector is computed in each iteration step. After a nearest neighbor communication in order to change the storage type of the vector (CG → norm → AdditiveToUnique → MPI_Isend), the norm is first computed on each process (CG → norm) and then summed up globally (CG → norm → allreduce → MPI_Allreduce).

This way our expectations for the code kernels of the multigrid solver are confirmed and we have strong evidence that only logarithmical complexity with respect to the core count (or better) occurs.

Table 1 Skin 3d study. Models for self-time and bytes sent for selected kernels of the geometric multigrid method with outer CG iteration. $|1 - R^2|$, the absolute difference between R^2 and the optimum scaled by 10^{-3}, can be considered a normalized error

Kernel	Time		Bytes sent	
	Model	$|1 - R^2|$	Model	$|1 - R^2|$
	time $= f(p)$[ms]	$[10^{-3}]$	bytes $= f(p)$	$[10^{-3}]$
CG $\rightarrow$ GMG $\rightarrow$ PreSmooth $\rightarrow$ Jacobi $\rightarrow$ apply $\rightarrow$ step	$18.9 + 0.4 \cdot \log p$	42.6	0	0.0
CG $\rightarrow$ GMG $\rightarrow$ PreSmooth $\rightarrow$ Jacobi $\rightarrow$ apply $\rightarrow$ AdditiveToConsistent $\rightarrow$ MPI_Isend	$1.51 + 1.12 \cdot \log p$	36.0	$5.77 \cdot 10^5 + 0.95 \cdot 10^5 \cdot \log p$	53.2
CG $\rightarrow$ GMG $\rightarrow$ Restrict $\rightarrow$ init	510	0.0	0	0.0
CG $\rightarrow$ GMG $\rightarrow$ Restrict $\rightarrow$ apply	$51.0 + 0.05 \cdot \log p$	378	0	0.0
CG $\rightarrow$ norm	$3.52 + 0.002 \cdot \log^2 p$	544	0	0.0
CG $\rightarrow$ norm $\rightarrow$ AdditiveToUnique $\rightarrow$ MPI_Isend	$0.52 + 0.45 \cdot \log p$	38.9	$1.95 \cdot 10^5 + 0.34 \cdot 10^5 \cdot \log p$	45.5
CG $\rightarrow$ norm $\rightarrow$ allreduce $\rightarrow$ MPI_Allreduce	$1.67 + 0.92 \cdot \log^2 p$	7.5	$\mathcal{O}(\text{MPI\_Allreduce})$	0.0

5 Related Work

Numerous analytical and automated performance modeling approaches have been proposed and developed. The field ranges from manual models [5, 23], over source-code annotations [34] to specialized languages [32]. Automated modeling methods are developed based on machine-learning approaches [19], and via extrapolating trace measurements in [42] (extrapolating from single-node to parallel architectures), in [41] (predicting communication costs at large core counts) and in [9] (extrapolating based on a set of canonical functions). The Dimemas simulator provides tools for performance analysis in message-passing programs [13].

Various frameworks to solve partial differential equations use multigrid methods. Highly scalable multigrid methods are presented in [4, 14, 29, 33], and [39] for geometric multigrid, and in [1, 2], and [3] for algebraic multigrid methods. Work on performance modeling for multigrid can be found in [11, 38] for geometric and in [12] for algebraic multigrid. For an overview for the numerical treatment of skin permeation, we refer to [21] and the references therein.

6 Conclusion

The numerical simulation framework UG4 consists of half a million lines of code and is used to address problems formulated in terms of partial differential equations employing multigrid methods to solve arising large sparse matrix equations. In order to analyze, predict and improve the scaling behavior of UG4 we have conducted a performance modeling at code kernel granularity. Inspecting automated performance models we validated the scalability of entire simulations and presented the close to optimal weak scaling properties for the components of the employed geometric multigrid method that only depend logarithmically on the core count.

Acknowledgements Financial support from the DFG Priority Program 1648 *Software for Exascale Computing* (SPPEXA) is gratefully acknowledged. The authors also thank the Gauss Centre for Supercomputing (GCS) for providing computing time on the GCS share of the supercomputer JUQUEEN at Jülich Supercomputing Centre (JSC).

References

1. Baker, A., Falgout, R., Kolev, T., Yang, U.: Multigrid smoothers for ultra-parallel computing. SIAM J. Sci. Comput. **33**, 2864–2887 (2011)
2. Baker, A.H., Falgout, R.D., Gamblin, T., Kolev, T.V., Schulz, M., Yang, U.M.: Scaling algebraic multigrid solvers: on the road to exascale. In: Competence in High Performance Computing 2010, pp. 215–226. Springer, Berlin/New York (2012)
3. Bastian, P., Blatt, M., Scheichl, R.: Algebraic multigrid for discontinuous Galerkin discretizations of heterogeneous elliptic problems. Numer. Linear Algebra **19**(2), 367–388 (2012)

4. Bergen, B., Gradl, T., Rude, U., Hulsemann, F.: A massively parallel multigrid method for finite elements. Comput. Sci. Eng. **8**(6), 56–62 (2006)
5. Boyd, E.L., Azeem, W., Lee, H.H., Shih, T.P., Hung, S.H., Davidson, E.S.: A hierarchical approach to modeling and improving the performance of scientific applications on the KSR1. In: Proceedings of the 1994 International Conference on Parallel Processing, St. Charles, vol. III, pp. 188–192. IEEE (1994)
6. Braess, D.: Finite Elemente. Springer, Berlin (2003)
7. Calotoiu, A., Hoefler, T., Poke, M., Wolf, F.: Using automated performance modeling to find scalability bugs in complex codes. In: Proceedings of the ACM/IEEE Conference on Supercomputing (SC13), Denver. ACM (2013)
8. Calotoiu, A., Hoefler, T., Wolf, F.: Mass-producing insightful performance models. In: Workshop on Modeling & Simulation of Systems and Applications, Seattle, Aug 2014
9. Carrington, L., Laurenzano, M., Tiwari, A.: Characterizing large-scale HPC applications through trace extrapolation. Parallel Process. Lett. **23**(4), 1340008 (2013)
10. Ciarlet, P.G., Lions, J.: Finite Element Methods (Part 1). North-Holland, Amsterdam (1991)
11. Gahvari, H., Gropp, W.: An introductory exascale feasibility study for FFTs and multigrid. In: International Symposium on Parallel & Distributed Processing (IPDPS), pp. 1–9. IEEE, Piscataway (2010)
12. Gahvari, H., Baker, A.H., Schulz, M., Yang, U.M., Jordan, K.E., Gropp, W.: Modeling the performance of an algebraic multigrid cycle on HPC platforms. In: Proceedings of the International Conference on Supercomputing, pp. 172–181. ACM, New York (2011)
13. Girona, S., Labarta, J., Badia, R.M.: Validation of dimemas communication model for MPI collective operations. In: Proceedings of the 7th European PVM/MPI Users' Group Meeting on Recent Advances in Parallel Virtual Machine and Message Passing Interface, pp. 39–46. Springer, London (2000). http://dl.acm.org/citation.cfm?id=648137.746640
14. Gmeiner, B., Köstler, H., Stürmer, M., Rüde, U.: Parallel multigrid on hierarchical hybrid grids: a performance study on current high performance computing clusters. Concurr. Comput.: Pract. Exp. **26**(1), 217–240 (2014)
15. Hackbusch, W.: Multi-grid Methods and Applications, vol. 4. Springer, Berlin/New York (1985)
16. Hackbusch, W.: Theorie und Numerik elliptischer Differentialgleichungen: mit Beispielen und Übungsaufgaben. Teubner (1996)
17. Heppner, I., Lampe, M., Nägel, A., Reiter, S., Rupp, M., Vogel, A., Wittum, G.: Software framework ug4: parallel multigrid on the hermit supercomputer. In: High Performance Computing in Science and Engineering '12, pp. 435–449. Springer, Cham (2013)
18. Ierusalimschy, R., De Figueiredo, L.H., Celes Filho, W.: Lua-an extensible extension language. Softw. Pract. Exp. **26**(6), 635–652 (1996)
19. Lee, B.C., Brooks, D.M., de Supinski, B.R., Schulz, M., Singh, K., McKee, S.A.: Methods of inference and learning for performance modeling of parallel applications. In: Proceedings of the 12th ACM SIGPLAN Symposium on Principles and Practice of Parallel Programming (PPoPP '07), pp. 249–258. ACM, New York (2007)
20. Nägel, A., Heisig, M., Wittum, G.: A comparison of two- and three-dimensional models for the simulation of the permeability of human stratum corneum. Eur. J. Pharm. Biopharm. **72**(2), 332–338 (2009)
21. Nägel, A., Heisig, M., Wittum, G.: Detailed modeling of skin penetration—an overview. Adv. Drug Deliv. Rev. **65**(2), 191–207 (2013)
22. ParMetis (Nov 2015), http://glaros.dtc.umn.edu/gkhome/metis/parmetis/overview
23. Petrini, F., Kerbyson, D.J., Pakin, S.: The case of the missing supercomputer performance: achieving optimal performance on the 8,192 processors of ASCI Q. In: Proceedings of the ACM/IEEE Conference on Supercomputing (SC'03), pp. 55ff. ACM, New York (2003)
24. Picard, R.R., Cook, R.D.: Cross-validation of regression models. J. Am. Stat. Assoc. **79**(387), 575–583 (1984)
25. Reiter, S.: Efficient algorithms and data structures for the realization of adaptive, hierarchical grids on massively parallel systems. Ph.D. thesis, University of Frankfurt, Germany (2014)

26. Reiter, S.: Promesh (Nov 2015), http://wwww.promesh3d.com
27. Reiter, S., Vogel, A., Heppner, I., Rupp, M., Wittum, G.: A massively parallel geometric multigrid solver on hierarchically distributed grids. Comput. Vis. Sci. **16**(4), 151–164 (2013)
28. Saad, Y.: Iterative Methods for Sparse Linear Systems. SIAM, Philadelphia (2003)
29. Sampath, R., Biros, G.: A parallel geometric multigrid method for finite elements on octree meshes. SIAM J. Sci. Comput. **32**, 1361–1392 (2010)
30. Shudler, S., Calotoiu, A., Hoefler, T., Strube, A., Wolf, F.: Exascaling your library: will your implementation meet your expectations? In: Proceedings of the International Conference on Supercomputing (ICS), Newport Beach, pp. 1–11. ACM (2015)
31. Siebert, C., Wolf, F.: Parallel sorting with minimal data. In: Recent Advances in the Message Passing Interface, pp. 170–177. Springer, Berlin/New York (2011)
32. Spafford, K.L., Vetter, J.S.: Aspen: a domain specific language for performance modeling. In: Proceedings of the International Conference on High Performance Computing, Networking, Storage and Analysis, SC '12, pp. 84:1–84:11. IEEE Computer Society Press, Los Alamitos (2012)
33. Sundar, H., Biros, G., Burstedde, C., Rudi, J., Ghattas, O., Stadler, G.: Parallel geometric-algebraic multigrid on unstructured forests of octrees. In: Proceedings of the International Conference on High Performance Computing, Networking, Storage and Analysis, p. 43. IEEE Computer Society Press, Los Alamitos (2012)
34. Tallent, N.R., Hoisie, A.: Palm: easing the burden of analytical performance modeling. In: Proceedings of the International Conference on Supercomputing (ICS), pp. 221–230. ACM, New York (2014)
35. UG4 (Nov 2015), https://github.com/UG4
36. Vogel, A.: Flexible und kombinierbare Implementierung von Finite-Volumen-Verfahren höherer Ordnung mit Anwendungen für die Konvektions-Diffusions-, Navier-Stokes- und Nernst-Planck-Gleichungen sowie dichtegetriebene Grundwasserströmung in porösen Medien. Ph.D. thesis, Universität Frankfurt am Main (2014)
37. Vogel, A., Reiter, S., Rupp, M., Nägel, A., Wittum, G.: UG 4: a novel flexible software system for simulating PDE based models on high performance computers. Comput. Vis. Sci. **16**(4), 165–179 (2013)
38. Vogel, A., Calotoiu, A., Strube, A., Reiter, S., Nägel, A., Wolf, F., Wittum, G.: 10,000 performance models per minute—scalability of the UG4 simulation framework. In: Träff, J.L., Hunold, S., Versaci, F. (eds.) Euro-Par 2015: Parallel Processing, Theoretical Computer Science and General Issues, vol. 9233, pp. 519–531. Springer, Heidelberg (2015)
39. Williams, S., Lijewski, M., Almgren, A., Straalen, B.V., Carson, E., Knight, N., Demmel, J.: s-step Krylov subspace methods as bottom solvers for geometric multigrid. In: 28th International Parallel and Distributed Processing Symposium, pp. 1149–1158. IEEE, Piscataway (2014)
40. Wolf, F., Bischof, C., Hoefler, T., Mohr, B., Wittum, G., Calotoiu, A., Iwainsky, C., Strube, A., Vogel, A.: Catwalk: a quick development path for performance models. In: Euro-Par 2014: Parallel Processing Workshops. Lecture Notes in Computer Science, pp. 589–600. Springer, Cham (2014)
41. Wu, X., Mueller, F.: ScalaExtrap: trace-based communication extrapolation for SPMD programs. In: Proceedings of the 16th ACM Symposium on Principles and Practice of Parallel Programming (PPoPP '11), pp. 113–122. ACM, New York (2011)
42. Zhai, J., Chen, W., Zheng, W.: Phantom: predicting performance of parallel applications on large-scale parallel machines using a single node. Sigplan Not. **45**(5), 305–314 (2010)

Part XII
GROMEX: Unified Long-Range Electrostatics and Dynamic Protonation for Realistic Biomolecular Simulations on the Exascale

Accelerating an FMM-Based Coulomb Solver with GPUs

Alberto Garcia Garcia, Andreas Beckmann, and Ivo Kabadshow

Abstract The simulation of long-range electrostatic interactions in huge particle ensembles is a vital issue in current scientific research. The Fast Multipole Method (FMM) is able to compute those Coulomb interactions with extraordinary speed and controlled precision. A key part of this method are its shifting operators, which usually exhibit $O(p^4)$ complexity. Some special rotation-based operators with $O(p^3)$ complexity can be used instead. However, they are still computationally expensive. Here we report on the parallelization of those operators that have been implemented for a GPU cluster to speed up the FMM calculations.

1 Introduction

The simulation of dynamical systems of N particles subject to physical potentials, such as gravitation or electrostatics, is a crucial issue in scientific research. This problem is commonly referred as the N-body problem, which has no analytical solution for $N > 3$. However, using an iterative numerical approach, the dynamical behavior of such systems can be simulated. Therefore, the total force exerted on each particle is computed at discrete time intervals, so that the velocities and positions of the particles can be updated.

A typical example is the simulation of a system of particles with electric charges q_i. The Coulomb force $\mathbf{F}_{ij}$ of a particle j with charge q_j acting on a particle i with charge q_i is defined by the following expression:

$$\mathbf{F}_{ij} = \frac{q_i q_j}{|\mathbf{r}_{ij}|^3} \mathbf{r}_{ij}, \tag{1}$$

A. Garcia Garcia • A. Beckmann • I. Kabadshow (✉)
Jülich Supercomputing Centre (JSC), Forschungszentrum Jülich GmbH, Jülich, Germany
e-mail: a.garcia@fz-juelich.de; a.beckmann@fz-juelich.de; i.kabadshow@fz-juelich.de

© Springer International Publishing Switzerland 2016   485
H.-J. Bungartz et al. (eds.), *Software for Exascale Computing – SPPEXA 2013-2015*, Lecture Notes in Computational Science and Engineering 113, DOI 10.1007/978-3-319-40528-5_22

where $\mathbf{r}_{ij}$ is the distance vector between particles i and j. Given that, the total force $\mathbf{F}_i$ acting on each particle i can be expressed as the following summation:

$$\mathbf{F}_i = \sum_{j=1}^{N} \frac{q_i q_j}{|\mathbf{r}_{ij}|^3} \mathbf{r}_{ij} \quad (j \neq i). \tag{2}$$

As we can observe, calculating the forces acting on each particle has a computational complexity of $O(N)$ since we have to compute all pairwise interactions of the current particle with the rest of the system. Therefore, a naive algorithm for computing all forces $\mathbf{F}_i$ exhibits $O(N^2)$ complexity. The update step has a complexity of $O(N)$ since computing the velocities from the forces just needs to iterate once over each particle. The same applies for the position update step. In this regard, the quadratic complexity may be negligible for a small number of particles, but interesting and useful simulations often involve huge particle ensembles, so the simulation will be considerably slowed down to a point in which it is non-viable to apply this kind of summation method. Fortunately, due to the increasing importance of N-body simulations for research purposes, fast summation methods have been developed throughout the latter years [1–3, 5, 7, 8].

In this work, we will focus on the Fast Multipole Method (FMM). The main goal is to develop a CUDA accelerated implementation of a rotation operator that is used during the FMM passes to reduce the computational complexity of the typical FMM mathematical operators, used for shifting and converting the multipole expansions, from $O(p^4)$ to $O(p^3)$. In contrast to other GPU implementations[13, 16], we focus our efforts on achieving good performance for a high multipole order ($p > 10$) required for MD simulations.

This document is structured as follows: Sect. 2 introduces the FMM, its core aspects and the role of the M2M operator as well as the functioning of the rotation-based operators. Section 3 describes the existing CPU/sequential implementation of the FMM $O(p^3)$ and $O(p^4)$ operators and sets baseline timings for all of them. Section 4 explains the changes in the application layout and the included abstraction layers to support the future GPU implementation. Section 5 shows the implementation details of the GPU-accelerated version using CUDA. In Sect. 6 we draw conclusions about this work and outline possible future improvements.

2   Theoretical Background

In this section we provide a brief description of the FMM, reviewing its core aspects and its mathematical foundations. We also briefly describe the role of the mathematical operators used for shifting. At last, we explain how the application of rotation-based operators to those expansions is capable to reduce the complexity of the aforementioned operators from $O(p^4)$ to $O(p^3)$.

2.1 The FMM Workflow

The FMM is a fast summation method which is able to provide an approximate solution to the calculation of forces, potentials or energies within a given precision goal, namely ΔE. The FMM developed at JSC is capable of automatically tuning [4] the FMM parameters for a given energy threshold ΔE. The method exhibits linear computational complexity $O(N)$, resulting from a sophisticated algorithmic structure. The core aspects of the FMM are: spatial grouping of particles, hierarchical space subdivision, multipole expansion of the charges and a special interaction scheme.

The main idea behind the FMM is based on the following intuitive property of the Coulomb and gravitational potential: the effect of particles close to the observation point (called target), on the target particle is dominant compared to the effect produced by remote particles. As opposed to a cutoff scheme, the FMM takes into account the effects of all particles no matter how remote. Cutoff methods have a $O(N)$ complexity, but ignore interactions beyond a cutoff completely.

Consider the particle distribution shown in Fig. 1, for which the remote interactions between two clusters shall be computed: target, with m particles, and source, with n particles. The FMM is based on the idea that a remote particle from a spatial cluster will have almost the same influence on the target particle as another one from the same cluster, given that the inter-cluster distance is large enough. The FMM therefore groups all particles in the remote cluster into a pseudo-particle. By doing this, the amount of interactions is effectively reduced to m. This grouping scheme is also used in reverse, by grouping the target cluster thus requiring n interactions. When grouping both source and target clusters, the computation reduces to a single however more complex interaction.

To implement spatial grouping, the simulation space is subdivided to generate particle groups. The FMM decomposes space recursively in cubic boxes, generating eight different child boxes from each parent box. This hierarchy of cubes is arranged in a tree, called *octree* of depth d. Figure 2 shows an example of this recursive subdivision visualized in a 2D plane.

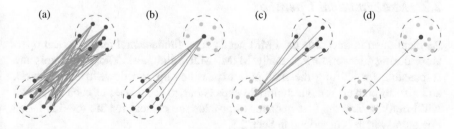

Fig. 1 From *left* to *right*: (**a**) Direct interactions of the particles of one cluster with all particles in the other cluster. (**b**) Interaction via source pseudo-particle. (**c**) Interaction via target pseudo-particle. (**d**) Interaction with both source and target pseudo-particles

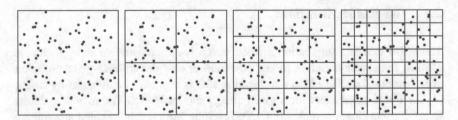

Fig. 2 Space subdivision using an octree. From *left* to *right*: Trees with depth $d = 0$, $d = 1$, $d = 2$, and $d = 3$

Given a certain separation criterion ws, the multipole order p and the depth of the tree d, the FMM consists of the following steps, called passes:

- **Pass 1:** Expand charges into spherical multipole moments ω_{lm} on the lowest level for each box, and translate multipole moments ω_{lm} of each box up the tree
- **Pass 2:** Transform remote multipole moments ω_{lm} into local moments μ_{lm} for each box on every level
- **Pass 3:** Translate local moments μ_{lm} down the tree towards the leaf nodes
- **Pass 4:** Compute far field contributions: potentials Φ_{FF}, forces $\mathbf{F}_{FF}$, and energy E_{FF} on the lowest level
- **Pass 5:** Compute near field contributions: potentials Φ_{NF}, forces $\mathbf{F}_{NF}$, and energy E_{NF} on the lowest level

This algorithm exhibits a linear computational complexity. Its derivation is beyond the scope of this work, and can be found in [11]. The first pass is performed by the P2M operator, which is often considered a preprocessing step, and the M2M operator, while the second one is done via the M2L operator and the third one with the L2L operator. This work focuses on the M2M operator. The extension to the remaining operators M2L and L2L is straightforward and can be implemented following the same strategies.

2.2 Mathematical Operators

As mentioned in Sect. 2.1, the FMM needs three fundamental mathematical operators during its workflow, namely M2M, M2L, and L2L. Those operators are responsible for shifting the multipole expansions up and down the tree levels, and also to convert remote multipole expansions to local ones at each level. We will briefly review the first operator to provide the context for the rotation-based operators, which is described in Sect. 2.3.

2.2.1 Multipole-to-Multipole (M2M) Operator

The M2M is a vertical operator which shifts the multipole coefficients up to higher
levels of the tree structure. Each box of the 3D tree has eight child boxes in the
next lower level. The M2M operator sums up all the moments of the multipole
expansions of the child boxes at the center of the parent box. This operator is applied
to each level up to the root of the tree. By doing this, each box on every level has
a multipole expansion. This operator is applied in the first pass, and is also known
as A.

From a mathematical perspective, each child multipole expansion ω^i at the center
a_i of that child box i is shifted up to the center $a + b$ of its parent box (see Fig. 3).
Equation (3) shows how the moments $\omega^i_{jk}(a_i)$ of each child multipole expansion
are shifted by the A operator to produce the moments $\omega^i_{lm}(a_i + b_i)$ of the parent's
expansion:

$$\omega^i_{lm}(a_i + b_i) = \sum_{j=0}^{l} \sum_{k=-j}^{j} A^{lm}_{jk}(b_i)\omega^i_{jk}(a_i) . \tag{3}$$

All the shifted moments of the eight child boxes are finally added up to conform
the multipole expansion at the center of the parent box

$$\omega_{lm}(a + b) = \sum_{i=1}^{8} \omega^i_{lm}(a_i + b_i) . \tag{4}$$

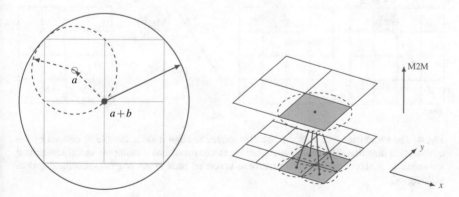

Fig. 3 The *left panel* shows the analytical domain of the M2M operator in a 2D tree. The centers
(*blue dots*) of a sample child box and the parent are shown. The *right panel* depicts the functioning
of the M2M operator for a 2D system. The operator has $O(p^4)$ complexity

2.3 Rotation-Based Operators

A set of more efficient operators with $O(p^3)$ computational complexity scaling were proposed by White and Head-Gordon [15]. The reduced complexity is achieved by rotating the multipole expansions so that the translations or shifts are performed along the quantization axis of the boxes (see Fig. 4). This reduces the 3D problem to a 1D one.

The multipole moments of an expansion with respect to a coordinate system which has been rotated twice, first by an angle ϕ about the z-axis and then by θ about the y-axis, can be expressed as a linear combination of the moments with respect to the original coordinate system. The rotated multipole expansion (see Fig. 4) can be expressed as ω'_{lm} as shown in Eq. (5):

$$\omega'_{l,m}(\theta, \phi) = \sum_{k=-l}^{l} \frac{\sqrt{(l-k)!(l+k)!}}{\sqrt{(l-m)!(l+m)!}} d_l^{m,k}(\theta) e^{ik\phi} \omega_{l,k} . \tag{5}$$

In the last equation, $d_l^{m,k}$ represents Wigner small d-rotation coefficients whose computation falls beyond the scope of this work. A detailed explanation and implementation on how to compute them can be found in [9]. The term $e^{ik\phi}$ represents a factor that is needed for each moment to compute the rotation. Usually, the operator will compute both terms on the fly, adding a prefactor to the $O(p^3)$ complexity of this operator. However, that prefactor can be removed by precomputing and reusing the constants.

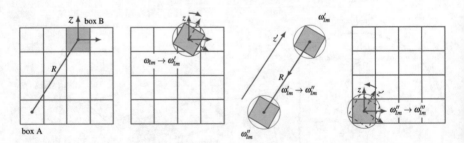

Fig. 4 The coordinate system of the box B is rotated to align it along the z'-axis defined by the quantization direction. The multipole expansion is translated into a multipole expansion around the center of A. The new multipole expansion is rotated back to the original coordinate system yielding ω'''_{lm}

3 Existing Implementation

In this section we will describe the existing C++ implementation of the FMM, which has been implemented within the GROMEX SPPEXA project. This project addresses the development, implementation, and optimization of a unified electrostatics algorithm that will account for realistic, dynamic ionization states (λ-dynamics) [6] and at the same time overcome scaling limitations on current architectures.

We will show benchmarks carried out to determine a baseline for future optimizations, i.e., the parallel implementation. In addition, this baseline will prove the effectiveness of the $O(p^3)$ operators.

From an application point of view, the FMM is implemented in a set of abstraction layers, each on top of another, with different responsibilities. By using a layered approach, the internal functionality of a layer can be changed and optimized at any time without having to worry about the other layers. This design provides flexibility, and it is implemented with the help of templates in the different layers (see Fig. 5).

As shown in the figure, the implementation is composed of four well distinguished layers: (1) the algorithm, (2) data structures, (3) allocator and (4) memory. The top layer contains the FMM logic itself, i.e., the implementations of the described passes. Here, we keep the focus on the M2M operator, for which templates allow us to choose between the $O(p^4)$ or $O(p^3)$ version.

Those implementations need data structures to store the information that is being processed. In this regard, the algorithm layer leverages to the data structures one. This layer contains the data types needed for the algorithm, e.g., coefficient matrices (ω), rotation matrices (R), and other simple data structures, including their internal

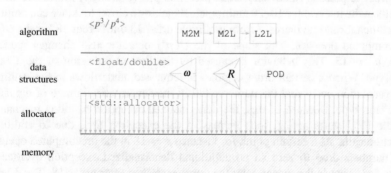

Fig. 5 Layout of the existing FMM implementation for CPUs. There are four different abstraction layers: the algorithm, the data structures, the allocator and the memory. The algorithm layer is templated to choose between the p^3 or p^4 operators. The data structures are also templated so that the underlying data type precision can be chosen. The allocator is templated as well, so that it can make use of custom or predefined memory allocators

logic. The templated design allows to choose the precision of the underlying data types.

The allocator layer enables the data structures to allocate memory for storing their information. The data structures delegate the memory allocation to an allocator performing the corresponding calls for allocating and deallocating memory. This layer is also templated, so any allocator can perform this task.

This existing CPU implementation was tested to establish a set of baseline performance results. A baseline helps to determine multiple facts: the actual effectiveness of the reduced complexity operators, the impact of the precomputed constants, and precision bounds. It will also serve as a starting point to compare the performance of the GPU implementation.

The benchmarks were carried out on the JUHYDRA cluster at the JSC, featuring an Intel Xeon E5-2650 CPU. Both versions of the M2M operator, $O(p^4)$ and $O(p^3)$, were compared. Our benchmarks are focused on this operator since it is the one that we decided to parallelize on the GPU as a starting point, given the fact that the optimizations performed over the M2M phase can be easily applied to the M2L and L2L ones. In addition, it can even be argued that porting the M2M phase to a GPU implementation is harder than the M2L one, due to less workload and parallelism. Note that the $O(p^3)$ operators employ some prefactors for the rotation steps. Those prefactors as well as the Wigner d-matrices are computationally expensive but can be precomputed to reduce the runtime. We tested both variants of the rotation-based operators: on the fly and precomputed. The benchmarks were carried out for both single and double precision floating point datatypes.

As seen in Fig. 6, the M2M $O(p^3)$ on the fly operator is even slower than the $O(p^4)$ version. However, when all the constants are precomputed the complexity reduction pays off because most of its prefactor penalty is removed. Nevertheless, there is still a small prefactor which makes it slower than its $O(p^4)$ counterpart when the order of poles is small. The single precision plot (Fig. 6 top) shows unexpected results in the interval for 10–20 multipoles. By taking a closer look, we can point out a significant runtime increase from multipole order 13 until order 18 for the $O(p^3)$ precomputed operator. The slope of the $O(p^4)$ operator also changes suddenly after $p = 15$. This behavior is caused by the limited precision of the float datatype. When a certain order of poles is requested, underflows in the multipole representation occur and the numbers fall in the denormalized range of the single precision type. Because of this, the denormalized exception handling mechanism of the FPU starts acting, thus increasing the execution time due to additional function calls. At a certain point, for instance $p = 18$ in the precomputed operator, the numbers drop to zero so no additional denormalized exception overhead is produced. That is the reason why the curve *stabilizes* after order 18. The float implementation achieves a 2.5× speedup before denormalization overhead starts at $p = 13$.

If we look at the double precision plot (Fig. 6 bottom), the unexpected slope does not occur since the double representation is able to handle the required precision. The runtime is reduced by one order of magnitude (13.8× speedup) when using order 50.

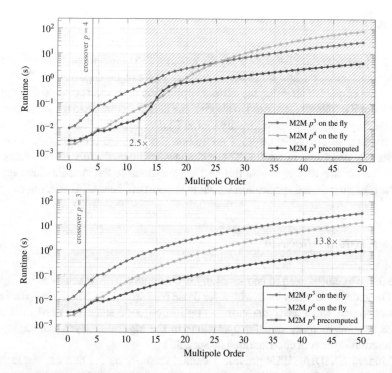

Fig. 6 Runtimes of a full M2M execution from the lowest level of an FMM tree with tree depth $d = 4$ to the highest one, i.e., 4096 M2M operator runs, shifting all boxes into a single one at the top level. Both M2M variants with $O(p^4)$ and $O(p^3)$ complexity were tested. The $O(p^3)$ operator was tested using non-precomputed constants which were calculated on the fly, and using those precomputed constants. The *upper panel* shows the timings using single precision floating point numbers and the *lower panel* shows those timings using double precision ones, varying the multipole order. For single-precision the speedup for the precomputed $O(p^3)$ compared to $O(p^4)$ was 2.5×, for double precision 13.8×. The kinks between multipole order four and six are only visible on the Sandy Bridge architecture, on Ivy Bridge the kinks are smaller, on Haswell the effect is not visible

In conclusion, a significant benefit is obtained by using the optimized operators. However, the complexity reduction implies a more sophisticated implementation and also a computationally more expensive prefactor, which should therefore be precomputed. These benchmarks establish a baseline for future improvements. In the following section, we will discuss the required steps prior to the CUDA-optimized implementation that will be deployed on a GPU.

4 Application Layout

Having a performance baseline using the existing CPU implementation, we start the code transformation into a CUDA-based one that can be deployed on GPU. The first steps consist of adapting the current application to support computations on the GPU. For that purpose, we need to allocate the data structures into the GPU memory so that they can be accessed directly by the CUDA kernels. Since we want to keep the changes in our codebase to a minimum, we leverage the application layout, previously described in Sect. 3. Those changes will heavily rely on templates and additional indirection layers. In this section, we describe the modifications applied to the aforementioned application layout to support efficient GPU execution.

4.1 Custom Allocator

The data structures need to be moved to GPU memory. This is achieved by explicit CUDA memory transfer calls whenever those data structures are needed. However, this approach will clutter the current application code since we need to include those explicit memory transfer operations in the algorithm layer, sabotaging the abstraction layer concept described in Sect. 3.

Modern NVIDIA GPUs provide a unified memory model that fits our requirements. Since all data structures make use of the allocator abstraction layer, we can just modify that layer without affecting the rest. In this way, we do not add any additional logic to the algorithm or the data structures. In addition, the allocator layer is templated so that we are flexible enough to choose between an allocator for CPU or GPU memory easily.

In this regard, we developed a custom CUDA managed allocator, which inherits from the `std::allocator` class and overrides the allocation and deallocation methods. It can be plugged in as a template parameter to our allocator abstraction layer.

In order to ease development, we decided to make use of unified memory despite its reduced efficiency when compared to other memory transfer operations, especially when data can be batched. Nevertheless, our current design provides complete control over the internal memory management mechanisms, so it can be easily extended to support other memory models or techniques such as overlapping memory transfers with computation.

4.2 Pool Allocator

The raw CUDA managed allocator has drawbacks if a considerable number of allocations has to be done. For big problem sizes bad allocations occur due to the limited amount of memory map areas a process can provide. The Linux kernel value `vm.max_map_count` limited our allocations to 65,536. Since most of our data structures contain other nested ones, our implementation performs many allocation calls and for big problem sizes we eventually reach that limit.

There are several ways to solve this problem. In our case, we resorted to using a pool allocator with a reasonable chunk size to decrease the number of allocation calls that reach the operating system. This memory management scheme is integrated in our application as a new abstraction layer between the data structures and the actual allocator.

The pool allocator allocates chunks of a predefined size and then serves parts of those chunks to the allocation calls performed by the data structures. The improvement compared to the previous layout is twofold, (i) it decreases the execution time since each allocation call has a significant latency penalty, and (ii) allows to fully utilize the GPU memory without hitting the allocation calls limit.

Thanks to the decoupled design and the templated layers, introducing this middle level is straightforward. Neither algorithm logic nor data structures code has to be changed to include a new memory management strategy. We carried out a set of benchmarks that confirmed that adding this intermediate layer has no performance impact.

4.3 Merging the CPU and GPU Codebases

The pool allocator enables the application to efficiently deal with big problem sizes. However, two distinct implementations of the same routines exist for CPU and GPU architecture. As a result code cannot be reused at the algorithm level and if one implementation changes, the other has to be changed manually.

CPU kernels for the different operators and the rotation steps take references as input arguments by design. Additionally these kernels are usually implemented by a set of nested loops which iterate over all the elements of the coefficient matrix in a sequential manner. The GPU kernels make use of pointers to those data structures, and the loop starting points and strides are different since the threads will no longer iterate sequentially over them but rather choose the data elements to compute depending on their identifiers or positions in the block/grid.

To merge both implementations into a single codebase, the data structures are converted from references to pointers for the GPU kernel wrappers or launchers. The GPU kernels access the corresponding elements of the pointers to the data structures and call the operator or rotation kernels which make use of references. These operator and rotation kernels are used by both the CPU and the GPU. Figure 7 shows the final layout of our application with all the aforementioned layers.

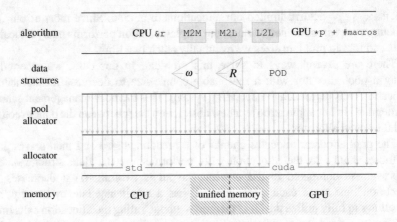

Fig. 7 Final application layout after merging the CPU and GPU logic for M2M, M2L, and L2L operators. The CPU uses references directly and the GPU launchers wrap them as pointers for the kernels. Also, preprocessor macros allow us to determine the loop starting points and strides, and specific features depending on the architecture

5 CUDA Implementation

As a starting point for the CUDA-optimized implementation, we focus on the M2M kernel. As we previously stated in Sect. 3, the optimizations performed over the M2M phase can be easily reused later for the M2L and L2L ones. Furthermore, it can even be argued that porting the M2M phase to a GPU implementation is harder than the M2L one, due to less floating point operations. Hence, an efficient implementation of M2M automatically enables an even better performance for the M2L operator, which has a significantly increased workload (see Fig. 8).

Now we will describe the parallelization of the $O(p^3)$ M2M operator, including both rotation steps, forwards and backwards. We will first focus on how to distribute the work to expose enough parallelism for the kernel functions to ensure a high GPU utilization. Then we will take an in-depth look at the different optimization strategies and CUDA techniques applied to each of the kernels. We will close showing the results of the accelerated operator and the speedup with respect to the CPU version.

5.1 Exposing Parallelism

A possible way to expose parallelism in a simple manner is to make each thread compute the whole operator for a single box, i.e., the rotation forward, M2M operator, and rotation backwards. This naive approach will spawn as many threads as boxes have to be processed. In the best case scenario we will have $(2^d)^3$ boxes in

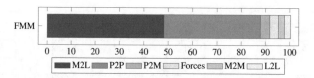

Fig. 8 Relative time distribution for the different passes of the FMM. Relative timings obtained after a full FMM run with 103k particles, $d = 4$, $p = 10$ and $ws = 1$. The CUDA parallelized versions of the $O(p^4)$ operators [12] were executed on a K40m

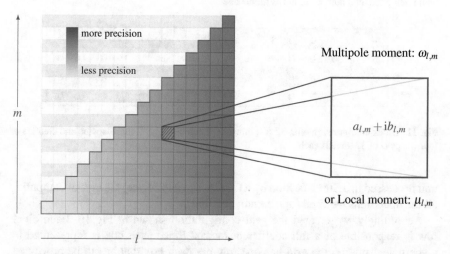

Fig. 9 Representation of the coefficient matrix datatype with an exemplary number of 15 poles. As the coefficient matrix grows, the precision increases, so a higher number of poles leads to more accurate simulations. Both axes are in the range $[0, p]$, this produces a coefficient matrix of $(p + 1)(p + 2)/2$ coefficients. The coefficients, which are represented as *squares* in the picture, are multipole or local moments depending on the type of the expansion. Each one of them holds a complex number

the lowest level of the tree. This means that for $d = 3$ we will launch 512 threads, and even for $d = 4$ only 4096 threads will be launched. Even for small block sizes, the grids will be composed of only a few blocks, preventing us to achieve a high GPU utilization.

Since there are not enough boxes to be processed, we have to take another approach to expose more parallelism. Before getting into any more detail, it is worth taking a look at the main data structure that is processed by the rotation and operator steps: the coefficient matrix. Figure 9 shows the representation of the coefficient matrix using only its upper part. It consists of a set of coefficients, representing the local or multipole moments, which are distributed in a triangular shape along the horizontal l and the vertical m axes. Each coefficient is represented by a complex number, and the rotation and operator steps usually iterate over all those coefficients to apply certain transformations (rotations, shifts, or translations). More parallelism can be exposed by assigning each warp (group of 32 threads which is the minimum

	gridDim.x									
$b_{0,0}$	$b_{1,0}$	$b_{2,0}$	$b_{3,0}$	$b_{4,0}$	$b_{5,0}$	$b_{6,0}$	$\cdots$	$b_{p,0}$
$b_{0,1}$	$b_{1,1}$	$b_{2,1}$	$b_{3,1}$	$b_{4,1}$	$b_{5,1}$	$b_{6,1}$	$\cdots$	$b_{p,1}$
$\vdots$	$\vdots$	$\vdots$	$\vdots$	$\vdots$	$\vdots$	$\vdots$	$\ddots$	$\vdots$
$b_{0,t}$	$b_{1,t}$	$b_{2,t}$	$b_{3,t}$	$b_{4,t}$	$b_{5,t}$	$b_{6,t}$	$\cdots$	$b_{p,t}$

(grid rows labelled along gridDim.y)

Fig. 10 Grid configuration. Each row is composed by $p + 1$ blocks b. The grid consists of t rows, with t being the total number of boxes minus one

w_3	t_0	t_1	t_2	t_3	t_4	t_5	t_6	t_7	t_8	t_9	t_{10}	t_{11}	$\cdots$	t_{29} t_{30} t_{31}
w_2	t_0	t_1	t_2	t_3	t_4	t_5	t_6	t_7	t_8	t_9	t_{10}	t_{11}	$\cdots$	t_{29} t_{30} t_{31}
w_1	t_0	t_1	t_2	t_3	t_4	t_5	t_6	t_7	t_8	t_9	t_{10}	t_{11}	$\cdots$	t_{29} t_{30} t_{31}
w_0	t_0	t_1	t_2	t_3	t_4	t_5	t_6	t_7	t_8	t_9	t_{10}	t_{11}	$\cdots$	t_{29} t_{30} t_{31}

Fig. 11 Block configuration with 32×4 threads t. Each block is 32 threads wide and consists of four warps w of 32 threads each

unit processed in a SIMT fashion by a CUDA-capable device) the task of computing all the operations required for a certain coefficient.

Accordingly, we created the grid configuration shown in Fig. 10. Each block row is responsible of a full coefficient matrix. Since each box is represented by a coefficient matrix, the grid has one row per each box that has to be processed. Note that by using grid-strided loops [10], we can launch less blocks and distribute the work accordingly. The block configuration is shown in Fig. 11. Each consists of four warps of 32 threads thus creating a 32×4 2D structure.

Since each row of the grid is responsible of a full coefficient matrix, i.e., the `blockIdx.y` determines which coefficient matrix the block processes each block of the row is assigned to a certain column of the corresponding coefficient matrix. In other words, the `blockIdx.x` gets mapped to the l axis as shown in Fig. 12.

Once the blocks are mapped to the coefficient matrix, the next step does the same with the threads inside those blocks. Since we have groups of 32 threads inside each block which share the same y position, i.e., each warp has the same `threadIdx.y`, we can map the warps to individual coefficients or cells of the assigned column. This means that the `threadIdx.y` variable will be mapped to the m dimension of the coefficient matrix. Figure 13 shows the warp distribution for an arbitrary block. However, the distribution is not trivial since each column has a different height and after the fourth column there are more coefficients to process than warps in the threads.

To overcome this, the warps are reassigned to the remaining coefficients in a round-robin way. By doing this, warp zero will be always assigned to $m \in \{0, 4, 8, \ldots\}$, warp one to $m \in \{1, 5, 9, \ldots\}$, warp two to $m \in \{2, 6, 10, \ldots\}$ and warp three to $m \in \{3, 7, 11, \ldots\}$ taking into account the block configuration shown in Fig. 11. It is important to remark that, even considering that the warps will be

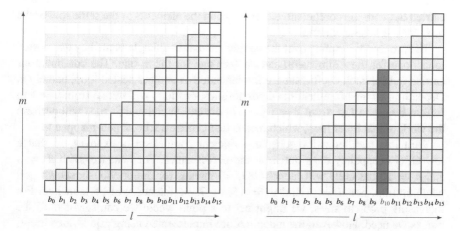

Fig. 12 Grid block row distribution. `blockIdx.x` is mapped to the l dimension of the corresponding coefficient matrix which is selected by the `blockIdx.y`, i.e., the y position of the block in the grid. The *left panel* shows the block mapping for an exemplary coefficient matrix with $p = 15$, so each grid row is composed of 16 blocks. The *right panel* shows an example of work assigned to a block. In this case the block b_{10} with `blockIdx.x` $= 10$ will have to compute all the coefficients of the highlighted column $l = 10$ of its corresponding coefficient matrix

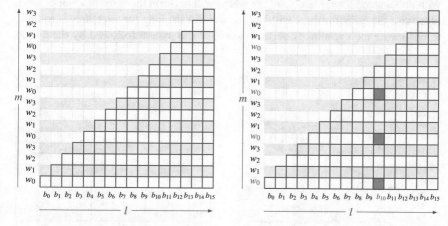

Fig. 13 Block warp distribution. Individual elements of the corresponding column of the coefficient matrix, depending on `threadIdx.y`, are mapped to warps in a round-robin fashion. The *left panel* shows the warp distribution in an exemplary coefficient matrix with $p = 15$. The *right panel* shows an example with the coefficients assigned to the first warp w_0 of a block b_{10} highlighted. That warp will compute the elements $(10, 0)$, $(10, 4)$, and $(10, 8)$. Although it would be assigned to the element $(10, 12)$ too, it will not compute it because it is outside the boundaries of the coefficient matrix ($m > l$)

theoretically assigned to a certain m, they will not process that coefficient if it is not part of the coefficient matrix. Basically, warps will only compute if $m \leq l$.

The next step is mapping the threads depending on their `threadIdx.x` value which identifies the 32 threads inside each warp. Depending on the step to be performed (rotation, operator or rotation backwards) different computations will be carried out with the coefficient, either by using the elements of the same row or the ones from the same column.

The process of iterating over the coefficient matrix is implemented by two nested loops, one for the l dimension and another one for the m one. The computations carried out for each coefficient are implemented as another nested loop, namely k. Depending on the aforementioned possibilities, this loop will iterate from $k = 1$ to l, from $k = m$ to l, or from $k = 1$ to m. In the end, individual threads will perform the work of that inner loop, which means that threadIdx.x gets mapped to k.

With this parallelization scheme, sufficient parallelism is exposed, so that a low GPU occupancy does not limit the performance. For instance, with $d = 3$ and $p = 15$ the GPU will launch 983,040 threads for the previously shown grid configuration. For $d = 4$ and $p = 15$, $\approx 7.9M$ threads will be launched. For increasing problem sizes, we might get to a point where we can't launch all the blocks we need. However, the mappings are implemented using grid-strided loops, so we can support any problem size by launching an arbitrary number of blocks and reusing them in a scalable manner.

Nevertheless, this approach has also some drawbacks. Due to the shape of the coefficient matrix and the way the m loop is mapped, a workload unbalance is produced among warps of the same block. Figure 14 shows two examples of warps with unbalanced load. Ideally, all the warps will perform the same amount of work, otherwise the early finishers will have to wait for the long running threads to finish to deallocate their resources. Also, because of the pattern followed by the k loop,

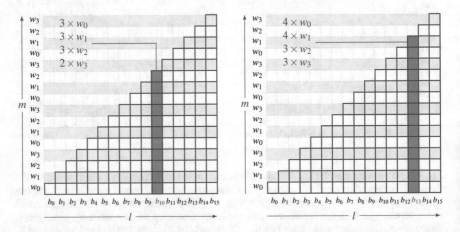

Fig. 14 Warp divergence due to different workload among the different warps of the same block. Blocks get scheduled and some of the warps finish earlier than others so they will be idle waiting for the others to end their workload. The resources allocated for the early finished warps are wasted since they will remain allocated until the longest running warp of the same block finishes. The *left panel* shows the workload for the different warps of block b_{10}, three warps will compute three coefficients each but the last warp will only compute two. The *right panel* shows another example of divergence in the block b_{13} since two warps will compute four coefficients each one and the other two will only calculate three each

not all the threads execute the same code path, so thread divergence may lead to performance degradation.

Despite the disadvantages, this strategy provides a starting point to start getting performance out of the GPU, although it can be improved to avoid the aforementioned pitfalls.

5.2 Results

We carried out a performance study to determine the improvement achieved by the CUDA-accelerated implementation. Figure 15 shows the results of that benchmark for single and double precision representations.

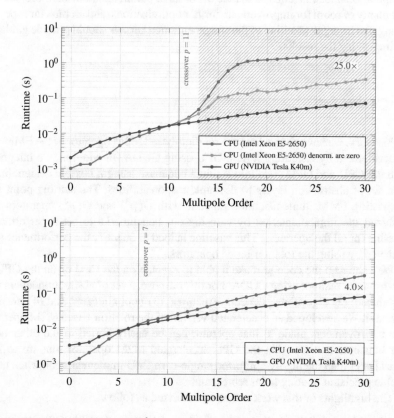

Fig. 15 Runtimes of full M2M $O(p^3)$ operator execution over the lowest level of an FMM tree with depth $d = 4$, i.e., 4096 M2M operator runs, shifting all boxes into a single one at the top level. Both CPU baseline implementation and GPU-CUDA-accelerated implementation are shown as a function of multipole order. The *upper panel* shows the results using single precision floating point numbers (with and without denormalization handling), the *lower panel* corresponds to double precision. All tests were executed on the JUHYDRA cluster at the JSC, the CPU tests ran on an Intel Xeon E5-2650 while the GPU ones used the NVIDIA Tesla K40m

For the floating point representation the same problem mentioned in Sect. 3 occurs again: the `float` datatype is not able to handle the required precision, leading to exception handling mechanisms of the CPU increasing the execution time. Currently, GPUs do not support denormalized numbers and truncate to zero immediately if an underflow occurs. For valid results, with $p < 10$, no gain is achieved by using the current GPU implementation.

The double precision benchmarks show that the extended representation is able to cope with the required precision. The crossover point is located at $p = 7$, from there the GPU implementation shows a faster execution time than the CPU one. A maximum speedup of $4.0\times$ is obtained at the biggest problem size tested, $p = 30$.

The results confirm that it is possible to improve the performance of the M2M operator by using a massively parallel device such as a GPU. However, a significant computational load is required to hide the costs of parallelism. Furthermore, there is still plenty of room for improvement, further optimizations, and architecture specific tuning. An in-depth profiling of the aforementioned kernels should provide guidance for improving the results.

6 Conclusion

In this work, we have shown how the rotation-based M2M operator of the FMM can be accelerated by executing it on a GPU using CUDA. In addition, we integrated both the CPU and GPU code into a single codebase using a flexible design, based on a set of abstraction layers to decouple responsibilities. The starting point was an existing FMM implementation pipeline with $O(p^3)$ and $O(p^4)$ operators. We analyzed the implementation by carrying out benchmarks to set a performance baseline for all the operators. This baseline helped to quantify the performance gain achieved by using the rotation-based operators.

We enhanced the code to make it able to execute on the CPU or on the GPU in a transparent manner using CUDA. For this purpose, a set of abstraction layers was introduced: (1) algorithms, (2) data structures, (3) pool allocator, and (4) memory allocator. We developed an accelerated version of the rotation-based M2M operator. The improvements made to that operator can be easily ported to the other ones. Our benchmarks show that the GPU-accelerated M2M operator runs up to four times faster than the highly optimized single-core CPU implementation when using double precision floating point representation.

The highlights of this work can be summarized as follows:

- a flexible application layout with a single codebase for the CPU/GPU implementation, based on a set of abstraction layers, atop of another:
  - an algorithm layer to hold the FMM logic
  - a data structures layer containing types, structures and their internal logic
  - a pool allocator layer for efficient memory management
  - an allocator layer for transparent memory space allocation

- a CUDA-accelerated version of the rotation-based M2M operator
  - flexible and scalable grid-strided loops
  - coalesced accesses to the data structures and fast warp reductions using CUB [14]
  - launch bounds to help the compiler optimize kernels
  - precomputed factors to save global memory round trips

Here we focused on building the abstraction layout and on accelerating the rotation-based M2M operator. The acceleration of the remaining operators M2L and L2L is straightforward since they share the same data representation and building blocks used for M2M. In addition, all the CUDA kernels can be further optimized to improve occupancy and reduce divergence. Furthermore, architecture specific tuning can be applied.

Acknowledgements This work is supported by the German Research Foundation (DFG) under the priority programme 1648 "Software for Exascale Computing—SPPEXA", project "GROMEX".

References

1. Appel, A.W.: An efficient program for many-body simulation. SIAM J. Sci. Stat. Comput. **6**(1), 85–103 (1985)
2. Barnes, J., Hut, P.: A hierarchical O (N log N) force-calculation algorithm. Nature **324**, 446–449 (1986)
3. Brandt, A.: Multi-level adaptive solutions to boundary-value problems. Math. Comput. **31**(138), 333–390 (1977)
4. Dachsel, H.: An error-controlled fast multipole method. J. Chem. Phys. **132**(11), 244102 (2009)
5. Darden, T., York, D., Pedersen, L.: Particle mesh Ewald: an N log(N) method for Ewald sums in large systems. J. Chem. Phys. **98**(12), 10089–10092 (1993)
6. Donnini, S., Ullmann, R.T., Groenhof, G., Grubmüller, H.: Charge-neutral constant ph molecular dynamics simulations using a parsimonious proton buffer. J. Chem. Theory Comput. **12**(3), 1040–1051 (2016)
7. Eastwood, J.W., Hockney, R.W., Lawrence, D.N.: P3M3DP-the three-dimensional periodic particle-particle/particle-mesh program. Comput. Phys. Commun. **19**(2), 215–261 (1980)
8. Greengard, L., Rokhlin, V.: A fast algorithm for particle simulations. J. Comput. Phys. **73**(2), 325–348 (1987)
9. Gumerov, N.A., Duraiswami, R.: Recursive computation of spherical harmonic rotation coefficients of large degree. CoRR abs/1403.7698 (2014)
10. Harris, M.: CUDA Pro Tip: Write Flexible Kernels with Grid-Stride Loops. http://devblogs. nvidia.com/parallelforall/cuda-pro-tip-write-flexible-kernels-grid-stride-loops/
11. Kabadshow, I.: Periodic boundary conditions and the error-controlled fast multipole method, vol. 11. Forschungszentrum Jülich (2012)
12. Kohnke, B., Kabadshow, I.: FMM goes GPU: a smooth trip or a bumpy ride? (2015), GPU Technology Conference

13. Lashuk, I., Chandramowlishwaran, A., Langston, H., Nguyen, T.A., Sampath, R., Shringarpure, A., Vuduc, R., Ying, L., Zorin, D., Biros, G.: A massively parallel adaptive fast multipole method on heterogeneous architectures. Commun. ACM **55**(5), 101–109 (2012)
14. Merrill, D.: CUB – collective software primitives (2013), GPU Technology Conference
15. White, C.A., Head-Gordon, M.: Rotating around the quartic angular momentum barrier in fast multipole method calculations. J. Chem. Phys. **105**(12), 5061–5067 (1996)
16. Yokota, R., Barba, L.: Treecode and fast multipole method for N-body simulation with CUDA. ArXiv e-prints (2010)

Part XIII
ExaSolvers: Extreme Scale Solvers for Coupled Problems

Space and Time Parallel Multigrid for Optimization and Uncertainty Quantification in PDE Simulations

Lars Grasedyck, Christian Löbbert, Gabriel Wittum, Arne Nägel,
Volker Schulz, Martin Siebenborn, Rolf Krause, Pietro Benedusi, Uwe Küster,
and Björn Dick

Abstract In this article we present a complete parallelization approach for simulations of PDEs with applications in optimization and uncertainty quantification. The method of choice for linear or nonlinear elliptic or parabolic problems is the geometric multigrid method since it can achieve optimal (linear) complexity in terms of degrees of freedom, and it can be combined with adaptive refinement strategies in order to find the minimal number of degrees of freedom. This optimal solver is parallelized such that weak and strong scaling is possible for extreme scale HPC architectures. For the space parallelization of the multigrid method we use a tree based approach that allows for an adaptive grid refinement and online load balancing. Parallelization in time is achieved by SDC/ISDC or a space-time formulation. As an example we consider the permeation through human skin which serves as a diffusion model problem where aspects of shape optimization, uncertainty quantification as well as sensitivity to geometry and material parameters are studied. All methods are developed and tested in the UG4 library.

L. Grasedyck (✉) • C. Löbbert
IGPM, RWTH Aachen, Aachen, Germany
e-mail: lgr@igpm.rwth-aachen.de; loebbert@igpm.rwth-aachen.de

G. Wittum • A. Nägel
G-CSC, University of Frankfurt, Frankfurt, Germany
e-mail: wittum@gcsc.uni-frankfurt.de; naegel@gcsc.uni-frankfurt.de

V. Schulz • M. Siebenborn
University of Trier, Trier, Germany
e-mail: volker.schulz@uni-trier.de; siebenborn@uni-trier.de

R. Krause • P. Benedusi
ICS, University of Lugano, Lugano, Germany
e-mail: rolf.krause@usi.ch; pietro.benedusi@usi.ch

U. Küster • B. Dick
HLRS, University of Stuttgart, Stuttgart, Germany
e-mail: dick@hlrs.de; kuester@hlrs.de

© Springer International Publishing Switzerland 2016   507
H.-J. Bungartz et al. (eds.), *Software for Exascale Computing – SPPEXA*
2013-2015, Lecture Notes in Computational Science and Engineering 113,
DOI 10.1007/978-3-319-40528-5_23

1 Introduction

From the very beginning of computing, numerical simulation has been the force driving the development. Modern solvers for extremely large scale problems require extreme scalability and low electricity consumption in addition to the properties solvers are always expected to exhibit—like optimal complexity and robustness. Naturally, the larger the system becomes, the more crucial is the asymptotic complexity issue. In this article, in order to get the whole picture, we give a brief review of recent developments towards optimal parallel scaling for the key components of numerical simulation. We consider parallelization in space in Sect. 2, in time in Sect. 4, and with respect to (uncertain) parameters in Sect. 6. These three approaches are designed to be perfectly compatible with each other and can be combined in order to multiply the parallel scalability. At the same time they are kept modular and could in principle also be used in combination with other methods. We address the optimal choice of CPU frequencies for the components of the multigrid method in Sect. 3. This serves as a representative first step for the general problem of finding an energy optimal solver, or respectively energy optimal components. Finally, in Sect. 5 the whole simulation tool is embedded in a typical optimization framework.

2 Parallel Adaptive Multigrid

To accommodate parallel adaptive multigrid computation, we developed the simulation system UG [2], which is now available in the fourth version, UG4 [21, 28]. UG4 is a solver for general systems of partial differential equations. It features hybrid unstructured grids in one, two and three space dimensions, a discretization toolbox using finite element and finite volume schemes of arbitrary order and geometric and algebraic multigrid solvers. It allows for adaptive grid refinement. Furthermore, UG4 features a flexible and self adaptive graphical user interface based on VRL [13].

In our first test we investigate the scaling properties of the geometric multigrid method in UG4 by a weak scaling test for the simple 3d-Laplace model problem (cf. Sect. 5 for strong scaling tests). As can be seen from Table 1 and Fig. 1 we achieve almost perfect weak scaling.

In our second numerical test we consider the weak scaling efficiency for a slightly more involved structural mechanics problem, 3d linear elasticity. The results in Table 2 show the same almost perfect weak scaling.

Adaptivity is a key tool for HPC, the larger the problem becomes, the more important adaptive grid resolution becomes. This can be seen from the following numerical test example computed by Arne Nägel, Sebastian Reiter and Andreas Vogel, see also [29]. To compute diffusion across human skin, we model the main barrier, i.e. the uppermost skin layer, the stratum corneum. The stratum corneum

Table 1 Weak scaling on JUQUEEN. 3d-Laplacian, uniform grid, finite volumes with linear ansatz functions, geometric multigrid V-cycle, damped Jacobi smoother ($\nu_1 = \nu_2 = 2$). We denote by p the number of processors, by dofs the number of degrees of freedom, by N_{iter} the number of multigrid iterations, and by $T_{\text{ass}}, T_{\text{setup}}$, and T_{solve} the elapsed time for the assembly, setup, and solve, respectively

p	L	dofs	N_{iter}	T_{ass}	(eff.)	T_{setup}	(eff.)	T_{solve}	(eff.)
64	8	4,198,401	10	4.46	–	2.22	–	3.04	–
256	9	16,785,409	10	4.47	99.6	2.17	102.2	3.08	98.6
1,024	10	67,125,249	10	4.46	99.9	2.32	95.6	3.13	97.0
4,096	11	268,468,225	10	4.40	101.3	2.26	98.3	3.17	95.8
16,384	12	1,073,807,361	10	4.42	100.9	2.38	98.3	3.27	93.0
65,536	13	4,295,098,369	10	4.42	100.9	2.47	89.7	3.40	89.5
262,144	14	17,180,131,329	10	4.47	99.7	2.62	84.9	3.55	85.5

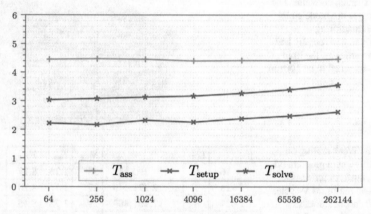

Fig. 1 Weak scaling on JUQUEEN. 3d-Laplacian, uniform grid, finite volumes with linear ansatz functions, geometric multigrid V-cycle, damped Jacobi smoother ($\nu_1 = \nu_2 = 2$). Plotted is the elapsed time $T_{\text{ass}}, T_{\text{setup}}$, and T_{solve} for the assembly, setup and solve phase, respectively, for $p = 64, \ldots, 262{,}144$ processors (From [21])

consists of dead horn cells, the corneocyctes, which are glued together by lipid bilayers. As geometry model we use the so-called cuboid model as shown in Fig. 2. To compute diffusion of a substance across stratum corneum, we use a diffusion equation with constant diffusivities in the two different materials, corneocytes and lipids, and add a transmission condition for the interior material boundaries

$$\frac{\partial c(\mathbf{x}, t)}{\partial t} = \operatorname{div}\left(k(\mathbf{x}) \nabla c(\mathbf{x}, t)\right)$$

with the diffusivities

$$k(\mathbf{x}) = \begin{cases} k_{lip}, & \mathbf{x} \text{ in lipid layer} \\ k_{cor}, & \text{otherwise} \end{cases}$$

Table 2 Weak scaling on JUQUEEN. 3d linear elasticity, uniform grid, finite volumes with linear ansatz functions, geometric multigrid V-cycle, damped Jacobi smoother ($\nu_1 = \nu_2 = 2$). Legend as in Table 1

p	L	dofs	$T_{\text{ass}} + T_{\text{setup}} + T_{\text{solve}}$	(eff.)
1	3	14,739	7.33	–
8	4	107,811	7.42	98.8
64	5	823,875	7.58	96.8
512	6	6,440,067	7.79	94.2
4,096	7	50,923,779	7.90	92.8
32,768	8	405,017,091	8.08	90.7
262,144	9	3,230,671,875	8.21	89.4

Fig. 2 Cuboid model of human stratum corneum. The corneocytes are modeled by cuboids, measuring $30 \times 30\mu$m horizontally and 1μm in vertically, the lipid layer is assumed to be 100 nm thick

Table 3 Weak scaling on JUQUEEN, skin problem 3d cuboid, uniform refinement, geometric multigrid V-cycle, damped Jacobi smoother ($\nu_1 = \nu_2 = 3$, $\omega = 0.6$), base level 4, base solver LU. Legend as in Table 1

p	L	dofs	N_{iter}	T_{ass}	T_{setup}	T_{solve}
16	6	290,421	25	1.76	8.17	20.23
128	7	2,271,049	27	1.77	8.20	22.31
1,024	8	17,961,489	29	1.78	8.45	24.10
8,192	9	142,869,025	29	1.78	8.48	23.35
65,536	10	1,139,670,081	29	1.79	8.59	24.79

and the transmission condition

$$K_{cor/lip} \cdot c_{lip}(\mathbf{x}, t) \big|_{\mathbf{n}_-} = c_{cor}(\mathbf{x}, t) \big|_{\mathbf{n}_+}$$

for the interior material boundaries. The transmission condition describes the so-called partitioning effect caused by the lipophilicity, respectively hydrophilicity, of the diffusing substance. The problem here is the extreme anisotropy, which is combined with the jumping diffusivities of the material. Together, these features cause an optimal barrier effect as described in [12, 20]. We used this model for a scaling study with uniform refinement. The results are shown in Table 3. The same model was used to study the influence of adaptive grid refinement in parallel. To that end, we did a weak scaling study of this problem with adaptive refinement using a residual error estimator as refinement criterion and compared this with the uniform refinement results. Plotting this into one graph, we obtain the results in Fig. 3.

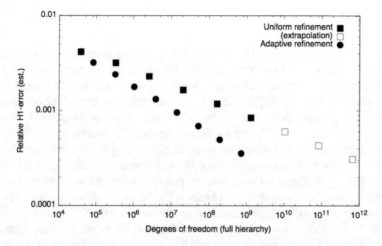

Fig. 3 Error reduction per degrees of freedom for uniform and adaptive refinement in parallel (From [29])

From that we conclude:

- Using the full machine with the adaptive approach, we gain an accuracy comparable to the one with a computer 512 times as large.
- Adaptivity is a leading method for power saving. To reach the same error with the adaptive method, you need just 1024 CPUs instead of 65,536 CPUs in the uniform case or using 65,536 CPUs for the adaptive computation, we would need a computer 512 times larger, i.e. with 33,554,432 CPUs, to reach the same accuracy on a uniform grid. This means saving 99.5 % in CPU time and in power consumption.

3 Empirically Determined Energy Optimal CPU Frequencies

Besides improved scaling properties also energy efficiency poses a challenge that needs to be tackled in order to enable exascale computing. This is due to rising energy costs, a limited availability of electric power at many sites as well as challenges for heat dissipation. From our point of view, increasing energy efficiency requires approaches on multiple fields. The most important one will be efficient algorithms as addressed in the previous and following sections. Furthermore, efficient implementations of these algorithms are required and the resulting codes need to be executed on energy efficient hardware. Moreover, the CPU's clock frequency may be adjusted according to the current load. In this section, we will give an overview of our approach to do the latter. A detailed description of this method has already been published in [6]. Hence, we just give a summary here and refer to the aforementioned document regarding further details.

3.1 Approach

Our purpose is to figure out the *maximum* energy saving potentials and corresponding runtime impacts achievable by adjusting the CPU's clock frequency. We hence try to minimize the energy required to solve a given problem. This energy can be determined by $E = \int_{t_1}^{t_2} P(t)\, dt$ with $P(t)$ denoting the present power consumption of the corresponding code, running from time t_1 to t_2. According to [4], P can be approximated by $P = CV^2(t)f(t)$ where C denotes the semiconductor's capacity and $V(t)$ respectively $f(t)$ denote the time dependent supply voltage as well as clock frequency of the CPU. Hence, reducing $f(t)$ and $V(t)$ by so-called dynamic voltage and frequency scaling (DVFS), decreases power but may increase the runtime and therefore energy consumption.[1] In phases with intensive memory access, however, one may observe only a slight increase of runtime because the CPU is anyway forced to wait on the memory subsystem most of the time. However, predicting memory access characteristics in complex codes is a challenging task. Thus, it is also hard to predict the optimal clock frequency and we therefore deploy an empirical approach. Linux also does this in its standard configuration but clock frequency decisions are based on an idle time analysis. In contrast to this, we take advantage of knowledge about potential phase boundaries and adjust the clock frequency immediately to the optimal value instead of spending time for a runtime analysis first.

In order to do so, we employ preparatory measurements to figure out energy optimal clock frequencies and utilize them in subsequent production runs. The overhead induced by this method is negligible if it is possible to determine optimal frequencies within a single node or timestep and use them within plenty of those. The core of our approach is to run the entire target code at a *fixed* clock frequency, measure the resulting power consumption over time, and repeat this procedure with all the available frequencies. The resulting energy consumption of a routine can be determined by integrating the measured power over the routine's runtime. Since all routines have been run and profiled at *all* available frequencies, one can now—per routine—pick out the optimal ones in terms of energy. Hence, phases with varying memory access characteristics can be reflected by adapting the clock frequency per routine to its optimal value.

We emphasize that—while minimizing energy—this method may significantly increase the runtime. Nevertheless we deploy it because we are interested in the maximum energy saving potentials and corresponding runtime impacts of DVFS, as stated above.

---

[1] We use the commonly utilized term "energy consumption" despite the fact that electrical energy is *converted* to thermal energy.

3.2 Implementation Details

In order to implement the aforementioned approach, a measurement method is required that yields highly reliable and time correlated results. We measure the actual supply voltage V and current I of the used CPU and its associated memory modules as close to these components as possible. The current flow I can be determined according to Ohm's law as $I = \frac{V_R}{R}$ with V_R denoting the voltage drop over a high precision shunt $R = 0.01\Omega$ in the CPU's supply line. Based on these values, one can calculate the present power consumption by $P = VI$. The required measurements are performed by an A/D converter in a separate machine with high accuracy ($\epsilon_{relative} < \pm 1.5\%$) and a time resolution of 80 μs.

As already mentioned before, integrating P over time yields the energy spent in a particular routine. The corresponding time interval is determined by calls to gettimeofday() from within the code to be evaluated. This method requires a precise synchronization between the real time clocks of the compute node and the measurement hardware. The Precision Time Protocol (PTP) is employed for this purpose via a separate ethernet link. By this method, an average time deviation of about 20 μs can be achieved, which is below the time resolution of the used A/D converter and therefore admissible.

Unfortunately it is not possible to separate the power consumption of distinct components—especially CPU cores—with the described method. Parallel runs on multiple cores without tight synchronization and perfect load balancing will hence blur the measured power consumption. We therefore restrict our method to serial runs for a start. Parallel runs might nevertheless be regarded in future research.

In order to compensate for OS jitter as well as other transients, five runs of every test case are evaluated and their median is used in further processing.

Since the time resolution of our measurement system is 80 μs, it is not possible to make reasonable statements on the energy consumption of routines with a runtime in or below this order of magnitude. We therefore take into account only routines with a runtime of at least 1 ms. Since the used profiler solely provides accumulated times, $\frac{t_{ac}}{n} \geq 1$ ms is used as the selection criterion, where t_{ac} denotes the accumulated runtimes (including subroutine calls) of all routine calls and n denotes their number. In analogy to this, optimal clock frequencies are selected based on the average energy consumption of entire calls (i.e., including subroutine calls) to the respective routine. These criteria are just one of many possible choices and other ones will be investigated in future research.

According to [14], the frequency transition latency of current CPUs is substantial. In case of a rapid series of frequency transition requests it is hence not reasonable to immediately set the new frequency in the target production runs. As a consequence, we wait for a period of 10 μs after the first request of the series, track further ones, and serve only the latest. The choice of this value will also be subject to future research.

Unfortunately, the library call used for setting the frequencies[2] is blocked until completion of the transition (which may be a substantial amount of time), although the CPU can be used in the usual manner during this period. Because of the possibly large number of frequency transitions, this may decrease the overall performance and at the same time raise the energy consumption to an extent that may diminish the benefits of optimal frequency usage. We therefore trigger the actual frequency transition from within a concurrent helper thread, cf. [6].

3.3 Evaluation

In order to evaluate the described approach, it has been applied to the already mentioned numerical simulation code UG4. As a representative application, UG4 has been deployed to solve a time dependent convection–diffusion problem using the vertex centered finite volume method on a two dimensional grid with the geometric multigrid solver and several combinations of setup parameters, particularly different smoothers. Every timestep involves several phases with differing characteristics, i.e., discretization and system assembly (memory-bounded), system solution (CPU-bounded), as well as output of results (potentially I/O-bounded), which may be exploited by the approach by means of differing clock frequencies. The corresponding runs have been executed on an Ivy Bridge compute node, cf. [6] for technical details.

To quantify the effects of the approach, we compared the resulting runtime and energy consumption of entire runs to those resulting from Linux' default clock frequency management. In our first experimental measurements (cf. [6]) we have found an average energy saving potential of about 10 %, which was, however, contrasted by an average runtime penalty of about 19 %. By further investigating our method since publishing those data, we have found that results are not fully reproducible in between different runs of the approach. Despite this fact, it still seems to be possible to reduce the energy requirements by allowing an increased runtime.

One shortcoming of the approach is the limitation to systems with special measurement equipment. Hence, it will be important to investigate the precision of power estimation by means of hardware performance counters with respect to our approach, in order to use them for the preparatory measurements on conventional systems.

In future research, we will tackle these problems in order to enable full reproducibility. We will, moreover, try to reduce the induced runtime penalty and expand the method to multiple active cores within a socket.

---

[2]`cpufreq_set_frequency()`

4 Parallel in Time Multigrid

Firstly we mention an inexact variant of the well known time integrator Spectral Deferred Correction (SDC); SDC is commonly used as a smoother for multilevel algorithms in time. The Inexact SDC (ISDC) algorithm can reduce significantly the computational cost of SDC. In fact in SDC, a full system solution is required for an implicit, or semi-implicit strategy. On the other hand in ISDC few multigrid V-cycles are used to get an approximate solution. The effectiveness of this technique is due to the iterative nature of SDC that provides an accurate initial guess for the multigrid cycles. This method has been tested on the heat equation (see Table 4) and Viscous Burgers' equations in [27].

The natural usage of the ISDC time stepper is in the context of multilevel time-parallel algorithms, e.g. MLSDC [26] or PFASST [7], based originally on SDC. Both schemes perform SDC sweeps in a hierarchy of levels and use a FAS correction term for the spatial representation of the problem on different levels. ISDC can further improve parallel efficiency of those parallel-in-time methods [16].

Secondly we mention the results in the context of a multigrid space–time solution method for the Navier-Stokes equations with periodic boundary conditions in time [3]. The Navier-Stokes equations are discretized in space–time with high order finite differences on a staggered grid. The discretization leads to a large, ill-conditioned, non-linear system that has to be solved in parallel. Picard iterations are used to treat the non-linearity and we design a block Gauss-Seidel smoother for a space–time multigrid algorithm. A local Fourier analysis is used to analyze the smoothing property of such a method on a staggered grid. The space–time domain is fully decomposed resulting in a parallel-in-time method. Convergence and weak/strong scaling were successfully tested (see Fig. 4).

Table 4 Accumulated Multigrid V-cycles over all sweeps to reduce the SDC or ISDC residual below $5 \cdot 10^{-8}$ for different values of the diffusion coefficient k in the heat equation and the number of quadrature nodes M

k	M	SDC	ISDC	Savings (%)
1	3	16	12	25
	5	23	20	13
	7	32	28	13
k	M	SDC	ISDC	Savings (%)
10	3	36	20	44
	5	61	40	34
	7	79	47	41
k	M	SDC	ISDC	Savings (%)
100	3	106	52	51
	5	150	104	31
	7	187	167	11

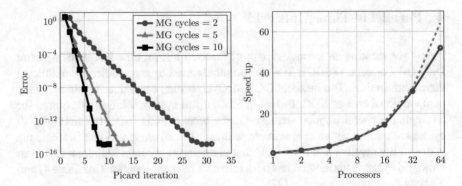

Fig. 4 *Left*: convergence of the Picard iteration with different numbers of multigrid cycles per iteration. *Right*: strong scaling results with processors equally distributed in space and time

5 Scalable Shape Optimization Methods for Structured Inverse Modeling in 3D Diffusive Processes

We consider the inverse modeling of the shape of cells in the outermost layer of the human skin, the so-called stratum corneum. For this purpose we present a novel algorithm combining mathematical shape optimization and high performance computing. In order to show the capabilities of this method, we assume that we have an experiment providing a time-series of data describing the spatial distribution of a tracer in a skin sample. Based on this information, we aim at identifying the structure and the parameters matching the experimental results best. The starting point is a common computational model for the so-called stratum corneum based on tightly coupled tetrakaidecahedrons. For a review, the reader is referred, e.g., to [17, 18].

From a computational point of view, this means to evaluate the model equations, compute the defect to the measurements, evaluate sensitivities of this defect with respect to the shape of the parameter distribution and finally update the shape in order to minimize the defect. A special focus is on the scalability of the optimization algorithm for large scale problems. We therefore apply the geometric multigrid solver UG4 [28].

In this particular application, we are dealing with flows dominated by diffusion. We thus choose the classical parabolic model equation for the simulation together with standard finite elements. By c we denote the concentration of the quantity of interest in the domain $\Omega = \Omega_1 \cup \Omega_2$ over the time interval $[0, T]$. At the initial time $t = 0$, the concentration c is fixed to homogeneously zero in the entire domain and one at the upper boundary Γ_{top}. The other boundaries denoted by Γ_{out} are modeled such that there is no flux across them. The permeability of the domain Ω is given by a jumping coefficient k taking two distinct values k_1 and k_2 in Ω_1 and Ω_2. It can thus be thought of as a homogeneous material with inclusions of different permeability separated by the interior boundary Γ_{int}. The underlying model is given by

$$\min \; J(\Omega) := \frac{1}{2} \sum_{i=1}^{M} \int_{\Omega} (c(t_i) - \bar{c}(t_i))^2 dx + \mu \int_{\Gamma_{int}} 1 \, ds \tag{1a}$$

$$\text{s.t.} \; \frac{\partial c}{\partial t} - \operatorname{div}(k\nabla c) = f \quad \text{in } \Omega \times (0, T] \tag{1b}$$

$$c = 1 \quad \text{on } \Gamma_{top} \times (0, T] \tag{1c}$$

$$[\![c]\!] = 0, \quad \left[\!\!\left[k\frac{\partial c}{\partial \mathbf{n}} \right]\!\!\right] = 0 \quad \text{on } \Gamma_{int} \times (0, T] \tag{1d}$$

$$\frac{\partial c}{\partial \mathbf{n}} = 0 \quad \text{on } \Gamma_{out} \times (0, T] \tag{1e}$$

$$c = c_0 \quad \text{in } \Omega \times \{0\} . \tag{1f}$$

The first term in (1a) tracks the observations and the second term is a perimeter regularization. Thus, the optimization tends to shapes Γ_{int} with minimal surface area. Equations (1d) describe the continuity of the concentration and of the flux across Γ_{int}.

The corresponding adjoint equation, which is obtained by deriving the Lagrangian (cf. [25]) of problem (1a), (1b), (1c), and (1d) with respect to the state c, then reads as

$$-\frac{\partial p}{\partial t} - \operatorname{div}(k\nabla p) = \begin{cases} -(c - \bar{c}) & \text{in } \Omega \times \{t_1, \ldots, t_M\} \\ 0 & \text{in } \Omega \times [0, T) \setminus \{t_1, \ldots, t_M\} \end{cases} \tag{2a}$$

$$p_2 = k_1 \frac{\partial p}{\partial n}, \quad p = 0 \quad \text{in } \Omega \times \{T\} \tag{2b}$$

$$[\![p]\!] = 0, \quad \left[\!\!\left[k\frac{\partial p}{\partial n} \right]\!\!\right] = 0 \quad \text{on } \Gamma_{int} \times [0, T) \tag{2c}$$

$$p_1 = -k_1 p, \quad \frac{\partial p}{\partial n} = 0 \quad \text{on } \Gamma_{out} \times [0, T) \tag{2d}$$

$$p = 0 \quad \text{on } \Gamma_{top} \times [0, T) . \tag{2e}$$

In order to derive the derivative with respect to the shape, first the space of feasible shapes has to be defined. For more details on the connection of shape calculus and shape manifolds, see [22]. We consider the manifold

$$B_e(S^2, \mathbb{R}^3) := \operatorname{Emb}(S^2, \mathbb{R}^3) / \operatorname{Diff}(S^2) \tag{3}$$

of smooth embeddings $\operatorname{Emb}(S^2, \mathbb{R}^3)$ of the unit sphere S^2 into $\mathbb{R}^3$. Let $b \in B_e(S^2, \mathbb{R}^3)$ be a feasible shape, then the tangent space to the manifold in b is given by all smooth deformations in normal direction

$$T_b B_e = \{h \mid h = \alpha \mathbf{n}, \ \alpha \in C^\infty(S^2, \mathbb{R})\} . \tag{4}$$

Additionally, we need to equip the tangential space with an inner product. Here we take the so-called Sobolev metric for a constant $\gamma > 0$ given by

$$g^1 : T_b B_e \times T_b B_e \to \mathbb{R}, \ (u, v) \mapsto \int_b \langle (\mathrm{id} - \gamma \Delta_b) u, v \rangle \, ds . \tag{5}$$

The symbol Δ_b denotes the tangential Laplace or Laplace-Beltrami operator along b. This inner product determines the representation of the shape gradient which is then the actual update to the shape in each optimization step. The Sobolev inner product with the Laplace-Beltrami operator and a proper parameter γ ensure smooth shape deformations such that the optimized shape remains in B_e.

The shape derivative in direction of a smooth vector field $V : \Omega \to \mathbb{R}^3$ is defined as

$$dJ(\Omega)[V] := \lim_{h \to 0} \frac{J(\Omega_h) - J(\Omega)}{h} \tag{6}$$

where $\Omega_h = \{x + h \cdot V(x) \mid x \in \Omega\}$ is perturbed according to V. For the underlying model equations the shape derivative is derived in [25] and is given by

$$dJ(\Omega)[V] = \int_{\Gamma_{int}} \left(\int_0^T \langle V, \mathbf{n} \rangle \left[-2k \frac{\partial c}{\partial \mathbf{n}} \frac{\partial p}{\partial \mathbf{n}} + k \nabla c^T \nabla p \right] \, dt + \langle V, \mathbf{n} \rangle \, \mu \kappa \right) ds \tag{7}$$

where $\kappa : \Gamma_{int} \to \mathbb{R}$ denotes the sum of the principle curvatures of the variable surface Γ_{int}.

In most applications, the measurements $\bar{c}$ are not available as a continuous function. There is rather a set of discrete measurements in space. We thus apply radial basis functions in order to interpolate $\bar{c}$ to the finite element nodes where c is given.

The next step is to obtain a descent direction which can be applied as a deformation to the mesh. On each triangle $\tau \subset \Gamma_{int}$ we evaluate the quantity

$$\delta_0 := \left[-2k \frac{\partial c}{\partial \mathbf{n}} \frac{\partial p}{\partial \mathbf{n}} + k \nabla c^T \nabla p \right] \tag{8}$$

i.e., the jump of the value in brackets between in two opposing tetrahedra on Γ_{int} sharing a common triangle. Rescaling $\mathbf{n}$, we define the vector

$$\mathbf{g}_0 := \delta_0 \mathbf{n}. \tag{9}$$

For linear finite elements, both δ_0 and $\mathbf{g}_d$ are piecewise constant on each surface triangle. Thus, in order to be consistent with the curvature, which is available in each surface node, we project $\mathbf{g}_d$ onto a vector $\mathbf{g}_c$ in the space of piecewise linear basis functions via the L_2 projection, i.e.,

$$\int_{\Gamma_{int}} \mathbf{g}_c \, \mathbf{v} \, ds = \int_{\Gamma_{int}} \mathbf{g}_d \, \mathbf{v} \, ds \tag{10}$$

for all piecewise linear trial functions $\mathbf{v}$ on Γ_{int}. By solving $(\mathrm{id} - \gamma \Delta_b)\mathbf{g} = \mathbf{g}_c$ with a discretization of the Laplace-Beltrami operator as derived in [15] we finally obtain the representation of the shape gradient $\mathbf{g}$.

One optimization iteration can be summarized in the following steps:

1. Evaluate measurements on current grid via radial basis function representation,
2. solve parabolic and its adjoint PDE with geometric multigrid,
3. compute δ_0 and integrate over time,
4. L^2 projection of piecewise constant gradient to linear basis function space and add curvature for regularization,
5. solve Laplace-Beltrami equation for the representation of the gradient in the Sobolev metric,
6. solve linear elasticity equations with $\mathbf{g}$ as Dirichlet condition on Γ_{int} and deform the mesh.

The algorithm described here is implemented within the software toolbox UG4 [28]. This software is known to be scalable and features parallel multigrid solvers [21]. Numerical experiments were conducted on the HERMIT[3] supercomputer.

The investigation of the scalability is depicted in Fig. 5a–d. The computations shown are based on a coarse grid with 9923 elements and, due to uniform refinements, a fine grid with 325,156,864 elements on the 5th level. For strong scalability (cf. Fig. 5a–c), one observes that most timings decrease, when p increases. All operations not involving any solver, show this decrease. We can explain the saturation for larger number of cores by the time the coarse grid solver requires, which is a natural behavior. Also the weak scalability, which can be seen in Fig. 5d, reflects our expectations. In the decreasing times, for the gradient computation one clearly sees the difference in the asymptotic behavior of volume cells and surface cells. A more detailed analysis can be found in [19].

In our future work we will focus especially on two issues of the presented method. First, due to the incorporation of the shape gradient as a Dirichlet condition in the mesh deformation, the iterated finite element grids tend to have overlapping elements. In [23] we present shape metrics which circumvent this issue and additionally lead to good mesh qualities. Second, the scalability of the presented

---

[3]HLRS, Stuttgart, Germany, http://www.hlrs.de/systems/platforms/cray-xe6-hermit/

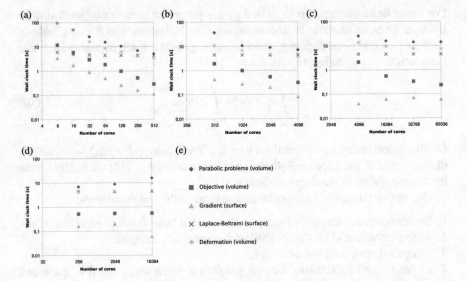

Fig. 5 Scaling of different components of the algorithm in first optimization iteration. (**a**) Strong scaling on level 3. (**b**) Strong scaling on level 4. (**c**) Strong scaling on level 5. (**d**) Weak scaling, increment factor 8 for cells and processors. (**e**) Legend (From [19])

approach is affected by the necessity to solve PDEs on surfaces only. In [24] equivalent formulations for (7) using volume formulations are investigated which overcome the effect on the scalability.

6 Uncertainty Quantification

As a typical parameter-dependent extreme scale problem we consider a PDE involving diffusion coefficients that are parametrized by $\mathbf{p} \in P = [0, 1]^d$,

$$\text{div } (k(\mathbf{x}, \mathbf{p})\nabla c(\mathbf{x}, \mathbf{p})) = f(\mathbf{x}, \mathbf{p}), \quad \mathbf{x} \in \Omega(\mathbf{p}) \quad + b.c. \tag{11}$$

and assume that for fixed parameters $\mathbf{p} \in P$ a solution $c(\mathbf{x}, \mathbf{p})$ is computed in parallel by the UG4 library. The parameters could, e.g., be piecewise diffusion coefficients in each corneocyte. However, we are not interested in the whole solution c of (11) itself but rather in a quantity $\phi : P \to \mathbb{R}$, e.g. the integral mean of the solution $c(\cdot, \mathbf{p})$ over a subset Ω_ϕ:

$$\phi(\mathbf{p}) = \frac{1}{|\Omega_\phi|} \int_{\Omega_\phi} c(\mathbf{x}, \mathbf{p}) \, d\mathbf{x} . \tag{12}$$

Since the parameter set P is d-dimensional, even a discretization with 5 parameter values for each component gives rise to 5^d possible combinations, which exceeds the estimated number of particles in the observable universe already for $d \approx 150$ parameters. Therefore, the full tensor cannot be stored or computed, but rather an extremely data-sparse approximation of it. This approximation is sought in the hierarchical low rank Tucker format [8, 11].

In [1, 10] we have devised a strategy for parallel sampling of tensors in the hierarchical Tucker format, i.e. we compute a few of the values $\phi(\mathbf{p})$ and derive all others from these—based on the assumption that $\phi(\cdot)$ can be approximated in the data sparse low rank hierarchical Tucker format (cf. [5]). In this context sampling means that only certain entries of the tensor are required as opposed to intrusive methods that require us to solve the underlying system of PDEs in the tensor format. The sampling strategy that we propose is guided by the idea that samples are taken one after the other and that later samples can be adapted to the already obtained information of prior samples. This is in contrast to tensor completion strategies [9] where the samples are taken randomly (perfectly parallelizable) and the tensor is completed afterwards.

As a result of [10], parallelization of the (adaptive) sampling process is possible with an almost optimal speedup. Since the method is only a heuristic, it would be helpful to obtain an a posteriori estimate of the approximation quality. For this, we require

- a representation of the underlying (discrete) operator A, the right-hand side b, and the solution c in the hierarchical Tucker format,
- to approximately compute the (discrete) residual $r = b - Ac$,
- to estimate the accuracy by relating it to the residual.

This, however, is still under development. As a first step into this direction, we have distributed the hierarchical low rank tensor according to the dimension tree layout over $2d - 1$ nodes (here we use a complete binary tree and consider only powers of 2 for d). For such a distributed tensor the parallel tensor arithmetic has to be developed. One key ingredient is the evaluation of the tensor, i.e. extracting a single entry from the compressed representation. This procedure has been parallelized and gives the results in Table 5.

d	Parallel time (s)	Serial time (s)	Speedup
4	0.127	0.246	1.9
8	0.261	0.777	3.0
16	0.433	1.880	4.3
32	0.627	4.206	6.7
64	0.882	8.673	9.8
128	0.869	18.82	21.6
256	1.057	38.09	36.0

Table 5 Parallel weak scaling of the tensor evaluation for distributed tensors. The tensor is of size $100,000^d$, the number of processors used is $2d - 1$, the internal rank is $k = 500$ for every node in the dimension tree

We observe that the parallel speedup is roughly 36 for a tensor in dimension $d = 256$ with 511 processors. The loss is due the fact that the nodes in the dimension tree have to be processed sequentially one level after the other (which was expected). In addition to distributing the data of the tensor over several nodes, we also gain a considerable speedup.

7 Conclusion

We have presented the development of a parallel multigrid based solver for complex systems and tasks such as shape optimization or uncertainty quantification within the unified UG4 software library. The modular parallelization in space, time, and with respect to parametric dependencies allows us to provide the software for computing way beyond exascale.

Acknowledgements All ten authors gratefully acknowledge support from the DFG (Deutsche Forschungsgemeinschaft) within the DFG priority program on software for exascale computing (SPPEXA), project Exasolvers.

References

1. Ballani, J., Grasedyck, L.: Hierarchical tensor approximation of output quantities of parameter-dependent PDEs. SIAM/ASA J. Uncertain. Quantif. **3**(1), 852–872 (2015)
2. Bastian, P., Wittum, G.: Robustness and adaptivity: the UG concept. In: Hemker, P., Wesseling, P. (eds.) Multigrid Methods IV, Proceedings of the Fourth European Multigrid Conference. Birkhäuser, Basel (1994)
3. Benedusi, P., Hupp, D., Arbenz, P., Krause, R.: A parallel multigrid solver for time-periodic incompressible Navier–Stokes equations in 3d. In: Karasözen, B., Manguoglu, M., Tezer-Sezgin, M., Göktepe, S., Ugur, Ö. (eds.) Numerical Mathematics and Advanced Applications – ENUMATH 2015. Springer, Ankara (2016)
4. Corporation, I.: Enhanced Intel® SpeedStep® Technology for the Intel® Pentium® M Processor. White Paper (2004). http://download.intel.com/design/network/papers/30117401.pdf
5. Dahmen, W., DeVore, R., Grasedyck, L., Süli, E.: Tensor-sparsity of solutions to high-dimensional elliptic partial differential equations. Found. Comput. Math. 1–62 (2015). http://dx.doi.org/10.1007/s10208-015-9265-9
6. Dick, B., Vogel, A., Khabi, D., Rupp, M., Küster, U., Wittum, G.: Utilization of empirically determined energy-optimal CPU-frequencies in a numerical simulation code. Comput. Vis. Sci. **17**(2), 89–97 (2015). http://dx.doi.org/10.1007/s00791-015-0251-1
7. Emmett, M., Minion, M.L.: Toward an efficient parallel in time method for partial differential equations. Commun. Appl. Math. Comput. Sci. **7**, 105–132 (2012)
8. Grasedyck, L.: Hierarchical singular value decomposition of tensors. SIAM J. Matrix Anal. Appl. **31**, 2029–2054 (2010)
9. Grasedyck, L., Kluge, M., Krämer, S.: Variants of alternating least squares tensor completion in the tensor train format. SIAM J. Sci. Comput. **37**(5), A2424–A2450 (2015)

10. Grasedyck, L., Kriemann, R., Löbbert, C., Nägel, A., Wittum, G., Xylouris, K.: Parallel tensor sampling in the hierarchical tucker format. Comput. Vis. Sci. **17**(2), 67–78 (2015)
11. Hackbusch, W., Kühn, S.: A new scheme for the tensor representation. J. Fourier Anal. Appl. **15**(5), 706–722 (2009)
12. Heisig, M., Lieckfeldt, R., Wittum, G., Mazurkevich, G., Lee, G.: Non steady-state descriptions of drug permeation through stratum corneum. I. The biphasic brick-and-mortar model. Pharm. Res. **13**(3), 421–426 (1996)
13. Hoffer, M., Poliwoda, C., Wittum, G.: Visual reflection library: a framework for declarative gui programming on the java platform. Comput. Vis. Sci. **16**(4), 181–192 (2013)
14. Mazouz, A., Laurent, A., Benoît, P., Jalby, W.: Evaluation of CPU frequency transition latency. Comput. Sci. **29**(3–4), 187–195 (2014). http://dx.doi.org/10.1007/s00450-013-0240-x
15. Meyer, M., Desbrun, M., Schröder, P., Barr, A.H.: Discrete differential-geometry operators for triangulated 2-manifolds. In: Hege, H.C., Polthier, K. (eds.) Visualization and Mathematics III, pp. 35–57. Springer, Berlin (2003)
16. Minion, M.L., Speck, R., Bolten, M., Emmett, M., Ruprecht, D.: Interweaving PFASST and parallel multigrid. SIAM J. Sci. Comput. **37**, S244–S263 (2015)
17. Mitragotri, S., Anissimov, Y.G., Bunge, A.L., Frasch, H.F., Guy, R.H., Hadgraft, J., Kasting, G.B., Lane, M.E., Roberts, M.S.: Mathematical models of skin permeability: an overview. Int. J. Pharm. **418**(1), 115–129 (2011)
18. Naegel, A., Heisig, M., Wittum, G.: Detailed modeling of skin penetration – an overview. Adv. Drug Delivery Rev. **65**(2), 191–207 (2013). http://www.sciencedirect.com/science/article/pii/S0169409X12003559. Modeling the human skin barrier – towards a better understanding of dermal absorption
19. Nägel, A., Schulz, V., Siebenborn, M., Wittum, G.: Scalable shape optimization methods for structured inverse modeling in 3D diffusive processes. Comput. Vis. Sci. **17**(2), 79–88 (2015)
20. Nägel, A., Heisig, M., Wittum, G.: A comparison of two- and three-dimensional models for the simulation of the permeability of human stratum corneum. Eur. J. Pharm. Biopharm. **72**(2), 332–338 (2009)
21. Reiter, S., Vogel, A., Heppner, I., Rupp, M., Wittum, G.: A massively parallel geometric multigrid solver on hierarchically distributed grids. Comput. Vis. Sci. **16**(4), 151–164 (2013). http://dx.doi.org/10.1007/s00791-014-0231-x
22. Schulz, V.: A Riemannian view on shape optimization. Found. Comput. Math. **14**, 483–501 (2014)
23. Schulz, V., Siebenborn, M.: Computational comparison of surface metrics for PDE constrained shape optimization. Comput. Methods Appl. Math. (submitted) (2015). arxiv.org/abs/1509.08601
24. Schulz, V., Siebenborn, M., Welker, K.: A novel Steklov-Poincaré type metric for efficient PDE constrained optimization in shape spaces. SIAM J. Optim. (submitted) (2015). arxiv.org/abs/1506.02244
25. Schulz, V., Siebenborn, M., Welker, K.: Structured inverse modeling in parabolic diffusion problems. SIAM J. Control Optim. **53**(6), 3319–3338 (2015). arXiv.org/abs/1409.3464
26. Speck, R., Ruprecht, D., Emmett, M., Minion, M.L., Bolten, M., Krause, R.: A multi-level spectral deferred correction method. BIT Numer. Math. **55**, 843–867 (2015)
27. Speck, R., Ruprecht, D., Minion, M., Emmett, M., Krause, R.: Inexact spectral deferred corrections. In: Domain Decomposition Methods in Science and Engineering XXII. Lecture Notes in Computational Science and Engineering, vol. 104, pp. 127–133. Springer, Cham (2015)
28. Vogel, A., Reiter, S., Rupp, M., Nägel, A., Wittum, G.: UG4: a novel flexible software system for simulating PDE based models on high performance computers. Comput. Vis. Sci. **16**(4), 165–179 (2013). http://dx.doi.org/10.1007/s00791-014-0232-9
29. Wittum, G.: Editorial: algorithmic requirements for HPC. Comput. Vis. Sci. **17**(2), 65–66 (2015)

Part XIV
Further Contributions

Domain Overlap for Iterative Sparse Triangular Solves on GPUs

Hartwig Anzt, Edmond Chow, Daniel B. Szyld, and Jack Dongarra

Abstract Iterative methods for solving sparse triangular systems are an attractive alternative to exact forward and backward substitution if an approximation of the solution is acceptable. On modern hardware, performance benefits are available as iterative methods allow for better parallelization. In this paper, we investigate how block-iterative triangular solves can benefit from using overlap. Because the matrices are triangular, we use "directed" overlap, depending on whether the matrix is upper or lower triangular. We enhance a GPU implementation of the block-asynchronous Jacobi method with directed overlap. For GPUs and other cases where the problem must be overdecomposed, i.e., more subdomains and threads than cores, there is a preference in processing or scheduling the subdomains in a specific order, following the dependencies specified by the sparse triangular matrix. For sparse triangular factors from incomplete factorizations, we demonstrate that moderate directed overlap with subdomain scheduling can improve convergence and time-to-solution.

1 Introduction

Sparse triangular solves are an important building block when enhancing Krylov solvers with an incomplete LU (ILU) preconditioner [28]. Each iteration of the solver requires the solution of sparse triangular systems involving the incomplete factors. Exact solves with sparse triangular matrices are difficult to parallelize due to the inherently sequential nature of forward and backward substitution.

H. Anzt (✉) • J. Dongarra
University of Tennessee, Knoxville, TN, USA
e-mail: hanzt@icl.utk.edu; dongarra@icl.utk.edu

E. Chow
Georgia Institute of Technology, Atlanta, GA, USA
e-mail: echow@cc.gatech.edu

D.B. Szyld
Temple University, Philadelphia, PA, USA
e-mail: szyld@temple.edu

© Springer International Publishing Switzerland 2016   527
H.-J. Bungartz et al. (eds.), *Software for Exascale Computing – SPPEXA*
2013-2015, Lecture Notes in Computational Science and Engineering 113,
DOI 10.1007/978-3-319-40528-5_24

Level scheduling strategies [28] aim at identifying sets of unknowns that can be computed in parallel (called "levels"), but these sets are often much smaller than the parallelism provided by the hardware. Particularly on manycore architectures like graphics processing units (GPUs), level-scheduling techniques generally fail to exploit the concurrency provided.

At the same time, the incomplete factorizations are typically only a rough approximation, and exact solutions with these factors may not be required for improving the convergence of the Krylov solver. Given this situation, interest has developed in using "approximate triangular solves" [7]. The concept is to replace the exact forward and backward substitutions with an iterative method that is easy to parallelize. Relaxation methods like the Jacobi method provide parallelism across vector components, and can be an attractive alternative when running ILU-preconditioned Krylov methods on parallel hardware. For problems where only a few steps of the iterative method applied to the sparse triangular systems are sufficient to provide the same preconditioning quality to the outer Krylov method, the approximate approach can be much faster [14, 15]. A potential drawback of this strategy, however, is that disregarding the dependencies between the vector components can result in slow information propagation. This can, in particular, become detrimental when using multiple local updates for better cache reuse [6]. In this paper, we investigate improving the convergence of approximate sparse triangular solves by using overlap strategies traditionally applied in domain decomposition methods. Precisely, we enhance a block-iterative method with restricted Schwarz overlap, and analyze the effect of non-uniform overlap that reflects the information propagation dependencies. The findings gained are then used to realize overlap in a GPU implementation of block-asynchronous Jacobi.

The remainder of the paper is structured as follows. Section 2 provides some background on sparse triangular solves, block-asynchronous Jacobi, and different types of Schwarz overlap with the goal of setting the larger context for this work. In Sect. 3, the benefits of restricted additive Schwarz and directed overlap are investigated for different synchronization strategies, and with specific focus on sparse triangular systems arising from incomplete factorization preconditioners. Section 4 gives details about how we realized overlap in the GPU implementation. In Sect. 5, we analyze the convergence and performance improvements we achieved by enhancing block-asynchronous Jacobi with restricted overlap. We conclude in Sect. 6.

2 Background and Related Work

2.1 Sparse Triangular Solves

Due to their performance-critical impact when used in preconditioned Krylov methods, much attention has been paid to the acceleration of sparse triangular solves. The traditional approach tries to improve the exact solves. The most common strategies are based on level scheduling or multi-color ordering [2, 21–23, 30]. A more disruptive approach is to use partitioned inverses [1, 27], where the triangular matrix is written as a product of sparse triangular factors, and each triangular solve becomes a sequence of sparse matrix vector multiplications. Also, the use of sparse approximate inverses for the triangular matrices were considered [9, 17]. The solution of the triangular systems is then replaced by the multiplication with two sparse matrices that are approximating the respective inverses of the triangular factors. With the increase in parallelism that is available in hardware, *iterative* approaches to solving triangular systems become tantalizing, as they provide much finer grained parallelism. In situations where an approximate solution is acceptable, which often is the case for incomplete factorization preconditioning, iterative triangular solves can accelerate the overall solution process significantly, even if convergence is slightly degraded [7, 14]. With regard to the increased parallelism expected for future hardware systems, iterative triangular solves are also attractive from the standpoint of fault-tolerance [8].

2.2 Jacobi Method and Block-Asynchronous Iteration

Classical relaxation methods like Jacobi and Gauss-Seidel use a specific update order of the vector components, which implies synchronization between the distinct iterations. The number of components that can be computed in parallel in an iteration depends on whether the update of a component uses only information from the previous iteration (Jacobi type) or also information from the current iteration (Gauss-Seidel type). Using newer information generally results in faster convergence, which however reduces the parallelism: Gauss-Seidel is inherently sequential and requires a strict update order; for Jacobi, all components are updated simultaneously within one iteration [7]. If no specific update order is enforced, the iteration becomes "chaotic" or "asynchronous" [13, 18]. In this case, each component update takes the newest available values for the other components. The asymptotic convergence of asynchronous iterations is guaranteed if the spectral radius of the positive iteration matrix, $\rho(|M|)$, is smaller than unity [18]. This is a much stronger requirement than needed for Jacobi, however it is always fulfilled for sparse triangular systems [7, 14]. The fine-grained parallelism and the lack of synchronization make asynchronous methods attractive for graphics processing units (GPUs), which themselves operate in an asynchronous-like fashion within one kernel operation [24]. In particular, the special case where subsets of the iteration vector are iterated in synchronous Jacobi fashion and asynchronous updates are used in-between the subsets can efficiently be realized on GPUs [3]. The potential of this "block-asynchronous Jacobi" on GPU hardware was investigated in [6]. Block asynchronous Jacobi was also considered as smoother for geometric multigrid methods [5], and evaluated in a mixed-precision iterative refinement framework [4]. In [7], block-asynchronous Jacobi was employed as an iterative solver for sparse triangular systems arising from incomplete factorization preconditioning. Precisely, the benefits of replacing exact sparse triangular solves with approximate triangular solves were demonstrated for an ILU-preconditioned Krylov solver running on GPUs. This work ties on the findings presented therein by enhancing the block-asynchronous Jacobi method with overlap strategies.

2.3 Overlapping Domains and Restricted Additive Schwarz

The idea of improving information propagation by overlapping blocks originated with domain decompositions methods. In these methods, a large domain is split into subdomains, where local solution approximations are computed. A global solution is generated by iteratively updating the local parts and communicating the components of the solutions in the domain intersections. Increasing the size of these intersections usually accelerates the information propagation [32, 34]. In the alternating Schwarz method, the subdomains are processed sequentially in a fixed order. The additive Schwarz method performs subdomains solves in parallel. To avoid write conflicts in the overlap region, the update of the global solution must be implemented carefully. One approach that has proven to be very effective is "Restricted additive Schwarz" (RAS) proposed in [12], where each processor restricts the writing of the solution to the local subdomain, and discards the part of the solution in the region that overlaps other subdomains. The convergence properties of RAS are analyzed in [19]. An initial asynchronous approach for an additive Schwarz method is presented in [20]. The underlying idea is very similar to the work presented in this paper, however it starts with a physical domain decomposition problem and then allows for asynchronism in the update order. In contrast, the approach we present starts from an arbitrary linear system that is interpreted as a domain decomposition problem with domain sizes induced by the GPU thread block size used in the block-asynchronous Jacobi method. Some theoretical convergence results for asynchronous iterations with overlapping blocks can also be found in [33].

3   Random-Order Alternating Schwarz

3.1   Domain Overlap Based on Matrix Partitioning

In domain decomposition methods for solving partial differential equations (PDEs), the subdomains are usually contiguous, physical subdomains of the region over which a solution is sought. In this work, however, we adopt a black-box-solver setting, where no details about the physical problem or domain are available. This is a reasonable premise as many software packages must accommodate this situation. We note that in this case, incomplete factorization preconditioners are often employed, as they work well for a large range of problems.

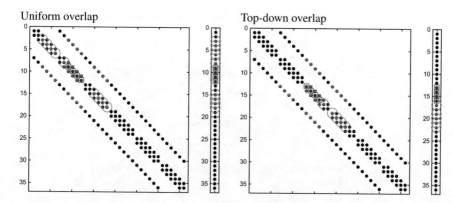

Fig. 1 Algebraic Schwarz overlap for a 5-point stencil discretization of the Laplace problem in 2D. The unknowns of the original subdomain are marked as *blue squares*, the overlap of level one is derived from the matrix structure and indicated by *red circles*. The *left figure* shows uniform overlap, the *right figure* shows top-down overlap

For a matrix problem, we call a subdomain the set of unknowns corresponding to a partition of unknowns (e.g., partitioning of the unknown vector) which may not correspond to a physical subdomain even if the problem comes from a PDE. Here, the overlap for the distinct subdomains cannot be derived from the physical domain, but has to be generated from the dependency graph of the matrix. This strategy was first proposed in [11]. We use the terminology introduced therein by calling this kind of overlap "algebraic Schwarz overlap". Algebraic Schwarz overlap of level 1 is generated by including all unknowns that are distant by one edge to the subdomain when solving the local problem. Recursively applying this strategy results in overlap of higher levels: for level o overlap, all unknowns distant by at most o edges are considered. See Fig. 1 for an illustration of algebraic Schwarz overlap.

In a first experiment, we analyze the effect of overlap for a block-iterative solver. The target problem is a finite difference discretization of the Laplace problem in 3D. The discretization uses a 27-point stencil on a $8 \times 8 \times 8$ grid, resulting in a symmetric test matrix where 10,648 edges connect 512 unknowns. The block-iterative method splits the iteration vector into 47 blocks that we call "subdomains" to be consistent with domain decomposition terminology. On each subdomain, the local problem is solved via 2 Jacobi sweeps. Subdomain overlap is generated as algebraic Schwarz overlap. Restricted Schwarz overlap only updates the components part of the original subdomain. The motivation for restricting the results also in sequential subdomain updates is the GPU architecture we target in the experimental part of the paper. There, multiple subdomains are updated in parallel. All experiments in this section are based on a MATLAB implementation (release R2014a) running in double precision.

Figure 2 shows how restricted alternating Schwarz using level 1 overlap improves the convergence rate when solving the above Laplace problem. Each subdomain is updated once in a global iteration, and the results are averaged over 100 runs.

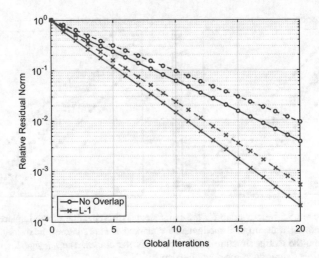

Fig. 2 Convergence of the restricted alternating Schwarz using 2 Jacobi sweeps as local solver on the subdomains applied to the Laplace test problem. The subdomain updates are scheduled in Gauss-Seidel order (*solid lines*), or in a random order (*dashed lines*)

The solid lines are for sequential top-down subdomain scheduling (Gauss-Seidel), the dashed lines are for a random update order. In the remainder, we call the latter "random-order restricted alternating Schwarz". Random subdomain scheduling results in slower average convergence, but the improvement obtained from restricted Schwarz overlap is of similar quality. We note that extending the original subdomains with overlap increases the computational cost, as on each subdomain a larger local problem has to be solved. For this test problem, level 1 overlap increases the total number of floating point operations for one global sweep by a factor of 6.68. This issue is not reflected in Fig. 2 showing the convergence with respect to global iterations. The increase in the computational cost does however typically not reflect the increase in execution time, as in a parallel setting also the communication plays an important role. In the end, the interplay between hardware characteristics, the linear system, and the used decomposition into subdomains determines whether the time-to-solution benefits from using overlapping subdomains [31].

3.2 Directed Overlap

The purpose of using overlap is to propagate information faster across the local problems. More overlap usually results in faster convergence. However, it can be expected that not all the information propagated provides the same convergence benefits:

1. It can be expected that propagating information in the dependency direction of the unknowns may provide larger benefit.

2. For non-parallel subdomain updates, propagating information from subdomains already updated in the current iteration may provide larger benefit than information from the previous iterate.

A non-directed or bidirected dependency graph (structurally symmetric matrix) makes it impossible to benefit from scheduling the subdomains in dependency order. For each dependency that is obeyed, the opposite dependency is violated. In this case, the optimization of the information propagation boils down to propagating primarily information from subdomains that have already been updated in the current iteration. For the sequential subdomain scheduling in top-down Gauss-Seidel order, domain overlap pointing opposite the subdomain scheduling order propagates information from already updated subdomains. Overlap pointing in the scheduling direction propagates information of the previous iteration. Hence, bottom-up overlap may carry "more valuable" information than overlap pointing top-down. For the remainder of the paper we use the term "directed overlap" if the original subdomain is extended only in a certain direction:

- "Top-down overlap" means that the original subdomain is extended by unknowns adjacent in the graph representation of the matrix that have larger indexes, i.e., are located closer to the end of the iteration vector.
- "Bottom-up overlap" means that the original subdomain is extended by unknowns adjacent in the graph representation of the matrix that have smaller indexes, i.e., are located closer to the top of the iteration vector.

We note, that in case the problem originates from a discretization of a partial differential equation, the concept of directed overlap does in general not correspond to a direction in the physical domain. An example where the directed overlap has a physical representation is a 1-dimensional physical domain in combination with consecutive numbering of the unknowns. More generally, if a domain in an n-dimensional space is divided into (possibly overlapping) subdomains with boundaries which do not intersect each other, consecutive numbering allows for a physical interpretation. For a visualization of directed overlap, see the right side of Fig. 1.

The advantage of extending the subdomains only in one direction compared to uniform overlap is that the computational cost of solving the local problems grows slower with the overlap levels.

Figure 3 compares the convergence of block-Jacobi using different restricted Schwarz overlap strategies for a sequential top-down subdomain scheduling. All overlap strategies result in faster convergence than the overlap-free block-Jacobi. However, significant difference in the convergence rate can be observed: the top-down overlap fails to propagate new information, and propagating information from subdomains not yet updated in the global iteration provides only small convergence improvement. The uniform overlap treats information from adjacent unknowns equally, independent of whether the respective subdomain has already been updated in the current iteration or not. The resulting convergence improvement comes at a 6.68 times higher computational cost, as elaborated previously. Using directed

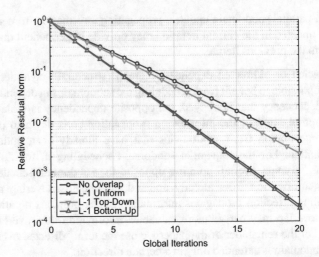

Fig. 3 Convergence of the sequential restricted alternating Schwarz using top-down subdomain scheduling and different overlap strategies. The test matrix is the Laplace problem introduced in the beginning of the section

overlap pointing bottom-up increases the computational cost only by a factor 3.34. For this test case, the bottom-up overlap provides the same convergence improvement like the uniform overlap. The lower computational cost makes this strategy superior. Although disregarding overlap in direction of "old" neighbors may in general result in a lower convergence rate than uniform overlap, this test validates the expectation that propagating new information provides higher benefits when solving a symmetric problem.

For a random update order, it is impossible to define an overlap direction that propagates information only from already updated subdomains. On average, the effects of using overlap pointing bottom-up and overlap pointing top-down equalize, and the resulting convergence rate is lower than when using uniform overlap, see Fig. 4.

The situation changes as soon as we look into structurally non-symmetric matrices with a directed dependency graph. While propagating information from freshly updated subdomains may still be preferred, the dependencies have a much more important role for convergence. Obviously, these dependencies should also be considered in the subdomain scheduling. Then, it is possible to choose directed overlap that benefits twofold: information gets propagated in dependency directions; and this information comes from subdomains that have already been updated in the current iteration.

For a non-symmetric matrix, it is impossible to always find a subdomain update order that obeys all dependencies. A scenario where this is possible, however, is the solution of sparse triangular systems and sequential component updates. The resulting algorithm is nothing other than forward and backward

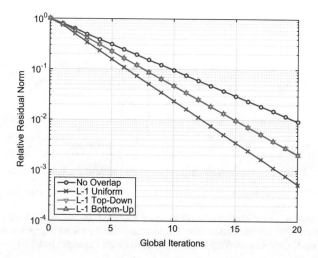

Fig. 4 Convergence of the sequential random-order restricted alternating Schwarz for different overlap strategies. The test matrix is the Laplace problem introduced in the beginning of the section

substitution. Subdomains containing multiple unknowns are available if components are independent and only depend on previously updated components. The strategy of identifying these subdomains and updating the components in parallel is known as level scheduling [29]. Unfortunately, the subdomains are often too small to allow for efficient parallel execution. Also, the components are usually not adjacent in the iteration vector, which results in expensive memory access patterns. But also when using subdomains coming from a decomposition of the iteration vector, it is often possible to identify an overall dependency direction. Some dependencies might be violated, but aligning the subdomain scheduling to the triangular system's dependency direction usually results in faster convergence [7].

In the scenario of random subdomain scheduling, the idea of orienting overlap opposite the update order fails, but overlap may still be adjusted to the dependency graph. Bottom-up overlap propagates information in dependency direction for a lower triangular system, top-down overlap obeys the dependencies for an upper triangular system.

In Fig. 5, the convergence of the different overlap strategies is compared for a lower and an upper sparse triangular system. The systems arise as incomplete LU factorization without fill-in (ILU(0)) for the previously considered Laplace problem. On each subdomain, two Jacobi sweeps are used to solve the local problem. The sequential subdomain updates are scheduled in a random order, and the results are averaged over 100 runs. Despite ignoring the benefits of orienting the overlap towards already updated subdomains, the convergence of the restricted alternating Schwarz using directed overlap opposite the dependency direction matches the convergence of uniform overlap. Propagating information opposite the dependency

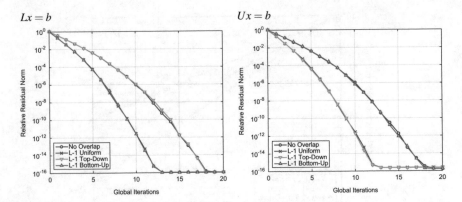

Fig. 5 Convergence of the sequential random-order restricted alternating Schwarz using 2 Jacobi sweeps as local solver on the subdomains. The test cases are the sparse triangular systems arising as incomplete LU factorization without fill-in (ILU(0)) of the Laplacian problem

direction does not provide noticeable benefits compared to the non-overlapping block-iterative solver.

4   Restricted Overlap on GPUs

We now want to evaluate whether using restricted Schwarz overlap can improve iterative sparse triangular solves on GPUs. The hardware characteristics however require some modification of the approach suggested in Sect. 3. On GPUs, operation execution is realized via a grid of thread blocks, where the distinct thread blocks apply the same kernel to different data [24]. The distinct thread blocks all have the same "thread block size", which is the number of compute threads in one thread block. Multiple thread blocks can be executed in concurrent fashion. The number of thread blocks handled in parallel depends on the thread block size, requested shared memory, and the characteristics of the used GPU hardware. This setting suggests to assign each thread block to one subdomain. If the subdomain size matches the thread block size, all the unknowns in a subdomain can be handled in parallel by the distinct compute threads, which is the desired setting for using Jacobi sweeps as local solver. For non-overlapping subdomains of equal size, this mapping works fine. Extending the subdomains with algebraic Schwarz overlap however becomes difficult, as the overlap for distinct subdomains may differ in size. Fixing the thread block size to the largest subdomain results in idle threads assigned to smaller subdomains; smaller thread block sizes fail to realize the Jacobi sweeps in parallel fashion. An additional challenge comes from the solution of the local problem being based on local memory. On GPUs, this is the shared memory of the distinct multiprocessors. Due to the limited size, it is impossible to keep the complete iteration vector in shared memory, but the unknowns of the local problem (original subdomain and overlap) have to be stored in consecutive fashion. As these components are in general not adjacent in the global iteration vector, an additional mapping is required. This increases pressure on the memory bandwidth, the typically performance-limiting instance in this algorithm. Also, the scattered access to the components of the algebraic Schwarz overlap in the global iteration vector results in expensive memory reads [24].

Given this background, we relax the mathematical validity of the overlap in favor of higher execution efficiency. Precisely, replace the algebraic Schwarz overlap with "block-overlap". For a given decomposition of the iteration vector, block overlap is generated by extending the subdomains in size such that the subdomains adjacent in the iteration vector overlap. Similar to the algebraic Schwarz overlap, unknowns can be part of multiple subdomains. However, it also is possible that not all components in a subdomain are connected in the dependency graph. The restricted Schwarz setting avoids not only write conflicts, but also ensures that structurally disconnected overlap is not written back to the global iteration vector. Compared to the algebraic Schwarz overlap, two drawbacks can be identified:

- Block overlap can miss important dependencies if the respective unknowns are not included in the block-extension of the subdomain.
- Components part of the block overlap but not structurally connected to the original subdomain increase the cost of the local solver, but do not contribute to the global approximation.

These handicaps become relevant for matrices with significant entries distant to the main diagonal. For matrices where most entries are reasonably close to the diagonal, the higher execution efficiency on GPUs may outweigh the drawbacks. Sparse triangular systems as they arise in the context of approximate incomplete factorization preconditioning often have a Reverse Cuthill-McKee (RCM) ordering, as this ordering helps in producing accurate incomplete factorization preconditioners [10, 16]. At the same time, RCM ordering reduces the matrix bandwidth, which makes block overlap more attractive (more matrix entries captured in the overlap regions).

To match the thread block size of the non-overlapping block-iterative solver, we shrink the original subdomains, and fill up the thread block with overlap. We note that shrinking the original subdomain size and restricting the writes requires a higher number of subdomains for covering the iteration vector. The corresponding higher number of thread blocks reflects the increased computational cost when using block overlap. If subdomains adjacent in the iteration vector are scheduled to the same multiprocessor, overlapping subdomains allows for temporal and spacial cache reuse [15]. The data is loaded into the fast multiprocessor memory only once, and can then be reused for the overlap of an adjacent subdomain. Figure 6 visualizes the strategy of mapping thread blocks to subdomains for the case of non-overlapping subdomains (*left*), and the case of directed bottom-up overlap (*right*), respectively. Note that in the latter, the overlap threads of each thread block only read the data for the local problem (**r**), but do not write back the local solution to the global iteration vector.

No overlap   Bottom-up overlap (30 %)

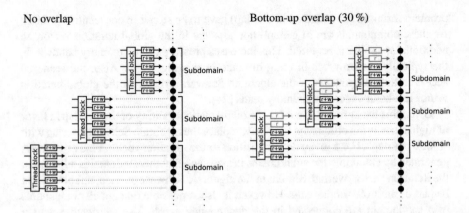

Fig. 6 Mapping thread blocks to the subdomains in case of non-overlapping subdomains (*left*) and bottom-up overlap (*right*). In the latter case, only the threads at the lower end of each thread block write back the solution update (**r+w**), the threads at the upper end only read the data in for the local problem (**r**)

On GPUs, multiple thread blocks are executed in concurrent fashion. For large problems there however exist more subdomains than can be handled in parallel. We call this case an "overdecomposition": not all subdomains are updated simultaneously, and the update order impacts the convergence of the block-iterative method. Unfortunately, GPUs generally do not allow insight or modifications to the thread block execution order. However, backward-engineering experiments reveal that for the used GPU architecture, the thread blocks are usually scheduled in consecutive increasing order [15]. For sparse triangular solves, this property can be exploited to improve the information propagation by numbering the thread blocks in dependency direction [7]. The fact that this scheduling order cannot be guaranteed gives the solver a block-asynchronous flavor, and requires to report all experimental results as average over multiple runs. As the block overlap is mathematically inconsistent with algebraic Schwarz overlap, we avoid the term "restricted additive Schwarz", but refer to the implementation as "block-asynchronous Jacobi with restricted overlap".

5 Experimental Results

5.1 Test Environment

The experimental results were obtained using a Tesla K40 GPU (Kepler microarchitecture) with a theoretical peak performance of 1,682 GFlop/s (double precision). The 12 GB of GPU main memory, accessed at a theoretical bandwidth of 288 GB/s, was sufficiently large to hold all the matrices and all the vectors needed in the iteration process. Although all operations are handled by the accelerator, we mention

Table 1 Characteristics of the sparse lower triangular ILU(0) factors employed in the experimental tests

	Matrix	Description	Size n	Nonzeros n_z	n_z/n	Condition number
UFMC	BCSSTK38	Stiffness matrix, airplane engine component	8,032	116,774	14.54	6.87e+08
	CHP	Convective thermal flow (FEM)	20,082	150,616	7.50	7.90e+05
	CONSPH	Concentric spheres (FEM)	83,334	1,032,267	12.39	6.81e+06
	DC	Circuit simulation matrix	116,835	441,781	3.78	6.54e+10
	M_T1	Structural problem	97,578	4,269,276	43.74	4.78e+10
	STO	3D electro-physical duodenum model	213,360	1,660,005	7.78	1.38e+07
	VEN	Unstructured 2D Euler solver (FEM)	62,424	890,108	14.26	1.85e+07
	LAP	3D Laplace problem (27-pt stencil)	262,144	3,560,572	13.58	9.23e+06

for completeness that the host was being an Intel Xeon E5 processor (Sandy Bridge). The implementation of all GPU kernels is realized in CUDA [25], version 7.0 [26], using a thread block size of 256. For non-overlapping subdomains, this thread block size corresponds to the size of the subdomains; for overlapping subdomains the size is split into subdomain and overlap. Double precision computations were used.

For the experimental evaluation, we consider solving with the incomplete factorizations of different test matrices, including all problems tested in [7] to show the potential of iterative sparse triangular solves. The test matrices are general sparse matrices from the University of Florida matrix collection (UFMC), and a finite difference discretization of the 3D Laplace problem with Dirichlet boundary conditions. For the discretization, a 27pt stencil on a $64 \times 64 \times 64$ mesh is used, resulting in structurally and numerically symmetric matrix. We consider all matrices in RCM ordering to reduce the matrix profile, as this increases the effectiveness of overlapping matrix rows that are nearby, as well as the effectiveness of incomplete factorizations. Table 1 lists the characteristics of the lower triangular matrices from the incomplete factors. The sparsity plots of these triangular matrices are shown in Fig. 7. We report all experimental results as the average over 100 runs to account for nondeterministic scheduling effects on the GPU in the block-asynchronous Jacobi solver.

5.2 Sparse Triangular Solves

Figures 8 and 9 show convergence and timing results for the lower sparse triangular factors coming from incomplete factorizations of the selected UFMC matrices and the Laplace matrix. The figures on the left side show convergence of the residual norm with respect to iterations, while the figures on the right side relate the residual norm to the execution time. The GPU thread block scheduling was set to promote

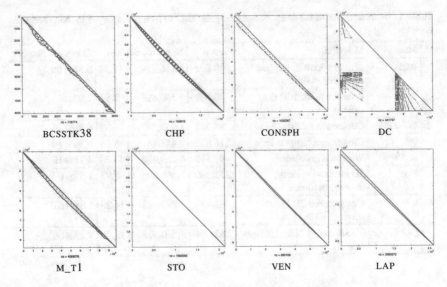

Fig. 7 Sparsity plots of the sparse lower triangular factors listed in Table 1

top-down subdomain scheduling for information propagation in the dependency direction. In addition, bottom-up overlaps of different sizes were used, which also accounts for the dependency direction of lower triangular matrices. The notation used in the figures relates the size of the overlap to the thread block size, i.e., 25 % overlap means that each thread block of size 256 contains a subdomain of size 192 and 64 overlap components. For 25 % overlap, the number of thread blocks necessary to cover the complete iteration vector increases by one third compared to a non-overlapping decomposition using a subdomains size of 256. The computational cost, i.e., the number of thread blocks being scheduled, increases by the same factor.

We first make some overall observations. From Figs. 8 and 9, we observe a range of behaviors for the different problems. In most of the cases, the use of overlap improves the convergence rate. The exceptions are the STO and VEN problems, for which there is very little effect due to overlap, and M_T1 where overlap can actually make convergence worse. For the problems where overlap improves convergence rate, there is still the question of whether or not computation time is improved, since overlap increases the amount of work. The best timings may come from a small or a moderate amount of overlap (rather than a large amount of overlap), balancing the extra computational effort with improved convergence rate.

For the CHP problem, a small convergence improvement can be achieved by using overlap, and this improvement grows, as expected, with the size of the overlap. However, when considering the GPU execution time, overlap is always worse than non-overlap for this problem. On the other hand, for the DC problem, overlap can improve convergence rate significantly. In addition, overlap does not significantly increase computational cost, as this matrix is very sparse compared to the other test matrices.

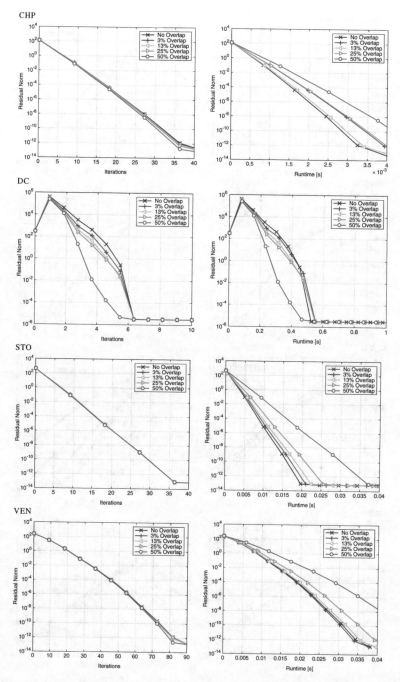

Fig. 8 Convergence (*left*) and residual-performance (*right*) of the block-asynchronous Jacobi using 2 Jacobi sweeps as local solver on the subdomains. The test cases are the sparse lower triangular systems arising as incomplete LU factorization without fill-in (ILU(0)) for the UFMC problems

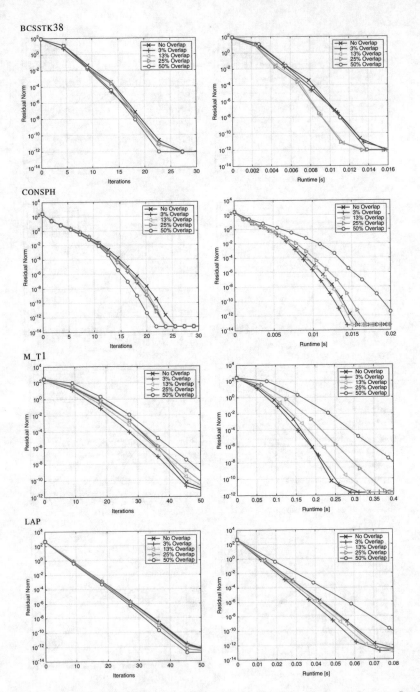

Fig. 9 Convergence (*left*) and residual-performance (*right*) of the block-asynchronous Jacobi using 2 Jacobi sweeps as local solver on the subdomains. The test cases are the sparse lower triangular systems arising as incomplete LU factorization without fill-in (ILU(0)) for the UFMC problems and the LAP problem

For the STO and VEN problems, overlap makes little or no improvement to convergence, as already mentioned. The STO matrix is a large matrix compared to the others, and overlap of adjacent rows in the matrix may introduce few additional couplings in the dependency direction, i.e., many off-diagonal entries are too far from the main diagonal for block overlap to include them. For these large matrices, a decomposition into physical subdomains would be better.

For the BCSSTK38 problem, overlap accelerates the overall solution process. The best results are achieved for 13 % and 25 % overlap. Overlap is beneficial also for the CONSPH problem, but the convergence improvement (left side) is not reflected in the time-to-solution metric. Only moderate overlap (see results for 3 % and 13 %, respectively) accelerate the solution process. From the figures, we see that for most problems, 50 % overlap has the worst execution time, but the best execution time is given by 3–25 % overlap.

The M_T1 problem is an example for which overlap degrades the convergence rate. For this problem, the matrix has a block structure that is easily captured by non-overlapping subdomains, but this structure is not well-matched by overlapping subdomains. Again for this problem, a decomposition into physical subdomains would be better.

Finally, for the LAP problem, convergence can be improved by using overlap. Again, the best solver is not the one using the most overlap, but the one using 3 % overlap. This is a typical pattern for block overlap: moderate overlap helps in faster information propagation, but large overlap includes too many structurally disconnected components that increase the computational cost, but do not aid in accelerating convergence.

6 Summary and Future Work

We investigate the potential of enhancing block-iterative methods with restricted Schwarz overlap. For systems carrying a nonsymmetric dependency structure, pointing the overlap opposite the dependencies propagates the information in dependency direction. This improves the convergence rate. We propose a GPU implementation where we relax the consistency to algebraic Schwarz overlap in favor of higher execution efficiency. For sparse triangular factors arising as incomplete LU factors we analyze the convergence and performance benefits achieved by enhancing block-asynchronous Jacobi with directed overlap of different size. Depending on the matrix structure, restricted overlap can improve time-to-solution performance.

In the future, we will look into optimizing the block overlap used in the GPU implementation to the matrix structure. Adapting the overlap size to the location of the most significant off-diagonal entries improves convergence at moderate cost increase. This optimization makes the block overlap more similar to algebraic Schwarz overlap, and will in particular work for problems with a finite element origin.

Acknowledgements This material is based upon work supported by the U.S. Department of Energy Office of Science, Office of Advanced Scientific Computing Research, Applied Mathematics program under Award Numbers DE-SC-0012538 and DE-SC-0010042. Daniel B. Szyld was supported in part by the U.S. National Science Foundation under grant DMS-1418882. Support from NVIDIA is also gratefully acknowledged.

References

1. Alvarado, F.L., Schreiber, R.: Optimal parallel solution of sparse triangular systems. SIAM J. Sci. Comput. **14**, 446–460 (1993)
2. Anderson, E.C., Saad, Y.: Solving sparse triangular systems on parallel computers. Int. J. High Speed Comput. **1**, 73–96 (1989)
3. Anzt, H.: Asynchronous and multiprecision linear solvers – scalable and fault-tolerant numerics for energy efficient high performance computing. Ph.D. thesis, Karlsruhe Institute of Technology, Institute for Applied and Numerical Mathematics (2012)
4. Anzt, H., Luszczek, P., Dongarra, J., Heuveline, V.: GPU-accelerated asynchronous error correction for mixed precision iterative refinement. In: Euro-Par 2012 Parallel Processing. Lecture Notes in Computer Science, pp. 908–919. Springer, Berlin/New York (2012)
5. Anzt, H., Tomov, S., Gates, M., Dongarra, J., Heuveline, V.: Block-asynchronous multigrid smoothers for GPU-accelerated systems. In: Ali, H.H., Shi, Y., Khazanchi, D., Lees, M., van Albada, G.D., Dongarra, J., Sloot, P.M.A. (eds.) ICCS. Procedia Computer Science, vol. 9, pp. 7–16. Elsevier, Amsterdam (2012)
6. Anzt, H., Tomov, S., Dongarra, J., Heuveline, V.: A block-asynchronous relaxation method for graphics processing units. J. Parallel Distrib. Comput. **73**(12), 1613–1626 (2013)
7. Anzt, H., Chow, E., Dongarra, J.: Iterative sparse triangular solves for preconditioning. In: Träff, J.L., Hunold, S., Versaci, F. (eds.) Euro-Par 2015: Parallel Processing. Lecture Notes in Computer Science, vol. 9233, pp. 650–661. Springer, Berlin/Heidelberg (2015)
8. Anzt, H., Dongarra, J., Quintana-Ortí, E.S.: Tuning stationary iterative solvers for fault resilience. In: Proceedings of the 6th workshop on latest advances in scalable algorithms for large-scale systems, ScalA'15, pp. 1:1–1:8. ACM, New York (2015)
9. Benzi, M., Tůma, M.: A comparative study of sparse approximate inverse preconditioners. Appl. Numer. Math. **30**, 305–340 (1999)
10. Benzi, M., Szyld, D.B., van Duin, A.: Orderings for incomplete factorization preconditionings of nonsymmetric problems. SIAM J. Sci. Comput. **20**, 1652–1670 (1999)
11. Benzi, M., Frommer, A., Nabben, R., Szyld, D.B.: Algebraic theory of multiplicative Schwarz methods. Numer. Math. **89**, 605–639 (2001)
12. Cai, X.C., Sarkis, M.: A restricted additive Schwarz preconditioner for general sparse linear systems. SIAM J. Sci. Comput. **21**, 792–797 (1999)
13. Chazan, D., Miranker, W.: Chaotic relaxation. Linear Algebra Appl. **2**(7), 199–222 (1969)
14. Chow, E., Patel, A.: Fine-grained parallel incomplete LU factorization. SIAM J. Sci. Comput. **37**, C169–C193 (2015)
15. Chow, E., Anzt, H., Dongarra, J.: Asynchronous iterative algorithm for computing incomplete factorizations on GPUs. In: Kunkel, J., Ludwig, T. (eds.) Proceedings of 30th International Conference, ISC High Performance 2015. Lecture Notes in Computer Science, vol. 9137, pp. 1–16. Springer, Cham (2015)
16. Duff, I.S., Meurant, G.A.: The effect of ordering on preconditioned conjugate gradients. BIT **29**(4), 635–657 (1989)
17. Duin, A.C.N.V.: Scalable parallel preconditioning with the sparse approximate inverse of triangular matrices. SIAM J. Matrix Anal. Appl. **20**, 987–1006 (1996)
18. Frommer, A., Szyld, D.B.: On asynchronous iterations. J. Comput. Appl. Math. **123**, 201–216 (2000)

19. Frommer, A., Szyld, D.B.: An algebraic convergence theory for restricted additive Schwarz methods using weighted max norms. SIAM J. Numer. Anal. **39**, 463–479 (2001)
20. Frommer, A., Schwandt, H., Szyld, D.B.: Asynchronous weighted additive Schwarz methods. Electron. Trans. Numer. Anal. **5**, 48–61 (1997)
21. Hammond, S.W., Schreiber, R.: Efficient ICCG on a shared memory multiprocessor. Int. J. High Speed Comput. **4**, 1–21 (1992)
22. Mayer, J.: Parallel algorithms for solving linear systems with sparse triangular matrices. Computing **86**(4), 291–312 (2009)
23. Naumov, M.: Parallel solution of sparse triangular linear systems in the preconditioned iterative methods on the GPU. Technical Report, NVR-2011-001, NVIDIA (2011)
24. NVIDIA Corporation: CUDA C best practices guide. http://docs.nvidia.com/cuda/cuda-c-best-practices-guide/
25. NVIDIA Corporation: NVIDIA CUDA Compute Unified Device Architecture Programming Guide, 2.3.1 edn. (2009)
26. NVIDIA Corporation: NVIDIA CUDA TOOLKIT V7.0 (2015)
27. Pothen, A., Alvarado, F.: A fast reordering algorithm for parallel sparse triangular solution. SIAM J. Sci. Stat. Comput. **13**, 645–653 (1992)
28. Saad, Y.: Iterative Methods for Sparse Linear Systems. SIAM, Philadelphia (2003)
29. Saad, Y., Zhang, J.: Bilum: block versions of multi-elimination and multi-level ILU preconditioner for general sparse linear systems. SIAM J. Sci. Comput. **20**, 2103–2121 (1997)
30. Saltz, J.H.: Aggregation methods for solving sparse triangular systems on multiprocessors. SIAM J. Sci. Stat. Comput. **11**, 123–144 (1990)
31. Shang, Y.: A parallel finite element variational multiscale method based on fully overlapping domain decomposition for incompressible flows. Numer. Methods Partial Differ. Equ. **31**, 856–875 (2015)
32. Smith, B.F., Bjørstad, P.E., Gropp, W.D.: Domain Decomposition: Parallel Multilevel Methods for Elliptic Partial Differential Equations. Cambridge University Press, New York (1996)
33. Szyld, D.B.: Different models of parallel asynchronous iterations with overlapping blocks. Comput. Appl. Math. **17**, 101–115 (1998)
34. Toselli, A., Widlund, O.B.: Domain Decomposition Methods – Algorithms and Theory. Springer Series in Computational Mathematics, vol. 34. Springer, Berlin/Heidelberg (2005)

Asynchronous OpenCL/MPI Numerical Simulations of Conservation Laws

Philippe Helluy, Thomas Strub, Michel Massaro, and Malcolm Roberts

Abstract Hyperbolic conservation laws are important mathematical models for describing many phenomena in physics or engineering. The Finite Volume (FV) method and the Discontinuous Galerkin (DG) method are two popular methods for solving conservation laws on computers. In this paper, we present several FV and DG numerical simulations that we have realized with the OpenCL and MPI paradigms. First, we compare two optimized implementations of the FV method on a regular grid: an OpenCL implementation and a more traditional OpenMP implementation. We compare the efficiency of the approach on several CPU and GPU architectures of different brands. Then we present how we have implemented the DG method in the OpenCL/MPI framework in order to achieve high efficiency. The implementation relies on a splitting of the DG mesh into subdomains and subzones. Different kernels are compiled according to the zone properties. In addition, we rely on the OpenCL asynchronous task graph in order to overlap OpenCL computations, memory transfers and MPI communications.

1 Introduction

Hyperbolic conservation laws are a particular class of Partial Differential Equations (PDE) models. They are present in many fields of physics or engineering. It is thus very important to have efficient software tools for solving such systems. The unknown of a system of conservation laws is a vector $\mathbf{W}(\mathbf{x}, t) \in \mathbb{R}^m$ that depends on a space variable $\mathbf{x} = (x^1, \ldots, x^d)$ and time t. The vector $\mathbf{W}$ is called the vector of

P. Helluy (✉)
IRMA, Université de Strasbourg and Inria Tonus, Strasbourg, France
e-mail: helluy@unistra.fr

T. Strub
AxesSim Illkirch, Illkirch-Graffenstaden, France
e-mail: thomas.strub@axessim.fr

M. Massaro • M. Roberts
IRMA, Université de Strasbourg, Strasbourg, France
e-mail: massaro@math.unistra.fr; roberts@math.unistra.fr

© Springer International Publishing Switzerland 2016   547
H.-J. Bungartz et al. (eds.), *Software for Exascale Computing – SPPEXA*
2013-2015, Lecture Notes in Computational Science and Engineering 113,
DOI 10.1007/978-3-319-40528-5_25

conservative variables. In this work we shall consider a space dimension $d = 2$ or $d = 3$. Generally, the space variable $\mathbf{x}$ belongs to a bounded domain $\Omega \subset \mathbb{R}^d$. The system of conservation reads

$$\partial_t \mathbf{W} + \partial_k \mathbf{F}^k(\mathbf{W}) = 0 . \tag{1}$$

In this formula, we use the following notations:

- The partial derivative operators are denoted by

$$\partial_t = \frac{\partial}{\partial_t}, \quad \partial_k = \frac{\partial}{\partial x^k} . \tag{2}$$

- We adopt the Einstein sum-on-repeated-indices convention

$$\partial_k \mathbf{F}^k(\mathbf{W}) = \sum_{k=1}^{d} \partial_k \mathbf{F}^k(\mathbf{W}) . \tag{3}$$

- The functions $\mathbf{F}^k(\mathbf{W}) \in \mathbb{R}^m$, $k = 1 \ldots d$, characterize the physical model that we wish to represent. It is classic to consider a space vector $\mathbf{n} = (n_1 \ldots n_d) \in \mathbb{R}^d$ and to also define the *flux* of the system

$$\mathbf{F}(\mathbf{W}, \mathbf{n}) = \mathbf{F}^k(\mathbf{W}) n_k . \tag{4}$$

System (1) is supplemented by an initial condition

$$\mathbf{W}(\mathbf{x}, 0) = \mathbf{W}_0(\mathbf{x}) , \tag{5}$$

at time $t = 0$, and conditions on the boundary $\partial\Omega$ of Ω. For example, one can prescribe the value of $\mathbf{W}$ on the boundary

$$\mathbf{W}(\mathbf{x}, t) = \mathbf{W}_b(\mathbf{x}, t), \quad \mathbf{x} \in \partial\Omega . \tag{6}$$

Generally, the system (1), (5), (6) admits a unique solution if it satisfies the hyperbolicity condition: the Jacobian matrix of the flux

$$\nabla_{\mathbf{W}} \mathbf{F}(\mathbf{W}, \mathbf{n}) \tag{7}$$

is diagonalizable with real eigenvalues for all values of $\mathbf{W}$ and $\mathbf{n}$.

The above mathematical framework is very general. It can be applied to electromagnetism, fluid mechanics, multiphase flows, magneto-hydro-dynamics (MHD), Vlasov plasmas, etc. Let us just give two examples:

1. The Maxwell equations describe the evolution of the electric field $\mathbf{E}(\mathbf{x}, t) \in \mathbb{R}^3$ and the magnetic field $\mathbf{H}(\mathbf{x}, t) \in \mathbb{R}^3$. The conservative variables are the

superimposition of these two vectors $\mathbf{W} = (\mathbf{E}^T, \mathbf{H}^T)^T$ (thus $m = 6$) and the Maxwell flux is given by

$$\mathbf{F}(\mathbf{W}, \mathbf{n}) = \begin{bmatrix} 0 & -\mathbf{n} \times \\ \mathbf{n} \times & 0 \end{bmatrix} \mathbf{W}. \tag{8}$$

In Sect. 3 we present numerical results obtained with the Maxwell equations.

2. In fluid mechanics, the Euler equations describe the evolution of a compressible gas of density ρ, velocity $\mathbf{u} = (u^1, u^2, u^3)^T$ and pressure p. The conservative variables are given here by

$$W = (\rho, \rho\mathbf{u}^T, p/(\gamma - 1) + 1/2\rho\mathbf{u} \cdot \mathbf{u})^T \tag{9}$$

and the flux by

$$\mathbf{F}(\mathbf{W}, \mathbf{n}) = (\rho\mathbf{u} \cdot \mathbf{n}, \rho\mathbf{u} \cdot \mathbf{n}\mathbf{u}^T + \tag{10}$$

$$p\mathbf{n}^T, \{\gamma p/(\gamma - 1) + 1/2\rho\mathbf{u} \cdot \mathbf{u}\} \mathbf{u} \cdot \mathbf{n})^T, \tag{11}$$

where $\gamma > 1$ is the polytropic exponent of the gas. The MHD equations are a generalization of the Euler equations for taking into account magnetic effects in conductive compressible gas. The MHD system is a complicated system of conservation laws, with $m = 9$. It is not the objective of this work to detail the MHD equations. For this we refer for instance to [12]. In Sect. 2, we present numerical results obtained with the MHD equations.

Because of their numerous fields of application, many numerical methods have been developed for the resolution of hyperbolic conservation laws. For instance the finite volume (FV) and discontinuous Galerkin (DG) method are very popular. They are easy to program on a standard parallel computer thanks to subdomain decomposition. However, on new hybrid architectures, the efficient implementation of those methods is more complex. It appears that there is possibility of optimizations. In this paper, we explore several numerical experiments that we have made for solving conservation laws with the FV and DG methods on hybrid computers. OpenCL and MPI libraries are today available on a wide range of platforms, making them a good choice for our optimizations. It is classic to rely on OpenCL for local computations and on MPI for communications between accelerators. In addition, in our work we will see that it is interesting to also use the OpenCL asynchronous task graph in order to overlap OpenCL computations, memory transfers and MPI communications.

In the first part of this paper, we compare a classic OpenMP optimization of a FV solver to an OpenCL implementation. We show that on a standard multicore CPU, we obtain comparable speedups between the OpenMP and the OpenCL implementation. In addition, using several GPU accelerators and MPI communications between them, we were able to make computations that would be unattainable with more classic architectures.

Our FV implementation is limited to regular grids. In the second part of the paper, we thus describe an efficient implementation of the DG algorithm on unstructured grids. Our implementation relies on several standard optimizations: local memory prefetching, exploitation of the sparse nature of the tensor basis, and MPI subdomain decomposition. Other optimizations are less common: idling work-item for minimizing cache prefetching and asynchronous MPI/OpenCL communication.

2   Comparison of an OpenCL and an OpenMP Solver on a Regular Grid

2.1   FV Approximation of Conservation Laws

The FV and DG method construct a discontinuous approximation of the conservative variables W. In the case of the FV method, the approximation is piecewise constant. In the case of the DG method, the approximation is piecewise polynomial. It is therefore necessary to extend the definition of the flux $\mathbf{F}(\mathbf{W}, \mathbf{n})$ at a discontinuity of the solution. We consider thus a spatial discontinuity Σ of $\mathbf{W}$. The discontinuity is oriented by a normal vector $\mathbf{n}_{LR}$. We use the following convention: the "left" (L) of Σ is on the side of $-\mathbf{n}_{LR} = \mathbf{n}_{RL}$ and the "right" (R) is on the side of $\mathbf{n}_{LR}$. We denote by $\mathbf{W}_L$ and $\mathbf{W}_R$ the values of $\mathbf{W}$ on the two sides of Σ. The numerical flux is then a function

$$\mathbf{F}(\mathbf{W}_L, \mathbf{W}_R, \mathbf{n}_{LR}) \ . \tag{12}$$

A common choice is to take the Lax-Friedrichs flux (see for instance [11] and included references)

$$\mathbf{F}(\mathbf{W}_L, \mathbf{W}_R, \mathbf{n}) = \frac{\mathbf{F}(\mathbf{W}_L, \mathbf{n}) + \mathbf{F}(\mathbf{W}_R, \mathbf{n})}{2} - \frac{s}{2}(\mathbf{W}_R - \mathbf{W}_L) \ , \tag{13}$$

where s is called the numerical viscosity. It is a supremum of all the wave speeds of the system. For more simplicity, in this section we consider the two-dimensional case $d = 2$ and a square domain $\mathbf{x} = (x^1, x^2) \in \Omega =]0, L[\times]0, L[$. The space step of the grid is $\Delta x = L/N$ where N is a positive integer. The grid cells are squares of size $h \times h$. The cell centers are defined by $\mathbf{x}_{i,j} = ((i + \frac{1}{2})\Delta x, (j + \frac{1}{2})\Delta x)$. We also consider a time step Δt and the times $t^n = n\Delta t$. We look for an approximation $\mathbf{W}_{i,j}^n$ of $\mathbf{W}$ at the cell centers $\mathbf{x}_{i,j}$ and at time t^n

$$\mathbf{W}_{i,j}^n \simeq \mathbf{W}(\mathbf{x}_{i,j}, t^n) \ . \tag{14}$$

Let $\boldsymbol{v}^1$ and $\boldsymbol{v}^2$ be normal vectors pointing in the x^1 and x^2 direction, respectively, so that

$$\boldsymbol{v}^1 = (1,0)^T, \quad \boldsymbol{v}^2 = (0,1)^T . \tag{15}$$

We adopt a Strang dimensional splitting strategy: for advancing the numerical solution from time step t^n to time step t^{n+1}, we first solve the finite volume approximation in direction x^1

$$\frac{\mathbf{W}_{i,j}^* - \mathbf{W}_{i,j}^n}{\Delta t} + \frac{\mathbf{F}(\mathbf{W}_{i,j}^n, \mathbf{W}_{i+1,j}^n, \boldsymbol{v}^1) - \mathbf{F}(\mathbf{W}_{i-1,j}^n, \mathbf{W}_{i,j}^n, \boldsymbol{v}^1)}{\Delta x} = 0 , \tag{16}$$

and then in direction x^2

$$\frac{\mathbf{W}_{i,j}^{n+1} - \mathbf{W}_{i,j}^*}{\Delta t} + \frac{\mathbf{F}(\mathbf{W}_{i,j}^n, \mathbf{W}_{i,j+1}^n, \boldsymbol{v}^2) - \mathbf{F}(\mathbf{W}_{i,j-1}^n, \mathbf{W}_{i,j}^n, \boldsymbol{v}^2)}{\Delta x} = 0 . \tag{17}$$

On the boundary cells, we simply replace, in the previous formulas, the missing values of $\mathbf{W}$ by the boundary values (6).

2.2 OpenMP Implementation of the FV Scheme

The chosen numerical scheme is very simple. We apply the FV scheme to the ideal MHD system with divergence correction. The MHD system models the coupling of a compressible fluid with a magnetic field. It contains $m = 9$ conservative variables and the numerical flux can be a rather complex function. For more details and bibliography on the MHD equations, we refer to [12].

We have first written a C/OpenMP implementation of the algorithm. It adopts a tiling strategy in order to avoid cache misses on large grids with sizes bigger than 1024×1024 points. More details are given in [12]. For later comparison with GPU computations, we only consider results with single precision. We use the optimized tiled OpenMP implementation as our reference for comparisons with OpenCL implementations (see Table 1 where the different implementations are compared).

2.3 OpenCL Implementation of the FV Scheme

2.3.1 OpenCL

It is necessary to adapt our code to new SIMD accelerators, such as GPUs, in order to decrease computation cost. For this, we have chosen OpenCL [14], which is a programming framework, similar to CUDA, for driving such accelerators. A feature

Table 1 Comparison of the different implementations of the FV scheme on a structured grid. Hardware : 2× Intel(R) Xeon(R) E5-2630 (6 cores, 2.3 GHz), AMD Radeon HD 7970, NVidia K20m. On Intel CPUs hyperthreading was deactivated

Implementation	Time	Speedup
OpenMP (Intel CPU 12 cores)	717 s	1
OpenCL (Intel CPU 12 cores)	996 s	0.7
OpenCL (NVIDIA K20)	45 s	16
OpenCL (AMD HD7970)	38 s	19
OpenCL + MPI (4 x NVIDIA K20)	12 s	58

of OpenCL is that multicore CPUs are also considered as accelerators. The same program can thus be run without modification on a CPU or a GPU.

2.3.2 Implementation

For the OpenCL version of our FV algorithm, we organize the data in a (x_1, x_2) grid: each conservative variable is stored in a two-dimensional (i, j) array. For advancing from time step t^n to time step t^{n+1}:

1. In principle, we associate an OpenCL thread (also called a work-item) to each cell of the grid and a thread block (also called a work-group) to each row. But OpenCL drivers generally impose a maximal work-group size. Thus when the row is too long it is also necessary to split the row and distribute it on several work-groups.
2. We compute the flux balance in the x_1-direction for each cell of each row of the grid (see formula (16)).
3. We then transpose the grid, which amounts to exchanging the x_1 and x_2 coordinates. The $(i, j) \rightarrow (j, i)$ transposition is performed on the two-dimensional array of each conservative variable. For ensuring coalescent memory access we adopt an optimized memory transfer algorithm [15] (see also [13]).
4. We can then compute the flux balance in the x_2-direction (17) for each row of the transposed grid. Because of the previous transposition, memory access is coalescent.
5. We again transpose the grid.

Let us mention that other strategies are possible. For instance in [13] the authors describe GPU computations of scalar ($m = 1$) elastic waves. The algorithm is based on two-dimensional tiling of the mesh into cache memory and registers in order to ensure fast memory access. However the tile size is limited by the cache size and the number of unknowns m in each grid cell. In our case for the MHD system we have $m = 9$ and the adaptation of the algorithm given in [13] is inefficient because, as of today (January 2016), GPU cache sizes are too small.

We have tested this OpenCL implementation in several configurations. See Table 1. We can run the OpenCL code on a two-CPU SMP computer or GPUs of different brands, without modification. In addition, we obtain interesting speedups on SMP architectures. The OpenCL speedup for CPU accelerator is approximately 70 % of the OpenMP speedup. It remains very good considering that the transposition algorithm probably deteriorates the memory access efficiency on CPU architectures. The fact that OpenCL is a possible alternative to OpenMP on multicore CPU has already been discussed in [16].

On AMD or NVIDIA GPUs, the same version of our code achieves good performance. If we replace the optimized transposition by a naive unoptimized transposition algorithm the code runs approximately 10 times slower on GPUs. The coalescent memory access is thus an essential ingredient of the efficiency.

2.4 OpenCL/MPI FV Solver

We now modify the OpenCL implementation in order to address several GPU accelerators at the same time. This could theoretically be achieved by creating several command queues, one for each GPU device. However, as of today, when GPUs are plugged into different nodes of a supercomputer, the current OpenCL drivers are not able to drive at the same time GPUs of different nodes. Therefore, we have decided to rely on the MPI framework for managing the communications between different GPUs. This strategy is very common (see for instance [1, 4, 7] and included references).

We split the computational domain Ω into several subdomains in the x^1 direction. An example of splitting with four subdomains is presented on Fig. 1. Then, each subdomain is associated to one MPI node and each MPI node drives one GPU. For applying the finite volume algorithm on a subdomain, it is necessary to exchange two layers of cells between the neighboring subdomains at the beginning of each time step. The layers are shaded in grey in Fig. 1. On each MPI node, an exchange thus requires a GPU to CPU memory transfer of the cell layers, a MPI send/recv communication and a CPU to GPU transfer for retrieving the neighbor layers. The exchanged cells represent a small amount of the total grid cells, however, the transfer and communication time represent a non-negligible amount of the computation cost.

In our first OpenCL/MPI implementation, the exchange task is performed in a synchronous way: we wait for the exchange to be finished before computing the flux balance in the subdomains. This explains why the speedup between the OpenCL code and the OpenCL/MPI code with four GPUs is approximately 3.5 (the ideal speedup would be 4). See Table 1.

Despite the synchronous approach, the OpenCL/MPI FV solver on structured grid is rather efficient. It has permitted us to perform computations on very fine grids that would be unreachable with standard parallel computers. For instance, we have performed two-fluid computations of shock-bubble interaction with grid size up to 40,000 × 20,000 in [6].

Our FV solver has several drawbacks: the FV method is limited to first or second order approximation and in some applications, it is important to have access to higher order schemes; MPI and host/GPU communications take time, so it is also

Fig. 1 Subdomain MPI
decomposition

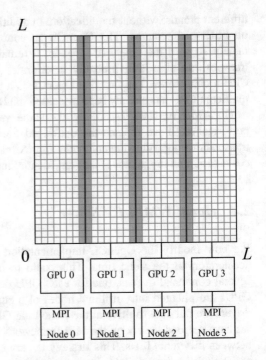

important to provide asynchronous implementations for scalability with more MPI
nodes; finally, the previously described approach is limited to structured grids and
we wish also to extend the method to arbitrary geometries.

In the next section we describe a Discontinuous Galerkin (DG) solver that
allows to achieving higher order, addressing general geometries, and overlapping
computations and communications.

3 Asynchronous OpenCL/MPI Discontinuous Galerkin Solver

We now present the Discontinuous Galerkin Method and explain our software
design for keeping high performance in the GPU implementation.

3.1 The DG Method

3.1.1 Interpolation on Unstructured Hexahedral Meshes

The DG method is a generalization of the FV method. We suppose that dimension
$d = 3$. We consider a mesh of the computational domain Ω made of cells L_i,

$i = 1 \ldots N_c$. In a cell L of the mesh, the field is approximated by a linear combination of basis functions ψ_j^L

$$\mathbf{W}(\mathbf{x}, t) = \mathbf{W}_L^j(t)\psi_j^L(\mathbf{x}), \quad \mathbf{x} \in L. \tag{18}$$

Each cell L of the mesh is obtained by a geometrical mapping τ_L that transforms a reference element $\hat{L}$ into L. In theory the shape of the reference element $\hat{L}$ may be arbitrary. A classic choice is to consider tetrahedra [9]. In this work we prefer hexahedra, as in [5]. Building a tetrahedral mesh of Ω is generally easier. The nodal basis functions of a hexahedral cell are constructed from tensor products of one-dimensional functions. The tensor nature of the basis allows many optimizations of the algorithm that are not possible with tetrahedra. The chosen basis is made of Lagrange polynomials of order D associated to Gauss-Legendre quadrature points. This choice is classic and described in details in [8]. In Fig. 2 we have represented the Gauss-Legendre points for an order $D = 2$. The volume Gauss points (which are also the chosen interpolation points) are blue and the face Gauss points are green. Because we have chosen Gauss-Legendre quadrature, an extrapolation is needed at face Gauss points for computing surface integrals. This would not be necessary with Gauss-Lobatto quadrature points.

3.1.2 DG Formulation

The numerical solution satisfies the DG approximation scheme

$$\forall L, \forall i \quad \int_L \partial_t \mathbf{W}\psi_i^L - \int_L \mathbf{F}(\mathbf{W}, \mathbf{W}, \nabla\psi_i^L)$$

$$+ \int_{\partial L} \mathbf{F}(\mathbf{W}_L, \mathbf{W}_R, \mathbf{n}_{LR})\psi_i^L = 0 . \tag{19}$$

Fig. 2 Volume and face Gauss-Legendre points in the reference cube

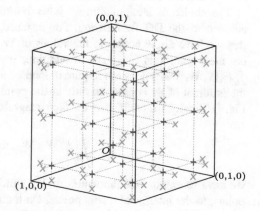

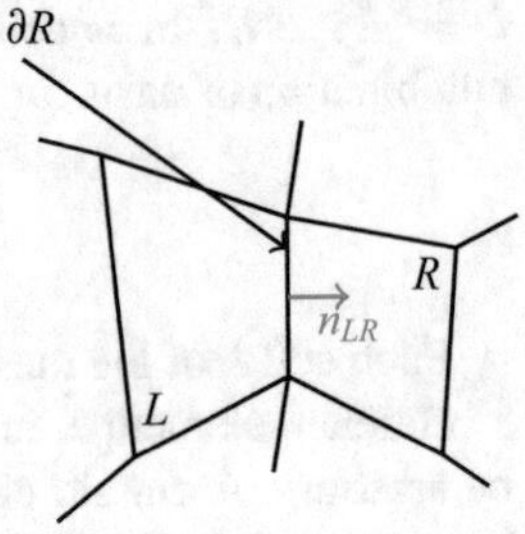

Fig. 3 Mesh: notation
conventions

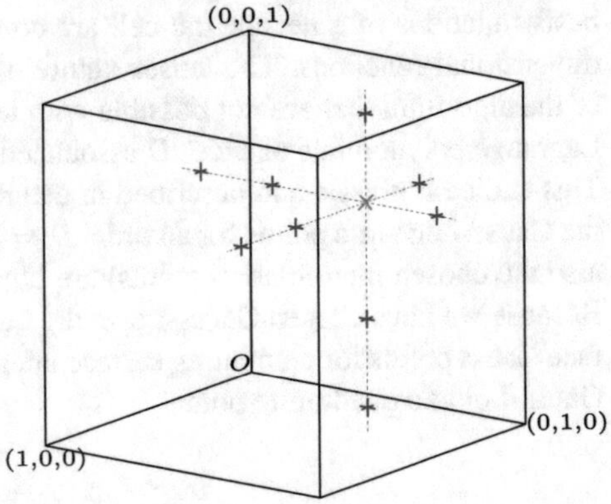

Fig. 4 Non-zero values of
the basis functions. The
gradient of the basis function
associated to the *red point* is
nonzero only on the *blue
points*

In this formula,

- R denotes the neighbor cells along ∂L,
- $\mathbf{n}_{LR}$ is the unit normal vector on ∂L oriented from L to R. See Fig. 3,
- $\mathbf{F}(\mathbf{W}_L, \mathbf{W}_R, \mathbf{n})$ is the numerical flux, which satisfies $\mathbf{F}(\mathbf{W}, \mathbf{W}, \mathbf{n}) = F^k(\mathbf{W})\mathbf{n}_k$.

Inserting expansion (18) into (19) we obtain a system of differential equations satisfied by the $\mathbf{W}_L^j(t)$. This system of differential equations can be solved numerically with a standard Runge-Kutta method.

The choice of interpolation we have described in the previous section is well adapted to the DG formulation. For instance, the nodal basis property ensures that we have direct access to the values of $\mathbf{W}$ at the Gauss points. Consequently the mass matrix is diagonal. In addition, the computation of the volume term $\int_L F(\mathbf{W}, \mathbf{W}, \nabla \psi_i^L)$ does not require to loop on all the volume Gauss points. Indeed, the gradient of ψ_i is nonzero only at the points that are aligned with point i (see Fig. 4). Finally, for computing the face integrals

$$\int_{\partial L} F(\mathbf{W}_L, \mathbf{W}_R, \mathbf{n}_{LR})\psi_i^L \qquad (20)$$

we have to extrapolate the values of $\mathbf{W}$, which are known on the volume Gauss points, to the interfacial Gauss points. On tetrahedra, all the volume Gauss points would be involved in the interpolation. With our nodal hexahedral basis, only the volume Gauss points aligned with the considered interfacial Gauss point are needed (see Fig. 2: for computing $\mathbf{W}$ at a green point, we only need to know $\mathbf{W}$ at the blue points aligned with this green point).

In the end, exploiting the tensor basis properties, the DG formulation (19) in a cell L requires computations of complexity $\sim D^4$ instead of $\sim D^6$. For high orders, this is a huge improvement.

Beyond these useful optimizations that are also applied in sequential implementations, The DG method presents many advantages: it is possible to have different orders on different cells, no conformity is required between the cell and mesh refinement is thus simplified; the computations inside a cell only depend on the neighboring cells; the stencil is more compact than for high order FV methods, so memory accesses are well adapted to GPU computations; high order inside a cell implies a high amount of local computations, this property is well adapted to GPU computations; and finally, two levels of parallelism can be easily exploited: a coarse grain parallelism, at the subdomain level, well adapted to MPI algorithms and a fine grain parallelism, at the level of a single cell, well adapted to OpenCL or OpenMP.

But there are also possible issues that could make an implementation inefficient: first, we have to take care of memory bandwidth, because unstructured meshes may imply non coalescent memory access. Moreover, a general DG solver has to manage many different physical models, boundary conditions, interpolation basis, etc. If the implementation is not realized with care it is possible to end up with poorly coded kernels with many barely used variables or branch tests. Such wastage may remain unseen on standard CPUs with many registers and large cache memory, but is often catastrophic on GPUs. Finally, as we have already seen, MPI communications imply very slow GPU to Host and Host to GPU memory transfers. If possible, it is advised to hide communication latency by an overlapping with computations.

3.2 OpenCL Kernel for a Single GPU

We first wrote optimized OpenCL kernels for computing, on a single cell L, the terms appearing in the DG formulation (19). After several experiments, we have found that an efficient strategy is to write a single kernel for computing the ∂L and L integration steps.

More precisely we construct a kernel with two steps.

In the first step ("flux step"), we compute the fluxes at the face Gauss points and store those fluxes in the cache memory of the work-group. The work-items are distributed on the face Gauss points. In this stage, $6(D + 1)^2$ work-items are activated.

After a sync barrier, in the second stage ("collecting step"), we associate a work-item to each volume Gauss point and we collect the contributions of the other volume Gauss points coming from the numerical integration. We also collect the contribution from the face fluxes stored in the first step. In this stage, $(D + 1)^3$ work-items are activated.

We observe that when the order $D < 5$, which is always the case in our computations, $(D + 1)^3 < 6(D + 1)^2$ and then some work-items are idling in the collecting step.

We have also tried to split the computation into two kernels, one for the flux step and one for the collecting step, but it requires saving the fluxes into global memory, and in the end it appears that the idling work-items method is more efficient.

3.3 Asynchronous MPI/OpenCL Implementation for Several GPUs

3.3.1 Subdomains and Zones

We have written a generic C++ DG solver called CLAC ("Conservation Laws Approximation on many Cores") for solving large problems on general hexahedral meshes. Practical industrial applications require a lot of memory and computations. It is thus necessary to address several accelerators in an efficient way.

We describe some features of the CLAC implementation.

First, the physical models are localized in the code: in practice, the user has to provide the numerical flux plus a few functions for applying boundary conditions, source terms, etc. With this approach it is possible to apply CLAC to very different physics: Maxwell equations, compressible fluids, MHD, etc. This approach is similar to the approach of A. Klöckner in [10].

We also adopt a domain decomposition strategy. The mesh is split into several domains, each of which is associated to a single MPI node, and each MPI node is associated to an OpenCL device (CPU or GPU).

In addition to the domain decomposition, in each domain we split the mesh into zones. We consider volume zones made of hexahedral cells and also interface zones made of cell faces. The role of a volume zone is to apply the source terms and the flux balance between cells inside the zone. The interface zones are devoted to computing the flux balance between cells of different volume zones. When an interface zone is at the boundary of the computational domain, it is used for applying boundary conditions. When it is situated at an interface between two domains, it is also in charge of the MPI communications between the domains. Interface zones also serve to manage mesh refinements between two volume zones. A simple example of a mesh with two subdomains, three volume zones and five interface zones is given in Fig. 5 and a schematic view of the same mesh is represented in Fig. 6. We observe in this figure that simple non-conformities are allowed between volume zones (for instance neighboring volume zones 2 and 3 do not have the same refinement).

Finally, a zone possesses identical elements (same order, same geometry, same physical model). Thus, different computation kernels are compiled for each zone, in order to save registers and branch tests. We have observed that this aspect is very important for achieving high efficiency. For example, it is possible to

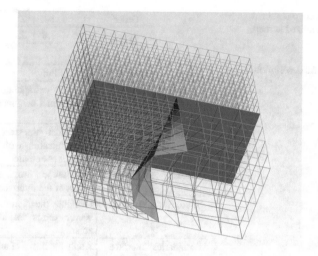

Fig. 5 A simple non conforming mesh

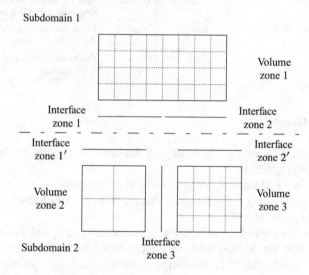

Fig. 6 Schematic view of the simple mesh

simplify the kernel that computes the flux balance at an interface zone between two volume zones with conforming meshes. At an interface between volume zones with different refinements, the kernel is more complex, because the Gauss integration points are not aligned (see Interface zone 3 on Fig. 6). The specialized kernels take advantage of the Gauss points alignment and store interpolation and geometrical data in constant arrays or preprocessor macros. The speedup obtained using the specialized kernels as opposed to the generic kernels is reported in Table 2 for different interpolation orders.

Table 2 Speedup obtained
with the specialized kernels

Order	0	1	2	3	4
Speedup	1.6	1.8	2.8	3.6	5.5

Table 3 Tasks description

Name	Attached to	Description
Extraction	Interface	Copy or extrapolate the values of W from a neighboring volume zone
Exchange	Interface	GPU/Host transfers and MPI communication with an interface of another domain
Fluxes	Interface	Compute the fluxes at the Gauss points of the interface
Sources	Volume	Compute the internal fluxes and source terms inside a volume zone
Boundaries	Interface	Apply the fluxes of an interface to a volume zone
Time	Volume	Apply a step of the Runge-Kutta time integration to a volume zone
Start	Volume	Fictitious task: beginning of the Runge-Kutta substep
End	Volume	Fictitious task: end of the Runge-Kutta substep

3.3.2 Task Graph

The zone approach is very useful to express the dependency between the different tasks of the DG algorithm.

We have identified tasks attached to volume or interface zones that have to be executed for performing a Runge-Kutta substep with the DG formulation. Those tasks are detailed in Table 3.

We express the dependencies between the tasks in a graph, and construct a task graph per subdomain. For instance, we have represented on Fig. 7 the task graph associated to Subdomain 2 of the simple mesh of Fig. 6. The volume tasks are represented in blue rectangles, the interface tasks in red ellipses. The interface tasks that require MPI communication are in red rhombuses.

We observe in these figures that it is possible to perform the exchange tasks and the internal computations at the same time. It is thus possible to overlap communications and GPU/Host transfers by computations.

OpenCL contains event objects for describing task dependencies between the operations sent to command queues. It is also possible to create user events for describing interactions between the OpenCL command queues and tasks that are executed outside of a call to the OpenCL library. We have decided to rely only on the OpenCL event management for constructing the task dependencies.

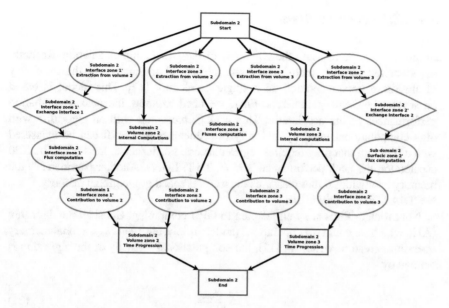

Fig. 7 Task graph for subdomain 2

Using asynchronous MPI communication requires calling `MPI_Wait` before launching tasks that depend on the completion of communication. We thus face a practical problem, which is to express the dependency between MPI and OpenCL operations in a non-blocking way. A possibility would have been to use an OpenCL "Native Kernel" containing MPI calls. A native kernel is a standard function compiled and executed on the host side, but that can be inserted into the OpenCL task graph. As of today, the native kernel feature is not implemented properly in all the OpenCL drivers. We thus had to adopt another approach in order to circumvent this difficulty.

Our solution uses the C++ standard thread class. It is also necessary to use an MPI implementation that provides the `MPI_THREAD_MULTIPLE` option. For programming the "Exchange" task, we first create an OpenCL user event. Then we launch a thread and return from the task. The main program flow is not interrupted and other operations can be enqueued. Meanwhile, in the thread, we start a blocking send/recv MPI operation for exchanging data between the boundary interface zones. Because the communication is launched in a thread, its blocking or non-blocking nature is not very important. When the communication is finished, we mark the OpenCL event as completed and exit the thread. The completion of the user event triggers the beginning of the enqueued tasks that depend on the exchange.

As we will see in the next section, this simple solution offers very good efficiency.

3.4 Efficiency Analysis

In this section we measure the efficiency of the CLAC implementation. Recently the so-called roofline model has been introduced for analyzing the efficiency of algorithm implementation on a single accelerator [17]. This model is based on several hardware parameters. First, we need to know the peak computation performances of the accelerator. This peak is measured with an algorithm with high computational intensity and very little memory access. It can be measured with a program that only requires register access. For instance, for a NVIDIA K20 accelerator, the peak performance is $P = 3.5$ TFLOP/s. Another parameter is the memory bandwidth B that measures the transfer speed of the global memory. For a NVIDIA K20 $B = 208$ GB/s.

Not all algorithms are well adapted to GPU computing. Consider an algorithm (A) in which we count N_{ops} operations (add, multiply, etc.) and N_{mem} global memory operations (read or write). In [17], the computational intensity of the algorithm is defined by

$$I = \frac{N_{ops}}{N_{mem}}. \tag{21}$$

The maximal attainable performance of one GPU for this algorithm is then given by the roofline formula:

$$P_A = \max(P, B \times I).$$

We have counted the computational and memory operations of our DG implementation. The counting method is simply based on source inspection because we have not been able to find an automatic and reliable tool for evaluating the amount of floating point and memory operations. We only count memory transfer to the global memory, floating point and integer operations: we neglected pointer arithmetic and register accesses. The results are plotted in Fig. 8. We observe that for order 1, the DG method is limited by the memory bandwidth. For higher orders, the method is limited by the peak performance of the GPU. The figure confirms that the DG method is well adapted to GPU architectures. We have also performed this analysis for the FV method described in Sect. 2. For large grids, the efficiency of the FV scheme is approximately 20 FLOP/B. The FV algorithm is thus also limited by the peak performance of the GPU. Our implementation of the FV scheme reaches approximately 800 GFLOP/s on a single K20 GPU.

In Table 5, we present the results that we have measured with the asynchronous MPI/OpenCL implementation with 1, 2, 4 and 8 GPUs. For comparison, we also give in Table 4 the results of the synchronous execution (we wait that each task is completed before launching the next one). The computational domain Ω is a cube. The chosen model is the Maxwell system ($m = 6$). The mesh is made of several subdomains of 90^3 cells. We perform single precision computations. The

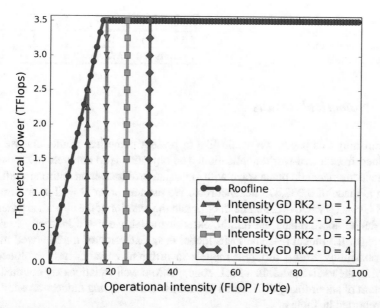

Fig. 8 Roofline model and DG method. *Abscissa*: computational intensity I (FLOP/B). *Ordinate*: Algorithm performance (TFLOP/s)

Table 4 Weak scaling of the synchronous MPI/OpenCL implementation

	1 GPU	2 GPUs	4 GPUs	8 GPUs
TFLOP/s	1.01	1.84	3.53	5.07
Speedup	1	1.83	3.53	5.01

Table 5 Weak scaling of the asynchronous MPI/OpenCL implementation

	1 GPU	2 GPUs	4 GPUs	8 GPUs
TFLOP/s	1.01	1.96	3.78	7.34
Speedup	1	1.94	3.74	7.26

interpolation is of order $D = 3$. The algorithm requires storing three time steps of the numerical solution. With these parameters the memory of each K20 board is almost entirely filled. Indeed the storage of the electromagnetic field on one subdomain requires approximately 3.4 GB.

We observe in Table 5 that the asynchronous implementation is rather efficient and that the communications are well overlapped by the GPU computations. In addition, we observe that with CLAC we attain approximately 30 % of the roofline limit. This result is not too bad, because CLAC handles unstructured meshes and some non-coalescent memory accesses are unavoidable.

Table 6 Additional cost for
5 and 10 PML expressed in
percentage of the total
computation time

Order	0	1	2	3	4
5 layers (%)	7.14	4.29	15.9	16.5	15.0
10 layers (%)	7.95	6.49	19.0	20.6	18.1

3.5 Numerical Results

For finishing this paper, we would like to present numerical results that we have
obtained from a real-world application. The objective is to compute the reflection
of an electromagnetic plane wave with Gaussian profile over an entire aircraft. The
mesh is made of 3,337,875 hexahedrons. We used an order $D = 2$ approximation
and 8 GPUs (NVIDIA K20). The interior and the exterior of the aircraft are meshed.
In order to approximate the infinite exterior model, we use a Perfectly Matched
Layers (PML) model [3]. The PML model is an extension of the Maxwell model.
The possibility to use different models in different zones is here exploited for
applying the PML model. In a PML zone, the Maxwell equations are coupled with
a system of six ordinary differential equations. This coupling induces an additional
cost reported in Table 6.

4 Conclusions

In this work we have reviewed several methods for solving hyperbolic conservation
laws. Such models are very useful in many fields of physics or engineering.
We have presented a finite volume OpenCL/MPI implementation. We have seen
that coalescent memory access is essential for obtaining good efficiency. The
synchronous MPI communication does not allow an optimal scaling with several
GPUs. However the MPI extension allows addressing computations that would not
fit into a single accelerator.

We have then presented a more sophisticated approach: the Discontinuous
Galerkin method on unstructured hexahedral meshes. We have also written an
OpenCL/MPI implementation of the method. Despite the unstructured mesh and
some non-coalescent memory accesses, we reach 30 % of the peak performance.

In future works we intend to change the description of the mesh geometry in
order to minimize the memory access: we can for instance share a higher order
geometrical transformation τ between several cells. We also plan to implement a
local-time stepping algorithm in order to be able to deal with locally refined meshes.
Finally, we would like to describe the task graph in a more abstract manner in
order to distribute the computation more effectively on the available resources. An
interesting tool for performing such distribution could be for instance the StarPU
environment [2].

Acknowledgements This work has benefited from several supports: from the French Defense Agency DGA, from the Labex ANR-11-LABX-0055-IRMIA and from the AxesSim company. We also thank Vincent Loechner for his helpful advice regarding the optimization of the OpenMP code.

References

1. Aubert, D.: Numerical cosmology powered by GPUs. Proc. Int. Astron. Union **6**(S270), 397–400 (2010)
2. Augonnet, C., Thibault, S., Namyst, R., Wacrenier, P.A.: StarPU: a unified platform for task scheduling on heterogeneous multicore architectures. Concurr. Comput.: Pract. Exper. **23**(2), 187–198 (2011)
3. Berenger, J.P.: A perfectly matched layer for the absorption of electromagnetic waves. J. Comput. Phys. **114**(2), 185–200 (1994)
4. Cabel, T., Charles, J., Lanteri, S.: Multi-GPU acceleration of a DGTD method for modeling human exposure to electromagnetic waves. Research report, vol. RR-7592, p. 27. INRIA. http://hal.inria.fr/inria-00583617 (2011)
5. Cohen, G., Ferrieres, X., Pernet, S.: A spatial high-order hexahedral discontinuous Galerkin method to solve Maxwell's equations in time domain. J. Comput. Phys. **217**(2), 340–363 (2006)
6. Helluy, P., Jung, J.: Interpolated pressure laws in two-fluid simulations and hyperbolicity. In: Finite Volumes for Complex Applications VII-Methods and Theoretical Aspects: FVCA 7, Berlin, June 2014, pp. 37–53. Springer, Cham (2014)
7. Helluy, P., Jung, J.: Two-fluid compressible simulations on GPU cluster. ESAIM Proc. Surv. **45**, 349–358 (2014)
8. Hesthaven, J.S., Warburton, T.: Nodal Discontinuous Galerkin Methods: Algorithms, Analysis, and Applications. Texts in Applied Mathematics, vol. 54. Springer, New York (2008)
9. Klöckner, A., Warburton, T., Bridge, J., Hesthaven, J.S.: Nodal discontinuous Galerkin methods on graphics processors. J. Comput. Phys. **228**(21), 7863–7882 (2009). http://dx.doi.org/10.1016/j.jcp.2009.06.041
10. Kloeckner, A.: Hedge: Hybrid and Easy Discontinuous Galerkin Environment. http://mathema.tician.de/software/hedge/ (2010)
11. LeVeque, R.J.: Finite volume methods for hyperbolic problems. Cambridge Texts in Applied Mathematics, vol. 31. Cambridge University Press, Cambridge (2002)
12. Massaro, M., Helluy, P., Loechner, V.: Numerical simulation for the MHD system in 2D using OpenCL. ESAIM Proc. Surv. **45**, 485–492 (2014)
13. Michéa, D., Komatitsch, D.: Accelerating a three-dimensional finite-difference wave propagation code using GPU graphics cards. Geophys. J. Int. **182**(1), 389–402 (2010)
14. OpenCL: The open standard for parallel programming of heterogeneous systems. https://www.khronos.org/opencl. Accessed 23 Feb 2016
15. Ruetsch, G., Micikevicius, P.: Optimizing matrix transpose in CUDA. Nvidia CUDA SDK Application Note (2009)
16. Shen, J., Fang, J., Sips, H., Varbanescu, A.L.: Performance gaps between OpenMP and OpenCL for multi-core CPUs. In: 2012 41st International Conference on Parallel Processing Workshops (ICPPW), pp. 116–125. IEEE (2012)
17. Williams, S., Waterman, A., Patterson, D.: Roofline: an insightful visual performance model for multicore architectures. Commun. ACM **52**(4), 65–76 (2009)

Editorial Policy

1. Volumes in the following three categories will be published in LNCSE:

i)   Research monographs
ii) Tutorials
iii) Conference proceedings

Those considering a book which might be suitable for the series are strongly advised to contact the publisher or the series editors at an early stage.

2. Categories i) and ii). Tutorials are lecture notes typically arising via summer schools or similar events, which are used to teach graduate students. These categories will be emphasized by Lecture Notes in Computational Science and Engineering. **Submissions by interdisciplinary teams of authors are encouraged.** The goal is to report new developments – quickly, informally, and in a way that will make them accessible to non-specialists. In the evaluation of submissions timeliness of the work is an important criterion. Texts should be well-rounded, well-written and reasonably self-contained. In most cases the work will contain results of others as well as those of the author(s). In each case the author(s) should provide sufficient motivation, examples, and applications. In this respect, Ph.D. theses will usually be deemed unsuitable for the Lecture Notes series. Proposals for volumes in these categories should be submitted either to one of the series editors or to Springer-Verlag, Heidelberg, and will be refereed. A provisional judgement on the acceptability of a project can be based on partial information about the work: a detailed outline describing the contents of each chapter, the estimated length, a bibliography, and one or two sample chapters – or a first draft. A final decision whether to accept will rest on an evaluation of the completed work which should include

– at least 100 pages of text;
– a table of contents;
– an informative introduction perhaps with some historical remarks which should be accessible to readers unfamiliar with the topic treated;
– a subject index.

3. Category iii). Conference proceedings will be considered for publication provided that they are both of exceptional interest and devoted to a single topic. One (or more) expert participants will act as the scientific editor(s) of the volume. They select the papers which are suitable for inclusion and have them individually refereed as for a journal. Papers not closely related to the central topic are to be excluded. Organizers should contact the Editor for CSE at Springer at the planning stage, see *Addresses* below.

In exceptional cases some other multi-author-volumes may be considered in this category.

4. Only works in English will be considered. For evaluation purposes, manuscripts may be submitted in print or electronic form, in the latter case, preferably as pdf- or zipped ps-files. Authors are requested to use the LaTeX style files available from Springer at http://www.springer.com/gp/authors-editors/book-authors-editors/manuscript-preparation/5636 (Click on LaTeX Template → monographs or contributed books).

For categories ii) and iii) we strongly recommend that all contributions in a volume be written in the same LaTeX version, preferably LaTeX2e. Electronic material can be included if appropriate. Please contact the publisher.

Careful preparation of the manuscripts will help keep production time short besides ensuring satisfactory appearance of the finished book in print and online.

5. The following terms and conditions hold. Categories i), ii) and iii):

Authors receive 50 free copies of their book. No royalty is paid.
Volume editors receive a total of 50 free copies of their volume to be shared with authors, but no royalties.

Authors and volume editors are entitled to a discount of 33.3 % on the price of Springer books purchased for their personal use, if ordering directly from Springer.

6. Springer secures the copyright for each volume.

Addresses:

Timothy J. Barth
NASA Ames Research Center
NAS Division
Moffett Field, CA 94035, USA
barth@nas.nasa.gov

Michael Griebel
Institut für Numerische Simulation
der Universität Bonn
Wegelerstr. 6
53115 Bonn, Germany
griebel@ins.uni-bonn.de

David E. Keyes
Mathematical and Computer Sciences
and Engineering
King Abdullah University of Science
and Technology
P.O. Box 55455
Jeddah 21534, Saudi Arabia
david.keyes@kaust.edu.sa

and

Department of Applied Physics
and Applied Mathematics
Columbia University
500 W. 120 th Street
New York, NY 10027, USA
kd2112@columbia.edu

Risto M. Nieminen
Department of Applied Physics
Aalto University School of Science
and Technology
00076 Aalto, Finland
risto.nieminen@aalto.fi

Dirk Roose
Department of Computer Science
Katholieke Universiteit Leuven
Celestijnenlaan 200A
3001 Leuven-Heverlee, Belgium
dirk.roose@cs.kuleuven.be

Tamar Schlick
Department of Chemistry
and Courant Institute
of Mathematical Sciences
New York University
251 Mercer Street
New York, NY 10012, USA
schlick@nyu.edu

Editor for Computational Science
and Engineering at Springer:
Martin Peters
Springer-Verlag
Mathematics Editorial IV
Tiergartenstrasse 17
69121 Heidelberg, Germany
martin.peters@springer.com

Lecture Notes
in Computational Science
and Engineering

24. T. Schlick, H.H. Gan (eds.), *Computational Methods for Macromolecules: Challenges and Applications.*

25. T.J. Barth, H. Deconinck (eds.), *Error Estimation and Adaptive Discretization Methods in Computational Fluid Dynamics.*

26. M. Griebel, M.A. Schweitzer (eds.), *Meshfree Methods for Partial Differential Equations.*

27. S. Müller, *Adaptive Multiscale Schemes for Conservation Laws.*

28. C. Carstensen, S. Funken, W. Hackbusch, R.H.W. Hoppe, P. Monk (eds.), *Computational Electromagnetics.*

29. M.A. Schweitzer, *A Parallel Multilevel Partition of Unity Method for Elliptic Partial Differential Equations.*

30. T. Biegler, O. Ghattas, M. Heinkenschloss, B. van Bloemen Waanders (eds.), *Large-Scale PDE-Constrained Optimization.*

31. M. Ainsworth, P. Davies, D. Duncan, P. Martin, B. Rynne (eds.), *Topics in Computational Wave Propagation.* Direct and Inverse Problems.

32. H. Emmerich, B. Nestler, M. Schreckenberg (eds.), *Interface and Transport Dynamics.* Computational Modelling.

33. H.P. Langtangen, A. Tveito (eds.), *Advanced Topics in Computational Partial Differential Equations.* Numerical Methods and Diffpack Programming.

34. V. John, *Large Eddy Simulation of Turbulent Incompressible Flows.* Analytical and Numerical Results for a Class of LES Models.

35. E. Bänsch (ed.), *Challenges in Scientific Computing - CISC 2002.*

36. B.N. Khoromskij, G. Wittum, *Numerical Solution of Elliptic Differential Equations by Reduction to the Interface.*

37. A. Iske, *Multiresolution Methods in Scattered Data Modelling.*

38. S.-I. Niculescu, K. Gu (eds.), *Advances in Time-Delay Systems.*

39. S. Attinger, P. Koumoutsakos (eds.), *Multiscale Modelling and Simulation.*

40. R. Kornhuber, R. Hoppe, J. Périaux, O. Pironneau, O. Wildlund, J. Xu (eds.), *Domain Decomposition Methods in Science and Engineering.*

41. T. Plewa, T. Linde, V.G. Weirs (eds.), *Adaptive Mesh Refinement – Theory and Applications.*

42. A. Schmidt, K.G. Siebert, *Design of Adaptive Finite Element Software.* The Finite Element Toolbox ALBERTA.

43. M. Griebel, M.A. Schweitzer (eds.), *Meshfree Methods for Partial Differential Equations II.*

44. B. Engquist, P. Lötstedt, O. Runborg (eds.), *Multiscale Methods in Science and Engineering.*

45. P. Benner, V. Mehrmann, D.C. Sorensen (eds.), *Dimension Reduction of Large-Scale Systems.*

46. D. Kressner, *Numerical Methods for General and Structured Eigenvalue Problems.*

47. A. Boriçi, A. Frommer, B. Joó, A. Kennedy, B. Pendleton (eds.), *QCD and Numerical Analysis III.*

48. F. Graziani (ed.), *Computational Methods in Transport.*

49. B. Leimkuhler, C. Chipot, R. Elber, A. Laaksonen, A. Mark, T. Schlick, C. Schütte, R. Skeel (eds.), *New Algorithms for Macromolecular Simulation.*

50. M. Bücker, G. Corliss, P. Hovland, U. Naumann, B. Norris (eds.), *Automatic Differentiation: Applications, Theory, and Implementations.*

51. A.M. Bruaset, A. Tveito (eds.), *Numerical Solution of Partial Differential Equations on Parallel Computers.*

52. K.H. Hoffmann, A. Meyer (eds.), *Parallel Algorithms and Cluster Computing.*

53. H.-J. Bungartz, M. Schäfer (eds.), *Fluid-Structure Interaction.*

54. J. Behrens, *Adaptive Atmospheric Modeling.*

55. O. Widlund, D. Keyes (eds.), *Domain Decomposition Methods in Science and Engineering XVI.*

56. S. Kassinos, C. Langer, G. Iaccarino, P. Moin (eds.), *Complex Effects in Large Eddy Simulations.*

57. M. Griebel, M.A Schweitzer (eds.), *Meshfree Methods for Partial Differential Equations III.*

58. A.N. Gorban, B. Kégl, D.C. Wunsch, A. Zinovyev (eds.), *Principal Manifolds for Data Visualization and Dimension Reduction.*

59. H. Ammari (ed.), *Modeling and Computations in Electromagnetics: A Volume Dedicated to Jean-Claude Nédélec.*

60. U. Langer, M. Discacciati, D. Keyes, O. Widlund, W. Zulehner (eds.), *Domain Decomposition Methods in Science and Engineering XVII.*

61. T. Mathew, *Domain Decomposition Methods for the Numerical Solution of Partial Differential Equations.*

62. F. Graziani (ed.), *Computational Methods in Transport: Verification and Validation.*

63. M. Bebendorf, *Hierarchical Matrices.* A Means to Efficiently Solve Elliptic Boundary Value Problems.

64. C.H. Bischof, H.M. Bücker, P. Hovland, U. Naumann, J. Utke (eds.), *Advances in Automatic Differentiation.*

65. M. Griebel, M.A. Schweitzer (eds.), *Meshfree Methods for Partial Differential Equations IV.*

66. B. Engquist, P. Lötstedt, O. Runborg (eds.), *Multiscale Modeling and Simulation in Science.*

67. I.H. Tuncer, Ü. Gülcat, D.R. Emerson, K. Matsuno (eds.), *Parallel Computational Fluid Dynamics 2007.*

68. S. Yip, T. Diaz de la Rubia (eds.), *Scientific Modeling and Simulations.*

69. A. Hegarty, N. Kopteva, E. O'Riordan, M. Stynes (eds.), *BAIL 2008 – Boundary and Interior Layers.*

70. M. Bercovier, M.J. Gander, R. Kornhuber, O. Widlund (eds.), *Domain Decomposition Methods in Science and Engineering XVIII.*

71. B. Koren, C. Vuik (eds.), *Advanced Computational Methods in Science and Engineering.*

72. M. Peters (ed.), *Computational Fluid Dynamics for Sport Simulation.*

73. H.-J. Bungartz, M. Mehl, M. Schäfer (eds.), *Fluid Structure Interaction II - Modelling, Simulation, Optimization.*

74. D. Tromeur-Dervout, G. Brenner, D.R. Emerson, J. Erhel (eds.), *Parallel Computational Fluid Dynamics 2008.*

75. A.N. Gorban, D. Roose (eds.), *Coping with Complexity: Model Reduction and Data Analysis.*

76. J.S. Hesthaven, E.M. Rønquist (eds.), *Spectral and High Order Methods for Partial Differential Equations.*

77. M. Holtz, *Sparse Grid Quadrature in High Dimensions with Applications in Finance and Insurance.*

78. Y. Huang, R. Kornhuber, O.Widlund, J. Xu (eds.), *Domain Decomposition Methods in Science and Engineering XIX.*

79. M. Griebel, M.A. Schweitzer (eds.), *Meshfree Methods for Partial Differential Equations V.*

80. P.H. Lauritzen, C. Jablonowski, M.A. Taylor, R.D. Nair (eds.), *Numerical Techniques for Global Atmospheric Models.*

81. C. Clavero, J.L. Gracia, F.J. Lisbona (eds.), *BAIL 2010 – Boundary and Interior Layers, Computational and Asymptotic Methods.*

82. B. Engquist, O. Runborg, Y.R. Tsai (eds.), *Numerical Analysis and Multiscale Computations.*

83. I.G. Graham, T.Y. Hou, O. Lakkis, R. Scheichl (eds.), *Numerical Analysis of Multiscale Problems.*

84. A. Logg, K.-A. Mardal, G. Wells (eds.), *Automated Solution of Differential Equations by the Finite Element Method.*

85. J. Blowey, M. Jensen (eds.), *Frontiers in Numerical Analysis - Durham 2010.*

86. O. Kolditz, U.-J. Gorke, H. Shao, W. Wang (eds.), *Thermo-Hydro-Mechanical-Chemical Processes in Fractured Porous Media - Benchmarks and Examples.*

87. S. Forth, P. Hovland, E. Phipps, J. Utke, A. Walther (eds.), *Recent Advances in Algorithmic Differentiation.*

88. J. Garcke, M. Griebel (eds.), *Sparse Grids and Applications.*

89. M. Griebel, M.A. Schweitzer (eds.), *Meshfree Methods for Partial Differential Equations VI.*

90. C. Pechstein, *Finite and Boundary Element Tearing and Interconnecting Solvers for Multiscale Problems.*

91. R. Bank, M. Holst, O. Widlund, J. Xu (eds.), *Domain Decomposition Methods in Science and Engineering XX.*

92. H. Bijl, D. Lucor, S. Mishra, C. Schwab (eds.), *Uncertainty Quantification in Computational Fluid Dynamics.*

93. M. Bader, H.-J. Bungartz, T. Weinzierl (eds.), *Advanced Computing.*

94. M. Ehrhardt, T. Koprucki (eds.), *Advanced Mathematical Models and Numerical Techniques for Multi-Band Effective Mass Approximations.*

95. M. Azaïez, H. El Fekih, J.S. Hesthaven (eds.), *Spectral and High Order Methods for Partial Differential Equations ICOSAHOM 2012.*

96. F. Graziani, M.P. Desjarlais, R. Redmer, S.B. Trickey (eds.), *Frontiers and Challenges in Warm Dense Matter.*

97. J. Garcke, D. Pflüger (eds.), *Sparse Grids and Applications – Munich 2012.*

98. J. Erhel, M. Gander, L. Halpern, G. Pichot, T. Sassi, O. Widlund (eds.), *Domain Decomposition Methods in Science and Engineering XXI.*

99. R. Abgrall, H. Beaugendre, P.M. Congedo, C. Dobrzynski, V. Perrier, M. Ricchiuto (eds.), *High Order Nonlinear Numerical Methods for Evolutionary PDEs - HONOM 2013.*

100. M. Griebel, M.A. Schweitzer (eds.), *Meshfree Methods for Partial Differential Equations VII.*

101. R. Hoppe (ed.), *Optimization with PDE Constraints - OPTPDE 2014.*

102. S. Dahlke, W. Dahmen, M. Griebel, W. Hackbusch, K. Ritter, R. Schneider, C. Schwab, H. Yserentant (eds.), *Extraction of Quantifiable Information from Complex Systems.*

103. A. Abdulle, S. Deparis, D. Kressner, F. Nobile, M. Picasso (eds.), *Numerical Mathematics and Advanced Applications - ENUMATH 2013.*

104. T. Dickopf, M.J. Gander, L. Halpern, R. Krause, L.F. Pavarino (eds.), *Domain Decomposition Methods in Science and Engineering XXII.*

105. M. Mehl, M. Bischoff, M. Schäfer (eds.), *Recent Trends in Computational Engineering - CE2014. Optimization, Uncertainty, Parallel Algorithms, Coupled and Complex Problems.*

106. R.M. Kirby, M. Berzins, J.S. Hesthaven (eds.), *Spectral and High Order Methods for Partial Differential Equations - ICOSAHOM'14.*

107. B. Jüttler, B. Simeon (eds.), *Isogeometric Analysis and Applications 2014.*

108. P. Knobloch (ed.), *Boundary and Interior Layers, Computational and Asymptotic Methods – BAIL 2014.*

109. J. Garcke, D. Pflüger (eds.), *Sparse Grids and Applications – Stuttgart 2014.*

110. H. P. Langtangen, *Finite Difference Computing with Exponential Decay Models.*

111. A. Tveito, G.T. Lines, *Computing Characterizations of Drugs for Ion Channels and Receptors Using Markov Models.*

112. B. Karazösen, N, Manguoglu, M. Tezer-Sezgin, S. Göktepe, U. Ömür (eds.), *Numerical Mathematics and Advanced Applications - ENUMATH 2015.*

113. H.-J. Bungartz, P. Neumann, W.E. Nagel (eds.), *Software for Exascale Computing – SPPEXA 2013-2015.*

For further information on these books please have a look at our mathematics catalogue at the following URL: www.springer.com/series/3527

Monographs in Computational Science and Engineering

1. J. Sundnes, G.T. Lines, X. Cai, B.F. Nielsen, K.-A. Mardal, A. Tveito, *Computing the Electrical Activity in the Heart.*

For further information on this book, please have a look at our mathematics catalogue at the following URL: www.springer.com/series/7417

Texts in Computational Science and Engineering

1. H. P. Langtangen, *Computational Partial Differential Equations.* Numerical Methods and Diffpack Programming. 2nd Edition

2. A. Quarteroni, F. Saleri, P. Gervasio, *Scientific Computing with MATLAB and Octave.* 4th Edition

3. H. P. Langtangen, *Python Scripting for Computational Science.* 3rd Edition

4. H. Gardner, G. Manduchi, *Design Patterns for e-Science.*

5. M. Griebel, S. Knapek, G. Zumbusch, *Numerical Simulation in Molecular Dynamics.*

6. H. P. Langtangen, *A Primer on Scientific Programming with Python.* 5th Edition

7. A. Tveito, H. P. Langtangen, B. F. Nielsen, X. Cai, *Elements of Scientific Computing.*

8. B. Gustafsson, *Fundamentals of Scientific Computing.*

9. M. Bader, *Space-Filling Curves.*

10. M. Larson, F. Bengzon, *The Finite Element Method: Theory, Implementation and Applications.*

11. W. Gander, M. Gander, F. Kwok, *Scientific Computing: An Introduction using Maple and MATLAB.*

12. P. Deuflhard, S. Röblitz, *A Guide to Numerical Modelling in Systems Biology.*

13. M. H. Holmes, *Introduction to Scientific Computing and Data Analysis.*

14. S. Linge, H. P. Langtangen, *Programming for Computations - A Gentle Introduction to Numerical Simulations with MATLAB/Octave.*

15. S. Linge, H. P. Langtangen, *Programming for Computations - A Gentle Introduction to Numerical Simulations with Python.*

For further information on these books please have a look at our mathematics catalogue at the following URL: www.springer.com/series/5151

Printed in the United States
By Bookmasters